Principles of
Biochemical
Toxicology

Fourth Edition

John A. Timbrell
Department of Pharmacy
King's College London
London, UK

informa
healthcare

New York   London

Informa Healthcare USA, Inc.
52 Vanderbilt Avenue
New York, NY 10017

International Standard Book Number-10: 0-8493-7302-6 (softcover : alk. paper)
International Standard Book Number-13: 978-0-8493-7302-2 (softcover : alk. paper)

Library of Congress Cataloging-in-Publication Data

Timbrell, John A.
   Principles of biochemical toxicology / by John A. Timbrell. —4th ed.
     p. ; cm.
   Includes bibliographical references and index.
   ISBN-13: 978-0-8493-7302-2 (softcover : alk. paper)
   ISBN-10: 0-8493-7302-6 (softcover : alk. paper)
   1. Biochemical toxicology. I. Title.
     [DNLM: 1. Poisoning—metabolism. 2. Poisoning—physiopathology.
   3. Poisons–metabolism. 4. Xenobiotics–adverse effects. QV 600 T583p
   2008]
    RA1219.5.T56 2008
    615.9—dc22

                                  2008035265

For Corporate Sales and Reprint Permissions call 212-520-2700 or write to: Sales Department, 52 Vanderbilt Avenue, 16th floor, New York, NY 10017.

Visit the Informa Web site at
www.informa.com

and the Informa Healthcare Web site at
www.informahealthcare.com

For
Anna, Becky, and Cathy
"... and she had never forgotten that, if you drink much from
a bottle marked 'poison,' it is almost certain to disagree with
you, sooner or later."
From Alice's Adventures in Wonderland, by Lewis Carroll

Preface

This fourth and probably final edition of *Principles of Biochemical Toxicology* has, like the previous editions, evolved against the background of my involvement with the teaching of toxicology on various courses at various levels and in various places.

The objective of the book has always been to form a sound introduction to the basic principles of the subject from a biochemical and mechanistic viewpoint. It is a testament to the vitality and progression of toxicology that the increasing sophistication, complexity, and expansion of the subject mean that revision of at least parts of this book is essential every few years. However, a book of this size cannot realistically cover all of the diverse aspects of toxicology in equal depth and detail and include all the new developments that are occurring, hence the extensive bibliography, which should be used to complement this text where more detail or other examples are wanted.

This is probably the most extensive revision, because I changed to a part-time teaching contract in order to do it. I hope that readers feel it has been worthwhile!

As previously, I have taken into account comments that have been made to me since the third edition was published. I have added new examples to broaden the scope as well as updated existing ones. I have also redone, and I hope improved, many of the diagrams as well as adding many new ones.

Special thanks to Anna for all her help with the diagrams.

Again, special thanks to Cathy, particularly this time for her critical comments and advice, the diagrams she has drawn or helped with, and, as always, for her patience, support, and, indeed, forebearance when I could not see the light at the end of the tunnel.

John A. Timbrell

April 2008

Contents

1 | Introduction

1.1 BACKGROUND

Toxicology is the subject concerned with the study of the noxious effects of chemical substances on living systems. It is a multidisciplinary subject, as it embraces areas of pharmacology, biochemistry, chemistry, physiology, and pathology; although it has sometimes been considered as a subdivision of some of these other subjects, it is truly a scientific discipline in itself.

Toxicology may be regarded as the science of poisons; in this context, it has been studied and practiced since antiquity, and a large body of knowledge has been amassed. The ancient Greeks used hemlock and various other poisons, and **Dioscorides** attempted a classification of poisons. However, the scientific foundations of toxicology were laid by **Paracelsus** (1493–1541), and this approach was continued by **Orfila** (1787–1853). Orfila, a Spanish toxicologist working in Paris, wrote a seminal work, Trait des Poisons, in 1814 in which he said toxicology should be founded on pathology and chemical analysis.

Today, highly sensitive and specific analytical methods are used and together with the new methods of molecular biology have a major impact on the development of the science.

Interactions between chemicals and living systems occur in particular phases. The first is the exposure phase where the living organism is exposed in some way to the chemical and which may or may not be followed by uptake or absorption of the chemical into the organism. This precedes the next phase in which the chemical is distributed throughout the organism. Both these phases may require transport systems. After delivery of the chemical to various parts of the organism, the next phase is metabolism, where chemical changes may or may not occur, mediated by enzymes. These phases are sometimes termed "toxicokinetics," whereas the next phase is the toxicodynamic phase in which the chemical and its metabolites interact with constituents of the organism. The metabolic phase may or may not be a prerequisite for the final phase, which is excretion.

This sequence may then be followed by a phase in which pathological or functional changes occur.

Ability to detect exposure and early adverse effects is crucial to the assessment of risk as will be apparent later in this book. Furthermore, understanding the role of metabolites rather than the parent chemical and the importance of concentration in toxic effects are essential in this process.

Therefore, toxicology has of necessity become very much a multidisciplinary science.

There are difficulties in reconciling the often-conflicting demands of public and regulatory authorities to demonstrate safety with pressure from animal rights organizations against the use of animals for this purpose.

Nevertheless, development of toxicology as a separate science has been slow, particularly in comparison with subjects such as pharmacology and biochemistry, and toxicology has a much more limited academic base. This may in part reflect the nature of the subject, which has evolved as a practical art, and also the fact that many practitioners were mainly interested in descriptive studies for screening purposes or to satisfy legislation.

Another reason may be that funding is limited because of the fact that toxicology does not generate novel drugs and chemicals for commercial use, rather it restricts them.

1.2   SCOPE

The interest in and scope of toxicology continue to grow rapidly, and the subject is of profound importance to human and animal health.

The increasing numbers **(at least 100,000)** of foreign chemicals (xenobiotics) to which humans and other organisms in the environment are exposed underlies this growth.

These include drugs, pesticides, environmental pollutants, industrial chemicals, and food additives about which we need to know much, particularly concerning their safety. Of particular importance, therefore, is the ability to predict, understand, and treat toxicity as shown by examples such as paracetamol hepatotoxicity (see chap. 7).

This requires a sound mechanistic base to be successful. It is this mechanistic base that comes within the scope of biochemical toxicology, which forms the basis for almost all the various branches of toxicology.

The development of toxicology has been hampered by the requirements of regulatory agencies, which have encouraged the "black box" approach of empiricism as discussed by Goldberg (see Bibliography). However, the black box has now been opened, and we are beginning to understand what is inside. But will it prove, especially in relation to risk assessment, to be Pandora's box? The routine gathering of data on toxicology, preferably of a negative nature, required by the various regulatory bodies of the industrial nations, has tended to constrain and regulate toxicology.

Furthermore, to paraphrase Zbinden (see Bibliography), misuse of toxicological data and adverse regulatory action in this climate of opinion have discouraged innovative approaches to toxicological research and have become an obstacle to the application of basic concepts in toxicology. However, the emphasis on and content of basic science at recent toxicology congresses is testimony to the progress that has taken place in the period since Goldberg and Zbinden wrote their articles (see Bibliography).

Ideally, basic studies of a biochemical nature should be carried out if possible before, but at least simultaneously with, toxicity testing, and a bridge between the biochemical and morphological aspects of the toxicology of a compound should be built. It is apparent that there are many gaps in our knowledge concerning this connection between biochemical events and subsequent gross pathological changes. Without an understanding of these connections, which will require a much greater commitment to basic toxicological research, our ability to predict toxicity and assess risk from the measurement of various biological responses will remain inadequate.

Thus, any foreign compound, which comes into contact with a biological system, will cause certain perturbations in that system. These biological responses, such as the inhibition of enzymes and interaction with receptors, macromolecules, or organelles, may not necessarily be toxicologically relevant. This point is particularly important when assessing *in vitro* data, and it involves the concept of a dose threshold or the lack of such a threshold, in the "one molecule, one hit" theory of toxicity.

1.3   BIOCHEMICAL ASPECTS OF TOXICOLOGY

Biochemical toxicology is concerned with the mechanisms underlying toxicity, particularly the events at the molecular level and the factors, which determine and affect toxicity.

The interaction of a foreign compound with a biological system is twofold: there is the effect of the organism on the compound and the effect of the compound on the organism.

It is necessary to appreciate both for a mechanistic view of toxicology. The first of these includes the absorption, distribution, metabolism, and excretion of xenobiotics, which are all factors of importance in the toxic process and which have a biochemical basis in many instances. The mode of action of toxic compounds in the interaction with cellular components, and at the molecular level with structural proteins and other macromolecules, enzymes, and receptors, and the types of toxic response produced are included in the second category of interaction. However, a biological system is a dynamic one, and therefore a series of events may follow the initial response. For instance, a toxic compound may cause liver or kidney damage and thereby limit its own metabolism or excretion.

The anatomy and physiology of the organism affect all the types of interaction given above, as can the site of exposure and entry of the foreign compound into the organism. Thus,

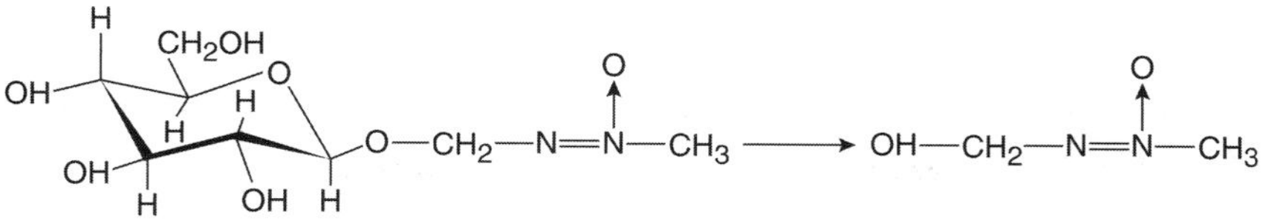

Figure 1.1　The bacterial metabolism of cycasin.

the gut bacteria and conditions in the gastrointestinal tract convert the naturally occurring compound **cycasin**, methylazoxymethanol glycoside, into the potent carcinogen **methylazoxymethanol** (Fig. 1.1). When administered by other routes, cycasin is not carcinogenic.

The distribution of a foreign compound and its rate of entry determine the concentration at a particular site and the number and types of cells exposed. The plasma concentration depends on many factors, not least of which is the metabolic activity of the particular organism. This metabolism may be a major factor in determining toxicity, as the compound may be more or less toxic than its metabolites.

The excretion of a foreign substance can also be a major factor in its toxicity and a determinant of the plasma and tissue levels. All these considerations are modified by species differences, genetic effects, and other factors. The response of the organism to the toxic insult is influenced by similar factors. The route of administration of a foreign compound may determine whether the effect is systemic or local.

For example, paraquat causes a local irritant effect on the skin after contact but a serious and often fatal lung fibrosis if it gains entry into the body and bloodstream. Normally, only the tissues exposed to a toxic substance are affected unless there is an indirect effect involving a physiological mechanism such as an immune response. The distribution and metabolism of a toxic compound may determine the target organ damaged, as does the susceptibility of the particular tissue and its constituent cells. Therefore, the effect of a foreign compound on a biological system depends on numerous factors, and an understanding and appreciation of them is a necessary part of toxicology.

The concept of toxicity is an important one: it involves a damaging, noxious, or deleterious effect on the whole or part of a living system, which may or may not be reversible. The toxic response may be a transient biochemical or pharmacological change or a permanent pathological lesion. The effect of a toxic substance on an organism may be immediate, as with a pharmacodynamic response such as a hypotensive effect, or delayed, as in the development of a tumor.

It has been said that there are "no harmless drugs only harmless ways of using them." It could equally be said, "There are no harmless substances, only harmless ways of using them," which underscores the concept of toxicity as a *relative* phenomenon. It depends on the dose and type of substance, the frequency of exposure, and the organism in question. There is no absolute value for toxicity, although it is clear that botulinum toxin has a much greater relative toxicity or potency than DDT(p,p'-dichlorodiphenyltrichloroethane) on a weight-for-weight basis (Table 1.1).

Table 1.1　Approximate Acute LD_{50} Values for a Variety of Chemical Agents

Agents	Species	LD_{50} (mg/kg body weight)
Ethanol	Mouse	10,000
Sodium chloride	Mouse	4,000
Morphine sulfate	Rat	900
Phenobarbital, sodium	Rat	150
DDT	Rat	100
Strychnine sulfate	Rat	2
Nicotine	Rat	1
Tetrodotoxin	Rat	0.1
Dioxin (TCDD)	Guinea pig	0.001
Botulinum toxin	Rat	0.00001

Abbreviations: LD, lethal dose; TCDD, 2,3,7,8-tetrachlorodibenzo-*p*-dioxin. *Source*: Data from Loomis TA, Hayes AW. Loomis's Essentials of Toxicology. 4th ed. San Diego: Academic Press, 1996.

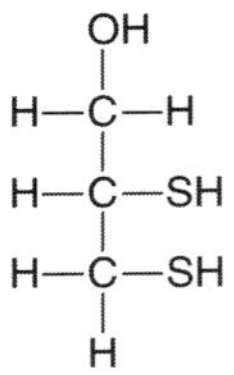

The structures of Lewisite and dimercaprol (British anti-Lewisite).

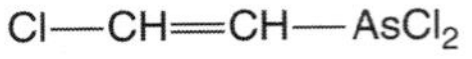

Figure 1.2 The structures of Lewisite and dimercaprol or British anti-Lewisite.

The derivation and meaning of LD_{50} will be discussed in detail in Chapter 2. However, the LD_{50} is now seldom regarded as a useful parameter of toxicity except in particular circumstances such as the design of pesticides.

There are many different types of toxic compounds producing the various types of toxicity detailed in chapter 6. One compound may cause several toxic responses. For instance, vinyl chloride is carcinogenic after low doses with a long latent period for the appearance of tumors, but it is narcotic and hepatotoxic after single large exposures (see chap. 7).

Investigation of the sites and modes of action of toxic agents and the factors affecting their toxicity as briefly summarized here is fundamental for an understanding of toxicity and also for its prediction and treatment.

For example, the elucidation of the mechanism of action of the war gas Lewisite (Fig. 1.2), which involves interaction with cellular sulfhydryl groups, allowed the antidote, British anti-Lewisite or **dimercaprol** (Fig. 1.2), to be devised. Without the basic studies performed by Sir Rudolph Peters and his colleagues, an antidote would almost certainly not have been available for the victims of chemical warfare.

Likewise, empirical studies with chemical carcinogens may have provided much interesting data but would have been unlikely to explain why such a diverse range of compounds causes cancer, until basic biochemical studies provided some of the answers.

SUMMARY

Toxicology, also called the science of poisons, is a multidisciplinary subject dealing with the noxious effects of chemicals on living systems. It has a long history in relation to the art of poisoning but has now become more scientifically based. The scientific foundations of toxicology were laid by Paracelsus and later by Orfila. Toxicology is interrelated with the activities of regulatory authorities, and its importance is a reflection of the large numbers of chemicals to which man and the environment are exposed. It relies on an understanding of the basic biochemistry and physiology of living systems and the relevant chemistry of toxic molecules. Thus, the interaction of a chemical with a living system occurs in phases and involves both an effect of the chemical on the biological system and of the biological system on the chemical. These interactions are affected by numerous factors.

The science of toxicology requires an appreciation of the fact that not all effects observed are toxicologically relevant. Toxicity is a damaging effect on whole or part of a living system.

An understanding of the mechanism of toxicity of a chemical is essential for a proper assessment of risk and can lead to the development of antidotes. There are no harmless chemicals, only harmless ways of using them.

REVIEW QUESTIONS

1. Which 16th century scientist was important in the development of toxicology and why?
2. Why is cycasin only carcinogenic when ingested by mouth?
3. How many times more toxic is botulinum toxin than nicotine in the rat?
4. What was the contribution of Orfila to the development of toxicology?

BIBLIOGRAPHY

Albert A. Selective Toxicity. London: Chapman & Hall, 1979.

Albert A. Xenobiosis. London: Chapman & Hall, 1987.

Aldridge WN. Mechanisms and Concepts in Toxicology. London: Taylor & Francis, 1996.

Ballantyne B, Marrs TC, Syversen T. Fundamentals of toxicology. In: Ballantyne B, Marrs TC, Syversen T, eds. General and Applied Toxicology. Vol. 1. London: Macmillan, 2000.

Barile FA. Principles of Toxicology Testing. Florida: CRC Press, 2007.

Beck BD, Calabrese EJ, Slayton TM, et al. The use of toxicology in the regulatory process. In: Hayes AW, ed. Principles and Methods of Toxicology. 5th ed. Florida: CRC Press, 2007.

Boerlsterli UA. Mechanistic Toxicology. 2nd ed. Boca Raton: CRC Press, 2007.

Derelanko MJ. Toxicologist's Pocket Handbook. Boca Raton: CRC Press, 2000.

Efron E. The Apocalyptics, Cancer and the Big Lie. New York: Simon & Schuster, 1984.

Fenton JJ. Toxicology: A Case Oriented Approach. Boca Raton: CRC Press, 2002.

Gallo MA. History and scope of toxicology. In: Klaassen CD, ed. Cassarett and Doull's Toxicology, The Basic Science of Toxicology. 6th ed. New York: McGraw Hill, 2001.

Gilbert SG. A Small Dose of Toxicology. Boca Raton: CRC Press, 2004.

Hellman B. General Toxicology. In: Mulder G, Dencker L, eds. Pharmaceutical Toxicology. London: Pharmaceutical Press, 2006.

Goldberg L. Toxicology: has a new era dawned? Pharmacol Rev 1979; 30:351.

Hayes AW, ed. Principles and Methods of Toxicology. 5th ed. Florida: CRC Press, 2007.

Hodgson E, Mailman RB, Chambers JE, eds. Dictionary of Toxicologyedition. 2nd ed. London: Macmillan, 1998.

Hodgson E, Smart RC. Biochemical toxicology: definition and scope. In: Hodgson E, Smart RC, eds. Introduction to Biochemical Toxicology. 2nd ed. New York: Wiley, 2001.

Hodgson E, Levi PE. A Textbook of Modern Toxicology. New York: Elsevier, 1987.

Lu FC, Kacew S. Toxicology: Principles and Applications. 4th ed. London: Taylor & Francis, 2002.

Koeman JH. Toxicology: history and scope of the field. In: Niesink RJM, de Vries J, Hollinger MA, eds. Toxicology: Principles and Practice. CRC Press and Open University of the Netherlands, 1996.

Lane RW, Borzelleca JF. Principles of toxicology. Harming and helping through time: the history of toxicology. In: Hayes AW, ed. Principles and Methods of Toxicology. 5th ed. Florida: CRC Press, 2007.

McClellan RO, ed. Critical Reviews in Toxicology. New York, NY: Taylor and Francis, 1971.

Otoboni A. The Dose Makes the Poison. 2nd ed. New York: Van Nostrand Reinhold, 1997.

Peters RA. Biochemical Lesions and Lethal Synthesis. Oxford: Pergamon Press, 1963.

Pratt WB, Taylor P, eds. Principles of Drug Action, The Basis of Pharmacology. 3rd ed. New York: Churchill Livingstone, 1990.

Shaw IC, Chadwick J. Principles of Environmental Toxicology. London: Taylor & Francis, 1998.

Stone T, Darlington G. Pills, Potions, Poisons. Oxford: Oxford University Press, 2000.

Stacey NH, ed. Occupational Toxicology. London: Taylor & Francis, 1993.

Timbrell JA. Study Toxicology Through Questions. London: Taylor & Francis, 1997.

Timbrell JA. Introduction to Toxicology. 3rd ed. London: Taylor & Francis, 2002.

Timbrell JA. The Poison Paradox. Oxford: Oxford University Press, 2005.

Walker CH, Sibly RM, Hopkin SP, et al. Principles of Ecotoxicology. Florida: CRC Press, 2005.

Wexler P. Information Resources in Toxicology, 2nd ed. New York: Elsevier, 1988.

Zbinden G. The three eras of research in experimental toxicology. Trends Pharmacol Sci 1992; 13: 221–223.

2 | Fundamentals of Toxicology and Dose-Response Relationships

2.1 INTRODUCTION

The relationship between the dose of a compound and its toxicity is central in toxicology. Paracelsus (1493–1541), who was the first to put toxicology on a scientific basis, clearly recognized this relationship. His well-known statement: "*All substances are poisons; there is none that is not a poison. The right dose differentiates a poison and a remedy*" has immortalized the concept. Implicit in this statement is the premise that there is a dose of a compound, which has no observable effect and another, higher dose, which causes the maximal response. The **dose-response relationship** involves quantifying the toxic effect or response and showing a correlation with exposure. The relationship underlies the whole of toxicology, and an understanding of it is crucial. Parameters gained from it have various uses in both investigational and regulatory toxicology. It should be appreciated, however, that toxicity is a *relative* phenomenon and that the ways of measuring it are many and various.

2.2 BIOMARKERS

To discuss the dose-response relationship, it is necessary to consider the dose of a chemical, the nature of the response to it, and what factors affect the response to the chemical. These considerations are also important in the process of risk assessment for any chemical.

Determination of the true exposure to a chemical substance and of the response of the organism to that chemical and its potential susceptibility to toxic effects are all crucial parameters in toxicology. Biomarkers are tools, which enable us to measure these things.

There are thus three types of biomarkers: biomarkers of exposure of the organism to the toxic substance, biomarkers of response of the organism to that exposure, and biomarkers of susceptibility of the organism to the chemical.

2.2.1 Biomarkers of Exposure

At its simplest, measurement of the dose is determination of the amount of chemical administered or the amount to which the animal or human is exposed (such as in air or water). However, it cannot be assumed that all of the dose will be absorbed, even in the case of a drug given to a patient (see chap. 3). Therefore, a more precise estimate of exposure is often needed. This is usually the blood level of the chemical. The level of a chemical in the blood approximates to the concentration in organs and tissues, which are perfused by that blood (see chap. 3), and so this is a true biomarker of exposure. It is assumed that the target for toxicity will be located in one or more of these organs or tissues. However, a metabolic breakdown product (see chap. 4) may be responsible for the toxicity, and therefore, measuring the parent chemical may not always be an appropriate biomarker of exposure. A more appropriate marker of exposure would be the metabolite itself, and this is termed a "biomarker of internal dose." Metabolites, especially if they are reactive, may interact with macromolecules such as proteins and nucleic acids (see chap. 6), often resulting in conjugates or adducts. These can also be measured. If the conjugate is part of the process of toxicity, measurement of such a conjugate in blood or other body fluid is a valuable biomarker of effective dose.

However, quantitative environmental exposure data for humans or other animals are often sadly lacking because studies are usually retrospective, and so, samples of body fluids will not have been taken. Biomarkers of exposure are relatively transient and even conjugate with hemoglobin in blood, which are the most persistent, and are generally only detectable for about three months after exposure. So unless a prospective study is being carried out or there is continuous or continual repeated exposure, measurement of biomarkers of exposure may not be possible. Biomarkers of effective dose were found to be extremely valuable in the study of aflatoxin-induced liver cancer in humans (see chap. 5).

For environmental exposure, therefore, the dose may have to be estimated from the amount in soil or water or air or food. If there is industrial exposure, workers in well-regulated industries are monitored such that their urine or blood may be sampled regularly and analyzed for biomarkers of exposure.

2.2.2 Biomarkers of Response
Living organisms can show many kinds of toxic or adverse response to a chemical exposure, ranging from biochemical or physiological to pathological. Consequently there are many biomarkers of response, which can be measured. These include markers such as enzymes, which appear in the blood when an organ is damaged, increases in enzymes or stress proteins resulting from induction (see chap. 5), changes in urinary constituents resulting from damage or metabolic dysfunction, increases or decreases in enzyme activity, and pathological changes detected at the gross, microscopic, and subcellular level. Indeed, a biomarker of response could be almost any indication of altered structure or function. The search for novel biomarkers now uses techniques in molecular biology, such as the study of changes in genes (genomics or transcriptomics), changes in the proteins produced from them (proteomics), and changes in the metabolites resulting from these proteins (metabonomics). However, although these new technologies have and certainly will have an increasingly important role, interpretation of the often large amount of data generated is a significant task requiring bioinformatic techniques such as pattern recognition. Furthermore, all biomarkers of response must be validated in relation to certain criteria. It cannot be assumed, because a gene is switched on or off, a protein is increased or decreased, or a metabolic pathway is influenced by a chemical, that the measurement is a usable biomarker, which reflects toxicity. Some changes are coincidental rather than causal, some reflect changes, which are inconsequential to the function of the organism, and some changes are transient and irrelevant.

2.2.3 Biomarkers of Susceptibility
Finally, biomarkers, which indicate variation in the susceptibility of the organism, can be determined, and again, these cover a range of types from deficiency in metabolic enzymes to variation in repair systems. These would typically be measured in individual members of a population. An example could be a genetic deficiency in a particular enzyme involved in detoxication or xenobiotic metabolism such as cyp 2D6 or *N*-acetyltransferase (see chap. 5). A less common type of susceptibility marker is that reflecting increased responsiveness of a receptor or resulting from a metabolic disorder, such as glucose 6-phosphate dehydrogenase deficiency, leading to increased susceptibility to toxicity. The interrelationships between these three types of biomarkers are indicated in Figure 2.1.

2.2.4 The Use of Biomarkers in Risk Assessment
Biomarkers are used at several stages in the risk assessment process. Biomarkers of exposure are important in risk assessment, as an indication of the internal dose is necessary for the proper description of the dose-response relationship. Similarly, biomarkers of response are necessary for determination of the no observed adverse effect level (NOAEL) and the dose-response relationship (see below). Biomarkers of susceptibility may be important for identifying especially sensitive groups to estimate an uncertainty factor.

Thus, biomarkers allow the crucial link between the response and exposure to be established (see the sect. 2.5).

2.3 CRITERIA OF TOXICITY

Clearly there are many different kinds of toxic effect as will be discussed in chapter 6 and also many different ways of detecting and measuring them. However, it is necessary at this stage to consider in general terms what is meant by the term "toxic response" or "toxic effect." This

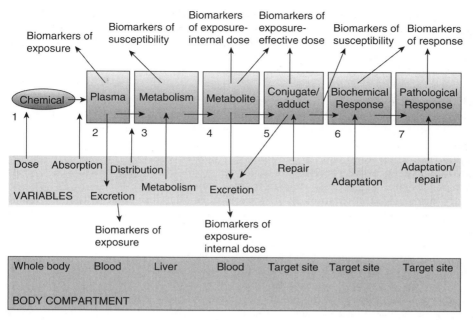

Figure 2.1 Types of biomarker and their relationship to the exposure-disease model of toxicity. *Source*: From Ref. 1.

may depend on the nature of the system being evaluated and the circumstances of exposure. Some biological effects of chemicals may be trivial, others serious and lethal.

Toxic effects or responses can be divided into those that are graded and those that are "all or none." Graded effects are those such as the inhibition of an enzyme, or a change in blood pressure, which can show some effect between zero and maximal, that is, an increase in severity will be seen. All-or-none responses are those that are only present or absent (on or off), such as death or the presence of a tumor. Both types of effect/response can be used to construct a dose-effect/dose-response curve, but there is a difference between them. With a graded effect, this is measured in each individual organism at a particular level of exposure to the chemical. Then usually the average effect for the organisms at a particular dose is plotted against the dose. If different doses have been investigated, then a dose-effect curve can be constructed. With an all-or-none response, the percentage of organisms responding to a particular dose is plotted, which might be 0% at a low dose and 100% at a high dose. This reflects an underlying frequency distribution as will be discussed later.

First, it is important to distinguish between toxic effects occurring at the point or site of exposure (e.g., the stomach), so-called local effects, and those toxic effects occurring at a site distant from the site of exposure, known as systemic effects. Local effects are usually limited to irritancy and corrosive damage such as from strong acids, which occur immediately but can be reversible. The one exception is sensitization, which involves the immune system but is often manifested at the site of the exposure (e.g., skin) although may be delayed.

It is also important at this point to identify that some effects may be reversible, whereas others are irreversible and that this can be for a number of reasons. Reversibility may be due to replacement of an inhibited enzyme, for example, or repair of damaged tissue. Irreversibility could be due to inability to repair a damaged organ or tissue. If this organ has a crucial function, then the organism may die.

In contrast to local irritants and corrosive acids and alkalis, other chemicals, such as the drug paracetamol (see chap. 7), cause systemic toxicity, damaging the liver, possibly irreversibly and with some delay after an oral overdose. Penicillin can also cause systemic toxicity as a result of an immune reaction, which may be immediate and serious, if it is anaphylaxis (see chap. 7). However, this effect, if not fatal, is reversible.

Some toxic chemicals cause both local and systemic effects. For example, cadmium fumes, which may occur in industrial environments, can cause lung damage when inhaled, but the cadmium absorbed will damage the kidneys.

Cancer caused by chemicals can occur at the site of action or systemically depending on the carcinogen. Thus, carcinogenic cigarette smoke constituents such as benzo[a]pyrene lead to cancer of the lung when inhaled but benzo[a]pyrene will also cause cancer of the skin if this is exposed to it.

Cancer is usually a delayed response, sometimes very much delayed, and is also often irreversible.

Later chapters will explore some of these examples and points in more detail.

Lethality has already been mentioned in the previous chapter, and at one time, this was considered the important measure of toxicity in experimental animals and was quantified as the LD_{50} (see below). However, it is an unnecessary and crude measure of the all-or-none type, which is dependent upon and can be influenced by many factors and so will show considerable variability. Lethality, therefore, is no longer an endpoint, which is used except in specific circumstances such as pesticide development.

As already discussed, information on toxicity is normally derived from experimental animals, starting with preliminary acute toxicity studies covering a range of doses, and these may be all that is necessary to *classify* a chemical [as described by van den Heuvel et al. (2)]. In such studies, careful observation of clinical signs and symptoms and their time course can be very important and could suggest underlying mechanisms. For example, a chemical, which caused a variety of symptoms such as constricted pupils, labored breathing, hind limb weakness, and diarrhea, might be acting by interfering with the cholinergic system as organophosphates do (see chap. 7). If the effects occur very rapidly, this might suggest that a major biochemical or physiological system is affected. For example, cyanide is rapidly lethal because the target is cytochrome aa_3 in the mitochondrial electron transport chain, which is vital to all cells (see chap. 7). The blockade of this enzyme will therefore stop cellular respiration in many different tissues.

At the end of toxicity studies during the postmortem, observations such as the occurrence of pathological change, for example, liver damage can be detected. In later, longer-term studies, a specific biochemical change, such as inhibition of an enzyme or a physiological change, might be detected in blood samples or specific physiological measurements. Some pathological damage can be detected from biomarkers in blood such as leakage of enzymes or changes in metabolites.

Simple changes can also be useful, and important indicators such as changes in body weight, which is quite a sensitive marker of dysfunction, or organ weight, which can be a sensitive indicator of pathological change. Generally, in studies of the toxicity of a chemical, a variety of biomarkers of effect or response will be measured, particularly biochemical markers, which are quantitative indicators of dysfunction.

The selection of a measurable index of toxicity in the absence of an obvious pathological lesion can be difficult, but information derived from preliminary toxicity studies may indicate possible targets.

Biomarkers, which are closely connected to the mechanism of toxicity, are preferable. This may, of course, require an underlying knowledge of the target site, which may be a *receptor, enzyme,* or *other macromolecule.*

For example, the industrial chemical hydrazine not only causes death, as a result of effects on the central nervous system, but also causes dysfunction in the liver, leading to the accumulation of fat. This effect is not related to the lethality, but it shows a clear relationship with dose. The response, that is, fatty liver, is a graded effect rather than an all-or-none response and can be quantitated either as an increase in liver weight (as percent of body weight) or by specific measurement of the triglycerides. Both measurements show a similar dose-effect curve (Fig. 2.2).

In contrast, the lung damage and edema (water accumulation) caused by ipomeanol, discussed in greater detail in chapter 7, is directly related to the lethality. This can be seen from the dose-response curve (chap. 7, Fig. 39) and can also be seen when the time course of death and lung edema, measured as the wet weight/ dry weight ratio, are compared (chap. 7, Fig. 38), strongly suggesting a causal relationship between them.

When information is derived from human epidemiological studies, it will normally be incidence of a particular disease or morbidity such as cancer or maybe the appearance of a novel disease, that is, all-or-none responses.

Toxicity studies can also be carried out *in vitro* and dose-response curves constructed.

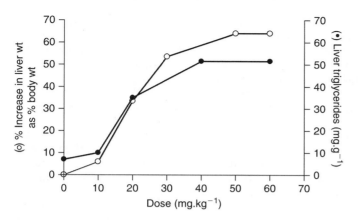

Figure 2.2 Increase in liver weight and liver triglycerides caused by **hydrazine**. *Source*: From Ref. 3.

In *in vitro* systems, criteria of toxicity will generally be measurements of either specific biochemical changes, such as ATP level or protein synthesis, or general indicators such as cell metabolic activity, viability, or membrane damage as indicated by dye uptake or enzyme leakage.

2.4 NEW TECHNOLOGIES

New techniques are being employed in toxicology in various ways, such as the development of new biomarkers. Thus, mechanisms underlying toxic effects are being unraveled using molecular biological techniques such as genomics and transcriptomics. Sophisticated separation techniques and mass spectrometry are being applied to identify protein targets and measure changes in them using proteomics. Separation and identification of metabolites using techniques such as high-pressure liquid chromatography and high-resolution nuclear magnetic resonance spectroscopy (NMR) are being used in metabonomics. When used together, these three techniques (genomics, proteomics, and metabonomics) show how chemicals cause metabolic alteration or dysfunction, starting in some cases with a change in gene expression and transcription or a mutation. The use of highly sensitive accelerator mass spectrometry (AMS) allows extremely small doses of a potential drug to be given to human volunteers and the drug and metabolites to be detected, so that some indication of potential metabolism is possible.

2.5 EVALUATION OF TOXICITY

The toxicity of a chemical can be determined in one of three ways:

1. By observing human, animal, or plant populations exposed to a chemical (epidemiology)
2. By administering the chemical to animals or plants under controlled conditions and observing the effects (*in vivo*)
3. By exposing cells, subcellular fractions, or single-celled organisms to the chemical (*in vitro*)

2.5.1 Human Toxicity Data

The exposure of humans to chemicals may occur accidentally through the environment, as part of their occupation or intentionally, as with drugs and food additives. Thus chemical accidents, if thoroughly documented, may provide important information about the toxicity of a chemical in humans. Similarly, monitoring of humans exposed to chemicals at work may, if well documented, provide useful evidence of toxicity. Thus, monitoring biochemical indices of pathological change may be carried out in humans during potential exposure (see the sect. 2.2). An example is the monitoring of agricultural workers for exposure to organophosphorus insecticides by measuring the degree of inhibition of the enzyme acetylcholinesterase in blood samples (see chap. 7). However, acquiring such human data is often difficult and is rarely complete or of a good-enough standard to be more than additional to animal studies. One problem is the lack of adequate exposure information. However, epidemiological data may at least indicate that a causal relationship exists between exposure to the chemical and an effect in humans. Studying particular populations of predatory birds and measuring certain parameters, such as eggshell thickness and pesticide level, is an ecotoxicological example of testing for toxicity in the field.

For a drug, detecting toxic effects is more straightforward. Experimental animals are given various doses of the new drug, and any toxic effects are evaluated. Then, before marketing, drugs are first given to a small number of human volunteers (5–10) in phase I clinical trials and then later to a relatively small number of patients (100–500) in phase II clinical trials, then to a larger number of patients (2000–3000) in phase III clinical trials. If it is licensed by the authorities, it is made available to the general public (phase IV clinical trials). Both during the early clinical trials and the eventual use by the general public, adverse reactions can be detected.

Data obtained from human exposure or clinical trials is analyzed by epidemiological techniques. Typically, effects observed will be compared with those in control subjects with the objective of determining if there is an association between exposure to the chemical and a disease or adverse effect.

There are four types of epidemiological study:

1. Cohort studies in which individuals exposed to the chemical of interest are followed overtime prospectively. This design is used in clinical trials of drugs. Controls are subjects selected out of the patient population and have the disease for which the drug is prescribed. The controls receive an inactive "placebo."
2. Case-control studies in which individuals who have been exposed and may have developed a disease are compared retrospectively with similar control subjects who have no disease.
3. Cross-sectional studies are those in which the prevalence of a disease in an exposed group is studied.
4. Ecological studies are those in which the incidence of a disease in one geographical area (where there may be hazardous chemical exposure) is compared with the incidence in another area without the hazardous chemical.

Epidemiological data can be analyzed in various ways to give measures of effect. The data can be represented as an odds ratio, which is the ratio of the risk of disease in an exposed group compared with a control group. Odds ratio can be calculated as

$$A \times B / C \times D,$$

where A = no. of cases of disease in exposed population; B = no. of unexposed controls without disease; C = no. of exposed subjects without disease; D = no. of unexposed controls with disease. The relative risk is determined as the ratio of the occurrence of the disease in the exposed to the unexposed population. The absolute excess risk is an alternative quantitative measure. Relative risk is calculated as A/B, where A = no. of cases of disease in total exposed group per unit of population; B = no. of cases of disease in total nonexposed control group per unit of population.

Absolute excess risk calculated as number of cases of disease per unit of exposed population minus number of cases of disease per unit of unexposed population.

When setting up epidemiological studies and when assessing their significance, it is important to be aware of confounding factors such as bias and the need for proper controls.

For further details on epidemiology, the reader is referred to the bibliography.

2.5.2 Animal Toxicity Data

Although human data from epidemiological studies are useful, the majority of data on the toxicity of chemicals are gained from experimental studies in animals. Some of these studies will be carried out to understand the mechanism behind the toxicity of a particular chemical, other studies will be carried for regulatory reasons. This data will be generated by toxicity studies, which are controlled and which generate histopathological, clinical, and biochemical data.

The data so acquired is used for the risk assessment and safety evaluation of drugs prior to human exposure, for food additives before use, and for industrial and environmental chemicals. In the case of drugs, this information is essential before the drug can be administered to patients in clinical trials, and similarly, for food additives and other chemicals, it is required to set a NOAEL (see below).

Because animal tests can be carefully controlled with the doses known exactly, the quality of the data is generally good. The number of animals used should be enough to allow statistical significance to be demonstrated. Humane conditions and proper treatment of animals are essential for scientific as well as ethical reasons, as this helps to ensure that the data is reliable and robust.

The problem of extrapolation between animal species and humans always has to be considered, but past data as well as theoretical considerations indicate that *in the majority* of cases, (but not all) toxic effects occurring in animals will also occur in humans.

The most common species used are rats and mice for reasons of size, the accumulated knowledge of these species, and cost. Normally, young adult animals of both sexes will be used. The exposure level of the chemical chosen will ideally span both nontoxic and maximally toxic doses.

Currently, mice have the advantage in being available as genetically modified varieties. They can, therefore, be engineered to express human enzymes, for example. Consequently, they can be used to evaluate the mechanism underlying a toxic effect or to simulate a human in terms of metabolism, for example.

To show and evaluate some types of toxic effect, a particular species might be required.

For veterinary drugs or environmental pollutants, the target species will normally be used.

2.5.3 *In Vitro* Toxicity Studies

It has become necessary to question the use of *in vivo* safety evaluation studies in animals because of the pressure from society to reduce the use of live animals in medical research. Consequently, there has been an increase in the exploration and use of various *in vitro* systems in toxicity testing. The current philosophy is embodied in the concept of the three R's: replacement, reduction, and refinement. Thus if possible, live animals should be replaced with alternatives. If this is not possible, then measures should be adopted to reduce the numbers used. Finally, research workers should also refine the methods used to ensure greater animal welfare and reduction in distress and improve the quality of the data derived, if possible.

In some areas, the use of *in vitro* systems has been successful. For example, the use of *in vitro* tests for the detection of genotoxicity is now well established. These tests include the well-known Ames test, which relies on detecting mutations in bacteria (such as *Salmonella typhimurium*). These are useful early screens for detecting potential genotoxicity.

Other microorganisms such as yeast may also be used. Mammalian cells can similarly be employed for tests for genotoxicity, typically mouse lymphoma or Chinese hamster ovary cell lines. Human lymphocytes can be used for the detection of chromosomal damage. Fruit flies are sometimes used for specific tests such as the detection of sex-linked recessive lethal mutations. However, there is only partial correlation between a positive result for mutagenicity in tests such as the bacterial test and carcinogenicity in an animal. That is, known animal carcinogens are not universally mutagenic in the bacterial tests and vice versa, some mutagenic chemicals are not carcinogenic in animals (see chap. 6 for more discussion). Therefore, although *in vitro* bacterial tests may be used to screen out potential genotoxic carcinogens, those compounds, which are not apparently mutagenic, may still have to be tested for carcinogencity *in vivo* at a later stage.

Unfortunately, there are a number of problems with many of the *in vitro* systems currently in use, which make the use of such systems for prediction and risk assessment difficult.

Thus primary cells (i.e., obtained freshly from a human or animal organ) may show poor viability in medium- to long-term experiments, and this can limit their usefulness to short-term exposures. There are also major biochemical changes, which occur with time in primary cells, starting from almost the moment of preparation of the tissue. Changes, such as in the level and proportions of isozymes of cytochrome P-450, for instance, which occur over the first 24 hours after isolation will influence the toxicity of chemicals in those cells if metabolic activation is a factor (see chaps. 4, 5, and 7).

An alternative *in vitro* system is the use of cell lines, immortal cells, which will continue to grow and can be frozen and used when needed. However, these cells, usually derived from tumors, are not the same as those in normal tissue.

Where comparisons have been made with *in vivo* data, in many cases, the *in vitro* system reacts differently to the tissue in the animal *in vivo*. This difference may be qualitative or quantitative.

Therefore, the data generated from them have to be viewed with caution. This is particularly the case if the data are being used as part of a risk assessment. Such *in vitro* data may underestimate the toxicity *in vivo*.

Thus, it is not yet possible to replace all animal experiments with *in vitro* systems even though considerable progress has been made. *In vitro* systems are particularly useful, however, for screening out toxic compounds, which might otherwise be developed, for mechanistic studies and for comparing different compounds within a group of analogues, for example.

2.6 INTERACTIONS

It is appropriate at this point to mention, in general terms, interactions, which may affect toxic responses. However, many specific factors will be discussed in detail later in this book and especially in chapter 5.

Although under experimental conditions, animals and humans are mostly exposed to only one chemical, in the environment, organisms of all kinds are potentially exposed to mixtures of chemicals. This is clearly the case with the administration of drugs, patients with some conditions receiving several drugs simultaneously. Also patients receiving only one drug may also be exposed to other chemicals in food or at their place of work. Similarly, pesticides may be mixtures, and wild organisms can be exposed to several different pesticides as they move through their environment.

Therefore, knowledge and understanding of interactive effects of chemicals are crucial for a number of reasons. For example

1. The design of antidotes to poisoning
2. The use of pesticides
3. The toxicity of drugs
4. The toxicity of environmental chemicals
5. The interaction of diet and drugs

In toxicology, we must be aware of this and the various possible consequences, which have a number of underlying mechanisms. The simplest situation is for a mixture of two chemicals.

When two chemicals (or more), which have the same toxic effect, are given to an animal, the resulting toxic response could simply be the sum of the individual responses. This situation, where there is no interaction, is known as an additive effect. Conversely, if the overall toxic response following exposure to two chemicals is more than the sum of the individual responses, the effect is called "synergism." There are a number of possible reasons for this. For example, both chemicals could interact with the same system differently, so as to enhance the effect such as by one increasing the sensitivity of the receptor for the other. An example of synergy is a combination of ethanol and carbon tetrachloride. Both are toxic to the liver, but together they are much more toxic than either separately. Another example is the combined effect of asbestos and cigarette smoking in humans, which both cause lung cancer. Asbestos, increases the incidence by fivefold, and smoking increases the incidence by 11-fold, but the combination of the two as in a smoker who works with asbestos, is a 55-fold increase in the incidence of cancer.

Potentiation is similar to **synergism** except that the two substances in question have different toxic effects, or perhaps, only one is toxic. For instance, when the drug **disulphiram** is given to alcoholics, subsequent intake of ethanol causes toxic effects to occur because of the interference in the metabolism of ethanol by disulphiram. However, disulphiram has no toxic effect at the doses administered. There are many other examples of potentiation and a number of them are covered in this book (see sects. 7.2.4, 7.2.5, and 7.2.1, chap. 7).

It should be noted that synergism and potentiation may be defined in the reverse way in some texts, and the term "insecticide synergists," as defined here, usually reflects potentiation. The definition used here is the same as that used in pharmacology.

It is also conceivable that the administration of two substances to an animal may lead to a toxic response, which is entirely different from that of either of the compounds.

This would be a **"coalitive"** effect. Alternatively, "antagonism" may occur in which one substance decreases the toxic effect of another toxic agent. Thus, the overall toxic effect of the two compounds together maybe less than additive.

2.6.1 Mechanisms

There are four basic mechanisms underlying interactions: functional, chemical, dispositional, and receptor.

Functional interactions are those in which both of the two chemicals affect a bodily system perhaps by different mechanisms, and either increase or decrease the combined effect. For example, both atropine and pralidoxime decrease the toxic effects of organophosphate compounds by different means, a combination of the two antidotes leads to a large increase in effectiveness (*synergism*).

Chemical interactions are those in which one chemical combines with another to become more or less toxic. For example, the chelating agent EDTA combines with toxic metals such as lead and decreases its toxicity (*antagonism*).

Dispositional interactions are those in which one chemical affects the disposition of the other, usually metabolism. Thus, one chemical may increase or inhibit the metabolism of another to change its toxicity. For example, 2,3-methylenedioxynaphthalene inhibits cytochrome P-450 and so markedly increases the toxicity of the insecticide carbaryl to flies (*potentiation*) (see chap. 5). Another example, which results in *synergy*, is the increased toxicity of the organophosphorus insecticide malathion (see chap. 5) when in combination with another organophosphorus insecticide, EPN. EPN blocks the detoxication of malathion. Many chemicals are either enzyme inhibitors or inducers and so can increase or decrease the toxicity of other chemicals either by synergism or potentiation (see chap. 5).

Receptor interactions are those where two chemicals both interact with the same receptor to change (usually decrease) the toxic effect of the combination. For example, naloxone binds to the same receptor as morphine and other opiates and so can be used as an antidote to excessive doses of opiates (*antagonism*). In other cases, such as when two organophosphates are used together, both acting on acetylcholinesterase, the combined effect would be as expected (*additive*).

These interactive effects may be visualized graphically as **isoboles** (Fig. 2.3) or alternatively, there are simple formulae, which may be used for detecting them:

$$V = \frac{\text{expected ED}_{50} \text{ of } (A + B)}{\text{observed ED}_{50} \text{ of } (A + B)}$$

If $V < 0.7$, there is antagonism; if $V = 0.7 - 1.3$, an additive effect occurs; if $V = 1.3 - 1.8$, the effect is more than additive; if $V > 1.8$, there is synergism or potentiation. For further discussion, see Brown (4) and references therein. Note that these interactive effects may occur with single acute doses or repeat dosing, and may depend on the timing of the doses relative to each other.

The response of an organism to a toxic compound may become modified after repeated exposure. For example, **tolerance** or reduced responsiveness can develop when a compound is repeatedly administered. This may be the result of increasing or decreasing the concentration of a particular enzyme involved or by altering the number of receptors. For example, repeated dosing of animals with phenobarbital leads to tolerance to the pharmacological response as a result of enzyme induction (see chap. 5). Conversely, tolerance to the hepatotoxic effect of a

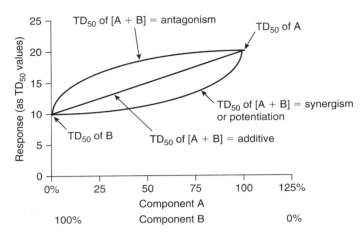

Figure 2.3 A set of isoboles for two toxic compounds A and B.

large dose of carbon tetrachloride results from the destruction of particular enzymes after small doses of the compound have been administered (for more details see chap. 7).

After repeated dosing of animals with β-agonist drugs such as clenbuterol, there is a decrease in both the density of β-receptors in muscle tissue and the stimulation of protein synthesis caused by this drug.

2.7 DOSE RESPONSE

It is clear from the earlier discussion that the measurable endpoint of toxicity may be a pharmacological, biochemical, or a pathological change, which shows percentage or proportional change. Alternatively, the endpoint of toxicity may be an all-or-none or quantal type of effect such as death or loss of consciousness. In either case, however, there will be a dose-response relationship. The basic form of this relationship is shown in Figure 2.4.

However, the dose response relationship is constructed, using either graded or all-or-none data, it is based on certain assumptions. Although not all toxic reactions can be ascribed to interactions with receptors with certainty, those due to pharmacological effects mostly can. Thus drugs show toxic reactions, which are commonly exaggerated pharmacological effects, which may represent effectively the top part of the dose-effect curve for therapeutic action. For example, if a drug causes lowering of blood pressure (via a receptor interaction), a high dose of the drug may lower this to a dangerous level, and this then is a toxic reaction. Alternatively, the drug may interact with another receptor, and this can lead to unwanted, adverse effects. It should be realized, however, that receptors are not the only structures specifically present in

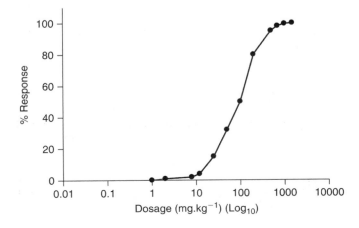

Figure 2.4 A dose-response curve.

cells, which can elicit a particular physiological or cellular response when a toxic chemical binds. Enzymes and other macromolecules such as DNA can also be considered as "receptors," binding to which may cause a particular response. However, for many toxic chemicals, the identity of the receptor or cellular target(s) is unknown.

Two examples of toxicity, where the target is known, are carbon monoxide, which interacts specifically with hemoglobin, and cyanide, which interacts specifically with the enzyme cytochrome a_3 of the electron transport chain (see chap. 7). The toxic effects of these two compounds are a direct result of these interactions and, it is assumed, depend on the number of molecules of the toxic compound bound to the receptors. However, the final toxic effects involve cellular damage and death and also depend on other factors. Other examples where specific receptors are known to be involved in the mediation of toxic effects are microsomal enzyme inducers, organophosphorus compounds, and peroxisomal proliferators (see chaps. 5–7).

The study of receptors has not featured as prominently in toxicology as in pharmacology. However, with some toxic effects such as the production of liver necrosis caused by paracetamol, for instance, although a dose-response relation can be demonstrated (see chap. 7), it currently seems that there may be no simple toxicant-receptor interaction in the classical sense. It may be that a specific receptor-xenobiotic interaction is not always a prerequisite for a toxic effect. Thus, the pharmacological action of volatile general anesthetics does not seem to involve a receptor, but instead the activity is well correlated with the oil-water partition coefficient. However, future detailed studies of mechanisms of toxicity will, it is hoped, reveal the existence of receptors or other types of specific targets where these are involved in toxic effects.

The dose-response relationship is predicated on certain assumptions, however:

1. That the toxic response is a function of the concentration of the compound at the site of action
2. That the concentration at the site of action is related to the dose
3. That the response is causally related to the compound given

Examination of these assumptions indicates that there are various factors that may affect the relationship. Furthermore, it is also assumed that there is a method for measuring and quantifying the toxic effect in question. As already indicated, there are many possible endpoints or criteria of toxicity, but not all are appropriate.

2.7.1 Toxic Response Is a Function of the Concentration at the Site of Action

The site of action may be an enzyme, a pharmacological receptor, another type of macromolecule, or a cell organelle or structure. The interaction of the toxic compounds at the site of action may be reversible or irreversible. The interaction is, however, assumed to initiate a proportional response. If the interaction is reversible, it may be described as follows:

$$R + T \underset{k_2}{\overset{k_1}{\rightleftharpoons}} RT \qquad (1)$$

where R = receptor; T = toxic compound, RT = receptor-compound complex, k_1 and k_2 = rate constants for formation and dissociation of the complex, then

$$\frac{[R][T]}{[RT]} = \frac{k_1}{k_2} = K_T \qquad (2)$$

where K_T = dissociation constant of the complex. If [Rt] is the total concentration of receptors and [Rt] = [R] + [RT], then

$$\frac{[RT]}{[R_t]} = \frac{[T]}{K_T + [T]} \qquad (3)$$

If the response of effect (e) is proportional to the concentration of RT, then

$$e = k_3[RT]$$

and the maximum response ($E_{max.}$) occurs when all the receptors are occupied:

$$E_{max.} = k_3[R_t]$$

then:

$$e = E_{max.} \cdot \frac{[RT]}{[R_t]}$$

This may be transformed into

$$e = \frac{E_{max.}[T]}{K_T + [T]} \qquad (4)$$

Thus, when $[T] = 0$, $e = 0$ and when $e = \frac{1}{2}E_{max.}$, $K_T = [T]$. Thus, Eq. 4 is analogous to the Michaelis–Menten equation describing the interaction of enzyme and substrate.

Thus, the more the molecules of the receptor that are occupied by the toxic compound, the greater the toxic effect. Theoretically, there will be a concentration of the toxic compound at which all of the molecules of receptor (r) are occupied, and hence, there will be no further increase in the toxic effect

i.e.,
$$\frac{[RT]}{[r]} = 1 \text{ or } 100\% \text{ occupancy}$$

The relationship described above gives rise to the classical **dose-response curve** (Fig. 2.4).

For more detail and the mathematical basis and treatment of the relationship between the receptor-ligand interaction and dose-response relationship, the reader is recommended to consult one of the texts indicated at the end of this chapter (5–7).

However, the mathematics describes an idealized situation, and the real situation *in vivo* may not be so straightforward. For example, with carbon monoxide, as already indicated, the toxicity involves a reversible interaction with a receptor, the protein molecule hemoglobin (see chap. 7 for further details of this example). This interaction will certainly be proportional to the concentration of carbon monoxide in the red blood cell. However, *in vivo* about 50% occupancy or 50% carboxyhemoglobin may be sufficient for the final toxic effect, which is cellular hypoxia and lethality. Duration of exposure is also a factor here because hypoxic cell death is not an instantaneous response. This time-exposure index is also very important in considerations of chemical carcinogenesis.

Therefore, *in vivo* toxic responses often involve several steps or sequelae, which may complicate an understanding of the dose-response relationship in terms of simple receptor interactions. Clearly, it will depend on the nature of response measured. Thus, although an initial biochemical response may be easily measurable and explainable in terms of receptor theory, when the toxic response of interest and relevance is a pathological change, which occurs over a period of time, this becomes more difficult.

The number of receptor sites and the position of the equilibrium (Eq. 1) as reflected in KT, will clearly influence the nature of the dose response, although the curve will always be of the familiar sigmoid type (Fig. 2.4). If the equilibrium lies far to the right (Eq. 1), the initial part of the curve may be short and steep. Thus, the *shape* of the dose-response curve depends on the type of toxic effect measured and the mechanism underlying it. For example, as already mentioned, cyanide binds very strongly to cytochrome a_3 and curtails the function of the electron transport chain in the mitochondria and hence stops cellular respiration. As this is a function vital to the life of the cell, the dose-response curve for lethality is very steep for cyanide. The intensity of the response may also depend on the number of receptors available. In some cases, a proportion of receptors may have to be occupied before a response occurs. Thus, there is a threshold for toxicity. With carbon monoxide, for example, there are no toxic effects below a carboxyhemoglobin concentration of about 20%, although there may be measurable physiological effects. A threshold might also occur when the receptor is fully occupied or saturated. For example, an enzyme involved in the biotransformation of the toxic compound may become saturated, allowing another metabolic pathway to occur, which is responsible for toxicity. Alternatively, a receptor involved in active excretion may become saturated, hence causing a disproportionate increase in the level of the toxic compound in the body when the dose is increased. Such saturable processes may determine the shape and slope of the dose-response curve.

However, when the interaction is irreversible, although the response may be proportional to the concentration at the site of action, other factors will also be important.

If the interaction is described as

$$R + T = RT \tag{5}$$
$$RT \rightarrow ?,$$

the fate of the complex RT in Eq. 5 is clearly important. The repair or removal of the toxin-receptor complex RT may therefore be a determinant of the response and its duration.

From this discussion, it is clear that the reversible and irreversible interactions may give rise to different types of response. With reversible interactions, it is clear that at low concentrations, occupancy of receptors may be negligible with no apparent response, and there may, therefore, be a threshold below which there is a "no-effect level." The response may also be very short, as it depends on the concentration at the site of action, which may only be transient. Also, repeated or continuous low-dose exposure will have no measurable effect.

With irreversible interactions, however, a single interaction will theoretically be sufficient. Furthermore, continuous or repeated exposure allows a cumulative effect dependent on the turnover of the toxin-receptor complex. An example of this is afforded by the organophosphorus compounds, which inhibit cholinesterase enzymes (see Aldridge (7) and chap. 7).

This inhibition involves reaction with the active site of the enzyme, which is often irreversible. Resynthesis of the enzyme is therefore a major factor governing the toxicity. Toxicity only occurs after a certain level of inhibition is achieved (around 50%). The irreversibility of the inhibition allows cumulative toxicity to occur after repeated exposures over an appropriate period of time relative to the enzyme resynthesis rate.

With chemical carcinogens, the interaction with DNA after a single exposure could be sufficient to initiate eventual tumor production with relatively few molecules of carcinogen involved, depending on the repair processes in the particular tissue. Consequently, chemical carcinogens may not show a measurable threshold, indicating that there may not be a no-effect level as far as the concentration at the site of action is concerned. Although the DNA molecule may be the target site or receptor for a carcinogen, it now seems as though there are many subsequent events or necessary steps involved in the development of a tumor. There may, therefore, be more than one receptor-carcinogen interaction, which will clearly complicate the dose-response relationship.

Also access to the target may be a factor. For example, DNA repair seems to be very important in some cases of chemical carcinogenesis and will contribute to the presence of a dose threshold.

The existence of "no-effect doses" for toxic compounds is a controversial point, but it is clear that to measure the exposure sufficiently accurately and to detect the response reliably are major problems (see below for further discussion). Suffice it to say that certain carcinogens are carcinogenic after exposure to concentrations measured in parts per million, and the dose-response curves for some nitrosamines and for ionizing radiation appear to pass through zero when the linear portion is extrapolated. At present, therefore, in some cases no-effect levels cannot be demonstrated for certain types of toxic effect.

With chemical carcinogens, time is also an important factor, both for the appearance of the effect, which may be measured in years, and for the length of exposure. It appears that some carcinogens do not induce tumors after single exposures or after low doses but others do. In some cases, there seems to be a relationship between exposure and dose, that is, low doses require longer exposure times to induce tumors than high doses, which is as would be expected for irreversible reactions with nucleic acids. For a further discussion of this topic, the reader is referred to the articles by Aldridge (7, chap. 6) and Zbinden (8).

2.7.2 Concentration at the Site of Action Is Related to the Dose

Although the concentration in tissues is generally related to the dose of the foreign compound, there are various factors, which affect this concentration. Thus, the absorption from the site of exposure, distribution in the tissues, metabolism, and excretion all determine the concentration at the target site. However, the concentration of the compound may not be directly proportional to the dose, so the dose-response relationship may not be straightforward, or marked thresholds may occur. For instance, if one or more of the processes mentioned is saturable or changed by dose, disproportionate changes in response may occur. For example, saturation of plasmaprotein binding sites may lead to a marked increase in the plasma and tissue levels of the free compound in question. Similarly, saturation of the processes of metabolism and excretion, or accumulation of the compound, will have a disproportionate effect. This may occur with acute dose-response studies and also with chronic dosing as, for example, with the drug chlorphentermine (chap. 3, Fig 19), which accumulates in the adrenals but not in the liver after chronic dosing.

The result of this is accumulation of phospholipids, or phospholipidosis, in the tissues where accumulation of the drug occurs. Active uptake of a toxic compound into the target tissue may also occur. For example, the herbicide paraquat is actively accumulated in the lung, reaches toxic concentrations in certain cells, and then tissue damage occurs (see chap. 7).

The relationship between the dose and the concentration of a compound at its site of action is also a factor in the consideration of the magnitude of the response and no-effect level.

The processes of distribution, metabolism, and excretion may determine that none of the compounds in question reaches the site of action after a low dose, or only does so transiently. For both irreversible and reversible interactions, but particularly for the latter, this may be the major factor determining the threshold and the magnitude and duration of the response. For example, the dose required for a barbiturate to induce sleep in an experimental animal and the length of time that that animal remains unconscious can be drastically altered by altering the activity of the enzymes responsible for metabolizing the drug. Changes in the level of a toxic compound in the target tissue may occur because of changes in the pH of the blood or urine, causing changes in distribution and excretion of the compound. This phenomenon is used in the treatment of poisoning to reduce the level of drug in the central nervous system after overdoses of barbiturates and salicylates (see chap. 7). Both of these examples involve alteration of the concentration of drug at the site of action.

2.7.3 Response Is Causally Related to the Compound

Although this may seem straightforward, in some cases, the response is only indirectly related and is therefore not a useful parameter of toxicity to use in a dose-response study. This may apply to situations where enzyme inhibition is a basic parameter but where it may not relate to the overall toxic effect. For example, inhibition by lead of aminolaevulinic acid dehydrase, an enzyme, which is involved in heme synthesis, can be readily demonstrated to be dose related, but is clearly not an appropriate indicator of lead-induced renal toxicity *in vivo*.

When more information has been gained about the toxicity or when the underlying mechanism of toxicity is understood, then more precise indicators of toxicity can be measured. Similarly, this criterion must be rigorously applied to epidemiological studies where a causal relationship may not be apparent or indeed may not even exist.

2.8   MEASUREMENT OF DOSE-RESPONSE RELATIONSHIPS

It should be clear from the earlier discussion that the measurable endpoint of toxicity could be a biochemical, physiological, or pathological change. This toxic effect will show a "graded" increase as the dose of toxicant increases. Alternatively, the toxic effect may be an all-or-none effect such as death or the presence or absence of a tumor (which can be considered in such a way). These are also called quantal effects. In this case, an increase in the dose will result in an increase in the *proportion of individuals* showing the response rather than an increase in the *magnitude* of the effect.

Therefore, we can identify two types of relationship with the dose of the toxicant: a dose-effect relationship (graded effect) and a dose-response relationship (all-or-none effect). However, the term "dose-response relationship" is often used to describe both types.

With graded effects, as the dose increases, the effect, such as inhibition of an enzyme, increases from zero to maximum. This results in a sigmoid curve when plotted (Fig. 2.4). There is, therefore, a threshold dose below which there is no effect but above which an effect is detectable. Clearly, there will also be maximal effect above which no further change is possible; it is impossible to inhibit an enzyme more than 100%, for instance.

However, the quantal type of relationship with dose will also show a sigmoid curve when appropriately plotted. In this case, the curve derives from a frequency distribution (Fig. 2.5), which is the familiar Gaussian curve.

Those animals or patients responding at the lowest doses (Fig. 2.5) are more sensitive (hypersensitive) and those responding at the highest doses are less sensitive than the average (hyposensitive). The median point of the distribution is the dose where 50% of the population has responded and is the midpoint of the dose-response curve (Fig. 2.6). If the frequency distribution of the response is plotted cumulatively, this translates into a sigmoid curve. The more perfect the Gaussian curve, the closer to a true sigmoid curve will the dose-response curve be.

A threshold also exists for quantal dose responses as well as graded, i.e., there will be a dose below which no individuals respond. However, the concept of a threshold also has to be considered in relation to the variation in sensitivity in the population, especially a human population with great variability. Thus, although there will be a dose at which the greatest number of individuals show a response (see point B in Fig. 2.5), there will be those individuals who are very much more sensitive (point A in Fig. 2.5) or those who are much less sensitive (point C in Fig. 2.5). This consideration is incorporated into risk assessment of chemicals such as food additives, contaminants, and industrial chemicals (see below).

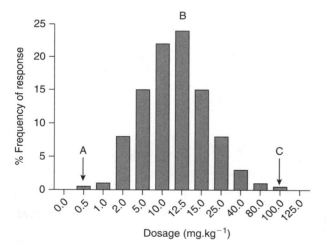

Figure 2.5 Dose-response relationship expressed as a frequency distribution.

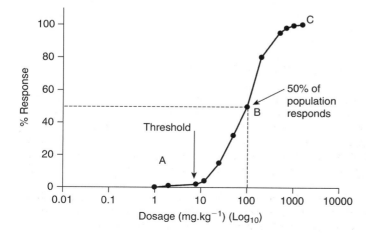

Figure 2.6 Illustration of the relation between a frequency distribution and dose-response curve. Points A, B and C relate to Figure 2.5.

For a drug, it may be that the therapeutic dose causes unacceptable side effects in such hypersensitive individuals and cannot be used.

The portion of the dose-response curve between 16% and 84% is the most linear and may be used to determine parameters such as ED_{50}, TD_{50}, or LD_{50}. These are the doses, which, from the dose-response curve, are estimated to cause a 50% response (either pharmacological, toxicological, or lethal) in 50% of the animals or a 50% inhibition of an enzyme, for example. The linearity of the dose-response curve may often be improved by plotting the $\log_{10}$ of the dose, although this is an empirical transformation. In some cases, dose-response curves may be linearized by applying other transformations. Thus, for the conversion of the whole sigmoid dose-response curve into a linear relationship, probit analysis may be used, which depends upon the use of standard deviation units. The sigmoid dose-response curve may be divided into multiples of the standard deviation from the median dose, this being the point at which 50% of the animals being used respond. Within one standard deviation either side of the median, the curve is linear and includes 68% of the individuals; within two standard deviations, fall 95.4% of the individuals.

Probit units define the median as probit five, and then each standard deviation unit is one probit unit above or below. The dose-response curve so produced is linear, when the logarithm of the dose is used (Fig. 2.7).

As well as mortality, other types of response can be plotted against dose. Similarly, a median effective dose can be determined from these dose-response curves such as the ED_{50} where a pharmacological, biochemical, or physiological response is measured or the TD_{50} where a toxic response is measured. These parameters are analogous to the LD_{50} (Fig. 2.8). The effective dose for 50% of the animals is used because the range of values encompassed is narrowest at this point compared with points at the extremities of the dose-response curve. A variation of the LD_{50} is the LC_{50}, which is the concentration of a substance, which is lethal to

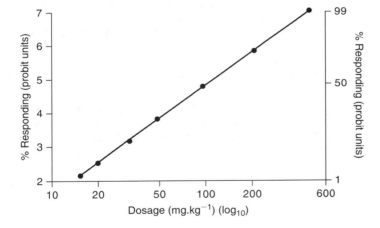

Figure 2.7 Dose-response relationship expressed as probit units.

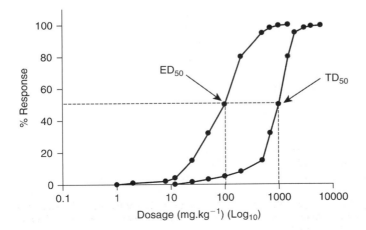

Figure 2.8 Dose-response curves for pharmacological effect and toxic effect, illustrating the ED_{50} and TD_{50}. The proximity of the curves for efficacy and toxicity indicates the margin of safety for the compound and the likelihood of toxicity occurring in certain individuals after doses necessary for the desired effect.

50% of the organisms, when exposed. This parameter is used in situations where an organism is exposed to a particular concentration of a substance in air or water, but the dose is unknown. Clearly, the exposure time in this case as well as the concentration must be indicated. The *slope* of the dose-response curve depends on many factors, such as the variability of measurement of the response and the variables contributing to the response. The greater the number of animals or individual measurements and the more precise the measurement of the effect, the more accurate are the parameters determined from the dose-response curve. The slope of the curve also reflects the type of response. Thus, when the response reflects a potent single effect, such as avid binding to an enzyme or interference with a vital metabolic function, as is the case with cyanide or fluoroacetate, for example, the dose-response curve will be steep and the value of the slope will be large. Conversely, a less specific toxic effect with more inherent variables results in a shallower curve with a greater standard deviation around the TD_{50} or LD_{50}. The slope therefore may give some indication of the mechanism underlying the toxic effect. Sometimes, two dose-response curves may be parallel. Although they may have the same mechanism of toxicity, this does not necessarily follow. The slope of the curve is also essential information for a comparison of the toxicity of two or more compounds and for a proper appreciation of the toxicity. The LD_{50} or TD_{50} value alone is not sufficient for this as can be seen from Figure 2.9.

The type of measurement made, and hence the type of data treatment, depends on the requirements of the test. Thus, measurement of the percent response at the molecular level may be important mechanistically and more precisely measured. However, for the assessment of toxicity, measurement of the population response may be more appropriate.

Apart from possibly giving an indication of the underlying mechanism of toxicity, one particular value of quantitation of toxicity in the dose-response relationship is that it allows comparison. Thus, for example, comparisons may be made between different responses, between different substances, and between different animal species.

Comparison of different responses underlies the useful parameter, **therapeutic index**, defined as follows:

$$\text{Therapeutic Index (TI)} = \frac{TD_{50}}{ED_{50}} \text{ or } \frac{LD_{50}}{ED_{50}}$$

It relates the pharmacologically effective dose to the toxic or lethal dose (Fig. 2.8). The therapeutic index gives some indication of the safety of the compound in use, as the larger the ratio, the greater the relative safety. However, as already indicated, simple comparison of parameters derived from the dose-response curve such as the LD_{50} and TD_{50} may be

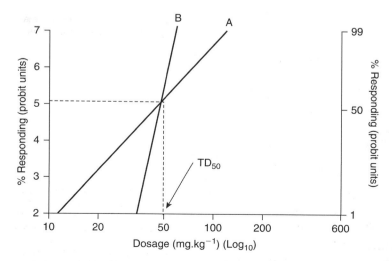

Figure 2.9 Comparison of the toxicity of two compounds, A and B. Although the TD_{50} is the same (50 mg/kg^{-1}) for A and B, toxicity occurs with A at a much lower dose than with B, but the minimum to maximum effect is achieved with B over a very much narrower dose range.

misleading without some knowledge of the shape and slope of the curve. A more critical index is the margin of safety:

$$\text{Margin of Safety} = \frac{TD_1}{ED_{99}} \text{ or } \frac{LD_1}{ED_{99}}$$

Similarly, comparison of two toxic compounds can be made using the LD_{50} (TD_{50}) (Fig. 2.9) and the dose-response curves, and this may also give information on possible mechanisms of toxicity. Thus, apart from the slope, which may be useful in a comparative sense, examination of ED_{50}, TD_{50}, and LD_{50} may also provide useful information regarding mechanisms. Comparison of the LD_{50} or TD_{50} values of a compound after various modes of administration (Table 1) may reveal differences in toxicity, which might indicate the factors, which affect the toxicity of that particular compound. Thus with the antitubercular drug isoniazid, there is little difference in toxicity after dosing by different routes of administration, whereas with the local anesthetic procaine, there is an 18-fold difference in the LD_{50} between intravenous and subcutaneous administration of the drug (Table 2.1). Shifts in the dose-response curve or parameters derived from it caused by various factors may give valuable insight into the mechanisms underlying toxic effects (Table 2.2). The dosage of the compound to which the animal is exposed is usually expressed as mg/kg body weight, or sometimes mg/m^2 of surface area. However, because of the variability of the absorption and distribution of compounds, it is preferable to relate the response to the plasma concentration or concentration at the target site.

This may be particularly important with drugs used clinically, which have a narrow therapeutic index or which show wide variation in absorption or where exposure is unknown (e.g., with industrial chemicals).

It will be clear from the discussion in the preceding pages, and should be noted, that the LD_{50} value is not an absolute biological constant as it depends on a large number of factors. Therefore, despite standardization of test species and conditions for measurement, the value for a particular compound may vary considerably between different determinations in different laboratories. Comparison of LD_{50} values must therefore be undertaken with caution and regard for these limitations.

Table 2.1 Effect of Route of Administration on the Toxicity of Various Compounds

Route of administration	Phenobarbital[a]		Isoniazid[a]		Procaine[a]		DFP[b]	
	LD_{50} (mg/kg)	Ratio to i.v.	LD_{50} (mg/kg)	Ratio to i.v.	LD_{50} (mg/kg)	Ratio to i.v.	LD_{50} (mg/kg)	Ratio to i.v.
Oral	280	3.5	142	0.9	500	11	4.0	11.7
Subcutaneous	130	1.6	160	1.0	800	18	1.0	2.9
Intramuscular	124	1.5	140	0.9	630	14	0.85	2.5
Intraperitoneal	130	1.6	132	0.9	230	5	1.0	2.9
Intravenous	80	1.0	153	1.0	45	1	0.34	1.0

[a]Mouse toxicity data.
[b]Rabbit toxicity data on di-isopropylfluorophosphate.
Abbreviation: i.v., intravenous.
Source: From Ref. 9.

Table 2.2 Effect of BDL on the Toxicity of Certain Compounds

Compound	LD$_{50}$ (mg/kg)		
	Sham operation	BDL	Sham-BDL ratio
Amitryptaline	100	100	1
Diethylstilboestrol	100	0.75	130
Digoxin	11	2.6	4.2
Indocyanine green	700	130	5.4
Pentobarbital	110	130	0.8

Abbreviation: BDL, bile duct ligation.
Source: From Ref. 10.

The value of the LD_{50} test and the problems associated with it have been reviewed (11).

Although the straightforward threshold model of the dose-response relationship as described here is the one originally conceived and the one for which there is clear mechanistic justification, other dose-response relationships have been suggested. The other dose-response relationships are substantially different and lead to different predictions in relation to toxicity. This becomes particularly important in risk assessment (see below).

2.9   LINEAR DOSE RESPONSE

When the effect of exposure to a chemical is the production of a cancer, it is sometimes assumed, for instance, by regulatory agencies such as the U.S. Environmental Protection Agency (EPA) that the dose-response curve passes through zero. Thus, it is not like the dose-response curve we have been discussing above where there is a threshold. The zero threshold dose response is predicated on the belief that the causation of cancer by a genotoxic mechanism is a stochastic (chance) event, in which a reactive chemical binds to and damages or alters DNA (see chap. 6).

Therefore, it is argued, there is no safe dose of such a chemical because one molecule could *theoretically* interact with the DNA in one cell, which *could* then become a tumor.

Therefore, the curve is not an S shape but the lowest portion is linearized and extrapolated to the origin (Fig. 2.10).

Now in practice it seems very likely that this is not the case and that an "effective" threshold exists. This can be justified on the following grounds:

First, the chemical must gain access to a cell. This requires, at the least, crossing biological membranes and entering an aqueous environment in which substances such as glutathione and vitamin C are present, which can detoxify reactive chemicals.

Furthermore, most carcinogenic chemicals of concern need to be metabolized to reactive intermediates. This requires interaction with an enzyme, which may only be present in certain types of cell. Therefore, the chemical may need to traverse several cells and the bloodstream to enter a metabolically competent cell such as those in the liver. As already discussed, by the laws of mass action, just as a single molecule is unlikely to interact with and affect a receptor, a single molecule is also unlikely to interact with the necessary enzyme and be converted into a reactive metabolite. Even if it did, protective systems such as gluthione exist to remove such reactive metabolites. In the event that a reactive metabolite formed reached the nucleus, entered and reacted with DNA, further protective systems exist. One is DNA repair, which removes and repairs damaged and altered DNA bases. Another is programmed cell death or apoptosis, which removes damaged cells. Finally, the cell with damaged DNA may not divide, an essential step in the production of a tumor. If these modifying factors were not significant, we would probably all get cancer early in life.

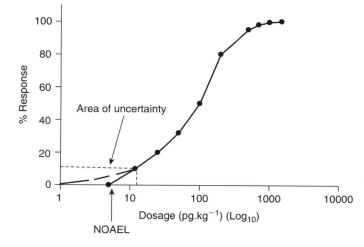

Figure 2.10 Dose-response curve showing the area of uncertainty around the low-dose, low-response area. Thus the dotted line shows the bottom portion of a dose-response curve where no threshold exists. *Abbreviation*: NOAEL, no observed adverse effect level.

For these reasons, a threshold will exist for genotoxic carcinogens in practice. However, one problem of demonstrating this experimentally is the absence of sufficiently sensitive biomarkers, which can detect effects at very low doses. Using the appearance of tumors as the endpoint is too insensitive, and therefore, the true nature of the dose-response curve at low doses is unknown. Thus, the bottom of the dose-response curve is an area of uncertainty, effectively a "black box" (Fig. 2.10). New and more sensitive biomarkers will help in this.

Furthermore, other types of toxic effect may also be stochastic events, if a reactive metabolite interacts with a critical protein or affects a gene involved with development of the embryo, for example.

2.10 HORMESIS

A third type of dose response relationship has been proposed, which is increasingly gaining acceptance, and this is the hormetic kind. This kind of dose response, for which there is experimental evidence, involves opposite effects at low doses, giving rise to a U-shaped or J-shaped curve (Fig. 2.11). That is, there may be positive or stimulatory beneficial effects at low doses. For example, some data indicate that at low doses of dioxin, the incidence of certain cancers in animals exposed is *less* than occurs in controls. Another example is alcohol (ethanol), for which there is evidence from a number of studies that low to moderate intake in man leads to lower levels of cardiovascular disease. Of course, high levels of intake of alcohol are well established to cause liver cirrhosis, various cancers, and also damage to the cardiovascular system.

However, it must be ascertained if the positive effect is directly related to the toxic effect and whether the same positive effect is observed using a variety of markers.

Mechanisms, which have been proposed to account for hormesis, are based on the premise that low doses stimulate repair or protective measures and that this is followed by overcompensation. Hence, there is a reduced level of pathological change. As the dose increases, the damage and dysfunction are less easily repaired or there is less reserve capacity, and as doses increase further, these processes are overwhelmed. This gives rise to a threshold for toxicity. Changes stimulated by low doses include increased DNA repair following a genotoxic insult, induction of stress proteins, and other endogenous protective substances such as metallothionein and glutathione and induction of enzymes such as cytochrome P-450 (see chaps. 5 and 6). However, these mechanisms can require exposure to more than one dose and thus may have a temporal component.

Because the positive effects will occur at low doses, showing these experimentally is difficult and it adds a layer of complexity to determining a dose-response relationship for a chemical.

Such effects may be confused or obscured by normal biological variation, as they are typically only 30% to 60% above the control. Furthermore, if the background level of tumor incidence (or other effect being measured) is low, it may be impossible to assess hormesis.

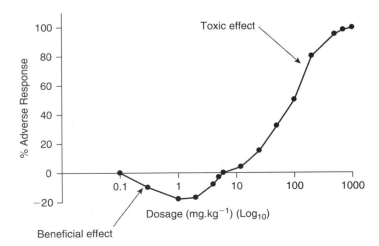

Figure 2.11 A hormetic type of dose-response curve where low doses give a positive, beneficial effect and high doses a negative, toxic effect.

Many toxicity studies, especially long-term bioassays carried out to determine potential carcinogenicity, use high-dose levels (e.g., maximum tolerated dose), and consequently, any hormetic response would be missed. To be properly evaluated, more doses and a wider dose response would have to be investigated.

Even if hormesis occurs, there could still be a threshold for a toxic or adverse effect, below which positive effects may occur. Therefore, the significance in toxicological risk assessment and even in toxicology is not entirely clear (12).

However, it does mean that extrapolation of a dose response in a linear fashion to zero could be too simplistic for some chemicals at least. Thus, mechanisms occurring after high-dose exposure may not be relevant to low-dose risk assessment.

It is currently uncertain if this phenomenon occurs across all chemical types, species of animal or cell, and type of toxic response.

This seems to be an area, which requires much further research using low doses and sensitive biomarkers to detect effects at these doses.

2.11 HAZARD AND RISK ASSESSMENT

2.11.1 Risk Assessment

An important role for the dose-response relationship and biomarkers is in risk assessment.

Risk is a mathematical concept, which refers to the likelihood of undesirable effects resulting from exposure to a chemical.

Risk is defined as the probability that a hazard will cause an adverse effect under specific exposure conditions and may also be defined in the following way:

$$Risk = hazard \times exposure$$

Hazard is defined as the capability of a substance to cause an adverse effect.

Conversely, safety may be defined as "the practical certainty that adverse effects will not occur when the substance is used in the manner and quantity proposed for its use."

As exposure increases so does the probability of harm, and therefore a reduction in exposure reduces the risk.

Risk assessment is carried out on chemicals for the following reasons:

1. The likelihood of being a hazard to humans in the environment
2. The likelihood of persistence of the chemical in the environment and bioaccumulation
3. The likelihood that sensitive human and ecological populations may be exposed to significant levels
4. An indication of hazard to human health
5. The likelihood of exposure via use or production

Risk assessment is the process whereby hazard, exposure, and risk are determined.

An underlying concept in risk assessment relies on the statement by Paracelsus (see above) and the fact that for most types of effect, there will be a dose-response relationship. Therefore, the corollary is that there should be a safe dose. Consequently, it should be possible to determine a level of exposure, which is without appreciable risk to human health or the ecosystem.

Risk assessment is a scientific process. The next stages are risk benefit analysis and risk management, which require a different type of approach.

Risk assessment is the process whereby the nature and magnitude of the risk is determined. It requires four steps:

1. Hazard identification. This is the evaluation of the toxic effects of the chemical in question.
2. Demonstration of a dose-response or dose-effect relationship. Evaluation of the causal relationship between the exposure to the hazard and an adverse effect in individuals or populations, respectively.

3. Exposure assessment. Determination of the level, frequency, and duration of exposure of human or other organisms to the hazardous substance.
4. Risk characterization. Estimation of the incidence of adverse effects under the various conditions of exposure.

Consider each of these in turn:

2.11.2 Hazard Identification

This is the evaluation of the potential of a chemical to cause toxicity and has been discussed earlier in this chapter. As indicated, the data used are normally derived from

1. human epidemiology,
2. animal toxicity studies, and
3. *in vivo* and *in vitro* mechanistic or other studies.

A chemical may constitute a number of hazards of different severity. However, the primary hazard (or critical effect) will be the one used for the subsequent stages of the risk assessment process. For example, a chemical may cause reversible liver toxicity at high doses but cause tumors in the skin at lower doses. The carcinogenicity is clearly the hazard of concern.

Therefore in practice, normally, animal toxicity data is required (see above). Of course, the differences between humans and other species must always be recognized and taken into account (see below). It may be possible to use *in vitro* data both from human cells and tissues as well as those from other animals to supplement the epidemiological and animal *in vivo* toxicity data. However, at present such data cannot replace experimental animal or human epidemiological data. The predictive use of structure-activity relationships is also possible, and it is an approach, which is becoming increasingly important.

For example, the Threshold of Toxicological Concern concept has been proposed, which reduces the amount of toxicological data necessary and therefore reduces the number of animals used in the assessment of hazard. This uses a tiered approach and excludes certain kinds of chemicals such as dioxins and organophosphates. It also makes use of structural alerts and chemical classes to select out chemicals, which are likely to be of little toxicological concern (13).

2.11.3 Dose Response Assessment

This stage quantitates the hazards already identified and estimates the relationship between the dose and the adverse effect in humans. However, this requires extrapolation from possibly high, experimental doses used in animals to levels likely to be encountered by humans.

The extrapolation from high to low doses will depend on the type of primary toxic effect. If this is a carcinogenic effect, then a threshold normally cannot be assumed, and a mathematical model is used to estimate the risk at low doses (see above). If the primary toxic effect is noncarcinogenic, then it will normally be assumed that a threshold exists.

Risk assessment of carcinogens is a two-step process involving first, a qualitative assessment of the data from the hazard identification stage (see above) and second, a quantitation of the risk for definite or probable human carcinogens.

There are several models, which can be used and which range from ultraconservative to least conservative:

1. The "one-hit" model. This is ultraconservative as it assumes that cancer involves only one stage, and a single molecular event is sufficient to induce a cellular transformation.
2. The linearized multistage model (used by the EPA). This determines the cancer slope factor, which can be used to predict cancer risk at a specific dose. It assumes a linear extrapolation to a zero-dose threshold (Fig. 2.10). This factor is an estimate (expressed in mg/kg/day) of the probability that an individual will develop cancer if exposed to the chemical for 70 years.

3. The multihit model, which assumes several interactions are necessary for transformation of a normal to a cancerous cell.
4. Probit model. This assumes a lognormal distribution for tolerance in the exposed population.

Another model, which is increasingly being used, is the physiologically based pharmacokinetic model. This uses data on the absorption, distribution, metabolism, tissue sequestration, kinetics, elimination, and mechanism to determine the target dose used for the extrapolation, but it requires extensive data.

The cancer risk values, which these models generate, are of course very different. For example, for the chemical chlordane, the lifetime risk for one cancer death in one million people ranges from exposures of 0.03 μg/L of drinking water for the one-hit model, 0.07 μg/L from the linearized multistage model to 50 μg/L for the probit model.

This problem of risk assessment of chemicals, which are mutagenic and potentially carcinogenic, relates to whether there is a threshold or not (see above). Although theoretically a single molecule of a genotoxic chemical could reach the DNA in a cell, the chances of this happening and causing a mutagenic change, which leads to a cancer, is extremely small. This is because there are many barriers, which stand in the way of such an event, and further factors, which modify the result even if it happens. As already discussed and will become clear in later chapters, the chemical has to pass many hurdles or barriers before it can initiate a potentially carcinogenic change.

For noncarcinogens, in which the dose response is believed to show a threshold, a dose can be determined at which there is no adverse effect, the NOAEL (Fig. 2.12). The effect will be one that is likely to occur in humans and which is the most sensitive toxic effect observed. If a NOAEL cannot be determined (if the data is insufficiently robust), then the "lowest adverse effect level (LOAEL)" is determined (Fig. 2.12).

2.11.4 Exposure Assessment

Exposure to a chemical converts it from being a hazard into a risk. Thus, determination of exposure is crucial to the whole process of risk assessment. This involves evaluation of the source of the exposure, the routes by which humans are exposed, and the level of exposure.

Of course in some situations of exposure to chemicals, such as around waste disposal areas or chemical factories, exposure is to a mixture of possibly many different chemicals. These may interact in a variety of ways (e.g., additivity, synergism, antagonism, potentiation, see above).

Exposure may be by more than one route (inhalation, skin contact, ingestion), and different types of organism may be exposed (human, animal, adult, infant). Therefore, the real-life situation

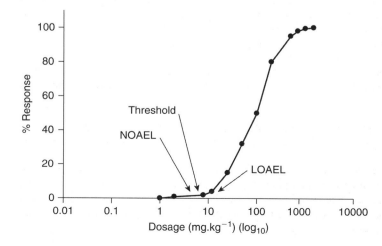

Figure 2.12 Dose-response curve showing the NOAEL, LOAEL, and threshold points. *Abbreviations*: NOAEL, no observed adverse effect level; LOAEL, lowest adverse effect level.

of exposure to chemicals in the workplace or environment can be immensely complex when these factors are taken into account. Risk assessment requires a consideration of these.

Actual exposure levels may not always be known, and therefore, such models may have to be used, which use knowledge of air dispersion or ground water movements.

The physicochemical characteristics of the chemical in question (i.e., lipid solubility, water solubility, vapor pressure, etc.) also will be important information.

However, the risk assessment process is more reliable if there is an indication of actual exposures for both the experimental animals and humans, which have provided the data on which it is based. As described, the exposure assessment may use biomarkers to improve the process.

2.11.5 Risk Characterization

The final stage involves integration of the results of the preceding stages to derive a probability of the occurrence of the adverse effect in humans exposed to the chemical. The biological, statistical, and other uncertainties will have to be taken into account.

For carcinogens, the risk is expressed in terms of increased risk of developing a cancer (e.g., 1 in 10^6). This is calculated from the cancer slope factor and the 70-year average daily intake in mg/kg/day.

From the NOAEL (or LOAEL if there is no reliable NOAEL), various parameters can be determined.

For food additives, this is normally the **acceptable daily intake** (**ADI**). The ADI is the amount of chemical to which a person can be exposed for a lifetime without suffering harmful effects. The determination of these intake values requires the use of a safety or uncertainty factor.

For food contaminants, the parameter is the **tolerable daily intake** (**TDI**). The tolerable daily intake (TDI) is an estimate of the daily intake of the chemical, which can occur over a lifetime without appreciable health risk.

Food may also contain veterinary drug residues and pesticide residues for which ADIs may be calculated.

Chemicals in water and air also have to be assessed for risk and guidance values set. As for ADI and TDI, the guidance values are determined from the NOAEL (or LOAEL).

For example, the guidance value for water is determined from the TDI and known daily intake of water by a standard adult of 60-kg weight drinking the water for 70 years.

For occupational exposure to chemicals as opposed to environmental exposure, other parameters such as **threshold limit values** (or maximum exposure limits) are determined in a similar way and are based on exposure for an eight-hour working day.

Exposure to multiple chemicals will be assumed to be additive.

The modifying or safety factors are as follows:

10× for human variability (intraspecies); this factor takes into account the variability in the human population, allowing for the most sensitive individuals (see 2.5 A)
10× for extrapolation from animals to humans (interspecies variability);
10×, if less than chronic doses have been used; and
10×, if the LOAEL rather than the NOAEL is used.

These uncertainty factors are combined and divided into the NOAEL (or LOAEL) to give the ADI or TDI. The modifying factor allows for judgment on the quality of the scientific data. Thus

$$TDI = \frac{NOAEL}{Uncertainty\ factor(s)}$$

$$ADI = \frac{NOAEL}{Uncertainty\ factor(s)}$$

Carcinogenic, non-threshold chemicals will be considered differently from noncarcinogenic chemicals, which is considered to have no threshold. In the case of carcinogens, a virtually safe dose (VSD) may be determined.

2.12 DURATION AND FREQUENCY OF EXPOSURE AND EFFECT

It is necessary to appreciate that exposures to chemicals vary in both duration and frequency, and therefore, the toxicities resulting from such exposures can also vary.

In the safety evaluation of chemicals, acute toxicity tests are those that evaluate effects occurring within about 7 days of a *single* dose, and sub-chronic toxicity tests evaluate toxicity resulting from short-term *repeated* dosing or exposure, often 28 days up to 90 days. Chronic toxicity tests are those that evaluate effects occurring after much longer exposures, at least six months and possibly up to the lifetime in experimental animals.

So exposure to a chemical can be an acute event, which is one dose, or repeated for a particular period of time, in which case it is either termed "chronic" or "sub-chronic" exposure, depending on the length of time. The effects of these types of exposure can be different although not necessarily so. For example, single, acute exposures usually lead to acute toxic effects, which occur within a few hours or days but then subside. For example, a reversible physiological change, such as bronchoconstriction, for example, caused by organophosphorus insecticides is due to interaction of the chemical at a receptor. This would be an acute effect. In contrast, sub-chronic or chronic exposures tend to cause chronic effects, that is, the effects persist at least for the length of the exposure and possibly longer (cigarette smoke–induced cancer, for example). However, sometimes single, acute exposures can lead to chronic effects, which may be delayed such as the peripheral neuropathy caused by tri-ortho-cresylphosphate (TOCP) (Fig. 2.13 and see chap. 7). Conversely, sub-chronic and chronic exposures can cause acute toxic effects, for example, due to the accumulation of the chemical or due to the accumulation of the effect. Such is the case with some organo-phosphorus compounds where one or both may occur.

Thus, the frequency of exposure is also an important factor because the concentration to which the organism, and more particularly the target site, is exposed can remain relatively constant or increase. As illustrated above, this is because repeated exposure may lead to accumulation of the chemical, depending on its half-life (see chap. 3), such that intake exceeds elimination. Thus the chemical can accumulate in the organism because of saturation of metabolism or elimination or because its physicochemical properties determine that the chemical becomes sequestered in tissues such as fat. Another factor in toxicity from chronic exposure can be the ability to repair damage or replace macromolecules and the speed with which this is done. If damage is not repaired or macromolecules are not replaced before the next exposure, then accumulation of damage or effect can also occur.

With humans and animals in the wild, environmental exposure can be discontinuous or erratic, making it difficult to predict the outcome in terms of toxicity.

Chronic toxicity may be quantitated in a similar manner to acute toxicity, using the TD_{50} concept. Measurement of chronic toxicity in comparison with acute toxicity measurements may reveal that the compound is accumulating *in vivo* and may therefore give a rough approximation of the probable whole-body half-life of the compound.

For chronic toxicity, the TD_{50} is measured for a specific period of time, such as 90 days of chronic dosing. The dose response is plotted as the percent response against the dose in (mg/kg)/day. If the TD_{50} values for acute and chronic toxicity are different, it may indicate that accumulation is taking place. This may be quantitated as the **chronicity factor**, defined as TD_{50} 1 dose/TD_{50} 90 doses, where the TD_{50} 90 dose is expressed as (mg/kg)/day. If this value is 90, the compound in question is absolutely cumulative, if more than 2, relatively cumulative, and if less than 2, relatively noncumulative.

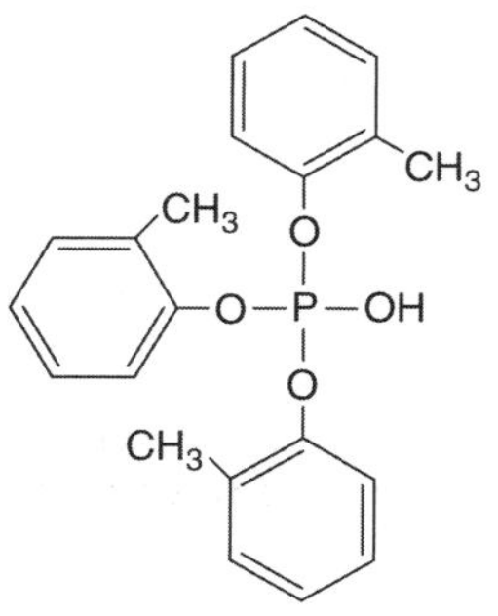

Figure 2.13 The structure of tri-*o*-cresyl phosphate (TOCP).

The chronicity factor could of course use dosing periods other than 90 days. The chronicity factor, however, should be viewed only as a crude indication of accumulation of the response. It does not indicate accumulation of the substance, and because it is based on the TD_{50} value, it takes no account of the shape and slope of the dose-response curve. Also the conditions for determination of the acute and chronic TD_{50} may be different, and this may introduce factors, which make comparison uncertain.

An example of absolutely cumulative toxicity is afforded by tri-o-cresyl phosphate or **TOCP** (Fig. 2.13). This compound is a cholinesterase inhibitor and neurotoxin. In chickens, an acute dose of 30 mg/kg has a severe toxic effect, which is produced to the same extent by a dose of 1 (mg/kg)/day given for 30 days. This effect may of course be produced by accumulation of the compound *in vivo* to a threshold toxic level, or it may result from the accumulation of the effect, as it probably does in the case of TOCP.

Thus, the inhibition of cholinesterase enzymes by organophosphorus compounds may last for several days or weeks, and repeated dosing at shorter intervals than the half-life of regeneration of the enzyme leads to accumulation of the inhibition until the toxic threshold of around 50% is reached.

SUMMARY

Exposure to toxic chemicals and the effect or response need to be quantitated to define the dose response relationship. These use what are called biomarkers, and new technology is constantly expanding the range of possible measurements. Susceptibility, important in risk assessment, can also be quantitated with biomarkers.

The dose-response relationship, which reflects the fact that toxicity is a relative phenomenon, was recognized by Paracelsus and is central to toxicology.

Toxic effects may be delayed or immediate, direct or indirect, local or systemic, and reversible or irreversible.

They may be described as graded or all or none (quantal).

Mixtures of toxic chemicals may give rise to the same toxicity, which is the sum of the components (additive) or it may be greater than the sum (potentiation; synergism) or less (antagonism). Alternatively, the toxicity may be different (coalitive).

Repeated exposure may lead to diminution of the toxic effect (tolerance).

Toxicity shows a dose-response relationship and may range from subtle biochemical changes to lethality and may involve receptor interactions. The dose-response relation depends on certain assumptions: the toxic response is a function of the concentration at the target site, the concentration at the target site is a function of dose, the toxic response is causally related to the compound.

Depending on the dose-response relationship, there may be a dose threshold.

As well as the traditional dose-response relationship, those based on hormesis may also occur. In this case, the dose response ranges from positive (helpful) effects to negative (harmful) effects. From the dose-response curve, it is possible to determine the LD_{50}, ED_{50}, TD_{50}, and NOEL. From a comparison of LD_{50} or TD_{50} and ED_{50}, the therapeutic index and margin of safety can be determined.

REVIEW QUESTIONS

1. Illustrate what is meant by the terms "therapeutic index" and "margin of safety."
2. Give an example of tolerance.
3. If the toxicity of two toxic chemicals together is greater than the sum of their individual toxicities, is this
   a. an additive effect,
   b. synergism,
   c. potentiation,
   d. a coalitive effect,
   e. antagonism, or
   f. an isobole?
4. On what assumptions is the dose-response relationship predicated?

5. Give an example of a quantal type of effect.
6. Define the term "LC_{50}."
7. Explain how the terms "NOEL" and "ADI" are derived.
8. Tri-o-cresyl phosphate shows absolutely cumulative toxicity. Explain this and its implications.
9. What are biomarkers used for?
10. What is hormesis?

REFERENCES

1. Waterfield CJ, Timbrell JA. Biomarkers-an overview. In: Ballantyne B, Marrs T, Syversen T eds. General and Applied Toxicology. Vol. 3. London: Macmillan, 2000.
2. Van Den Heuvel MJ, Dayan AD, Shillaker RO. Evaluation of the BTS approach to the testing of substances and preparations for their acute toxicity. Hum Toxicol 1987; 6:279.
3. Timbrell JA, Scales MDC, Streeter AJ. Studies on hydrazine hepatotoxicity.2. Biochemical findings. J Toxicol Environ Health 1982; 10:955.
4. Brown VK. Acute and Sub-acute Toxicology. London: Edward Arnold, 1988.
5. Hathway DE. Molecular Aspects of Toxicology. London: The Royal Society of Chemistry, 1984.
6. Pratt WB, Tayler P, eds. Principles of Drug Action: The Basis of Pharmacology. New York: Churchill Livingstone, 1990.
7. Aldridge WN. Mechanisms and Concepts in Toxicology. London: Taylor and Francis, 1996.
8. Zbinden G. The no-effect level, an old bone contention in toxicology. Arch Toxicol 1979; 43:3.
9. Loomis TA, Hayes AW. Loomis's Essentials of Toxicology. 4th ed. San Diego: Academic Press, 1996.
10. Klaassen CD. Comparison of the toxicity of chemicals in newborn rats to bile-duct ligated and sham operated rats and mice. Toxicol Appl Pharmacol 1974; 24:37.
11. Zbinden G, Flury-Roversi M. Significance of the LD_{50} test for the toxicological evaluation of chemical substances. Arch Toxicol 1981; 47:77.
12. Rodricks JV. Hormesis and toxicological risk assessment. Toxicol Sci 2003; 71:134–136.
13. Kroes R, Kleiner J, Renwick A. The threshold for toxicological concern concept in risk assessment. Toxicol Sci 2005; 86:226–230.

BIBLIOGRAPHY

Abdel-Rahman S, Kauffman RE. The integration of pharmacokinetics and pharmacodynamics: understanding dose response. Annu Rev Pharmacol Toxicol 2004; 44:111–136.
Ballantyne B, Marrs TC, Syversen T, eds. General and Applied Toxicology, Vol. 1. Part II , Techniques. This part has several chapters on the methods used in and design of toxicological studies. London: Macmillan, 2000.
Beck BD, Calabrese EJ, Slayton TM, et al. The use of toxicology in the regulatory process. In: Hayes AW, ed. Principles and Methods of Toxicology. 5th ed. Florida: CRC Press, 2007.
Calabrese EJ, Baldwin LA. Hormesis: the dose-response revolution. Annu Rev Pharmacol Toxicol 2003; 43:175–197.
DeCaprio AP. ed. Toxicologic Biomarkers. New York: Taylor and Francis, 2006.
Deichmann WB, Henschler D, Holmstedt B, et al. What is there that is not a poison: a study of the third defense by paracelsus. Arch Toxicol 1986; 58:207.
Eaton DL, Klaassen CD. Principles of toxicology. In: Klaassen CD, ed. Cassarett and Doull's Toxicology, The Basic Science of Toxicology. 6th ed. New York: McGraw Hill, 2001.
Faustman EM, Omenn GS. Risk assessment. In: Klaassen CD, ed. Cassarett and Doull's Toxicology, The Basic Science of Toxicology. 6th ed. New York: McGraw Hill, 2001.
Hellman B. General Toxicology. In: Mulder G, Dencker L, eds. Pharmaceutical Toxicology. London: Pharmaceutical Press, 2006.
Hengstler JG, Bogdanffy MS, Bolt HM, et al. Challenging dogma: thresholds for genotoxic carcinogens? The case of vinyl acetate. Annu Rev Pharmacol Toxicol 2003; 43:485–520.
Mailman RB, Lawler CP. Toxicant receptor interactions: fundamental principles. In: Hodgson E, Smart RC, eds. Introduction to Biochemical Toxicology. 3rd ed. New York: Wiley, 2001.
Musch A. Exposure: qualitative and quantitative aspects; and dose-time-effect relationships. In: Niesink RJM, de Vries J, Hollinger MA, eds. Toxicology: Principles and Practice. CRC Press and Open University of the Netherlands, 1996.
Olson H, Betton G, Robinson D, et al. Concordance of the toxicity of pharmaceuticals in humans and animals. Regul Toxicol Pharmacol 2000; 32:56–67.
Rhodes C. Principles of testing for acute effects. In: Ballantyne B, Marrs TC, Syversen T, eds. General and Applied Toxicology. Chapter 2, Vol. 1. London. Macmillan, 2000.

3 | Factors Affecting Toxic Responses: Disposition

3.1 INTRODUCTION

The disposition of a toxic compound in a biological system may be conveniently divided into four interrelated phases:

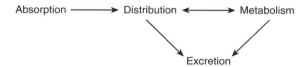

In this and the next chapter, each of these phases will be considered in turn.

3.2 ABSORPTION

It is clear that to exert a toxic effect a compound must come into contact with the biological system under consideration. It may exert a local effect at the site of administration on initial exposure, but it must penetrate the organism in order to have a **systemic** effect. The most common means of entry for toxic compounds are via the gastrointestinal tract and the lungs, although in certain circumstances, absorption through the skin may be an important route. Therapeutic agents may also enter the body by other routes such as injection.

The route and site(s) of exposure will be determined by the circumstances of exposure to a toxic chemical. Thus, most drugs are administered by mouth, a few topically to the skin, and a few by inhalation. Industrial chemicals in contrast are more likely to be inhaled or absorbed into and through the skin. With environmental chemicals, all the three routes of exposure will occur depending on the chemicals and the circumstances.

3.2.1 Transport Across Membranes

Although there are several sites of first contact between a foreign compound and a biological system, the absorption phase (and also distribution and excretion) necessarily involves the passage across cell membranes whichever site is involved. Therefore, it is important first to consider membrane structure and transport in order to understand the absorption of toxic compounds.

Membranes are basically composed of phospholipids and proteins with the lipids arranged as a bilayer interspersed with proteins as shown simply in Figure 3.1. A more detailed illustration is to be found in Figure 3.2, which shows that the membrane, on average about 70-Å (7 nm) thick, is not symmetrical and that there are different types of phospholipids and proteins as indicated in the figure. Furthermore, carbohydrates, attached to proteins (glycoproteins) and lipids (glycolipids), and cholesterol esters may also be constituents of the membrane. The presence of **cholesterol** is important as it contributes to the structural integrity of the membrane, affecting the fluidity, increasing the rigidity, and decreasing the permeability. It does this by interspersing with the phospholipids and interacting with them, and in the plasma membrane of liver cells and red blood cells, it constitutes 17% and 23%, respectively. However, some membranes have low levels of cholesterol, such as the mitochondrial membranes.

The particular proteins and phospholipids incorporated into the membrane, the proportions, and their arrangement vary depending on the cell type in which the membrane is located and also the part of the membrane. For example, the *ratio* of protein to lipid varies from 0.25:1 in the myelin membrane to 4.6:1 in the intestinal epithelial cell. Furthermore, the

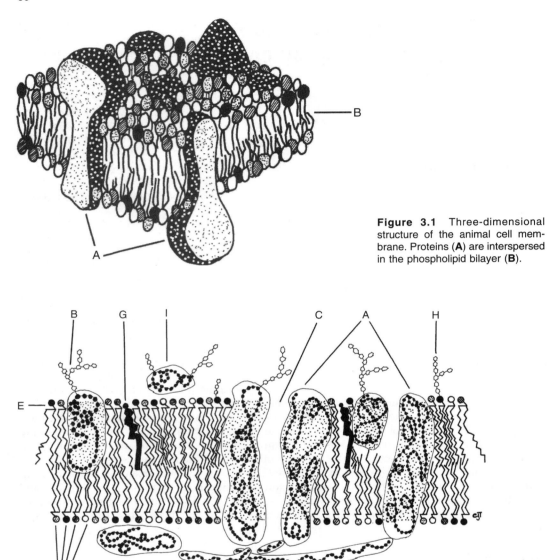

Figure 3.1 Three-dimensional structure of the animal cell membrane. Proteins (**A**) are interspersed in the phospholipid bilayer (**B**).

Figure 3.2 The molecular arrangement of the cell membrane: (**A**) integral proteins; (**B**) glycoprotein; (**C**) pore formed from integral protein; (**D**) various phospholipids with saturated fatty acid chains; (**E**) phospholipids with unsaturated fatty acid chains; (**F**) network proteins; (**G**) cholesterol; (**H**) glycolipid; (**I**) peripheral protein. There are four different phospholipids: phosphatidyl serine; phosphatidyl choline; phosphatidyl ethanolamine; and sphingomyelin represented as •, ○. The stippled area of the protein represents the hydrophobic portion. *Source*: From Ref. 1.

particular proteins and phospholipids on the inside may be different from those on the outside of the membrane and in different parts of the cell reflecting differences in function of these molecules. This leads to differences in charge between outside and inside. For example, the liver cell membrane with a protein-to-lipid ratio of about 1:1.4 has more phosphatidylcholine (neutral) in the exterior lipid layer than the interior layer, where there is more phosphatidylserine (negatively charged). Therefore, the external and internal surfaces of a membrane may have a different charge or one may be uncharged and so more hydrophobic and the other surface hydrophilic. Glycolipids are found only on the outer surface.

Although the sinusoidal and canalicular surfaces of the liver cell are similar, having phosphatidyl ethanolamine and sphingomyelin in the exterior surface, on the contiguous surface of the liver cell, the exterior layer is almost entirely composed of phosphatidylcholine.

As well as the four basic types of phospholipids (Fig. 3.2), there are also variations in the fatty acid content, which are very significant. The most common fatty acids in the phospholipids have 16 to 18 carbon atoms although C12 to C22 fatty acids may occur. However, not only does the chain length vary but so also does the extent of saturation. Thus, one or more double bonds may occur in the fatty acid chain, and the greater the unsaturation the greater will be the fluidity of the membrane. The character of the membrane may change between different tissues and cells and even within the same cell so that fluidity and function will vary. The presence of double bonds in the membrane phospholipids is also significant from a toxicological point of view, as these bonds are susceptible to peroxidation. Consequently, peroxidation of membrane phospholipids may occur following exposure to toxic chemicals such as carbon tetrachloride (see chaps. 6 and 7). The membranes of cells found in the nervous system may contain a high proportion of lipid (**type 1** membranes) and thereby allow the ready passage and accumulation of lipophilic substances. Membranes or regions of membranes, which have a specific role in transport other than passive diffusion (**type 2** and **type 3** membranes), contain specific carriers. Membranes containing many pores such as those in the kidney glomerulus and liver parenchymal cells are known as **type 4**.

The membrane proteins will have different characteristics and functions: structural, receptor, or enzymatic. For example, some proteins, which have a transport function, may also have ATPase activity. For example, the sodium–potassium pump, which maintains the osmotic balance of the cell, is an ATP-requiring system with ATPase activity. The transporter proteins known as ATP-binding cassette (ABC) [or also multidrug resistance (MDR)] are a family of proteins located in the plasma membrane and serve to pump drugs and other foreign chemicals out of cells in organs involved with absorption, metabolism, and elimination or those organs and tissues needing special protection. They are therefore extremely important in toxicology and will be discussed later in this chapter.

The different surfaces of the membrane of a cell may contain proteins that reflect the function at that surface. For example, the *sinusoidal surface* of the liver cell will have proteins such as transferases, which are involved in the transport of carbohydrates and amino acids, and receptors for hormones such as insulin, whereas the *bile canalicular membrane surface* will have specialized proteins for the transport of bile salts. Pores in the membrane will involve integral proteins, which span the entire membrane and may be hydrophobic, having outer hydrophilic regions and inner hydrophobic regions. Pores that can allow the passage of ions and charged compounds will have the hydrophilic side of the protein and the polar head groups of phospholipids exposed inside the pore. These pores will vary in both frequency and diameter ranging from 4 to perhaps 45 Å in the glomerulus of the kidney. The network proteins, such as **spectrin**, are involved in the cytoskeleton, which may also be a target for toxic substances (see chap. 6). The proteins of the outer surface are often associated with carbohydrates as glycoproteins, which may be involved in cell-cell interactions and may help to maintain the orientation of the proteins in the membrane. Membrane proteins may also be receptors and involved in cell signaling. For example, a particular transmembrane receptor protein when appropriately stimulated can interact with phosphoinositol kinase (PI 3 kinase) on the cytosolic side of the membrane, which phosphorylates a lipid, which then recruits particular proteins to the membrane as part of a signaling cascade.

For more details of plasma membrane structure, see the section "Bibliography."

Perhaps the most important feature of the plasma membrane is that it is *selectively* permeable. Overall permeability of the membrane to a chemical depends on the nature of the membrane, its surface charge and rigidity, and the chemical in question. Therefore, only certain substances are able to pass through the membrane, depending on particular physicochemical characteristics. It will be apparent throughout this book that the *physicochemical characteristics* of molecules are major determinants of their disposition and often of their toxicity.

Thus, with regard to the passage of foreign, potentially toxic molecules through membranes, the following physicochemical characteristics are important:

1. Size/shape
2. Lipid solubility/hydrophobicity

3. Structural similarity to endogenous molecules
4. Charge/polarity

It has been found that there will be poor absorption of a chemical if

1. The chemical has more than five hydrogen bond donor groups (e.g., NH or OH),
2. A molecular weight of more than 800,
3. A log P of more than 5, and
4. More than 10 hydrogen bond acceptor groups.

The role and importance of these characteristics will become apparent with the following discussion of the different ways in which foreign compounds may pass across membranes:

1. Filtration
2. Passive diffusion
3. Active transport
4. Facilitated diffusion
5. Phagocytosis/pinocytosis.

Filtration
This process relies on diffusion through pores in the membrane down a concentration gradient. Only small, hydrophilic molecules with a molecular weight of 100 or less, such as ethanol or urea, will normally cross membranes by filtration. Ionized compounds and even small ions such as sodium will not pass through pores, however; the latter will in fact be hydrated in aqueous environments and therefore too large for normal pores. Pore sizes vary between cells and tissues, and in the kidney, pores may be large enough (up to 45 Å) to allow passage of molecules with molecular weights of several thousand. The sinusoidal membrane in the liver is a specialized, discontinuous membrane, which also has large pores, allowing the ready passage of materials into and out of the bloodstream (see chap. 6).

Passive Diffusion
This is probably the most important mechanism of transport for foreign and toxic compounds. It does not show substrate specificity but relies on diffusion through the lipid bilayer. Passive diffusion requires certain conditions:

1. There must be a concentration gradient across the membrane.
2. The foreign compound must be lipid soluble.
3. The compound must be nonionized.

These conditions are embodied in the **pH-partition theory**: only nonionized lipid-soluble compounds will be absorbed by passive diffusion down a concentration gradient. Let us examine the three conditions in turn.

The concentration gradient.   This is normally in the direction external to internal relative to the cell or organism. The rate of diffusion is affected by certain factors: it is proportional to the concentration gradient across the membrane; the area and thickness of the membrane; and a diffusion constant, which depends on the physicochemical characteristics of the compound in question. This relationship is known as **Fick's Law**:

$$\text{Rate of diffusion} = \frac{KA(C_2 - C_1)}{d} \tag{1}$$

where A is the surface area, C_2 is the concentration of compound outside and C_1 the concentration on the inside of the membrane, d is the thickness of the membrane, and K is a constant, the diffusion coefficient. The concentration gradient is represented by C_2–C_1, and for the above relationship, it is assumed that the temperature is constant. In practice, the diffusion coefficient K for a particular compound will incorporate physicochemical characteristics such as lipophilicity, size, and shape. From this relationship, it is clear that passive diffusion is a *first-order* rate process, as the rate is directly proportional to the concentration of the compound at the membrane surface. This means that it is not a saturable process in contrast to active transport (see below) (Fig. 3.3).

As biological systems are *dynamic*, the concentration gradient will normally be maintained and an equilibrium will not be reached. Thus, the concentration on the inside of the membrane will be continuously *decreasing* as a result of ionization (see below), metabolism (see chap. 4), and removal by distribution into other compartments such as via **blood flow** (Fig. 3.4). It follows from Fick's Law that because A the surface area is an important term in the equation, the surface area of the site of exposure will have a major effect on the absorption of chemicals.

It is clear from this discussion that tissues such as the lungs, which have a large surface area, are served by an extensive vascular system, and with few cell membranes to cross, will allow rapid passage of suitable foreign compounds.

A factor of general importance is the ability of a chemical to interact with and bind to proteins in the blood, interstitial fluid, or other fluids. This binding maintains a concentration gradient similar to the way ionization of the chemical in the blood will maintain a gradient.

Lymph may also be of importance as a vehicle for facilitating absorption and transport of lipid-soluble substances but also of particles by specialized cells (M cells) in the gut-associated lymphoid tissue (GALT).

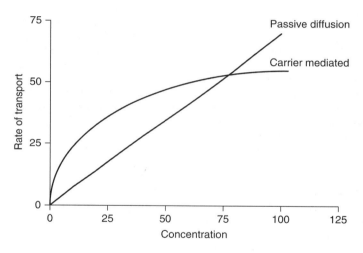

Figure 3.3 Comparison of the kinetics of carrier-mediated transport and passive diffusion.

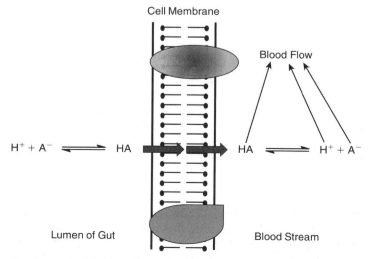

Figure 3.4 Role of blood flow and ionization in the absorption of foreign compounds. Both blood flow and ionization create a gradient across the membrane.

Table 3.1 Comparison Between Intestinal Absorption and Lipid: Water Partition of the Nonionized Forms of Organic Acids and Bases

Drug	Percentage absorbed	Partition coefficient ($K_{Chloroform}$)
Thiopental	67	100
Aniline	54	26.4
Acetanilide	43	7.6
Acetylsalicylic acid	21	2.0
Barbituric acid	5	0.008
Mannitol	<2	<0.002

$K_{Chloroform}$ is the partition coefficient determined between chloroform and an aqueous phase, the pH of which was such that the drug was largely in the nonionized form.
Source: From Ref. 2.

Lipid solubility. Passive diffusion relies on dissolution of the compound in the lipid component of the membrane, and therefore only lipid-soluble (lipophilic) compounds will pass through the membrane. This is illustrated in Table 3.1, which shows the absorption of various compounds through the intestinal wall in relation to their partition coefficient. Although there is often a good correlation between lipid solubility and ability to diffuse through membranes, very lipophilic compounds may become trapped in the membrane. Diffusion through the membrane will also depend on other factors such as the nature of the particular membrane or part of the membrane, especially the proportion of lipid and the presence of hydrophilic areas on integral proteins. Thus, a lipid-soluble foreign compound, which also has a polar but nonionized group, may diffuse through some membranes more efficiently than a highly lipophilic molecule. Also, some degree of water solubility may assist the passage through membranes, and in absorption from the gastrointestinal tract, especially, may indeed be an important factor (see below). The lipid solubility of a compound is an intrinsic property of that compound, dependent on the structure and usually denoted by the **partition coefficient** P (or log P). The larger the partition coefficient the greater is the lipophilicity of the compound.

Compounds of similar structure and ionization may have very different partition coefficients. For example, **thiopental** and **pentobarbital** are very similar in structure and acidity but have very different lipophilicity (Fig. 3.5), and hence their disposition *in vivo* is different (see below).

Solvents such as carbon tetrachloride, which are very lipid soluble, are rapidly and completely absorbed from most sites of application, whereas more polar compounds such as the sugar mannitol (Fig. 3.6) are very poorly absorbed as a consequence of limited lipid solubility (Table 1). For certain compounds, the lipid solubility as indicated by the log P (octanol-water) has been found to correlate well with the acute toxicity, for example, in some aquatic single-celled organisms.

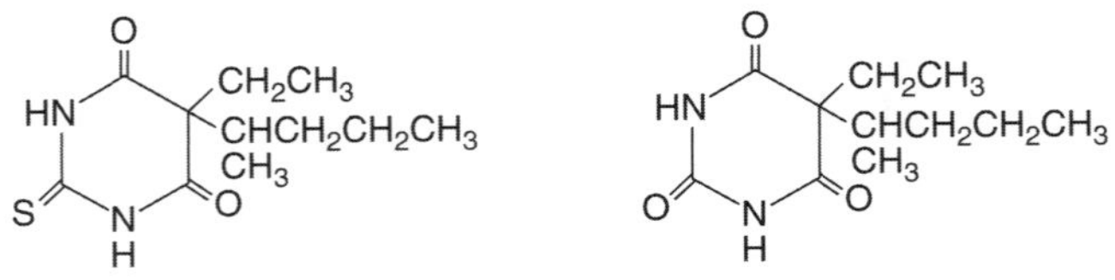

	Thiopental	Pentobarbital
pK$_a$	7.6	8.1
Fraction nonionised at pH 7.4	0.81	0.83
Partition coefficient (heptane/water)	3.3	0.05

Figure 3.5 Comparison of the structures and physicochemical characteristics of pentobarbital and thiopental.

(A)   **(B)**   **Figure 3.6**   Structures of (**A**) 5-fluorouracil and (**B**) mannitol.

The degree of ionization.   This determines the extent of absorption, as only the nonionized form will be able to pass through the lipid bilayer by passive diffusion. As already indicated, the lipid and water solubility of this nonionized form is also a major factor.

The degree of ionization of a compound can be calculated from the **Henderson–Hasselbach equation**:

$$pH = pK_a + \frac{Log[A^-]}{[HA]} \tag{2}$$

where pK_a is the dissociation constant for an acid HA and where:

$$HA \rightleftharpoons H^+ + A^-$$

For a base A,

$$pH = pK_a + \frac{Log[A]}{[HA^+]}$$

where

$$H^+ + A \rightleftharpoons HA^+$$

The pK_a of a compound, the pH at which it is 50% ionized, is a physicochemical characteristic of that compound.

Normally, only HA in the case of an acid or A in the case of a base will be absorbed (Fig. 3.14). Therefore, knowing the pH of the environment at the site of absorption and the pK_a of the compound, it is possible to calculate the amount of the compound, which will be in the nonionized form, and therefore estimate the likely absorption by passive diffusion.

For example, an acid with a pK_a of 4 can be calculated to be mainly nonionized in acidic conditionsat pH 1. Rearranging the Henderson–Hasselbach equation (Eq. 2),

$$pH - pK_a = \frac{Log[A^-]}{[HA]}$$

$$\text{anti-log } pH - pK_a = \frac{[A^-]}{[HA]}$$

For an acid with a pK_a 4 in an environment of pH 1,

$$\text{anti-log } 1 - 4 = \frac{[A^-]}{[HA]}$$

i.e.,

$$\text{anti-log } -3 = \frac{[A^-]}{[HA]} = 0.001$$

i.e., $A^-/HA = 1/1000$ or the acid is 99.9% nonionized.

Conversely, for a base pK_a 5 at pH 1,

$$pH - pK_a = \frac{Log[A]}{[HA^+]}$$

$$1 - 5 = \frac{Log[A]}{[HA^+]}$$

$$\text{anti-log} - 4 = \frac{[A]}{[HA^+]} = 0.0001$$

i.e., $A/HA^+ = 1/10,000$ or the base is 99.99% ionized.

The same calculations may be applied to calculate the degree of ionization of acids and bases under alkaline conditions. It can be easily seen that weak acids will be mainly nonionized and will therefore, if lipid soluble, be absorbed from an acidic environment, whereas bases will not, being mainly ionized under acidic conditions. Conversely, under alkaline conditions, acids will be mainly ionized, whereas bases will be mainly nonionized and will therefore be absorbed.

Because the situation *in vivo* is normally dynamic, continual removal of the nonionized form of the compound from the inside of the membrane causes continued ionization rather than the attainment of an equilibrium:

$$H^+ + A^- \rightleftharpoons HA \xrightarrow{membrane} HA \text{ removal}$$

If HA is continuously removed from the inside of the membrane, most of the compound will be absorbed from the site, provided its concentration at the site is not reduced by other factors.

Active Transport

This mechanism of membrane transport has several important features:

1. Specific membrane carrier system is required.
2. Metabolic energy is necessary to operate the system.
3. Transport occurs against a concentration gradient.
4. Metabolic poisons may inhibit the process.
5. The process may be saturated at high substrate concentration.
6. Substrates may compete for uptake.

As active transport uses a carrier system, it is normally specific for a particular substance or group of substances. Thus, the chemical structure of the compound and possibly even the spatial orientation are important. This type of transport is normally reserved for endogenous molecules such as amino acids, required nutrients, precursors, or analogues. For example, the anticancer drug **5-fluorouracil** (Fig. 3.6), an analogue of uracil, is carried by the pyrimidine transport system. The toxic metal lead is actively absorbed from the gut via the calcium transport system. Active uptake of the toxic herbicide paraquat into the lung is a crucial part of its toxicity to that organ (see chap. 7). Polar and nonionized molecules as well as lipophilic molecules may be transported. As active transport may be saturated, it is a *zero-order rate process* in contrast to passive diffusion (Fig. 3.3).

There are, however, various types of active transport systems, involving protein carriers and known as **uniports**, **symports**, and **antiports** as indicated in Figure 3.7. Thus, symports and antiports involve the transport of two different molecules in either the same or a different direction. Uniports are carrier proteins, which actively or passively (see section "Facilitated Diffusion") transport one molecule through the membrane. Active transport requires a source of energy, usually ATP, which is hydrolyzed by the carrier protein, or the cotransport of ions such as Na^+ or H^+ down their electrochemical gradients. The transport proteins usually seem to traverse the lipid bilayer and appear to function like membrane-bound enzymes. Thus, the protein carrier has a specific binding site for the solute or solutes to be transferred. For example, with the Na^+/K^+ ATPase antiport, the solute (Na^+) binds to the carrier on one side of

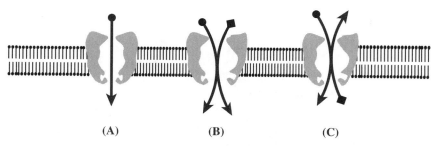

Figure 3.7 Membrane transporters involved in active transport. (**A**) a uniport, (**B**) a symport, and (**C**) an antiport.

the membrane, an energy-mediated *conformational change* occurs involving phosphorylation of the protein via ATP, and this allows the solute to be released on the other side of the membrane (Fig. 3.7). The second solute (K$^+$) then binds to the carrier protein, and this then undergoes a conformational change following release of the phosphate moiety, allowing the substance to be released to the other side of the membrane. **Glucose** is transported into intestinal cells via a symport along with Na$^+$, which enters the cell along its electrochemical gradient. The Na$^+$ is then transported out in the manner described above. Carrier-mediated transport can also involve **gated** channels, which may require the binding of a ligand to open the channel for instance.

Facilitated Diffusion
This type of membrane transport has some similar features to active transport:

1. It involves a specific carrier protein molecule.
2. The process may be saturated or competitively inhibited.

However:

3. Movement of the compound is only down a concentration gradient.
4. There is no requirement for metabolic energy.

Thus, a specific carrier molecule is involved, but the process relies on a concentration gradient, as does passive diffusion. The transport of glucose out of intestinal cells into the bloodstream occurs via facilitated diffusion and uses a uniport.

Phagocytosis/Pinocytosis
These processes are both forms of **endocytosis** and involve the invagination of the membrane to enclose a particle or droplet, respectively. The process requires metabolic energy and may be induced by the presence of certain molecules, such as ions, in the surrounding medium. The result is the production of a vesicle, which may fuse with a primary lysosome to become a secondary lysosome in which the enzymes may digest the macromolecule. In some cases, a particular part of the plasma membrane with specific receptors binds the macromolecule and then invaginates. Certain types of cells such as macrophages are especially important in the phagocytic process. Thus, large molecules such as carrageenens with a molecular weight of about 40,000 may be absorbed from the gut by this type of process. Insoluble particles such as those of uranium dioxide and asbestos are known to be absorbed by phagocytosis in the lungs (see below).

3.3 SITES OF ABSORPTION

The following are the major routes of entry for foreign compounds:

1. Skin
2. Gastrointestinal tract
3. Lungs/gills
4. Intraperitoneal (i.p.)
5. Intramuscular (i.m.)
6. Subcutaneous (s.c.)
7. Intravenous (i.v.)

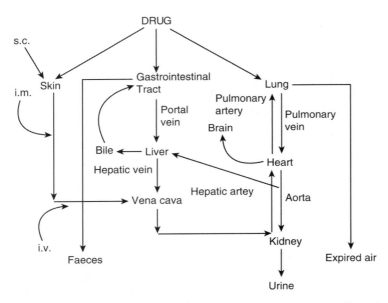

Figure 3.8 Blood flow and resulting distribution of a foreign compound from the three major sites of absorption or routes of injection. *Abbreviations*: i.v., intravenous injection; s.c., subcutaneous injection; i.m., intramuscular injection; i.p., intraperitoneal injection.

Routes 4 to 7, known as **parenteral** routes, are normally confined to the administration of therapeutic agents or used in experimental studies. The site of entry of a foreign, toxic compound may be important in the final toxic effect. Thus, the acid conditions of the stomach may hydrolyze a foreign compound, or the gut bacteria may change the nature of the compound by metabolism and thereby affect the toxic effect. The site of entry may also be important to the final disposition of the compound. Thus, absorption through the skin may be slow and will result in initial absorption into the peripheral circulation (Fig. 3.8). Absorption from the lungs, in contrast, is generally rapid and exposes major organs very quickly (Fig. 3.8). The compounds absorbed from the gastrointestinal tract first pass through the liver, which may mean that extensive metabolism takes place (Fig. 3.8). The toxicity of compounds after oral administration is therefore often less than that after i.v. administration (chap. 2, Table 1).

3.3.1 Skin
The skin is constantly exposed to foreign compounds such as gases, solvents, and substances in solution, and so absorption through the skin is potentially an important route. However, although the skin has a large surface area, some 18,000 cm^2 in humans, fortunately, it represents an almost continuous *barrier* to foreign compounds as it is not highly permeable. The outer layer of the nonvascularized **epidermis**, the stratum corneum, consists mainly of cells packed with **keratin**, which limits the absorption of compounds, and a few hair follicles and sebaceous glands (Fig. 3.9).

The underlying dermis is more permeable and vascularized, but to reach the systemic circulation through the skin, the toxic compound would have to traverse several layers of cells, in contrast to the situation in, for example, the gastrointestinal tract, where only two cells may separate the compound from the bloodstream.

Absorption through the skin is by passive diffusion mainly through the epidermis. Consequently, compounds that are well absorbed percutaneously are generally lipophilic, such as solvents like carbon tetrachloride, which may cause systemic toxicity (liver damage) following absorption by this route. Indeed, insecticides such as **parathion** have been known to cause death in agricultural workers following skin contact and absorption. However, lipophilicity as indicated by a large partition coefficient is not always a prerequisite for extensive absorption, and there is not necessarily a good correlation (Table 3.2). Thus, polar compounds, such as the small, water-soluble compound hydrazine, may also be absorbed

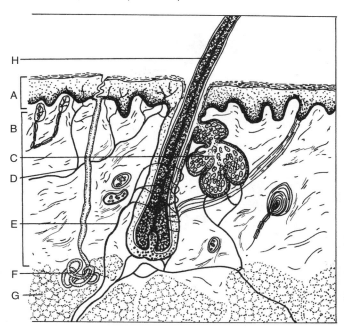

Figure 3.9 The structure of mammalian skin: (**A**) epidermis, (**B**) dermis, (**C**) sebaceous gland, (**D**) capillary, (**E**) nerve fibers, (**F**) sweat gland, (**G**) adipose tissue, (**H**) hair.

Table 3.2 Physicochemical Properties of Various Pesticides and Their Oral Absorption and Skin Penetration in Mice

Compound	Partition coefficient	Water solubility (ppm)	Penetration half-life		Penetrated (%)	
			Skin	Oral	Skin	Oral
DDT	1775	0.001	105	62	34	55
Parathion	1738	24	66	33	32	57
Chloropyrifos	1044	2	20	78	69	47
Permethrin	360	0.07	6	178	80	39
Nicotine	0.02	Miscible	18	23	71	83

Source: From Refs. 3–5.

through the skin sufficiently to cause a systemic toxic effect as well as a local reaction. The absorption of this compound may reflect its small molecular size (Fig. 5.31). Damage to the outer, horny layer of the epidermis increases absorption, and a toxic compound might facilitate its own absorption in this way. Absorption through the skin will, however, vary depending on the site, and hence nature, of the skin and thickness of the stratum corneum. Thus, penetration through the skin of the foot is at least an order of magnitude less than that through the skin of the scalp. It should also be noted that the epidermis has significant metabolic activity and so may metabolize substances, as they are absorbed.

3.3.2 Lungs

Exposure to and absorption of toxic compounds via the lungs is toxicologically important and more significant than skin absorption. The ambient air in the environment, whether it is industrial, urban, or household, may contain many foreign substances such as toxic gases, solvent vapors, and particles. The lungs have a large surface area, around 50 to 100 m² in humans; they have an excellent blood supply, and the barrier between the air in the alveolus and the bloodstream may be as little as two cell membranes thick (Fig. 3.10). Consequently, absorption from the lungs is usually *rapid* and *efficient*. The main process of absorption is passive diffusion through the membrane for lipophilic compounds (solvents such as chloroform), small molecules (gases such as carbon monoxide), and also solutions dispersed as aerosols. The substance will generally dissolve in the blood and may also react with plasma proteins or some other constituent. Therefore, as the blood flow is *rapid*, there will be a continuous removal of the substance and consequently a constant concentration gradient.

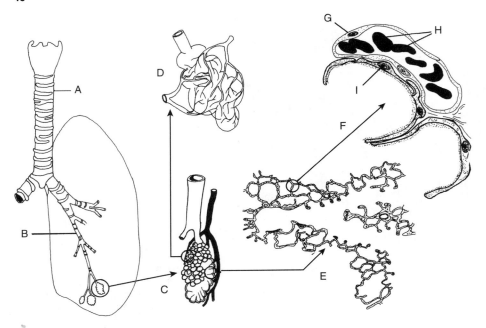

Figure 3.10 The structure of the mammalian respiratory system: (**A**) trachea, (**B**) bronchiole, (**C**) alveolar sac with blood supply, (**D**) arrangement of blood vessels around alveoli, (**E**) arrangement of cells and airspaces in alveoli showing the large surface area available for absorption, (**F**) cellular structure of alveolus showing the close association between (**G**) the endothelial cell of the capillary (**H**) with erythrocytes and (**I**) the epithelial cell of the alveolar sac. The luminal side of the epithelial cell is bathed in fluid, which also facilitates absorption and gaseous exchange. *Source*: From Ref. 1.

However, the solubility in the blood is a major factor in determining the rate of absorption. For compounds with low solubility, the rate of transfer from alveolus to blood will be mainly dependent on blood flow (*perfusion limited*), whereas if there is high solubility in the blood, the rate of transfer will be mainly dependent on respiration rate (*ventilation limited*).

As well as gases, vapors and aerosols, particles of toxic compounds may also be taken into the lungs. However, the fate of these particles will depend on a number of factors, but especially the size (Fig. 3.11). Thus, the larger particles will be retained in the respiratory tract initially to a greater extent than smaller particles because of rapid sedimentation under the influence of gravity, whereas small particles will be exhaled more easily. Overall, approximately 25% of particles will be exhaled, 50% retained in the upper respiratory tract, and 25% deposited in the lower respiratory tract. It can be seen from Figure 3.11 that the larger particles (20 μm) tend to be retained in the upper parts of the respiratory system, whereas the smaller particles (<6 μm) are confined to the alveolar ducts and terminal bronchioles. The optimum size for retention in the alveolar sacs is around 6 μm. Particles trapped by the mucus on the walls of the bronchi will be removed by the ciliary escalator, whereas those of around 1 μm or less, which penetrate the alveolus, may be absorbed by phagocytosis, and hence may remain in the respiratory system for a long time. For example, **asbestos** fibers are phagocytosed in the lungs, remain there, and eventually cause fibrosis and possibly lung tumors. It is known that **uranium dioxide** particles of less than 3-μm diameter can enter the bloodstream and cause kidney damage after inhalation. Similarly, particles of **lead** of about 0.25-μm diameter enter the bloodstream and cause biochemical effects following absorption via the lungs (see chap. 7). The lymphatic system also seems to be involved in the movement of phagocytosed particles.

3.3.3 Gastrointestinal Tract

Many foreign substances are ingested orally, either in the diet or as drugs, and poisonous substances taken either accidentally or intentionally. Most suicidal poisonings involve oral intake of the toxic agent. Consequently, the gastrointestinal tract is a very important site and perhaps the *major* route of absorption for foreign compounds.

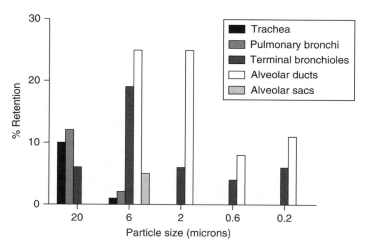

Figure 3.11 The retention of inhaled particles in various regions of the human respiratory tract in relation to size. *Source*: From Ref. 6.

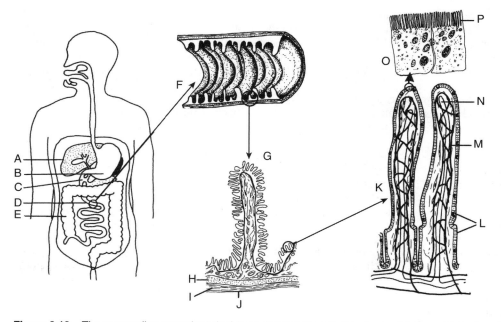

Figure 3.12 The mammalian gastrointestinal tract showing important features of the small intestine, the major site of absorption for orally administered compounds: (**A**) liver; (**B**) stomach; (**C**) duodenum; (**D**) ileum; (**E**) colon; (**F**) longitudinal section of the ileum showing folding, which increases surface area; (**G**) detail of fold showing villi with circular and longitudinal muscles, (**H**) and (**I**) respectively, bounded by (**J**) the serosal membrane; (**K**) detail of villi showing network of (**L**) epithelial cells, (**M**) capillaries, and (**N**) lacteals; (**O**) detail of epithelial cells showing brush border or (**P**) microvilli. The folding, vascularization, and microvilli all facilitate absorption of substances from the lumen. *Source*: From Ref. 1.

The internal environment of the gastrointestinal tract varies throughout its length, particularly with regard to the pH. Substances taken orally first come into contact with the lining of the mouth (buccal cavity), where the pH is normally around 7 in human, but more alkaline in some other species such as the rat. The next region of importance is the stomach, where the pH is around 2 in human and certain other mammals.

The substance may remain in the stomach for some time, particularly if it is taken in with food. In the small intestine, where the pH is around 6, there is a good blood supply and a large surface area because of the folding of the lining and the presence of villi (Fig. 3.12). The lining of the gastrointestinal tract essentially presents a continuous lipoidal barrier, passage through which is governed by the principles discussed above. Because of the change in pH in the gastrointestinal tract, different substances may be absorbed in different areas depending on their physicochemical characteristics, although absorption may occur along the whole length of the tract. Lipid-soluble, nonionized compounds may be absorbed anywhere in the tract, but ionizable substances will generally only be absorbed by passive diffusion if they are nonionized at the pH of the particular site and are also lipid soluble. However, despite the fact that the gastrointestinal tract is well adapted for the absorption of compounds and lipophilic substances should be readily and rapidly absorbed by passive diffusion, this is perhaps simplistic and not always the case as illustrated by comparative data from mice (Table 3.2).

Thus, in the acidic areas of the tract such as the stomach (pH 1–3), compounds that are lipid soluble in the nonionized form, such as weak acids, are absorbed, whereas in the more alkaline (pH 6) small intestine, weak bases are more likely to be absorbed. The fact that in practice weak acids are also absorbed in the small intestine despite being ionized (Fig. 3.15) is due to the following:

1. The large surface area of the intestine (Fig. 3.12)
2. Removal of compound by blood flow (Fig. 3.4)
3. Ionization of the compound in blood at pH 7.4 (Fig. 3.15).

These factors ensure that the concentration gradient is maintained, and so weak acids are often absorbed to a significant extent in the small intestine if they have not been fully absorbed in the stomach. Using the Henderson–Hasselbach equation, the degree of ionization can be calculated and the site and likelihood of absorption may be indicated.

Let us consider the situation in the gastrointestinal tract, using benzoic acid and aniline as examples.

The pH of gastric juice is 1 to 3. The pK_a of benzoic acid, a weak acid, is 4. Taking the pH in the stomach as 2, and using the Henderson–Hasselbach equation as described above, it can be calculated that benzoic acid is almost completely nonionized at this pH (Fig. 3.13):

$$\text{anti-log pH} - pK_a = \frac{[A^-]}{[HA]} = \text{anti-log2} - 4$$

$$\frac{[A^-]}{[HA]} = \frac{1}{100}$$

or 99% nonionized.

Benzoic acid should therefore be absorbed under these conditions and passed across the cell membranes into the plasma. Here, the pH is 7.4, which favors more ionization of benzoic acid.

COOH

COO⁻

+ H⁺

Benzoic Acid
100

1

NH₂

+ H⁺

NH₃⁺

Aniline
1

1000

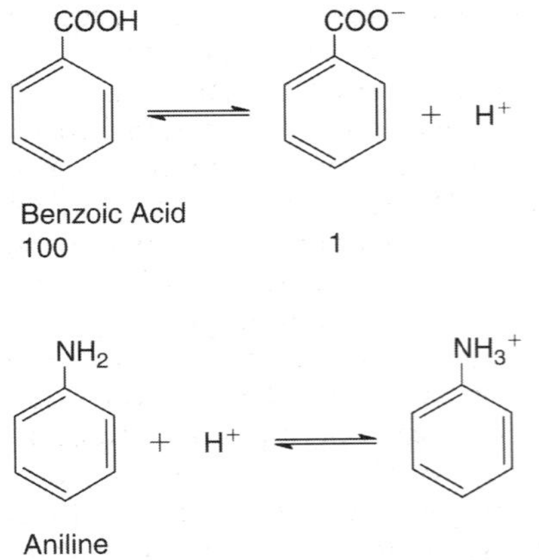

Figure 3.13   Ionization of benzoic acid and aniline at pH 2.

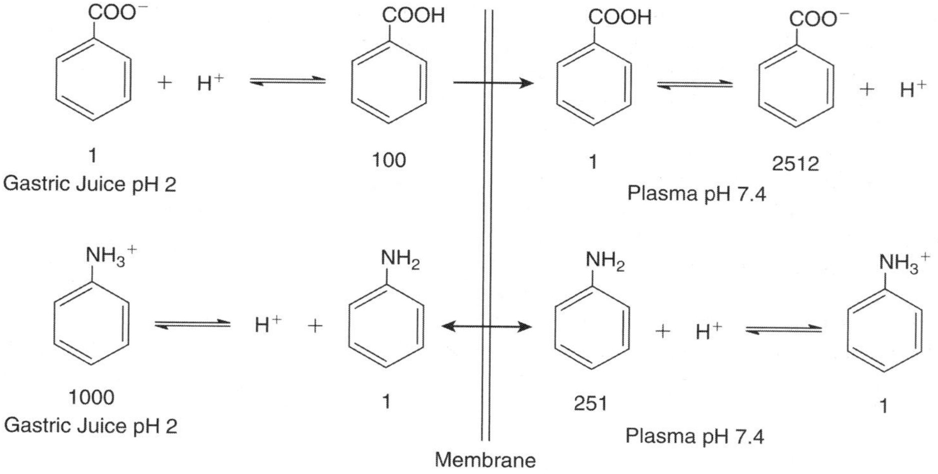

Figure 3.14 Disposition of benzoic acid and aniline in gastric juice and plasma. Figures represent proportions of ionized and nonionized forms.

Using the same calculation, at pH 7.4:

$$\frac{[A^-]}{[HA]} = \frac{1000}{1}$$

or 99.9% ionized.

Therefore, the overall situation is as shown in Figure 3.14.

The nonionized form of the benzoic acid crosses the membrane, but the continual removal by ionization in the plasma ensures that no equilibrium is reached. Therefore, the ionization in the plasma facilitates the absorption by removing the transported form.

Considering the situation for aniline in the same way (Fig. 3.13), for a base in the gastric juice of pH 2:

$$\text{anti-log2} - 5 = \frac{[A]}{[HA^+]} = 0.001$$

$$\frac{[A]}{[HA^+]} = \frac{1}{1000}$$

or 99.9% ionized.

Aniline is therefore not absorbed under these conditions (Fig. 3.14). Furthermore, the ionization in the plasma does not facilitate diffusion across the membrane, and with some bases, secretion from the plasma back into the stomach may take place. The situation in the small intestine, where the pH is around 6, is the reverse, as shown in Figure 3.15.

Therefore, it is clear that a weak base will be absorbed from the small intestine (Fig. 3.15) and, although the ionization in the plasma does not favor removal of the nonionized form, other means of redistribution ensure removal from the plasma side of the membrane.

With the weak acid, however, it can be appreciated that although most is in the ionized form in the small intestine, ionization in the plasma facilitates removal of the transported form, maintaining the concentration gradient across the gastrointestinal membrane (Fig. 3.15). Consequently, weak acids are generally fairly well absorbed from the small intestine.

In contrast, strong acids and bases are not usually appreciably absorbed from the gastrointestinal tract by passive diffusion. However, some highly ionized compounds are absorbed from the gastrointestinal tract, such as the quaternary ammonium compound pralidoxime, an antidote (Fig. 3.16), which is almost entirely absorbed from the gut, and paraquat, a highly toxic herbicide (Fig. 3.16). Sufficient paraquat is absorbed from the gastrointestinal tract after oral ingestion for fatal poisoning to occur, but the nature of the transport systems for both of these compounds is currently unknown, although

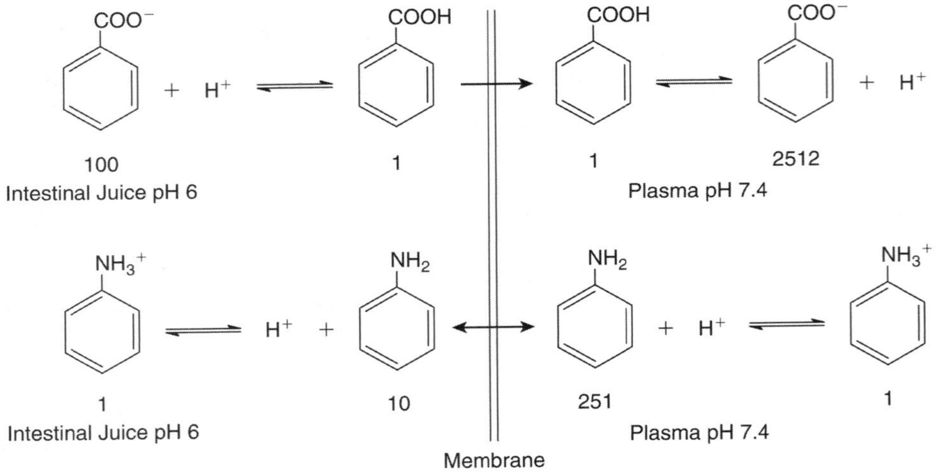

Figure 3.15 Disposition of benzoic acid and aniline in the small intestine and plasma. Figures represent proportions of ionized and nonionized forms.

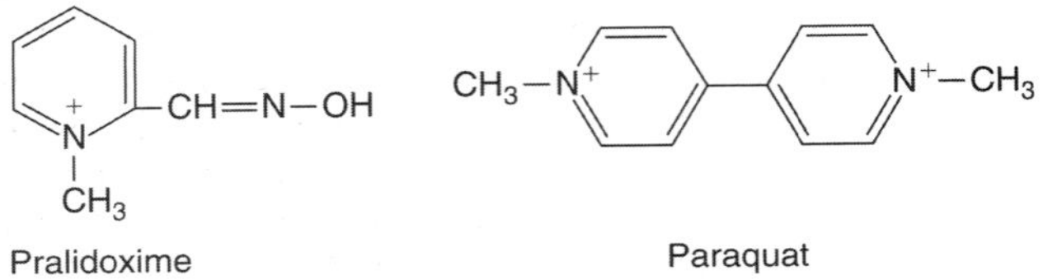

Figure 3.16 Structures of paraquat and pralidoxime.

carrier-mediated transport is perhaps the most likely (see chap. 7). Carrier-mediated transport systems, important for foreign, toxic compounds, are known to operate in the gastrointestinal tract. For example, cobalt is absorbed via the system that transports iron and lead by the calcium uptake system.

Large molecules and particles such as carrageenen and polystyrene particles of 22-μm diameter may also be absorbed from the gut, presumably by phagocytosis. The bacterial product **botulinum toxin**, a large molecule (molecular weight 200,000–400,000), is sufficiently well absorbed after oral ingestion to be responsible for toxic and often fatal effects.

There are a number of factors that affect the absorption of foreign compounds from the gut or their disposition; one factor, which is of particular importance, is the aqueous solubility of the compound in the nonionized form. With very lipid-soluble compounds, water solubility may be so low that the compound is not well absorbed (Table 2), because it is not dispersed in the aqueous environment of the gastrointestinal tract. In relation to this, a factor of particular importance in absorption of chemicals from the gut is the presence of bile, which is produced in the liver and secreted into the small intestine. This contains detergent-like substances, which will facilitate the dispersal of lipid-soluble chemicals in the aqueous medium of the intestine.

The gut also produces significant amounts of other secretions, including mucus, which coats the lining of both the stomach and intestine, thereby changing the charge of the gut surface. Therefore, the pH of the bulk of the site (e.g., stomach or intestine) may not be the same as at the surface. There will also be variations in the pH of the gastrointestinal tract between individual humans or animals. However, it must be appreciated that absorption is a dynamic process as is the ionization of the remaining drug.

Also, when drugs and other foreign compounds are administered, the vehicle used to suspend or dissolve the compound may have a major effect on the eventual toxicity by affecting the rate of absorption. Furthermore, the physical form of the substance may be important, for example, large particle size may decrease absorption. Similarly, when large masses of tablets are suicidally ingested, even those with reasonable water solubility, such as **aspirin** (acetylsalicylic acid), the bolus of tablets may remain in the gut for many hours after ingestion. Another factor, which may affect absorption from the gastrointestinal tract, is the presence of food. This may *facilitate* absorption if the substance in question dissolves in any fat present in the foodstuff. Alternatively, food may *delay* absorption if the compound binds to food or constituents, or if it is only absorbed in the small intestine, as food prolongs gastric-emptying time. Allied to this is gut motility, which may be altered by disease, infection, or other chemical substances present and hence change the absorption of a compound from the gut.

Apart from influencing the absorption of foreign compounds, the environment of the gastrointestinal tract may also affect the compound itself, making it more or less toxic. For example, **gut bacteria** may enzymically alter the compound, and the pH of the tract may affect its chemical structure.

The natural-occurring carcinogen cycasin, which is a glycoside of methylazoxymethanol (Fig. 1.1), is hydrolyzed by the gut bacteria after oral administration. The product of the hydrolysis is methylazoxymethanol, which is absorbed from the gut and which is the compound responsible for the carcinogenicity. Given by other routes, cycasin is not carcinogenic, as it is not hydrolyzed.

The gut bacteria may also reduce nitrates to nitrites, which can cause methemoglobinemia or may react with secondary amines in the acidic environment of the gut, giving rise to carcinogenic nitrosamines.

Conversely, the acidic conditions of the gut may inactivate some toxins, such as snake venom, which is hydrolyzed by the acidic conditions.

The absorption from the gastrointestinal tract is of particular importance, because compounds so absorbed are transported directly to the liver via the hepatic-portal vascular system (Fig. 3.8). Extensive metabolism in the liver may alter the structure of the compound, making it more or less toxic. Little of the parent compound reaches the systemic circulation in these circumstances. This **"first-pass"** effect is very important if hepatic metabolism can be saturated; it may lead to markedly different toxicity after administration by different routes. Highly cytotoxic compounds given orally may consequently selectively damage the liver by exposing it to high concentrations, whereas other organs are not exposed to such high concentrations, as the compound is distributed throughout body tissues after leaving the liver.

The gastrointestinal tract itself has significant metabolic activity and can metabolize foreign compounds en route to the liver and systemic circulation giving rise to a first-pass effect. For example, the drug **isoprenaline** undergoes significant metabolism in the gut after oral exposure, which effectively inactivates the drug. Therefore, administration by aerosol *into* the lungs, the target site, is the preferred route.

3.3.4 Specific Drug Transporter Proteins (ABC Proteins; MDR Proteins)

As indicated earlier in this chapter, these transporter proteins are found in a variety of cell types but especially in those organs exposed to chemicals from the environment (e.g., gastrointestinal tract), excretory organs (e.g., kidney), and sensitive organs (e.g., brain). The proteins are usually found on the luminal side of epithelial cells in organs of exposure, such as the small intestine, which allows the cells to pump out the potentially hazardous chemical. In sensitive organs such as the brain, the transporters are on that side of cells that will allow chemicals to be pumped back into the blood or interstitial fluid. In organs of excretion, such as the kidney, the transporters are located on the apical side of cells such as the proximal convoluted tubular cells.

Although there are at least 49 known ABC transporter proteins in humans, the most important in relation to foreign chemicals are **ABCB1, ABCC1, ABCC2,** and **ABCG2.** They require ATP and contain transmembrane domains and one or two nucleotide-binding domains for the binding/hydrolysis of ATP. These transport proteins are found in various mammalian species, and there is some homology between those in rodents, for example, and those found in human. However, they show different substrate specificity.

The level/expression of the transporters may be increased (upregulated) by chemicals or by disease and varies with age and during pregnancy. Other chemicals are known to inhibit certain transporter proteins. This may be due to competition between substrates (e.g., verapamil) or an effect on ATP binding (e.g., vanadate) or hydrolysis (e.g., cyclosporine). Other known inhibitors are flavonoids and organophosphate insecticides.

Cerivastatin, a drug used to reduce blood cholesterol, is known to be exported by the MDR1 transporter at the canalicular membrane in the liver cells, which functions to eliminate the drug into the bile. If there is competition with another drug, this elimination is decreased, and consequently, the blood level of the statin will be higher. It has been suggested that this may be part of the reason that unacceptable toxicity occurred with this drug (damage to muscle), which led to its withdrawal. The other reason may be the competition for metabolism by CYP3A4, which will also lead to higher blood levels (see chap. 5). Genetic polymorphisms in the genes coding for the transporters are known to occur.

The first transporter to be discovered was **p-glycoprotein** (now known as ABCB1 but also abbreviated as pGp or MDR1). This was found in tumor cells resistant to a wide range of structurally unrelated cytotoxic anticancer drugs, in which the transporter functions to pump the cytotoxic agent out of the cell, thus protecting it. It is also found in normal cells, especially in organs of elimination or absorption, and it functions as a route of detoxication in tissues and organs that are vulnerable such as brain and testis.

This transporter is particularly important in the small intestine, in the gut wall enterocytes, where its activity in humans is sevenfold higher than liver tissue. In the gut, pGp, acting in concert with cytochrome P-450 (CYP3A4) (see chap. 4), functions to keep chemicals, which may be potential toxicants, out of the body by pumping them back into the lumen of the gut. The CYP3A4 converts them into more polar compounds, which are less readily absorbed or further metabolized into water-soluble conjugates.

The chemical is removed before it can properly reach the cytoplasm or important organelles. The substrates for this transporter are structurally diverse but tend to be organic, weakly basic (cationic), or uncharged hydrophobic or amphipathic substances. Thus, the chemical *diffuses* into the cell and is then *pumped* out. Substrates include anions conjugated with glutathione (GSH), glucuronic acid, and sulfate.

MRP1 (ABCC1) transports cytotoxic drugs such as doxorubicin (see page 344). It also pumps out conjugates with glutathione, glucuronic acid, and sulfate. Nonconjugated chemicals are cotransported with GSH. Although there are a variety of substrates, it is still specific for these substrates. It is expressed in cells in many tissues on the basolateral membrane (in cells with polarity) of epithelial cells, allowing secretion into the interstitial space, so protecting the cell. MRP1 may interact with phase II conjugating enzymes to give increased resistance (synergy). In contrast to MRD1, it transports large organic anions. Other substrates are the GSH conjugates of aflatoxin and arsenic; as transport by this protein system is GSH dependent, depletion of GSH is possible.

MRP2 (MDR2; ABCC2) transports similar substrates (e.g., conjugates of anions) to MRP1 but has a more limited distribution in the body in excretory tissues and is involved in biliary excretion. Hence, it is also known as the canalicular multiorganic anion transporter (CMOAT). Acyl glucuronide conjugates, such as the conjugate of the drug benoxaprofen (see below) are transported into bile by this transporter. Concentrations of such compounds in bile may reach several thousand times those in the blood with potential toxic consequences, as with benoxaprofen. The food contaminant PhIP is transported back into the gut, and its conjugates are transported by this protein from the liver into the bile also.

BCRP (ABCG2) is similar to pGp but has only one ATP-binding domain. It is found in stem cells, where it may function to protect them against toxicants, and also in placenta and the canalicular membrane in hepatocytes. It is upregulated by low oxygen levels.

The location of the transporters is shown in Table 3.3.

Table 3.3 Distribution and Tissue Location of ABC Transporters in Human Tissues

	Transporter			
Tissue	ABCB1 (*p*-glycoprotein, MDR1)	ABCC1 (MRP1)	ABCC2 (MRP2, MDR2)	ABCG2 (BCRP)
Lung	Apical	Basolateral	n.d.	Apical
Intestine	Apical	Basolateral	Apical	Apical
Liver	Apical	Basolateral	Apical	Apical
Kidney	Apical	Basolateral	Apical	n.d
Brain	Apical	Apical	Apical	Apical
Testis	Apical	Basolateral	n.d.	n.d.
Placenta	Apical	Basolateral	Apical	Apical

Alternate designation is given in brackets.
Abbreviations: ABC, ATP-binding cassette; n.d., not detected.

These protein transporters are clearly important in the absorption, distribution, and excretion of xenobiotics. There are many other transporters, which may play a role at times in xenobiotic toxicity and detoxication.

3.4 DISTRIBUTION

Following absorption by one of the routes described, foreign compounds will enter the bloodstream. The part of the vascular system into which the compound is absorbed will depend on the site of absorption (Fig. 3.8). Absorption through the skin leads to the peripheral blood supply, whereas the major pulmonary circulation will be involved if the compound is absorbed from the air via the lungs. For most compounds, oral absorption will be followed by entry of the compound into the portal vein supplying the liver with blood from the gastrointestinal tract. Once in the bloodstream, the substance will distribute around the body and be diluted by the blood. Although only a small proportion of a compound in the body may be in contact with the receptor or target site, it is the distribution of the bulk of the compound that governs the concentration and disposition of that critical proportion. The plasma concentration of the compound is therefore very important, because it often directly relates to the concentration at the site of action. Blood circulates through virtually all tissues, and some equilibration between blood and tissues is therefore expected. The distribution of foreign toxic compounds throughout the body is affected by the factors already discussed in connection with absorption. This distribution involves the passage of foreign compounds across cell membranes. The passive diffusion of foreign compounds across membranes is restricted to the nonionized form, and the proportion of a compound in this form is determined by its pK_a and the pH of the particular tissue.

The passage of compounds out of the plasma through capillary membranes into the extravascular water occurs fairly readily, the major barrier being molecular size. Even charged molecules may therefore pass out of capillaries by movement through pores or epithelial cell junctions and driven by a concentration gradient. The passage of substances through pores in arterial capillaries may also be assisted by hydrostatic pressure. For lipid-soluble compounds, a major determinant of the rate of movement across capillaries will be the partition coefficient. Therefore, most small molecules, whether ionized or nonionized, pass readily out of the plasma, either through pores in the capillary membranes or by dissolving in the lipid of the membrane.

Large molecules pass out very slowly, possibly by pinocytosis. The pores in the capillary membranes vary considerably in size, and therefore some capillaries are more permeable than others. Pore sizes of about 30 Å correspond to a molecular weight of 60,000. The exceptions to this are the capillaries of the brain, which are relatively impermeable. Passage across cell membranes from extravascular or interstitial water into cells is, however, much more restrictive. Again, the physicochemical characteristics of the compound will be a crucial determinant of its disposition in conjunction with the particular environment. A particularly important interaction in the bloodstream, which the foreign compound may undergo, is reaction with **plasma proteins**. In some cases, such as with compounds of low water solubility, this interaction may be essential for the transport of the compound in the blood and may facilitate transport to the tissues, although usually it will restrict distribution. There are many different types of proteins, but the most abundant and important regarding binding of foreign compounds is **albumin** (Table 3.4). However, other plasma proteins may be important in binding foreign compounds, for example, the **lipoproteins**, which bind lipophilic compounds such as **DDT**. In general, the interactions are noncovalent, although some drugs, such as **captopril**, are known to bind covalently to plasma proteins and even to cause immune responses as a result of this interaction (see chap. 6). There may be a specific interaction in the plasma between particular foreign molecules and antibodies.

The noncovalent binding to plasma proteins may involve four types of interaction (Fig. 3.17):

1. Ionic binding, in which there is bonding between charged groups or atoms, such as metal ions and the opposite charge on the protein.
2. Hydrophobic interaction occurs when two nonpolar hydrophilic groups associate and mutually repel water.

Table 3.4 The Major Proteins Present in Human Plasma

Protein	g 100 g^{-1} plasma protein (% total)	Function
Albumin	50–65	Colloid osmotic pressure, binds drugs, fatty acids, hormones
A-acid glycoprotein	0.5–1.5	Tissue breakdown product
A-antitrypsin	1.9–4.0	Trypsin inhibitor
α-lipoprotein	4.5–8.0	Lipid transport
β-lipoprotein	0.5–1.5	Lipid transport
Transferrin	3–6.5	Iron transport
Fibrinogen	2.5–5.0	Blood clotting

Source: From Ref. 7.

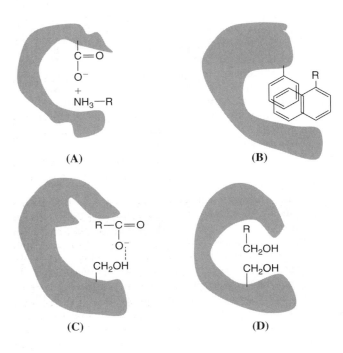

(A)   (B)

(C)   (D)

Figure 3.17 Types of bondings to plasma proteins that foreign compounds can undergo: (**A**) ionic bonding, (**B**) hydrophobic bonding, (**C**) hydrogen bonding, and (**D**) van der Waals forces. (**A**) Ionic binding, in which there is bonding between charged groups or atoms such as metal ions and the opposite charge on the protein. (**B**) Hydrophobic interactions occur when two nonpolar, hydrophilic groups associate and mutually repel water. (**C**) Hydrogen bonding, where a hydrogen atom attached to an electronegative atom (e.g., O) is shared with another electronegative atom (e.g., N). (**D**) Van der Waals forces are weak, acting between the nucleus of one atom and the electrons of another.

3. Hydrogen bonding, where a hydrogen atom attached to an electronegative atom (e.g., O) is shared with another electronegative atom (e.g., N).
4. Van der Waals forces are weak, acting between the nucleus of one atom and the electrons of another.

   The nature and strength of the binding will depend on the physicochemical characteristics of the foreign compound. For example, lipophilic substances such as DDT will bind to proteins that have hydrophobic regions such as lipoproteins and albumin. Ionized compounds may bind to a protein with available charged groups, such as albumin, by forming ionic bonds. The albumin molecule has approximately 100 positive and a similar number of negative charges at its isoelectric point (pH 5) and has a net negative charge at the pH of normal plasma (pH 7.4).
   The noncovalent binding to plasma proteins is a reversible reaction, which may be simply represented as:

$$T + P \underset{k_2}{\overset{k_1}{\rightleftharpoons}} TP$$

where T is the foreign compound and P is the protein and k_1 and1 k_2 are the rate constants for association and dissociation.

The overall dissociation constant K_d is derived from

$$\frac{1}{K_d} = \frac{[TP]}{[T][P]} \quad \text{or } K_d = k_2/k_1$$

When K_d is small, then binding is tight.

Plotting $1/[TP]$ versus $1/[T]$ may give some indication of the specificity of binding. Thus, if the plot passes through the origin, as is the case with DDT, then "infinite," nonsaturable binding is implied. If the concentration of bound/free compound is plotted against the concentration of bound compound, a straight line with a negative gradient may result. From this **Scatchard** plot, the **affinity constant** and the number of binding sites may be gained from the slope $(1/k_a)$ and intercept on the x axis (N), respectively. Binding can be described as (*i*) either specific and of low capacity and high affinity or (*ii*) nonspecific and of low affinity and high capacity. When there is a specific binding site, mathematical treatments derived from the above relationship may be applied to determine the nature of the site. Usually, however, binding to plasma proteins will involve several different binding sites on one protein and maybe several different proteins.

The binding of foreign compounds to plasma proteins has several important implications.

1. The *concentration* of the free compound in the plasma will be reduced. Indeed this removal of a portion of the compound from free solution will contribute to a concentration gradient.
2. *Distribution* to the tissues may be restricted. Although an equilibrium exists between the nonbound, free portion of the compound in the plasma and the bound portion, only the free compound will distribute into tissues.
3. Similarly, excretion by filtration and passive diffusion will be restricted to the free portion, and hence the half-life may be extended by protein binding. However, when a compound is excreted by an active process, then protein binding may have no significant effect. For example, **p-aminohippuric acid** (Fig. 3.18) is more than 90% bound to plasma proteins, yet it is cleared from the blood by a single pass through the kidney, being excreted by the organic acid transport system.
4. Saturation may occur. When a specific binding site is involved, there will be a limited number of sites. As the dose or exposure to the compound increases and the plasma level rises, these may become fully saturated. When this occurs, the concentration of the nonbound, free portion of the compound will rise. This may be the cause of a toxic dose threshold. The importance of this will depend partly on the extent of binding. Thus, with highly bound compounds, saturation will lead to a dramatic increase in the free concentration. For example, if a compound is 99% bound to plasma proteins, at a total concentration of say 100 mgL^{-1}, the free concentration of compound in the bloodstream will be only 1 mgL^{-1}. If all the plasma protein-binding sites are saturated at this concentration, then any increase in dosage can dramatically increase the free plasma concentration. Thus, doubling the dosage could increase the free concentration to 101 mgL^{-1}, if all other factors remained the same. Clearly, such a massive increase in the free concentration could result in the appearance of toxic effects.

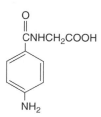

Figure 3.18 The structure of *p*-aminohippuric acid.

5. Displacement of one compound by another may occur. This may apply between two foreign compounds and between a foreign and an endogenous compound. For example, when some **sulfonamide** drugs bind to plasma proteins, they displace others such as **tolbutamide**, a hypoglycemic drug. The resultant increased plasma concentration of free drug can give rise to an excessive reduction in blood sugar. Some metals may compete for the binding sites on the protein **metallothionein**. Displacement of **bilirubin** in premature infants by **sulfisoxazole** competing for the same plasma protein-binding sites may lead to toxic levels of bilirubin entering the brain. The binding affinity will influence the consequences of this. Thus, if a toxic compound has relatively low binding (e.g., 30%), and 10% is displaced by another, the increase in free compound is only 3% (from 70% to 73%). If, however, the compound is 98% bound, then a similar displacement would cause a more dramatic increase in the free concentration (2–12%).

It is clear that the physicochemical characteristics of the compound and hence the extent and avidity of binding will determine the importance of plasma protein binding in the disposition of the compound.

3.4.1 Tissue Localization

The localization of a chemical in a particular tissue is an important factor in determining the toxicity of that chemical. The distribution of a foreign compound out of the bloodstream into tissues is, as already discussed, determined by its physicochemical characteristics. Thus, unless specific transport systems are available, such as for analogues of endogenous compounds, only nonionized, lipid-soluble compounds or small molecules will readily pass out of the bloodstream into cells throughout body tissues. The pores and spaces in the capillary membranes of certain tissues can be quite large, however, allowing much larger molecules to pass out of the bloodstream. For example, the sinusoids of the liver have a discontinuous basement membrane giving rise to large pores or fenestrations of about 2 μm in diameter. The **Kupffer** cells lining these sinusoids may also be involved in removal of foreign substances from the bloodstream by phagocytosis. The compounds that have some water solubility will distribute throughout body water, whereas highly lipophilic substances may become preferentially localized in fat tissue. For example, the drug **chlorphentermine** (Fig. 3.19) is found to accumulate in fatty tissues. Thus, although the blood: tissue ratio is 1:1 for the liver after both single and multiple doses, in the heart, lungs, and especially adrenal gland, the ratio becomes much greater than 1:1 after several doses. The accumulation of chlorphentermine results in a disturbance of lipid metabolism and a dramatic accumulation of phospholipids,

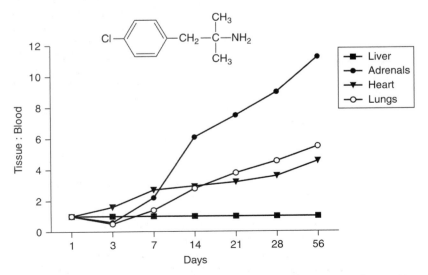

Figure 3.19 Tissue distribution of chlorphentermine on chronic dosing. *Source*: From Ref. 8.

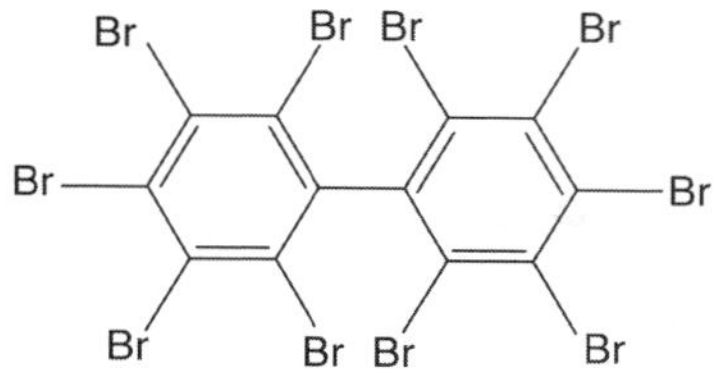

Figure 3.20　Structure of decabromobiphenyl.

especially in the adrenals. This toxic response, **phospholipidosis**, is due to the amphiphilic nature of the molecule. This is the possession of both lipophilic and hydrophilic groups in the same molecule. This characteristic is responsible for the disruption of lipid metabolism. Thus, it is believed that the hydrophobic portion of the chlorphentermine molecule associates with lipid droplets in target cells. The hydrophilic portion of the chlorphentermine molecule alters the surface charge of lipid droplets and so decreases their breakdown by lipases, giving rise to accumulation.

Many other drugs are now known to cause phospholipidosis, such as **chloroquine** and **amiodarone** (see chap. 6). Another, more dramatic, example of extensive tissue accumulation is the case of the **polybrominated biphenyls (PBBs)** (Fig. 3.20). These are industrial chemicals, which accidentally became added to animal feed in Michigan in 1973. Because of high lipophilicity, the PBBs became localized in the body fat of all the livestock primarily exposed and the humans subsequently exposed. There was little, if any, elimination of these compounds from the body, and so, the humans unfortunately exposed will be so for long, perhaps indefinite, periods of time. Sequestration of foreign compounds in particular tissues may in some cases be protective if the tissue is not the target site for toxicity. For example, highly lipophilic compounds such as DDT become localized in adipose tissue where they seem to exert little effect. Although this sequestration into adipose tissue reduces exposure of other tissues, mobilization of the fat in the adipose tissue may cause a sudden release of compound into the bloodstream with a dramatic rise in concentration and toxic consequences.

Lead is another example of a toxic compound that is localized in a particular tissue, bone, which is not the major target tissue (see chap. 7).

Some compounds accumulate in a specific tissue because of their affinity for a particular macromolecule. For instance, carbon monoxide specifically binds to hemoglobin in red cells, this being the target site (see chap. 7), whereas when cadmium binds to metallothionein in the liver and kidney, this is initially a detoxication process (see chap. 7). Sequestration into tissues may occur because of the action of an endogenous substance such as the protein **ligandin** (now identified as **glutathione transferase B**) (see chap. 4), which binds and transports both endogenous and exogenous compounds, such as carcinogenic azo dyes, into the liver.

Specific uptake systems may account for selective localization in tissues, and this may be the explanation for toxic effects in those particular tissues. For example, the herbicide paraquat (Fig. 3.16) is taken up by the polyamine transport system into the lungs and thereby reaches a toxic concentration (see chap. 7 for a detailed description of this system).

Therefore, foreign compounds may be distributed throughout all the tissues of the body, or they may be restricted to certain tissues. Two areas for special consideration are the fetus and the brain. Because of the organization of the placenta, the blood in the embryo and fetus is in intimate contact with the maternal bloodstream, especially at the later stages of pregnancy where the tissue layers between the two blood supplies may be only 2-μm thick (see chap. 6) (Fig. 3.21). Consequently, movement of nonionized lipophilic substances from the maternal bloodstream into the embryonic circulation by passive diffusion or filtration of small molecules through pores is facilitated. Specialized transport systems also exist for endogenous compounds and ions. ABC transporters, however, are located in the placenta and protect the embryo and fetus by pumping xenobiotics back in the bloodstream.

The placenta, through which the drugs will pass to the embryo or fetus, will vary in structure at different times in gestation and between species. For example, there may be six layers of cells between maternal and fetal blood in some species (e.g., pig), but in humans,

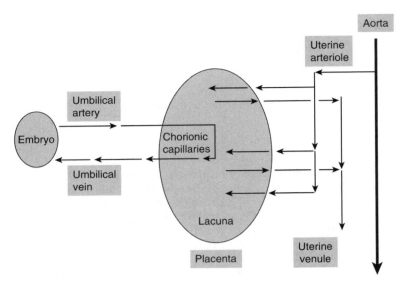

Figure 3.21 The blood supply to the mammalian embryo and placenta.

there are only three fetal layers. In other species there is only one layer (rat) at the end of gestation. However, there are also transporter proteins present in the placenta, which eject drugs and other foreign chemicals back into the maternal bloodstream. The placenta may also have metabolic activity and so be able to detoxify drugs.

The protein level in the embryonic or fetal bloodstream is also lower and different from that in the maternal circulation, and so the effect on the concentration gradient may well be less. So the concentration of drug in the embryonic or fetal blood may be lower than that in the maternal blood.

Consequently, many foreign compounds achieve the same concentration in fetal as in maternal plasma. However, if metabolism in utero converts the compound into a more polar metabolite, accumulation may occur in the fetus. Despite extensive blood flow (16% cardiac output; $0.5 \text{ mLmin}^{-1}\text{g}^{-1}$ of tissue), entry of foreign compounds into the brain takes place much less readily than passage into other tissues. Hence, the term "**blood-brain barrier.**" Ionized compounds will not penetrate the brain in appreciable quantities unless they are carried by active transport systems. The reasons for this are as follows:

1. The capillaries feeding the brain with blood are covered with the processes of brain cells called **glial cells or astrocytes**. This reduces the permeability of the basement membrane of the capillary because there are more layers of membranes to pass through.
2. There are **tight junctions** between the endothelial cells in the capillaries leaving few, if any, pores.
3. The endothelial cells of the capillaries in the brain have **transporter proteins** (ABC or MDR protein transporters such as *p*-glycoprotein), which export foreign substances (e.g., drugs) back into the blood.
4. There is a low-protein concentration in the interstitial fluid in the brain. This means that protein binding, which would contribute to a concentration gradient and also assist the transport of non–water-soluble drugs (paracellular), does not occur.
5. Endocytosis does not commonly occur.

The crucial role of this export system in protecting the brain can be illustrated by the use of specially bred mice in which the gene for *p*-glycoprotein is knocked out. When normal and knockout mice are exposed to the antihelminthic drug ivermectin, the null mice experience neurotoxic effects at doses 100 times less than the normal animals and also have levels of the drug in the brain almost 100 times those in the normal (wild type) mice.

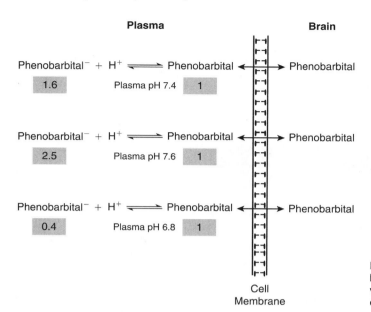

Figure 3.22 Disposition of phenobarbital in plasma at different pH values. Figures represent proportions of ionized and nonionized forms.

Lipid-soluble compounds, such as **methyl mercury**, which is toxic to the central nervous system (see chap. 7), can enter the brain readily, the facility being reflected by the partition coefficient. Another example, which illustrates the importance of the lipophilicity in the tissue distribution and duration of action of a foreign compound, is afforded by a comparison of the drugs **thiopental** and **pentobarbital** (Fig. 3.5). These drugs are very similar in structure, only differing by one atom. Their pK_a values are similar, and consequently, the proportion ionized in plasma will also be similar (Fig. 3.5). The partition coefficients, however, are very different, and this accounts for the different rates of anesthesia due to each compound. Thiopental being much more lipophilic enters the brain very rapidly and thereby causes its pharmacological effect, anesthesia. The duration of this, however, is short, as redistribution into other tissues, including body fat, causes a loss from the plasma, and hence by equilibration, the central nervous system also. Pentobarbital, conversely, distributes more slowly and so, although the concentration in the brain does not reach the same level as thiopental, it is maintained for a longer period.

Changes in plasma pH may also affect the distribution of toxic compounds by altering the proportion of the substance in the nonionized form, which will cause movement of the compound into or out of tissues. This may be of particular importance in the treatment of salicylate poisoning (see chap. 7) and barbiturate poisoning, for instance. Thus, the distribution of **phenobarbital**, a weak acid (pK_a 7.2), shifts between the brain and other tissues and the plasma, with changes in plasma pH (Fig. 3.22). Consequently, the depth of anesthesia varies depending on the amount of phenobarbital in the brain. Alkalosis, which increases plasma pH, causes plasma phenobarbital to become more ionized, alters the equilibrium between plasma and brain, and causes phenobarbital to diffuse back into the plasma (Fig. 3.22). Acidosis will cause the opposite shift in distribution. Administration of bicarbonate is therefore used to treat overdoses of phenobarbital. This treatment will also cause alkaline diuresis and therefore facilitate excretion of phenobarbital into the urine (see below).

3.4.2 Plasma Level

The plasma level of a toxic compound is a particularly important parameter, as (*i*) it reflects and is affected by the absorption, distribution, metabolism, and excretion (ADME) of the compound; (*ii*) it often reflects the concentration of compound at the target site more closely than the dose; it should be noted that this may not always be the case such as when sequestration in a particular tissue occurs which may or may not be the target tissue, for example, chlorphentermine (see Fig. 3.19), lead and, polybrominated biphenyls (Fig. 3.20);

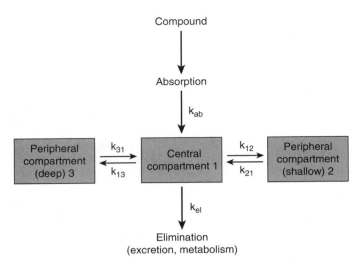

Figure 3.23 Distribution of a foreign compound into body compartments.

(*iii*) it may be used to derive other parameters; (*iv*) it may give some indication of tissue exposure and hence expected toxicity in situations of intentional or accidental overdosage; (*v*) it may indicate accumulation is occurring on chronic exposure; and (*vi*) it is central to any kinetic study of the disposition of a foreign compound.

Therefore, the pharmacokinetic parameters, which can be derived from blood level measurements, are important aids to the interpretation of data from toxicological dose-response studies. The plasma level profile for a drug or other foreign compound is therefore a composite picture of the disposition of the compound, being the result of various dynamic processes. The processes of disposition can be considered in terms of "compartments." Thus, absorption of the foreign compound into the central compartment will be followed by distribution, possibly into one or more peripheral compartments, and removal from the central compartment by excretion and possibly metabolism (Fig. 3.23). A very simple situation might only consist of one, central compartment. Alternatively, there may be many compartments. For such multicompartmental analysis and more details of pharmacokinetics and toxicokinetics, see references in the section "Bibliography." The central compartment may be, but is not necessarily, identical with the blood. It is really the compartment with which the compound is in rapid equilibrium. The distribution to peripheral compartments is reversible, whereas the removal from the central compartment by metabolism and excretion is irreversible.

The rates of movement of foreign compound into and out of the central compartment are characterized by **rate constants** k_{ab} and k_{el} (Fig. 3.23). When a compound is administered intravenously, the absorption is effectively instantaneous and is not a factor. The situation after a single, intravenous dose, with distribution into one compartment, is the most simple to analyze kinetically, as only distribution and elimination are involved. With a rapidly distributed compound then, this may be simplified further to a consideration of just elimination. When the plasma (blood) concentration is plotted against time, the profile normally encountered is an exponential decline (Fig. 3.24). This is because the rate of removal is proportional to the concentration remaining; it is a first-order process, and so a constant fraction of the compound is excreted at any given time. When the plasma concentration is plotted on a $\log_{10}$ scale, the profile will be a straight line for this simple, one compartment model (Fig. 3.25). The equation for this line is

$$\log_{10} C = \log_{10}C_0 - \frac{k_{el} \times t}{2.303}$$

where C is plasma concentration, t is time, C_0 is the intercept on the y axis, and the gradient or slope is $-k_{el}/2.303$. The unit of k_{el} is hr^{-1} or min^{-1}.

The elimination process is represented by the **elimination rate constant** k_{el}, which may be determined from the gradient of the plasma profile (Fig. 3.25). The reasons for the overall process of elimination being first order are that the processes governing it (excretion by various

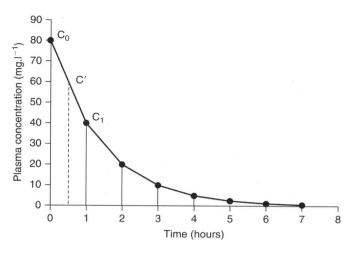

Figure 3.24 Profile of the decline in the plasma concentration of a foreign compound with time after intravenous administration. The AUC may be determined by dividing the curve into a series of trapezoids as shown and calculating the area of each. Thus, the $AUC_{C0 \to C1}$ is $(C' \times t_1 - t_0)$. The $AUC_{C0 \to C7}$ is then the sum of all the individual areas. *Abbreviation:* AUC, area under the curve.

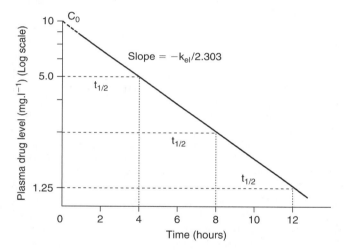

Figure 3.25 Log_{10} plasma concentration time profile for a foreign compound after intravenous administration. The plasma half-life $(t_{1/2})$ and the elimination rate constant (k_{el}) of the compound can be determined from the graph as shown.

routes, and metabolism) are irreversible processes and are also first order. When the latter is not the case, then this model does not apply.

3.4.3 Volume of Distribution

The total water in the body of an animal can be conveniently divided into three compartments: the plasma water, the interstitial water, and the intracellular water. The way a foreign compound distributes into these compartments will profoundly affect the plasma concentration. If a compound is only distributed in the plasma water (which is ~ 3 L in man), the plasma concentration will obviously be much higher than if it is distributed in all extracellular water (~ 14 L) or the total body water ~ 40 L. This may be quantified as a parameter known as the volume of distribution (V_D), which can be calculated as follows:

$$V_D = \frac{\text{dose(mg)}}{\text{Plasma conc.(mg L}^{-1})}$$

and is expressed in liters. This is the same principle as determining the volume of water in a jar by adding a known amount of dye, mixing, and then measuring the concentration of dye. The volume of distribution of a foreign compound is the volume of body fluids into which the compound is apparently distributed. It does not necessarily correspond to the actual body water compartments, as it is a mathematical parameter and does not have absolute physiological meaning. However, the V_D may yield important information about a compound. For example, a very high apparent V_D, perhaps higher than the total body water, indicates that the compound is localized or sequestered in a storage site such as fat or bone. If the value is low and similar to plasma water, it indicates that the compound is retained in the plasma. A compound with a high V_D will tend to be excreted more slowly than a compound with a low V_D. From the V_D and plasma concentration, the total amount of foreign compound in the body or the **total body burden** may be estimated:

$$\text{Total body burden(mg)} + \text{Plasma concentration(mg L}^{-1}) \times V_D \qquad (3)$$

The determination of V_D will depend on the type of model that describes the distribution of the particular compound. Thus, if the compound is given intravenously and is distributed into a one-compartment system, the V_D can be determined from the starting plasma concentration C_0. This may be determined from the graph of plasma concentration against time by extrapolation (Fig. 3.25). Thus,

$$V_D = \frac{\text{Dose}_{iv}}{C_0}$$

For compounds whose distribution fits more complex models, a more rigorous method uses the area under the plasma concentration versus time curve (AUC). Thus,

$$V_D = \frac{\text{Dose}_{iv}}{\text{AUC}_0 - \infty \times k_{el}}$$

The **AUC** is determined by plotting the plasma concentration versus time on normal rectilinear graph paper, dividing the area up into trapezoids, and calculating the area of each trapezoid (Fig. 3.24). The total area is then the sum of the individual areas. Although the curve theoretically will never meet the x axis, the area from the last plasma level point to infinity may be determined from C_t/k_{el}. The units are mgL^{-1}hr.

When a compound is administered by a route other than intravenously, the plasma level profile will be different, as there will be an absorption phase, and so the profile will be a composite picture of absorption in addition to distribution and elimination (Fig. 3.26). Just as first-order elimination is defined by a rate constant, so also is absorption k_{ab}. This can be determined from the profile by the method of *residuals*. Thus, the straight portion of the semilog plot of plasma level against time is extrapolated to the y axis. Then each of the actual plasma level points, which deviate from this during the absorptive phase, are subtracted from the equivalent time point on the extrapolated line. The differences are then plotted, and a straight line should result. The slope of this line can be used to calculate the absorption rate constant k_{ab} (Fig. 3.26). The volume of distribution should not really be determined from the plasma level after oral administration (or other routes except intravenous) as the administered dose may not be the same as the absorbed dose. This may be because of **first-pass metabolism** (see above), or incomplete absorption, and will be apparent from a comparison of the plasma

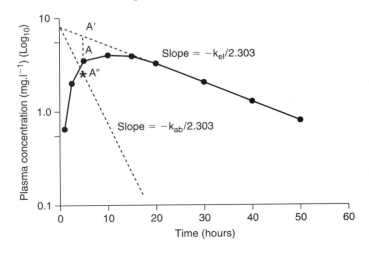

Figure 3.26 The $\log_{10}$ plasma concentration against time profile for a compound after oral administration. The absorption rate constant k_{ab} can be determined from the slope of the line (*dotted*) plotted using the method of residuals as described in the text. (Thus, A″ = A′−A, etc.)

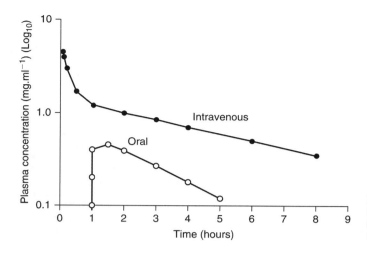

Figure 3.27 Effect of first-pass metabolism on plasma concentrations of a compound after oral and intravenous administration.

level profile of the compound after oral and intravenous administration (Fig. 3.27). First-pass metabolism may occur at the site of absorption such as in the gastrointestinal tract after oral administration, or it may occur in the liver during the passage of the substance through the portal circulation and the liver before it reaches the systemic circulation. The extent of this first-pass metabolism may be as high as 70% as with the drug **propranolol**, when given orally. This means that the systemic circulation is only exposed to 30% of the original dose as parent drug. The dose actually absorbed may be quantified as **bioavailability**:

$$\text{Bioavailability} = \frac{\text{AUC}_{\text{oral}}}{\text{AUC}_{\text{iv}}} \times 100$$

3.4.4 Plasma Half-Life

The plasma half-life is another important parameter of a foreign compound, which can be determined from the plasma level. It is defined as the time taken for the concentration of the compound in the plasma to decrease by half from a given point. It reflects the rates at which the various dynamic processes, distribution, excretion and metabolism, are taking place *in vivo*. The value of the half-life can be determined from the semilog plot of plasma level against time (Fig. 3.25), or it may be calculated:

$$\text{Half-life } (t_{1/2}) = \frac{0.693}{k_{\text{el}}}$$

The half-life will be independent of the dose, provided that the elimination is first order and therefore should remain constant. Changes in the half-life, therefore, may indicate alteration of elimination processes due to toxic effects because the half-life of a compound reflects the ability of the animal to metabolize and excrete that compound. When this ability is impaired, for example, by saturation of enzymic or active transport processes, or if the liver or kidneys are damaged, the half-life may be prolonged. For example, after overdoses of paracetamol, the plasma half-life increases severalfold as the liver damage reduces the metabolic capacity, and in some cases, kidney damage may reduce excretion (see chap. 7).

Another indication of the ability of an animal to metabolize and excrete, and therefore of the elimination of a foreign compound that can be gained from the plasma level data, is **total body clearance**. This may be calculated from the parameters already described:

$$\text{Total body clearance} = V_{\text{D}} \times k_{\text{el}}$$

or, a better calculation is

$$\text{Total body clearance} = \frac{\text{Dose mg}}{\text{AUC mg L}^{-1} \text{ hr}}$$

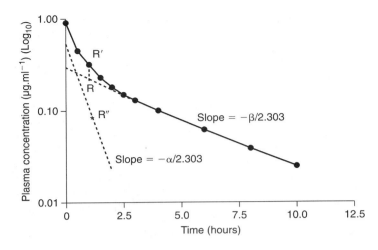

Figure 3.28 Log$_{10}$ plasma concentration against time profile for a foreign compound after intravenous administration. The distribution of this compound fits the two-compartment model. The dotted line is determined by the method of residuals as described in the text. (Thus, R″ = R′− R, etc.)

as it may be applied to multi- as well as single-compartment models. This is provided the absorbed dose is known or the drug is given i.v. The unit is L hr^{-1}.

If the semilog plot of the plasma level against time after an intravenous dose is not a straight line, then the compound may be distributing in accordance with a *two-compartment* or *multicompartment model* (Figs. 3.23 and 3.28). If a two-compartment model is appropriate, then the semilog plot can be resolved into two straight lines using the method of residuals ("feathering") already described. The first part of the plasma level profile is the α-phase and the second is the β-phase. The straight line that describes the α-phase is the difference between the observed points and the back-extrapolated β-phase line. The α-phase represents the initial, rapid distribution into the peripheral compartment, and the second, slower β-phase represents elimination from the central compartment.

The individual rate constants α and β can be determined from the graph. The rate constants for movement into the peripheral compartment k_{12} and k_{21} (Fig. 3.23) and the overall k_{el} can be determined as follows:

$$k_{21} = \frac{\alpha\beta + \beta\alpha}{A + B}$$

$$k_{el} = \frac{\alpha\beta}{k_{21}}$$

$$k_{12} = \alpha + \beta - k_{21} - k_{el}$$

Knowledge of these rate constants allows an assessment of the contribution of distribution and elimination.

For a further discussion of multicompartmental analysis, which is beyond the scope of this book, the reader is referred to the section "Bibliography."

The kinetics described so far have been based on first-order processes, yet often in toxicology, the situation after large doses are administered has to be considered when such processes do not apply. This situation may arise when excretion or metabolism is saturated, and hence the rate of elimination decreases.

This is known as **Michaelis–Menten** or saturation kinetics. The processes that involve specific interactions between chemicals and proteins such as plasma protein binding, active excretion from the kidney or liver via transporters, and metabolism catalyzed by enzymes can be saturated. This is because there are a specific number of binding sites that can be fully occupied at higher doses. In some cases, cofactors are required, and their concentration may be limiting (see chap. 7 for salicylate, paracetamol toxicity). These all lead to an increase in the free concentration of the chemical. Some drugs, such as phenytoin, exhibit saturation of metabolism and therefore nonlinear kinetics at therapeutic doses. Alcohol metabolism is also saturated at even normal levels of intake. Under these circumstances, the rate of metabolism is maximal, and so a constant amount of drug is removed by this process. Thus, the maximum rate of alcohol metabolism in humans is about 12 mL hr^{-1}, approximately the amount in one unit or standard measure of spirits, wine, or beer. As we will see in chapter 5, rates of metabolism vary with the individual. If metabolism is crucial to or an important factor in elimination from the body, repeat doses of a drug inevitably lead to accumulation. This is borne out by the experience of many "binge" drinkers.

When the concentration of foreign compound in the relevant tissue is lower than the k_m, then linear, first-order kinetics will apply, but when the concentration is greater, nonlinear, zero-order kinetics are observed. There are a number of consequences:there is an increase in half-life, the AUC is not proportional to the dose, a threshold for toxic effects may become apparent, a constant amount of compound is excreted (independent of the dose), and proportions of metabolites may change. When an arithmetic, as opposed to a semilog, plot of plasma concentration against time is drawn, this is linear. As the elimination changes with dose, a true half-life and k_{el} cannot be calculated. However, the consequences may not be easily predictable, as saturation of one process may lead to changes in others. For example, saturation of renal excretion would lead to a higher blood level of the chemical that could saturate plasma proteins. The resulting higher concentration of *free* chemical would be available for distribution, glomerular filtration into urine, and metabolism. The metabolism could lead to metabolites, which are excreted into bile. So, a number of processes might be changed as a result.

Saturation kinetics are important in toxicology because they may herald the point at which toxicity occurs.

The plasma level and half-life are also important parameters if a compound is to be administered chronically or there is repeated exposure. Thus, if the dosing interval is shorter than the half-life, the compound accumulates, whereas if the half-life is very short compared with the dosing interval, the compound does not accumulate. The effects on the plasma concentration of the compound of these two situations are shown in Figure 3.29.

This situation may also arise as a result of reduced excretion because of renal impairment due to age or disease. For example, following regular dosing with it, the drug gentamycin has been found to accumulate in patients with renal disease (see chap. 5). Similarly, the drug Opren accumulated with fatal consequences in some elderly patients with reduced renal function on repeated dosing because of the long half-life and the carrier-mediated transport into bile (see below and chap. 5).

In order to rapidly achieve a steady-state level of the compound in the plasma so that the organism is exposed to a fairly constant level, the dosage interval and half-life should be similar (Fig. 3.30). For steady-state conditions, the half-life determines the plasma level. That is, substrates with a long half-life attain a higher steady-state plasma level than compounds with shorter half-lives. It is obviously important to measure this plasma concentration for an assessment of chronic toxicity (Fig. 3.30).

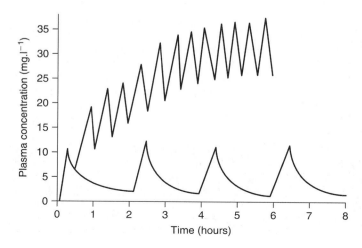

Figure 3.29 Effect of chronic dosing with a foreign compound on its plasma level. Half-life of compound is one hour. Upper curve shows the effect of dosing every 30 minutes and lower curve the effect of a dosage interval of two hours.

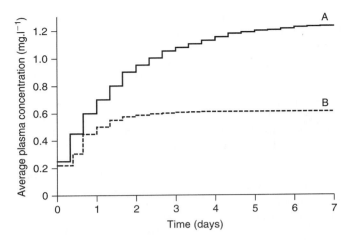

Figure 3.30 Average plasma concentrations of two foreign compounds after multiple dosing. Compound A, half-life 24 hours; compound B, half-life 12 hours; and dosage interval in each case is 8 hours. The accumulation plateau is directly proportional to the half-life, while the rate of accumulation is inversely proportional to the half-life. *Source*: From Ref. 9.

The average plateau concentration after repeated oral administration can be calculated:

$$C_{av} = \frac{f \times D}{V_D \times k_{el} \times t}$$

where f is the fraction absorbed, D is the oral dose, and t is the dosing interval. From this value of C_{av}, the **average body burden** can be calculated:

$$\text{Average body burden} = V_D \times C_{av}$$

3.5 EXCRETION

The elimination of toxic substances from the body is clearly a very important determinant of their biological effects. Rapid elimination will reduce both the likelihood of toxic effects occurring and their duration. Removal of a toxic compound may help to reduce the extent of damage. The elimination of foreign compounds is reflected in the parameters plasma half-life ($t\frac{1}{2}$), elimination rate constant (k_{el}), and total body clearance. The plasma half-life also reflects other processes as well as excretion; the whole-body half-life is the time required for half of the compound to be eliminated from the body and therefore reflects the excretion of the compound. It can be readily measured by administering a radiolabeled compound and determining amount excreted over time. The most important route of excretion for most nongaseous or nonvolatile compounds is through the kidneys into the urine. Other routes are secretion into the bile, expiration via the lungs for volatile and gaseous substances, and secretion into the gastrointestinal tract, or into fluids such as milk, saliva, sweat, tears, and semen.

3.5.1 Urinary Excretion

Many toxic substances and other foreign compounds are removed from the blood as it passes through the kidneys. The kidneys receive around 25% of the cardiac output of blood, and so they are exposed to and filter out a significant proportion of foreign compounds. However, excretion into the urine from the bloodstream applies to relatively small, water-soluble molecules; large molecules such as proteins do not normally pass out through the intact glomerulus, and lipid-soluble molecules such as bilirubin are reabsorbed from the kidney tubules (Fig. 3.31).

Excretion into the urine involves one of three mechanisms: filtration from the blood through the pores in the glomerulus, diffusion from the bloodstream into the tubules, and active transport into the tubular fluid. The principles governing these processes are essentially the same as already described and depend on the physicochemical properties of the compound in question. The structure of the kidney facilitates the elimination of compounds from the bloodstream (Fig. 3.31). The basic unit of the kidney, the nephron, allows most small molecules to pass out of the blood in the glomerulus into the tubular ultrafiltrate aided by large pores in the capillaries and the hydrostatic pressure of the blood. The **pores** in the glomerulus are relatively large (40 Å) and will allow molecules with a molecular weight of less than 60,000 to

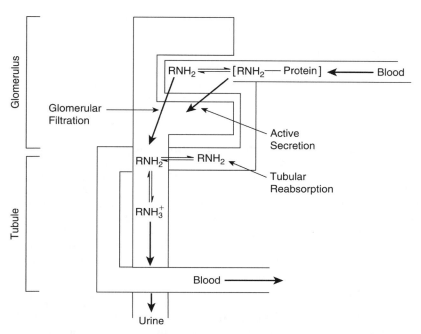

Figure 3.31 Schematic representation of the disposition of foreign compounds in the kidney.

pass through. This filtration therefore only applies to the nonprotein-bound form of the compound, and so the concentration of the compound in the glomerular filtrate will approximate to the free concentration in the plasma. Lipid-soluble molecules will passively diffuse out of the blood, provided there is a concentration gradient. However, such compounds, if they are not ionized at the pH of the tubular fluid, may be reabsorbed from the tubule by passive diffusion back into the blood, which flows through the vessels surrounding the tubule, because there will be a concentration gradient in the direction tubule → blood. Water-soluble molecules, which are ionized at the pH of the tubular fluid, will not be reabsorbed by passive diffusion and will pass out into the urine.

Certain molecules, such as **p-aminohippuric** acid (Fig. 3.18), a metabolite of *p*-aminobenzoic acid are actively transported from the bloodstream into the tubules by a specific anion transport system. Organic anions and cations appear to be transported by separate transport systems located on the proximal convoluted tubule. Active transport is an energy-requiring process and therefore may be inhibited by metabolic inhibitors, and there may be competitive inhibition between endogenous and foreign compounds. For example, the competitive inhibition of the active excretion of uric acid by compounds such as probenecid may precipitate gout.

All four kinds of ABC transporters are found in the kidney and may be involved in the export of some xenobiotics. Also, organic cations can be eliminated by transporters such as the organic cation transporter 2 (OCT2).

These transporters can be responsible for the toxicity of some xenobiotics. For example, the drug cephaloridine is toxic to the kidney as a result of accumulation in the proximal tubular cells, which form the cortex of the kidney. The drug is a substrate for OAT-1 on the basolateral surface and hence is transported into the proximal tubular cells. However, the transport out of these cells from the apical surface into the lumen of the tubule is restricted, probably because of the cationic group on the molecule (Fig. 7.34). The toxicity of cephaloridine is modulated by chemicals that inhibit the OAT-1 and cation transporters. The similar drug cephalothin is not concentrated in the cells and is not nephrotoxic (Table 3.5). See chapter 7 for more details.

Passive diffusion of compounds into the tubules is proportional to the concentration in the bloodstream, so the greater the amount in the blood, the greater will be the rate of elimination. However, when excretion is mediated via active transport or facilitated diffusion, which involves the use of specific carriers, the rate of elimination is constant, and the carrier

Table 3.5 Accumulation of Cephalosporins in the Kidney and Relation to Nephrotoxicity

Drug	Concentration in kidney cortex	Concentration in serum	Cortex/serum	Nephrotoxic dose (mg kg^{-1})
Cephaloridine	2576	167	15	90
Cephalothin	431	127	3	>1000

Source: From Ref. 10.

molecules may become saturated by large amounts of compound. This may have important toxicological consequences. As the dose of a compound is increased, the plasma level will increase. If excretion is via passive diffusion, the rate of excretion will increase, as this is proportional to the plasma concentration. If excretion is via active transport, however, increasing the dose may lead to saturation of renal elimination, and a toxic level of compound in the plasma and tissues may be reached. Another factor that may affect excretion is binding to plasma proteins. This may reduce excretion via passive diffusion, especially if binding is tight and extensive, as only the free portion will be filtered or will passively diffuse into the tubule. Protein binding does not affect active transport, however, and a compound such as *p*-aminohippuric acid (Fig. 3.18), which is 90% bound to plasma proteins, is cleared in the first pass of blood through the kidney.

One of the factors that affects excretion is the urinary pH. If the compound that is filtered or diffused into the tubular fluid is ionized at the pH of that fluid, it will not be reabsorbed into the bloodstream by passive diffusion. For example, an acidic drug such as phenobarbital is ionized at alkaline urinary pH, and a basic drug such as amphetamine is ionized at an acidic urinary pH. This factor is used in the treatment of poisoning with barbiturates and salicylic acid for example (see chap. 7). Thus, by giving sodium bicarbonate to the patient, the urine becomes more alkaline, and excretion of acidic metabolites is increased. The pH of urine may be affected by diet; high-protein diet, for instance, causes urine to become more acid. The rate of urine flow from the kidney into the bladder is also a factor in the excretion of foreign compounds; high-fluid intake and therefore production of copious urine will tend to facilitate excretion. Factors such as age and disease that affect kidney function will influence urinary excretion and may therefore increase toxicity by reducing elimination from the body. For example, sale of the nonsteroidal anti-inflammatory drug **Opren (benoxaprofen)** was stopped because of serious liver toxicity (cholestatic jaundice) and other adverse effects. The drug, used to treat arthritis has a very long half-life (>100 hours). The toxicity was probably partly caused by accumulation in the body following repeated administration to elderly patients in whom kidney function was reduced. An additional factor is the biliary secretion of the acyl glucuronide conjugate of the drug and resulting high concentrations in the biliary system (see below and chap. 5).

3.5.2 Biliary Excretion
Excretion into the bile is an important route for certain foreign compounds, especially large polar and amphipathic substances and may indeed be the predominant route of elimination for such compounds.

Bile production occurs in the liver, where it is secreted by the hepatocytes into the canaliculi, flows into the bile duct, and eventually into the intestine after storage in the gall bladder (Fig. 3.32). Consequently, compounds that are excreted into the bile are usually eliminated in the feces. The factors that affect biliary excretion are molecular weight, charge, and the species of animal (Table 6) (chap. 5, Table 8). Consequently, for polar compounds with a molecular weight of 300 or so, such as glutathione conjugates (see chap. 4), secretion into the bile can be a major route of excretion. Excretion into the bile is usually, although not exclusively, an active process, and there are three specific transport systems, one for neutral compounds, one for anions, and one for cations. For example, the biliary system also includes one of the ABC transporters, MRP2 or CMOAT. This is responsible for the transport of the acyl glucuronide conjugate of the drug benoxaprofen (Opren) into bile, which probably contributes to the hepatic toxicity of the drug. As a result of the specific transport process, such conjugates can accumulate to high concentrations in the lumen of the bile canaliculus and ducts. The relatively insoluble benoxaprofen acyl glucuronide may precipitate at this site. Furthermore,

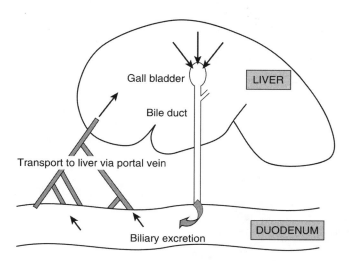

Figure 3.32 Enterohepatic circulation. The circulation of the compound is indicated by the arrows.

Table 3.6 Effect of Molecular Weight on the Route of Excretion of Biphenyls by the Rat

		Percentage of total excretion	
Compound	Molecular weight	Urine	Feces
Biphenyl	154	80	20
4-Monochlorobiphenyl	188	50	50
4,4'-Dichlorophenyl	223	34	66
2,4,5, 2'5'-Pentachlorobiphenyl	326	11	89
2,3, 6, 2',3',6'-Hexachlorobiphenyl	361	1	99

Source: From Ref. 11.

because of the alkaline conditions in the bile, the benoxaprofen conjugate could degrade and in the process interact with critical proteins in the bile duct, a process for which there is experimental evidence.

Quaternary ammonium compounds may be actively secreted into the bile by a separate process. Compounds that undergo biliary excretion have been divided into classes A, B, and C. Class A compounds are excreted by diffusion and have a bile-to-plasma ratio of around 1; class B compounds are actively secreted into bile and have a bile-to-plasma ratio of greater than 1; and class C compounds are not excreted into bile and have a bile-to-plasma ratio of less than 1. The latter type of compound is usually macromolecules such as proteins or phospholipids.

Clearly, large ionic molecules would be poorly absorbed from the lumen of the gastrointestinal tract if ionized at the prevailing pH. If such compounds are extensively secreted into the bile during the first pass through the liver, little may reach the systematic circulation after an oral dose. As with renal excretion via active transport, biliary excretion may be saturated, and this may lead to an increasing concentration of compounds in the liver. For example, the diuretic drug **furosemide** (Fig. 3.33) was found to cause hepatic damage in mice after doses of about 400 mg/kg. The toxicity shows a marked dose threshold, which is partly due to saturation of the biliary excretion and partly to saturation of plasma protein-binding

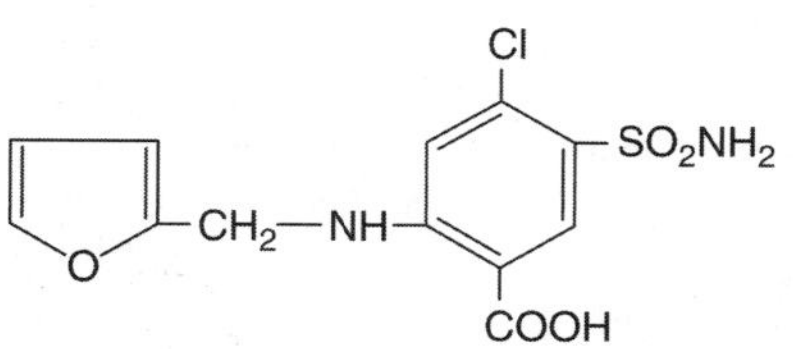

Figure 3.33 Structure of furosemide.

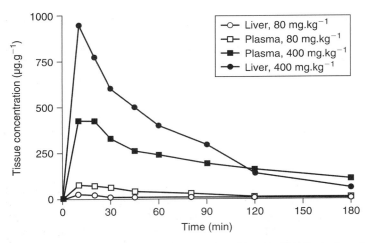

Figure 3.34 Changes in the tissue distribution of furosemide after toxic and nontoxic doses were administered to mice. Toxic dose (400 mg kg^{-1}), nontoxic dose (80 mg kg^{-1}). *Source*: From Ref. 12.

sites. The result is a disproportionate increase in the concentration of the drug in the liver (Fig. 3.34). As well as saturation of transporters, drug interactions can occur as a result of competition for, or inhibition of, specific transporters involved in excretion in the kidney and the biliary system. For example, the drugs cyclosporine and rifampicin may inhibit transport of drugs via OAT transporters and cephalosporins, and probenecids have the potential to inhibit OATP transport.

Another consequence of biliary excretion is that the compound comes into contact with the gut microflora. The bacteria may metabolize the compound and convert it into a more lipid-soluble substance, which can be reabsorbed from the intestine into the portal venous blood supply and so return to the liver. This may lead to a cycling of the compound known as **enterohepatic recirculation**, which may increase the toxicity (Fig. 3.32). If this situation occurs, the plasma level profile may show peaks at various times corresponding to reabsorption rather than the smooth decline expected. If the compound is taken orally, and therefore is transported directly to the liver and is extensively excreted into the bile, it may be that none of the parent compound ever reaches the systemic circulation. Alternatively, the gut microflora may metabolize the compound to a more toxic metabolite, which could be reabsorbed and cause a systemic toxic effect. An example of this is afforded by the hepatocarcinogen **2,4-dinitrotoluene** discussed in more detail in chapter 5. Compounds taken orally may also come directly into contact with the gut bacteria. For example, the naturally occurring glycoside cycasin is hydrolyzed to the potent carcinogen methylazoxymethanol by the gut bacteria when it is ingested orally. Biliary excretion may therefore (*i*) increase the half-life of the compound when followed by reabsorption from the gut, (*ii*) lead to the production of toxic metabolites in the gastrointestinal tract, (*iii*) increase hepatic exposure via the enterohepatic recirculation, and (*iv*) be saturated and lead to hepatic damage. The importance of biliary excretion in the toxicity of compounds can be seen from chapter 2, Table 2, which shows that ligation of the bile duct increases the toxicity of certain chemicals many times.

3.5.3 Excretion Via the Lungs

The lungs are an important route of excretion for volatile compounds and gaseous and volatile metabolites of foreign compounds. For example, about 50% to 60% of a dose of the aromatic hydrocarbon **benzene** is eliminated in the expired air. Excretion is by passive diffusion from the blood into the alveolus assisted by the concentration gradient. This is a very efficient, usually rapid, route of excretion for lipid-soluble compounds, as the capillary and alveolar membranes are thin and in very close proximity to allow for the normal gaseous exchange involved in breathing. There will be a continuous concentration gradient between the blood and air in the alveolus because of the rapid removal of the gas or vapor from the lungs and the rapid blood flow to the lungs. The rate of elimination also depends on the blood: gas solubility ratio. The effect of these factors on the rate of elimination may be crucial in the treatment of poisoning by such gases as the highly toxic carbon monoxide. Compounds may also be metabolized to gaseous toxic metabolites, for example, methylene chloride, which is metabolized to carbon monoxide and has been the cause of serious poisoning (see chap. 7).

3.5.4 Gastrointestinal Tract

Passive diffusion into the lumen of the gut may occur for compounds, such as weak bases, which are nonionized in the plasma but ionized in the stomach. This route may be of particular significance for highly lipid-soluble compounds.

3.5.5 Milk

Excretion into breast milk can be a very important route for certain types of foreign compounds, especially lipid-soluble substances, because of the high lipid content in milk. Clearly, newborn animals will be specifically at risk from toxic compounds excreted into milk. For example, nursing mothers exposed to DDT secrete it into their milk, and the infant may receive a greater dose, on a body-weight basis, than the mother. Also, because the pH of milk (6.5) is lower than the plasma, basic compounds may be concentrated in the fluid.

3.5.6 Other Routes

Foreign compounds may be excreted into other body fluids such as sweat, tears, semen, or saliva by passive diffusion, depending on the lipophilicity of the compound. Although these routes are generally of minor importance quantitatively, they may have a toxicological significance such as the production of dermatitis by compounds secreted into the sweat, for example.

SUMMARY

The disposition of toxic compounds in biological systems can be divided into four interrelated phases: ADME.

Absorption is necessary for the chemical to exert a systemic biological/toxic effect and involves crossing membranes. Membranes are semipermeable phospholipid/protein bilayers. The phospholipids and proteins are of variable structure, and the membrane is selectively permeable. The physicochemical characteristics of foreign molecules that are important include size/shape, lipid solubility, structure, and charge/polarity.

There are five types of transport (with their important features):

1. Filtration (small, water-soluble molecules)
2. Passive diffusion (lipid-soluble, noncharged molecules; concentration gradient; first order; pH partition theory)
3. Active transport (specific membrane carrier; energy required; operates against concentration gradient; inhibitable; saturable; possible competition; zero order)
4. Facilitated diffusion (specific carrier; saturable; possible competition; operates with concentration gradient; no energy required)
5. Endocytosis (phagocytosis/pinocytosis; requires energy; large, insoluble molecules).

There are three main sites of absorption: skin (large surface area; poorly vascularized; not readily permeable); gastrointestinal tract (major site; well vascularized; variable pH; large surface area; transport processes; food; gut bacteria); lungs (very large surface area; well vascularized; readily permeable). Compounds may be administered by direct injection (i.p., i.m., s.c., i.v.).

The end result of the absorptive phase is that the compound may pass through tissues and enter the blood.

Distribution is the phase in which the compound is carried to tissues by the bloodstream or lymphatic system. Compounds are usually first absorbed into the portal venous system after oral administration, directing them to the liver where they may be removed (extracted/metabolized) (first-pass effect). The blood (plasma) level reflects the concentration at the target/receptor and is governed by distribution. Distribution depends on passage through membranes (passive diffusion, carrier-mediated transport, etc.) and may be limited by binding to blood proteins. Protein binding may involve ionic, hydrophobic, hydrogen, or Van Der Waals bonding. It may show saturation, competitive inhibition, and displacement, therefore threshold effects may occur.

Chemicals may be sequestered and accumulated in tissues (compartments) depending on certain factors (e.g., lipid solubility; pK_a); distribution can change with pH of blood or tissue. Some tissues are poorly accessible (brain).

Blood level may be used to derive kinetic parameters such as half-life, elimination rate constant, AUC, and volume of distribution. Comparison of blood levels after oral and iv may be used to calculate bioavailability. Half-life can be used to predict the effect of repeated dosing.

Excretion is the elimination of the molecule from the organism by one of several routes. The urine is the major route, but expired air and bile may also be used. Urinary excretion involves filtration, passive diffusion, and active transport. Biliary excretion involves active transport, and there is a molecular weight threshold for compounds, above which excretion by this route becomes more important. Biliary excretion may lead to enterohepatic recirculation. Excretion by these routes can be saturated, leading to accumulation.

Volatile chemicals are transported by passive diffusion from the blood into the lungs prior to exhalation.

Excretion into milk is important for exposure of the newborn animal.

The concentration and physicochemical properties (ionization/charge, lipid solubility, size) of the molecule have a major impact on each of the phases of its disposition.

Membrane transporter proteins (MDR or ABC transporter proteins) such as p-glycoprotein are crucially important in the process of excretion and also in absorption and distribution and elimination of chemicals from cells. These transport organic anions or cations and neutral compounds across membranes, pump unwanted chemicals out of cells such as in gut, placenta, and brain, transport chemicals into bile from liver cells, and facilitate excretion from the kidney.

REVIEW QUESTIONS

1. Write short notes on three of the following:
   a. The pH partition theory
   b. Active transport
   c. Plasma protein binding
   d. Enterohepatic recirculation.
2. Explain the abbreviation ADME.
3. List five mechanisms for the transport of chemicals through biological membranes.
4. What physicochemical characteristics determine the absorption, distribution, and excretion of chemicals in biological systems?
5. How does the size of particles influence their absorption from the lungs?
6. Passive diffusion is a first-order rate process. Is this statement true or false?
7. What is the difference between the drugs thiopental and pentobarbital and how would it affect their absorption and distribution?
8. What is log P?
9. The three main sites of absorption for chemicals are the gastrointestinal tract, skin, and lungs. List the similarities and differences between these sites relevant to the absorptive process.
10. Explain the first-pass effect.
11. From which site would you expect a weak acid to be absorbed?
    - The mouth
    - The stomach
    - The small intestine
12. Which is the plasma protein most often involved in binding chemicals?
13. Explain why administration of sodium bicarbonate is used in the treatment of barbiturate poisoning.

14. Give two reasons why the plasma level of a chemical is an important piece of information for a toxicity study.
15. How is the volume of distribution determined?
16. How would you determine that a drug undergoes first-pass metabolism?
17. What factors influence the biliary excretion of chemicals?
18. What is the importance of ABC transporter proteins?

REFERENCES

1. Timbrell JA. Introduction to Toxicology. London: Taylor and Francis, 2002.
2. Hogben CAM, Tocco DJ, Brodie BB, et al. On the mechanism of intestinal absorption of drugs. J Pharmac Exp Therap 1959; 125:275–282.
3. Shah PV, Monroe RJ, Guthrie FE, et al. Comparative rates of dermal penetration of insecticides in mice. Toxicol Appl Pharmacol 1981; 59:414–423.
4. Ahdaya SM, Monroe RJ, Guthrie FE, Absorption and distribution of intubated insecticides in fasted mice. Pestic Biochem Physiol 1981; 16:38–46.
5. Hodgson E. Levi PE. A Textbook of Modern Toxicology. New York: Elsevier, 1987.
6. Hatch TF, Gross P. Pulmonary Deposition and Retention of Inhaled Aerosols. New York: Academic Press, 1964.
7. Documenta Geigy. In: Diem K, Lenter C, eds. Scientific Tables. Basle: CIBA-GEIGY, 1970.
8. Lullman H, Lullmann-Rauch R, Wassermann O, et al. Drug induced phospholipidosis. CRC Critical Rev Toxicol 1975; 4:185–218.
9. van Rossum JM. J Pharm Sci 1968; 57:2162.
10. Tune BM. . In: DeBroe ME, Porter GA, Bennett WM, et al., eds. Clinical Nephrotoxins: Renal Injury from Drugs and Chemicals. Dordrecht: Kluwer, 1998.
11. Matthews HB. In: Hodgson E, Guthrie FE, eds. Introduction to Biochemical Toxicology. New York: Elsevier-North Holland, 1980.
12. Mitchell JR, Potter WZ, Hinson JA, et al. Toxic Drug Reactions. In: Gillette JR, Mitchell JR eds. Handbook of Experimental Pharmacology, Vol. 28, Part 3. Concepts in Biochemical Pharmacology, Berlin: 1975.

BIBLIOGRAPHY

Absorption and Distribution

Adamson RH, Davies DS. Comparative aspects of absorption, distribution, metabolism and excretion of drugs. International Encyclopaedia of Pharmacology and Therapeutics, Section 85 (Comparative Pharmacology). Oxford: Pergamon Press, 1973:851.

Alberts B, Johnson A, Lewis J, et al. Molecular Biology of the Cell. New York: Garland Science, 2002.

Ayrton A, Morgan P. Role of transport proteins in drug absorption, distribution and excretion. Xenobiotica 2001; 31:469–497.

De Boer AG, van der Sandt, Gaillard PJ. The role of drug transporters at the blood brain barrier. Ann Rev Pharmacol Toxicol 2003; 43:629–656.

Brodie BB, Gillette JR, Ackerman HS, eds. Handbook of Experimental Pharmacology, Vol. 28, Part 1, Concepts in Biochemical Pharmacology. Berlin: Springer-Verlag, 1971.

Chasseaud LF. Processes of absorption, distribution and excretion. In: Hathway DE, Brown SS, Chasseaud LF, et al. Foreign Compound Metabolism in Mammals, Vol. 1. London: The Chemical Society, 1970.

Findlay JWA. The distribution of some commonly used drugs in human breast milk. Drug Metab Rev 1983; 14:653.

Florence AT, Atwood D. Physiochemical Principles of Pharmacy. 4th ed. London: Pharmaceutical Press, 2005.

Ginsburg J. Placental drug transfer. Annu Rev Pharmacol 1971; 11:387.

Hodgson E, Levi PE. A Textbook of Modern Toxicology. New York: Elsevier, 1987.

Jollow DJ. Mechanisms of drug absorption and drug solution. Rev Can Biol 1973; 32:7.

La Du BN, Mandel HG, Way EL, eds. Fundamentals of Drug Metabolism and Drug Disposition, Chapters 1–7. Baltimore: Williams & Wilkins, 1971.

Pratt WB. The entry, distribution and elimination of drugs. In: Pratt WB, Taylor P, eds. Principles of Drug Action. 3rd ed. New York: Churchill Livingstone, 1990.

Riviere JE. Absorption and distribution. In: Hodgson E, Levi PE, eds. Introduction to Biochemical Toxicology. 2nd ed. Connecticut: Appleton-Lange, 1994.

Rozman KK, Klaassen CD. Absorption, distribution and excretion of toxicants. In: Klaassen CD, ed. Cassarett and Doull's Toxicology, The Basic Science of Toxicology. 6th ed. New York: McGraw Hill, 2001.

Schanker LS. Drug absorption from the lung. Biochem Pharmacol 1978; 27:381.

Smyth RD, Hottendorf GH. Application of pharmacokinetics and biopharmaceutics in the design of toxicological studies. Toxic Appl Pharmacol 1980; 53:179.

Wilkinson GR. Plasma and tissue binding considerations in drug disposition. Drug Metab Rev 1983; 14:427.

Excretion

Chipman JK, Coleman R. Mechanisms and consequences of enterohepatic circulation (EHC). In: Hill MJ, ed. Role of Gut Bacteria in Human Toxicology and Pharmacology. London: Taylor and Francis, 1995.

Coleman R, Chipman JK. Factors governing biliary excretion. In: Hill MJ, ed. Role of Gut Bacteria in Human Toxicology and Pharmacology. London: Taylor and Francis, 1995.

Klaassen CD. Mechanisms of bile formation, hepatic uptake and biliary excretion. Pharmacol Rev 1984; 36:1.

Krishnamurthy P, Schuetz JD. Role ofABCG2/BCRP in biology and medicine. Ann Rev Pharmacol Toxicol 2006; 46:381–410.

Lee W, Kim RB. Transporters and renal drug elimination. Ann Rev Pharmacol Toxicol 2004; 44:137–166.

Leslie EM, Deeley R, Pole SPC. Multidrug resistance proteins: role of P-glycoprotein, MRP1, MRP2 and BCRP (ABCG2) in tissue defense. Toxicol Appl Pharmacol 2005; 204:216–237.

Levine WG. Excretion mechanisms. In: Caldwell J, Jakoby WB, eds. Biological Basis of Detoxication. New York: Academic Press, 1983.

Matthews HB. Excretion and elimination of toxicants and their metabolites. In: Hodgson E, Levi PE, eds. Introduction to Biochemical Toxicology. 2nd ed. New York: Elsevier-North Holland, 1994.

Pritchard JB, James MO. Metabolism and urinary excretion. In: Jakoby WB, Bend JR, Caldwell J, eds. Metabolic Basis of Detoxication. New York: Academic Press, 1982.

Pritchard JB, Millar DS. Mechanisms mediating renal secretion of organic anions and cations. Physiolog Rev 1993; 73:765–796.

Shitara Y, Sato H, Sugiyama Y. Evaluation of drug-drug interaction in the hepatobiliary and renal transport of drugs. Ann Rev Pharmacol Toxicol 2005; 45:689–723.

Smith RL. The Excretory Function of Bile. London: Chapman & Hall, 1973.

Stowe CM, Plaa GL. Extrarenal excretion of drugs and chemicals. Annu Rev Pharmacol 1968; 8:337.

Pharmacokinetics

Birkett DJ. Pharmacokinetics Made Easy. 2nd ed. Sydney: McGraw-Hill Australia, 2002.

Clark B, Smith DA. An Introduction to Pharmacokinetics. 2nd ed. Oxford: Blackwell, 1986.

Gibaldi M, Perrier D. Pharmacokinetics. 2nd ed. New York: Marcel Dekker, 1982.

Houston JB. Role of pharmacokinetics in rationalizing tissue distribution. In: Cohen GM, ed. Target Organ Toxicity, Vol. I. Boca Raton, Florida: CRC Press, 1986.

Krishnan K, Andersen ME. Physiologically based pharmacokinetic modelling in toxicology. In: Hayes AW, ed. Principles and Methods of Toxicology, 5th ed. Florida CRC Press, 2007.

Levy G, Gibaldi M. Pharmacokinetics. In: Gillette JR, Mitchell JR, eds. Handbook of Experimental Pharmacology, Vol. 28, Part 3, Concepts in Biochemical Pharmacology. Berlin: Springer, 1975.

Medinsky MA, Klaassen CD. Toxicokinetics. In: Klaassen CD, ed. Cassarett and Doull's Toxicology, The Basic Science of Toxicology. 5th ed. New York: McGraw Hill, 2001.

Neubig RR. The time course of drug action. In: Pratt WB, Taylor P,eds. Principles of Drug Action. 3rd ed. New York: Churchill Livingstone, 1990.

Renwick AG. Toxicokinetics. In: Ballantyne B, Marrs TC, Syversen T, eds. General and Applied Toxicology, Vol. 1. London: Macmillan, 2001.

Renwick AG. Toxicokinetics-pharmacokinetics in toxicology. In: Hayes AW, ed. Principles and Methods of Toxicologyedited. 5th ed. Florida: CRC Press, 2007.

Rowland M, Tozer TN. Clinical Pharmacokinetics. Concepts and Applications. 2nd ed. Philadelphia: Lea and Febiger, 1988.

Smith DA. Pharmacokinetics and pharmacodynamics in toxicology. Xenobiotica 1997; 27:513.

Zbinden G. Biopharmaceutical studies, a key to better toxicology. Xenobiotica 1988; 18(suppl 1):9.

4 | Factors Affecting Toxic Responses: Metabolism

4.1 INTRODUCTION

As discussed in the preceding chapter, foreign and potentially toxic compounds absorbed into biological systems are generally lipophilic substances. They are therefore not ideally suited to excretion, as they will be reabsorbed in the kidney or from the gastrointestinal tract after biliary excretion. For example, highly lipophilic substances such as polybrominated biphenyls and DDT are poorly excreted and therefore may remain in the animal's body for years.

In the **biotransformation** of foreign compounds, the body attempts to convert such lipophilic substances into more polar, and consequently, more readily excreted metabolites. The exposure of the body to the compound is hence reduced and potential toxicity decreased. This process of biotransformation is therefore a crucial aspect of the disposition of a toxic compound *in vivo*. Furthermore, as will become apparent later in the book, biotransformation may also underlie the toxicity of a compound.

The metabolic fate of a compound can therefore have an important bearing on its toxic potential, disposition in the body, and eventual excretion.

The primary results of biotransformation are therefore as follows:

1. The parent molecule is transformed into a more polar metabolite, often by the addition of ionizable groups.
2. Molecular weight and size are often increased.
3. The excretion is facilitated, and hence elimination of the compound from the tissues, and the body is increased.

The consequences of metabolism are as follows:

1. The biological half-life is decreased.
2. The duration of exposure is reduced.
3. Accumulation of the compound in the body is avoided.
4. The biological activity may be changed.
5. The duration of the biological activity may be affected.

For example, the analgesic drug **paracetamol** (see chap. 7) has a renal clearance value of 12 mL min^{-1}, whereas one of its major metabolites, the sulfate conjugate, is cleared at the rate of 170 mL min^{-1}. However, biotransformation may not always increase water solubility. For example, many **sulfonamide drugs** are acetylated *in vivo*, but the acetylated metabolites can be less water soluble (Table 4.1), precipitate in the kidney tubules, and cause toxicity. As the chemical structure is changed from that of the parent compound, there may be consequential changes to the pharmacological and toxicological activity of the compound. For some drugs, the pharmacological activity resides in the metabolite rather than the parent compound. A classic example of this is the antibacterial drug **sulfanilamide**, which is released from the parent compound, **prontosil**, by bacterial metabolism in the gut (Fig. 4.39). For other drugs, it is the parent compound that is active, as is the case with the muscle-relaxant drug succinylcholine (suxamethonium). The action of this drug normally only lasts a few minutes because metabolism rapidly cleaves the molecule to yield inactive products (chap. 7, Fig. 55). The duration of action is therefore determined by metabolism in this case. Although biotransformation is usually regarded as a detoxication process, this is not always so. Thus the pharmacological or toxicological activity of the metabolite may be greater or different from

Table 4.1 Solubility Data for Two Sulfonamides and Their Acetylated Metabolites

Solubility in urine	Drug (mg mL^{-1} at 37°C)	Urinary pH
Sulfisomidine	254	5.0
	282	6.8
Acetylsulfisomidine	9	5.0
	10	7.5
Sulfisoxazole	150	5.5
	1200	6.5
Acetylsulfisoxazole	55	5.5
	450	6.5

Source: From Ref. 2.

that of the parent compound. Later in the chapter, and especially in the final chapter, we will examine ways and consider examples in which toxicity is increased by metabolism.

Metabolism is therefore an important determinant of the *activity* of a compound, the *duration* of that activity, and the *half-life* of the compound in the body.

The metabolism of foreign compounds is catalyzed by enzymes, some of which are specific for the metabolism of xenobiotics. The metabolic pathways involved may be many and various but the major determinants of which transformations take place are

1. the structure and physicochemical properties of the compound in question and
2. the enzymes available in the exposed tissue.

Thus the *partition coefficient*, the *stereochemistry*, and the *functional groups* present on a molecule may all influence the particular metabolic transformation, which takes place, and these factors are discussed in more detail later.

The enzymes specifically involved in the metabolism of foreign compounds are necessarily often flexible, and the substrate specificity is generally broad. However, it follows from the above two conditions that if the structure of a foreign compound is similar to a normal endogenous molecule, then the foreign compound may be a suitable substrate for an enzyme primarily involved in intermediary metabolic pathways if the enzyme is present in the exposed tissue. Thus, foreign compounds are not exclusively metabolized by specific enzymes.

The organ most commonly involved in the biotransformation of foreign compounds is the liver because of its *position, blood supply*, and *function* (chap. 6, Fig. 2). Most foreign compounds are taken into the organism via the gastrointestinal tract, and the blood that drains the tract flows through the portal vein directly to the liver. Therefore, the liver represents a portal to the tissues of the body and is exposed to foreign compounds at higher concentrations than most other tissues. Detoxication in this organ and possible removal by excretion into the bile are therefore protective measures. The role of the liver in endogenous metabolism and its structure (see chap. 6) make it an ideal site for the biotransformation of xenobiotics. As already mentioned in the previous chapter, metabolism in the liver may be so extensive during the "first pass" of the compound through the organ that little or none of the parent compound reaches the systemic circulation. However, most other organs and tissues possess some metabolic activity with regard to foreign compounds and in some cases may be quantitatively more important than the liver.

The enzymes involved in biotransformation also have a particular subcellular localization: many are found in the **smooth endoplasmic reticulum (SER)**. Some are located in the cytosol, and a few are found in other organelles such as the mitochrondria. These subcellular localizations may have important implications for the mechanism of toxicity of compounds in some cases.

4.2 TYPES OF METABOLIC CHANGE

Metabolism can be simply and conveniently divided into two phases: **phase 1** and **phase 2**. Phase 1 is the alteration of the original foreign molecule so as to add on a functional group, which can then be conjugated in phase 2. This can best be understood by examining the example in Figure 7.1. The foreign molecule is **benzene**, a highly lipophilic molecule, which is

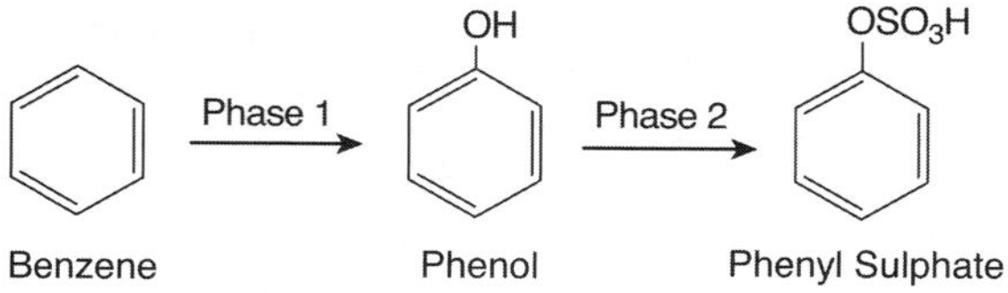

Figure 4.1 Metabolism of benzene to phenyl sulfate.

Table 4.2 The Major Biotransformation Reactions

Phase1	Phase 2	Phase 3
Oxidation	Sulfation	Further metabolism of glutathione conjugates
Reduction	Glucuronidation	
Hydrolysis	Hydration	
Hydration	Acetylation	
Dehalogenation	Amino acid conjugation	
	Methylation	

not readily excreted from the animal except in the expired air, as it is volatile. Phase 1 metabolism converts benzene into a variety of metabolites, but the major one is phenol. The insertion of a hydroxyl group allows a phase 2 conjugation reaction to take place with the polar sulfate group being added. Phenyl sulfate, the final metabolite is very water soluble and is readily excreted in the urine. Most biotransformations can be divided into phase 1 and phase 2 reactions, although the products of phase 2 biotransformations may be further metabolized in what is sometimes termed "**phase 3 reactions.**"

If the foreign molecule already possesses a functional group suitable for a phase 2 reaction, a phase 1 reaction will be unnecessary. Thus, if phenol is administered to an animal, then it may immediately undergo a phase 2 reaction, such as conjugation with sulfate. Alternatively it may undergo another phase 1 type of reaction. The major types of reactions are shown in Table 4.2.

Generally, therefore, the function of phase 1 reactions is to *modify* the *structure* of a xenobiotic so as to introduce a *functional group* suitable for conjugation with glucuronic acid, sulfate, or some other highly polar moiety, so making the entire molecule water soluble.

4.3 PHASE 1 REACTIONS

The major phase 1 reactions are oxidation, reduction, and hydrolysis.

4.3.1 Oxidation

For foreign compounds, the majority of oxidation reactions are catalyzed by **monooxygenase enzymes**, which are part of the mixed function oxidase (MFO) system and are found in the SER (and also known as **microsomal enzymes**). Other enzymes involved in the oxidation of xenobiotics are found in other organelles such as the mitochondria and the cytosol. Thus, amine oxidases located in the mitochondria, xanthine oxidase, alcohol dehydrogenase in the cytosol, the prostaglandin synthetase system, and various other peroxidases may all be involved in the oxidation of foreign compounds.

Microsomal oxidations may be subdivided into aromatic hydroxylation; aliphatic hydroxylation; alicyclic hydroxylation; heterocyclic hydroxylation; N-, S-, and O-dealkylation; N-oxidation; N-hydroxylation; S-oxidation; desulfuration; deamination; and dehalogenation.

Non-microsomal oxidations may be subdivided into amine oxidation, alcohol and aldehyde oxidation, dehalogenation, purine oxidation, and aromatization.

Microsomal Oxidations
Cytochromes P-450 MFO system. The majority of these reactions are catalyzed by one-enzyme system, the cytochromes P-450 monooxygenase system, which is located particularly

in the SER of the cell. The enzyme system is isolated in the so-called microsomal fraction, which is formed from the endoplasmic reticulum when the cell is homogenized and fractionated by differential ultracentrifugation. Microsomal vesicles are thus fragments of the endoplasmic reticulum in which most of the enzyme activity is retained.

The endoplasmic reticulum is composed of a convoluted network of channels and so has a large surface area. Apart from cytochromes P-450, the endoplasmic reticulum has many enzymes and functions, besides the metabolism of foreign compounds. These include the synthesis of proteins and triglycerides and other aspects of lipid metabolism and fatty acid metabolism. Specific enzymes present on the endoplasmic reticulum include cholesterol esterase, azo reductase, glucuronosyl transferase, NADPH cytochromes P-450 reductase and NADH cytochrome b_5 reductase and cytochrome b_5. A FAD-containing monooxygenase is also found in the endoplasmic reticulum, and this is discussed later in this chapter.

The cytochromes P-450 monooxygenase system is actually a collection of isoenzymes, all of which possess an iron protoporphyrin IX as the prosthetic group. The monomer of the enzyme has a molecular weight of 45,000 to 55,000. The enzyme is membrane bound within the endoplasmic reticulum. Cytochromes P-450 are closely associated with another vital component of the system, **NADPH cytochrome P-450 reductase**. This is a **flavoprotein**, which has 1 mol of FAD and 1 mol of FMN per mol of apoprotein. The monomeric molecular weight of the enzyme is 78,000. The enzyme transfers two electrons to cytochromes P-450, but one at a time. There only seems to be one reductase, which serves a group of isoenzymes of cytochromes P-450, and consequently, its concentration is 1/10 to 1/30 that of cytochromes P-450.

It is known that there is a binding site for NADPH cytochrome P-450 reductase and cytochrome b_5 and several substrate-binding sites. The heme-binding segment, which involves a cysteinyl residue, is common to cytochromes P-450 in many species.

Phospholipid is also required in the enzyme complex; seemingly this is important for the integrity of the overall complex and the interrelationship between the cytochromes P-450 and the reductase. The individual components can be separated, and reconstitution of these components results in a functional enzyme.

The overall reaction is

$$SH + O_2 + NADPH + H^+ \rightarrow SOH + H_2O + NADP^+$$

where S is the substrate. The reaction therefore also requires NADPH and molecular oxygen. The mechanism of action of cytochrome P-450 involves oxygen activation followed by abstraction of a hydrogen atom or an electron from the substrate and oxygen rebound (radical recombination). The overall result therefore is that a single atom of oxygen is inserted into the xenobiotic molecule.

The sequence of reactions is shown in Figure 7.2. It should be noted, however, that a number of different representations will be found in different texts and reviews. This can be extremely confusing for the student. Therefore, this figure has been kept relatively simple. More detail can be found in references in the Bibliography (3).

The catalytic cycle involves at least four distinct steps:

1. Addition of substrate to the enzyme
2. Donation of an electron
3. Addition of oxygen and rearrangement
4. Donation of a second electron and loss of water

Now let us look at these steps in more detail.

Step 1. Binding of the substrate to the cytochromes P-450. This takes place when the iron is in the oxidized, ferric state. There are three types of binding: type I, II, and reverse type I (or modified type II), which give rise to particular spectra. These are now known to be the result of perturbations in the spin equilibrium of cytochromes P-450. Thus, ethylmorphine gives rise to a type I spectrum, aniline, a type II, and phenacetin, a modified type II. There are two spin states: low spin, hexa coordinated or high spin, penta coordinated (Fig. 7.2). Most P-450s are in the low-spin state. When certain substrates bind (type I), they bind to a hydrophobic site on the protein close enough to the heme to interact with oxygen and cause a perturbation and a conformational change. The change results in a high-spin configuration. This change from low

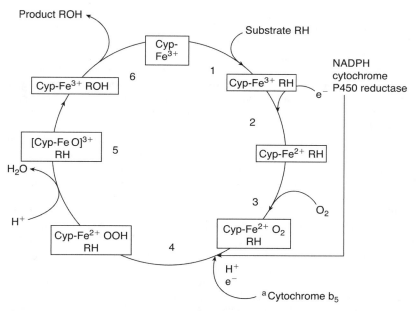

Figure 4.2 The catalytic cycle of the cytochrome(s) P-450 monooxygenase (MFO) system. [a]For explanation of step 4, see text. *Abbreviation*: Cyp, cytochromes P-450; MFO, mixed function oxidase. Source: From Ref. 1.

to high spin gives rise to spectral change with an increase in absorbance at 390 nm and a decrease at 420 nm. The difference spectrum, therefore, has a peak at 390 nm and a trough around 420 nm. The change in the spin state to high spin changes the redox potential such that the enzyme is now more easily reduced. This is therefore important as the next step is a one-electron reduction.

Type II substrates are compounds such as nitrogenous bases, with sp2 or sp3 nonbonded electrons. These bind to iron and give rise to a 6-coordinated, low-spin hemoprotein. Such compounds may also be inhibitors of cytochromes P-450. The spectrum shows a peak at 420 to 435 nm and a trough at 390 to 410 nm in the difference spectrum.

Step 2. First electron reduction of the substrate-enzyme complex. The iron atom is reduced from the ferric to the ferrous state. The reducing equivalents are transferred from NADPH via the reductase. The equilibrium between the high-spin and low-spin states may be important in this step, but it is not yet clear.

Thus the reaction is:

$$NADPH + H^+ \rightarrow FAD\ NADPH\ cytochrome\ FMN\ P\text{-}450\ reductase \rightarrow Cytochrome\ P\text{-}450$$

Both FAD and FMN are involved in the electron transfer in this and step 5.

In the reduced state, cytochromes P-450 may also bind certain ligands to give particular difference spectra. The most well known is that which occurs when carbon monoxide binds giving an absorption maximum at 450 nm. A type III spectrum gives two peaks at 430 and 455 nm after binding of certain compounds such as ethyl isocyanide or methylenedioxyphenyl compounds to the reduced enzyme. The latter form stable complexes with the enzyme and are also inhibitors.

Step 3. Addition of molecular oxygen and rearrangement of the ternary ferrous oxygenated cytochromes P-450-substrate complex. The reduced cytochromes P-450-substrate complex binds oxygen and undergoes a rearrangement. The oxidation state of the oxygen and the iron in the ternary complex is not entirely clear, but the oxygen may exist as hydroperoxide or superoxide. It has been shown by experiments using $^{18}O_2$ that the oxygen bound is molecular oxygen and is not derived from water.

Step 4. Addition of the second electron from NADPH via P-450 reductase. Step 4′. Alternatively, the electron may be donated by NADH via cytochrome b_5 reductase and cytochrome b_5. This alternative source of the second electron step is still controversial.

Step 5. The complex then rearranges with loss of water and insertion of one atom of oxygen into the substrate. The mechanism is obscure but seems to involve oxygen in an activated form. The other atom of oxygen is reduced to water, the other product.

Step 6. The final step is loss of the oxidized substrate.

It has been suggested that the hydroxylation of hydrocarbons by cytochromes P-450 may be a two-step mechanism involving a radical intermediate formed by removal of a hydrogen atom from the substrate and then transference to the oxygen bound to the iron. The hydroxyl may then react with the substrate radical to produce hydroxylated substrate. Under certain circumstances, cytochromes P-450 produces hydrogen peroxide. This seems to be when the cycle becomes uncoupled and the oxygenated P-450 complex breaks down differently to give the oxidized cytochrome-substrate complex and hydrogen peroxide. In some cases, with certain substrates, the catalytic cycle can become uncoupled or dissociated such that hydrogen peroxide is produced probably from the dissociation of the oxygen cytochrome P-450 complex. This hydrogen peroxide can potentially cause damage if it leaks out of the enzyme.

Cytochromes P-450 may be found in other organelles as well as the SER including the rough endoplasmic reticulum and nuclear membrane. In the adrenal gland, it is also found in the mitochondria, although here adrenodoxin and adrenodoxin reductase are additional requirements in the overall system. Although the liver has the highest concentration of the enzyme, cytochromes P-450 are found in most, if not all, tissues.

The cytochrome P-450 system is a remarkable enzyme system because it consists of so many (iso)enzymes or isoforms (probably at least 50 in man) and results from a gene family represented in all phyla and very many diverse species. More than 300 cDNAs have been cloned and more than 60 different types of chemical reaction are catalyzed. This multiplicity of isoforms accounts for the diversity of the reactions catalyzed and the substrates accommodated. The forms or isoenzymes can be separated using chromatographic and electrophoretic techniques and the DNA sequences determined using sophisticated molecular biological techniques. These isoenzymes may vary in distribution both within the cell and in the tissues of the whole organism. The proportions of the isoenzymes in any given tissue may change as a result of treatment with various compounds as described in the next chapter.

There are now a considerable number of isoenzymes, which have been identified, and some indication of the particular substrates and their characteristics is emerging. Those involved in the metabolism of xenobiotics can be seen in Table 4.3. The different isoenzymes are coded for by distinct genes, and the nomenclature used in this table is the current internationally accepted standard. This groups enzymes into gene families on the basis of primary amino acid sequence, resulting from sequencing of cytochromes P-450 proteins, cDNAs, and genes (5). Prior to the introduction of this nomenclature, a confusing array of names was in use based on substrates or inducers, and this may still be encountered.

Thus the **cytochromes P-450 gene superfamily** currently consists of 27 gene families. The various proteins and the genes coding for them are designated by CYP with the families indicated by Arabic numerals. There are three main gene families important in xenobiotic metabolism: CYP1, CYP2, and CYP3, and CYP4 is involved in fatty acid metabolism. However, the latter may be important for some xenobiotics, which have suitable carboxylic acids as part of the structure. The genes in these four families code for primarily hepatic, microsomal enzymes.

Within these families, there is one subfamily, i.e., CYP1A, CYP3A, CYP4A, except CYP2, which has five subfamilies A, B, C, D, and E. These may be further divided into genes coding for single distinct enzyme proteins such as CYP1A1 and CYP1A2. Subfamilies are designated by capital letters and proteins and genes within those by Arabic numerals. Thus there are multiple forms derived from distinct genes. CYP3A4 means the 4th gene to be sequenced in the CYP3A subfamily.

There may also be allelic variants giving rise to different proteins.

The four families CYP17, CYP19, CYP20, and CYP22 code for P-450s involved in steroid biosynthesis and found mainly in extrahepatic tissues. The mitochondrial P-450 is CYP11. The enzyme proteins CYP1A1, CYP1A2, and CYP2EI seem to be highly conserved and similar in all species. Humans have at least 40 different cytochrome P-450 enzymes. There are also other forms of P-450 to be found in insects, yeast, and bacteria. For further details on this topic, the reader is referred to the Bibliography.

As already indicated, however, there is overlap between some of the P-450s in terms of substrates and types of reaction catalyzed. These different isoenzymes may be separated on the

Table 4.3 Characteristics of Cytochromes P-450 Families 1–4

Isozyme	Substrate examples	Reactions
CYP1A1	benzo[a]pyrene	hydroxylation
	7-ethoxyresorufin	O-de-ethylation
CYP1A2	acetylaminofluorene	N-hydroxylation
	Phenacetin	O-de-ethylation
CYP2A1	testosterone	7-α-hydroxylation
CYP2A2	testosterone	15-α-hydroxylation
CYP2A3		
CYP2B1	hexobarbital	hydroxylation
	7-pentoxyresorufin	O-de-ethylation
CYP2B2	7-pentoxyresorufin	O-de-ethylation
	7,12-dimethylbenzanthracene	12-methyl hydroxylation
CYP2C	*S*-mephenytoin	hydroxylation
CYP2D	debrisoquine	alicyclic hydroxylation
CYP2E1	*p*-nitrophenol	hydroxylation
	Aniline	hydroxylation
CYP3A	ethylmorphine	N-demethylation
	aminopyrine	N-demethylation
CYP4A1	lauric acid	φ-hydroxylation
	lauric acid	φ-1-hydroxylation

It should be noted that this is not an exhaustive list and that in different species, different numbers of isozymes exist. It serves solely to illustrate the multiplicity of the forms of cytochrome P-450 generally involved with xenobiotic metabolism and the differences and similarities between them.
For a more detailed discussion see Ref. 4, or other references in the Bibliography.
Abbreviation: CYP, cytochrome P-450.

Table 4.4 The Relative Proportions of the Various Isozymes of Cytochrome P-450 with Representative Substrates

Human CYPs	Proportion in human liver	Representative substrates
3A4/5/7	30%	Nifedipine, erythromycin
2C8/9/18	20%	Warfarin, phenytoin
1A2	15%	Caffeine, theophylline
2E1	10%	Chlorzoxazone
2D6	5%	Debrisoquine, sparteine
2C19	5%	Mephenytoin, omeprazole
2A6	5%	Coumarin
1A1	Low level	
2B6	Low level	

Source: From Ref. 6.

basis of certain criteria. Thus, they may have different monomeric molecular weights, the carbon monoxide difference spectra may show different maxima, the amino acid composition and terminal sequences may be different, substrate specificities may be different, and they may be distinguished by specific antibodies.

The importance of these different isoenzymes is that they may catalyze different biotransformations, and this may be crucial to the toxicity of the compound in question. The isoforms found in human liver and their proportions are shown in Table 4.4. The most important subfamily is CYP3A, but there is variation in the expression of this and the other CYPs. For example, CYP3A5 is only found in 50% of human livers, and CYP3A7 is only found in the fetal liver and is switched off after birth. Even CYP3A4 shows significant interindividual variation (see also chap. 5).

There is considerable variation in CYP3A4 expression between individual humans; indeed, as much as one or two orders of magnitude. In part this may be due to the influence of factors such as exposure to chemicals, which may induce or inhibit the enzyme. However, there is also genetic variation resulting from mutations (see chap. 5). Other forms of cytochrome P-450 also show genetic variation as a result of mutations such as CYP2D6.

Variations in the proportions and presence of particular isoenzymes may underly differences in metabolism due to species, sex, age, nutritional status, and interindividual variability. However, the comparison and identification of particular forms (orthologues) of cytochrome P-450 between species have proved to be difficult in some cases. The presence or absence of a particular isoenzyme may be the cause of toxicity in one organ or tissue. The change in the proportion of isoenzymes, caused by exposure to substances in the environment, or drugs may explain changes in toxicity or other biological activity attributed to the compound of interest. These questions will be considered in greater detail in the following chapter.

One important feature of the cytochrome P-450 enzyme system is its broad and overlapping substrate specificity, which reflects the enormous variety of chemicals, which may be potential substrates. Furthermore, one substrate may be metabolized to more than one product by different forms of cytochrome P-450. For example, the drug propranolol can be metabolized by CYP2D6 and CYP2C19 to 4-hydroxypropranolol and naphthoxylacetic acid, respectively. Sometimes the same form of cytochrome P-450 may metabolize one drug to more than one product. For example, the drug methoxyphenamine can be metabolized by CYP2D6 either by O-demethylation or hydroxylation on the 5 position on the benzene ring. Despite this lack of substrate specificity, however, the enzyme may display significant stereoselectivity with chiral substrates (see below). There is an enormous variety of substrates for cytochromes P-450, and the only seemingly common factor is a degree of lipophilicity. Indeed there is a correlation between the metabolism of xenobiotics by microsomes and the lipophilicity of the compound. This is not surprising in that if the purpose of metabolism is to increase the water solubility of a foreign compound and hence its excretion, the compounds most needing this biotransformation are the lipophilic compounds. Furthermore, the lack of substrate specificity requires some control if many vital endogenous molecules are not to be wastefully metabolized. This control is exercised by the lipoidal character of the enzyme complex, which effectively excludes many endogenous molecules. Cytochromes P-450 may also be involved in the metabolism of endogenous compounds, particularly in some tissues where the appropriate isoenzyme is located, but these substrates again tend to be lipophilic. For example, in the kidney, fatty acidsare substrates, undergoing ω-1 hydroxylation, and prostaglandins also undergo this type of hydroxylation. In the adrenal cortex, steroids are hydroxylated by a mitochondrial cytochrome P-450.

Although in the majority of cases the MFO system catalyzes oxidation reactions, under certain circumstances the enzyme may catalyze other types of reaction such as reduction.

For example, carbon tetrachloride and halothane are both reduced in this way (see chap. 7). Although in the majority of cases, cytochromes P-450 catalyze oxidation.

Microsomal flavin-containing monooxygenases. As well as the cytochromes P-450 MFO system, there is also a system, which uses FAD. This flavin-containing monooxygenase or FMO enzyme system is found particularly in the microsomal fraction of the liver, and the monomer has a molecular weight of around 65,000. Each monomer has one molecule of FAD associated with it. The enzyme may accept electrons from either NADPH or NADH although the former is the preferred cofactor. It also requires molecular oxygen, and the overall reaction is as written for cytochromes P-450:

$$NADPH + H^+ + O_2 + S \rightarrow NADP^+ \rightarrow NADP^+ + H_2O + SO,$$

which includes the following steps:

$$Enzyme-FAD + NADPH\,H^+ \rightarrow NADP^+-Enzyme-FADH_2 \tag{1}$$

$$NADP^+-Enzyme-FADH_2 + O_2 \rightarrow NADP^+-Enzyme-FADH_2-O_2 \tag{2}$$

$$NADP^+-Enzyme-FADH_2-O_2 \rightarrow NADP^+-Enzyme-FAD-OOH \tag{3}$$

$$NADP^+-Enzyme-FADH_2-OOH + S \rightarrow NADP^+-Enzyme-FAD-OOH-S \tag{4}$$

$$NADP^+-Enzyme-FADH_2-OOH-S \rightarrow SO + NADP^+-Enzyme-FAD + H_2O \tag{5}$$

$$NADP^+-Enzyme-FAD \rightarrow Enzyme-FAD \tag{6}$$

Thus, step 1 involves addition of NADPH and reduction of the flavin, step 2 the addition of oxygen. At step 3, an internal rearrangement results in the formation of a peroxy complex, which then binds the substrate at step 4. The substrate is oxygenated and released at step 5.

Step 6 regenerates the enzyme.

In the absence of a suitable substrate, the complex at step 3 may degrade to produce hydrogen peroxide.

This enzyme system catalyzes the oxidation of various nitrogen-, sulfur -, and phosphorus-containing compounds, which tend to be nucleophilic, although compounds with an anionic group are not substrates. For example, the N-oxidation of trimethylamine (Fig. 4.19) is catalyzed by this enzyme, but also the hydroxylation of secondary amines, imines, and arylamines and the oxidation of hydroxylamines and hydrazines:

$$NAD(P)H + O_2 + R_1R_2X\text{-}N \rightarrow R_1R_2X\text{-}NO + H_2O + NAD(P)$$

Various sulfur-containing compounds, including thioamides, thioureas, thiols, thioethers and disulfides, are oxidized by this enzyme system. However, unlike cytochromes P-450, it cannot catalyze hydroxylation reactions at carbon atoms. It is clear that this enzyme system has an important role in the metabolism of xenobiotics, and examples will appear in the following pages. Just as with the cytochromes P-450 system, there appear to be a number of isoenzymes, which exist in different tissues, which have overlapping substrate specificities.

The FMO system is a multigene family, with five families (FMO1–5). The major form in human liver is FMO3, which is involved in the detoxication of drugs and other chemicals but may also be involved in activation to toxic products.

FMO is not inducible but can be affected by hormonal and dietary components. For example, 3-carbinol, found in certain vegetables, has been shown to switch off FMO1 in rats.

There is also a genetic deficiency in FMO3 in humans leading to Fish Odor syndrome, which results from the inability of the afflicted individuals to metabolism trimethylamine, which has a strong fishy smell, to the *N*-oxide, which has no smell.

Let us look at the major types of oxidation reaction catalyzed by the cytochromes P-450 system.

Aromatic hydroxylation. Aromatic hydroxylation such as that depicted in Figure 4.3 for the simplest aromatic system, benzene, is an extremely important biotransformation. The major products of aromatic hydroxylation are phenols, but catechols and quinols may also be formed, arising by further metabolism. One of the toxic effects of **benzene** is to cause **aplastic**

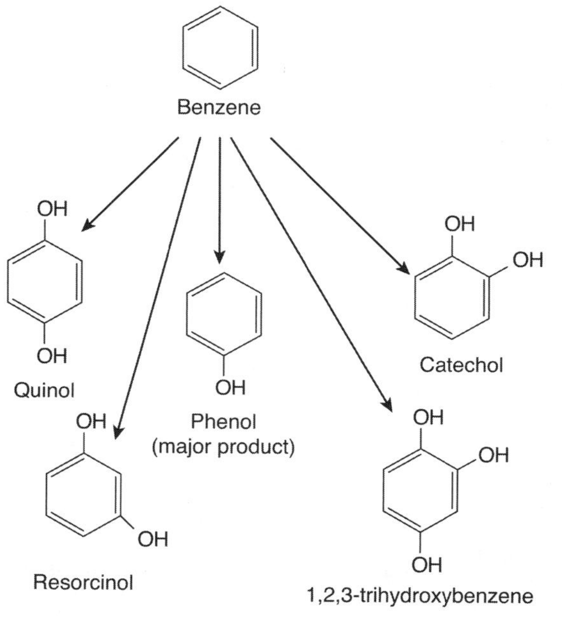

Figure 4.3 Aromatic products of the enzymatic oxidation of benzene. Phenol is the major metabolite.

anemia, which is believed to be due to an intermediate metabolite, possibly hydroquinone. As a result of further metabolism of epoxide intermediates (see below), other metabolites such as diols and glutathione conjugates can also be produced. Because epoxides exist in one of two enantiomeric forms, various isomeric metabolites can result. These may have significance in the toxicity of compounds such as the pulmonary damage caused by naphthalene and the carcinogenicity and the carcinogenicity of benzo[a]pyrene (see below and chap. 7 for more details). Consequently, a number of hydroxylated metabolites may be produced from the aromatic hydroxylation of a single compound (Figs. 4.4 and 4.5).

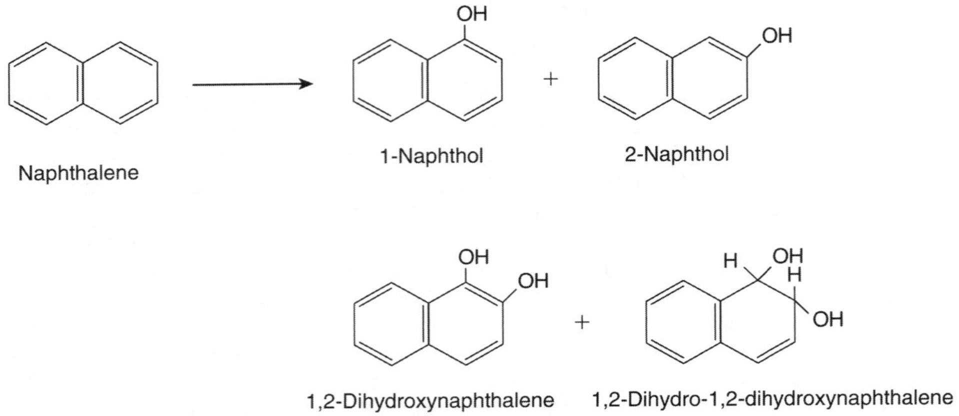

Naphthalene ⟶ 1-Naphthol + 2-Naphthol

1,2-Dihydroxynaphthalene + 1,2-Dihydro-1,2-dihydroxynaphthalene

Figure 4.4 Hydroxylated products of naphthalene metabolism.

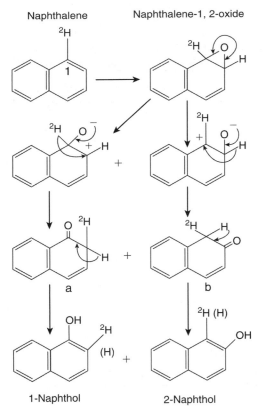

Naphthalene   Naphthalene-1, 2-oxide

1-Naphthol   2-Naphthol

Figure 4.5 Metabolism of deuterium-labeled naphthalene via the 1, 2-oxide (epoxide) intermediate, illustrating the NIH shift.

Aromatic hydroxylation generally proceeds via the formation of an epoxide intermediate. This is illustrated by the metabolism of **naphthalene**, labeled in the 1 position with deuterium (^{2}H), via the 1,2-oxide as shown in Figure 4.5. The shift in the deuterium atom that occurs during metabolism is the so-called **NIH shift**. This indicates that formation of an epoxide intermediate has occurred, and is one method of determining whether such an epoxide intermediate is involved. The phenolic products, 1- and 2-naphthols, retain various proportions of deuterium, however. The proposed mechanism involves the formation of an epoxide intermediate, which may break open chemically in two ways, leading to phenolic products (Fig. 4.5).

Each naphthol product may have deuterium or hydrogen in the adjacent position. The hydrogen and deuterium atoms (in a and b of Fig. 4.5) are equivalent, as the carbon atom to which they are attached is tetrahedral. Consequently, either hydrogen or deuterium may be lost, theoretically resulting in 50% retention of deuterium. However, in practice this may not be the case, as an isotope effect may occur. This effect results from the strength of the carbon-deuterium bond being greater than that of a carbon-hydrogen bond. Therefore, more energy is required to break the C-^{2}H bond than C-^{1}H bond, with a consequent effect on the rate-limiting chemical reactions involving bond breakage. Also, the direction of the opening of the epoxide ring is affected by the substituents, and the proportions of products therefore reflect this. The production of phenols occurs via a chemical rearrangement and depends on the stability of the particular epoxide.

The further metabolism of suitably stable epoxides may occur, with the formation of dihydrodiols as discussed later. Dihydrodiols may also be further metabolized to catechols. Other products of aromatic hydroxylation via epoxidation are glutathione conjugates. These may be formed by enzymic or nonenzymic means or both, depending on the reactivity of the epoxide in question.

The products of epoxidation *in vivo* depend on the reactivity of the particular epoxide. Stabilized epoxides react with nucleophiles and undergo further enzymic reactions, whereas destabilized ones undergo spontaneous isomerization to phenols. Epoxides are generally reactive intermediates, however, and in a number of cases are known to be responsible for toxicity by reaction with cellular constituents.

With carcinogenic polycyclic hydrocarbons, dihydrodiols are further metabolized to epoxides as shown in chapter 7, Figure 2. This epoxide-diol may then react with weak nucleophiles such as nucleic acids.

Unsaturated aliphatic compounds and heterocyclic compounds may also be metabolized via epoxide intermediates as shown in Figure 4.6 and chapter 5, Figure 14. Note that when the epoxide ring opens, the chlorine atom shifts to the adjacent carbon atom (Fig. 4.6). In the case of the furan ipomeanol and vinyl chloride, the epoxide intermediate is thought to be responsible for the toxicity (see below and chap. 7). Other examples of unsaturated aliphatic compounds, which may be toxic and are metabolized via epoxides, are diethylstilboestrol, **allylisopropyl acetamide**, which destroys cytochrome P-450, sedormid, and secobarbital.

Aromatic hydroxylation may also take place by a mechanism other than epoxidation. Thus, the *m*-hydroxylation of chlorobenzene is thought to proceed via a direct insertion mechanism (Fig. 4.7).

The nature of the substituent in a substituted aromatic compound influences the position of hydroxylation. Thus, *o-p*-directing substituents, such as amino groups, result in *o*-and

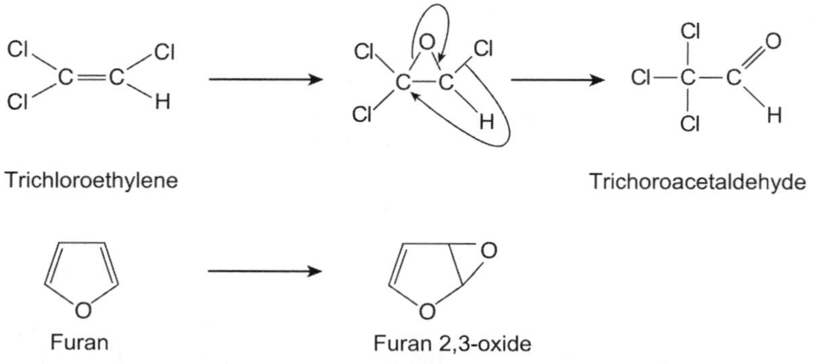

Trichloroethylene   Trichoroacetaldehyde

Furan   Furan 2,3-oxide

Figure 4.6   Metabolism of unsaturated aliphatic and heterocyclic compounds via epoxides.

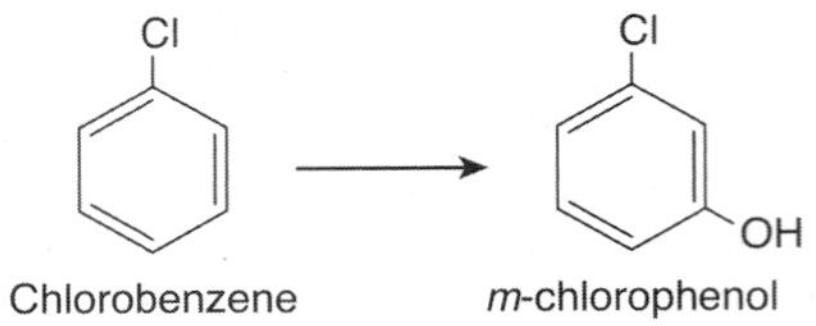

Chlorobenzene   *m*-chlorophenol   **Figure 4.7**   *m*-hydroxylation of chlorobenzene.

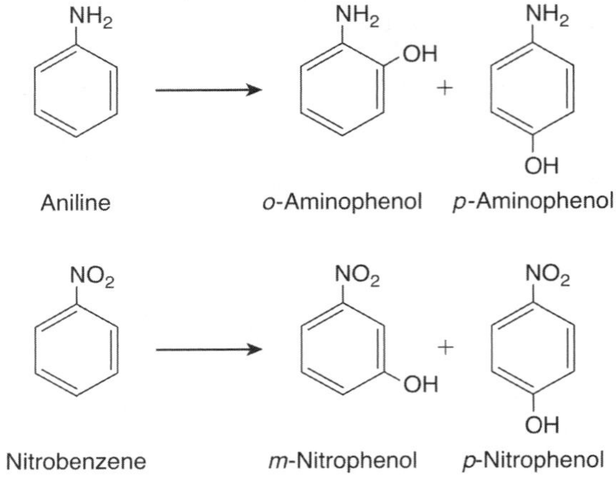

Aniline   *o*-Aminophenol   *p*-Aminophenol

Nitrobenzene   *m*-Nitrophenol   *p*-Nitrophenol

Figure 4.8   Hydroxylation of aniline and nitrobenzene.

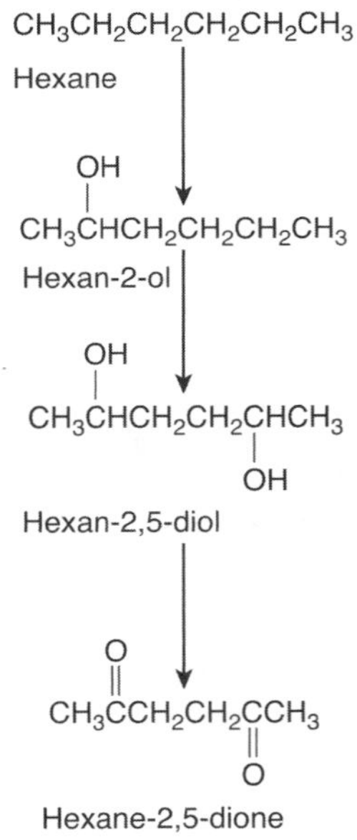

$CH_3CH_2CH_2CH_2CH_2CH_3$

Hexane

$CH_3\overset{\text{OH}}{\underset{\text{|}}{C}}HCH_2CH_2CH_2CH_3$

Hexan-2-ol

$CH_3\overset{\text{OH}}{\underset{\text{|}}{C}}HCH_2CH_2\overset{}{\underset{\text{|}}{C}}HCH_3$
$\quad\quad\quad\quad\quad\quad\quad OH$

Hexan-2,5-diol

$CH_3\overset{O}{\overset{\text{||}}{C}}CH_2CH_2\overset{O}{\overset{\text{||}}{C}}CH_3$

Hexane-2,5-dione   **Figure 4.9**   The aliphatic oxidation of hexane.

p-hydroxylated metabolites such as the *o*-and *p*-aminophenols from aniline (Fig. 4.8). Meta-directing substituents such as nitro groups lead to *m*-and *p*-hydroxylated products; for example, **nitrobenzene** is hydroxylated to *m*-and *p*-nitrophenols (Fig. 4.8).

Aliphatic hydroxylation. As well as unsaturated aliphatic compounds such as vinyl chloride mentioned above, which are metabolized by epoxidation, saturated aliphatic compounds also undergo oxidation. The initial products will be primary and secondary alcohols. For example, the solvent **n-hexane** is known to be metabolized to the secondary alcohol hexan-2-ol and then further to hexane-2,5-dione (Fig. 4.9) in occupationally exposed humans. The latter metabolite is believed to be responsible for the neuropathy caused by the solvent. Other toxicologically important examples are the nephrotoxic petrol constituents, 2,2,4- and 2,3,4-trimethylpentane, which are hydroxylated to yield primary and tertiary alcohols. However, aliphatic hydrocarbon chains are more readily metabolized if they are side chains on aromatic structures. Thus, **n-propylbenzene** may be hydroxylated in three positions, giving the primary alcohol 3-phenylpropan-1-ol and two secondary alcohols (Fig. 4.10). Further oxidation of the primary alcohol may take place to give the corresponding acid phenylpropionic acid, which may be further metabolized to benzoic acid, probably by oxidation of the carbon β to the carboxylic acid.

Alicyclic hydroxylation. Hydroxylation of saturated rings yields monohydric and dihydric alcohols. For instance, **cyclohexane** is metabolized to cyclohexanol, which is further hydroxylated to *trans*-cyclohexane-1,2-diol (Fig. 4.11). With mixed alicyclic/aromatic, saturated and unsaturated systems, alicyclic hydroxylation appears to predominate, as shown for the compound **tetralin** (Fig. 4.12).

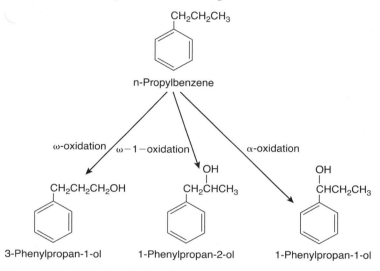

Figure 4.10   Aliphatic oxidation of *n*-propylbenzene.

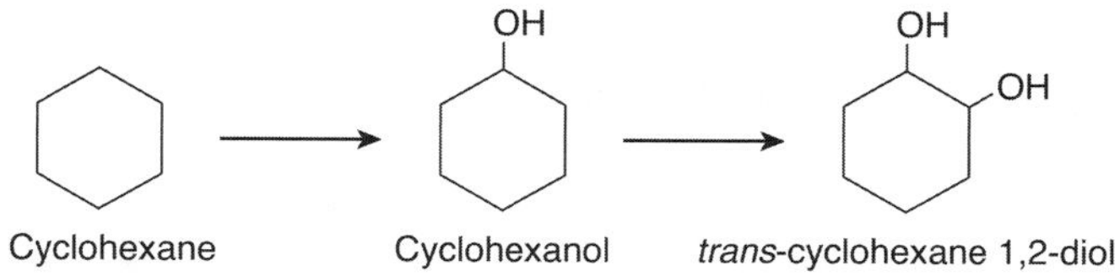

Figure 4.11   Hydroxylation of cyclohexane.

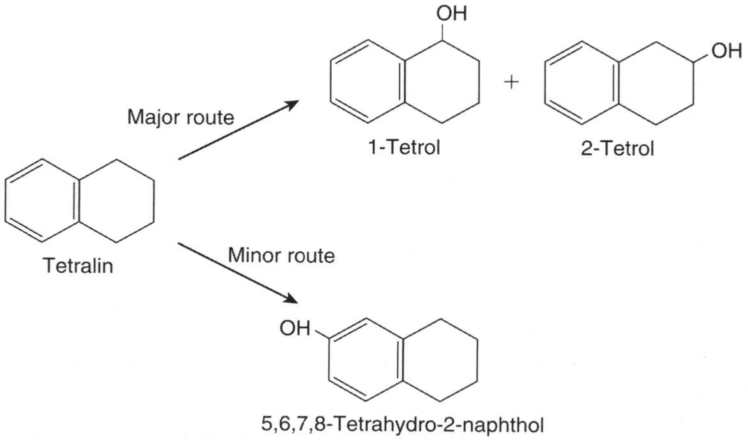

Figure 4.12   Hydroxylation of tetralin.

Heterocyclic hydroxylation. Nitrogen heterocycles such as **pyridine** and **quinoline** (Fig. 4.13) undergo microsomal hydroxylation at the 3 position. In quinoline, the aromatic ring is also hydroxylated in positions *o* and *p* to the nitrogen atom.

Aldehyde oxidase, a non-microsomal enzyme discussed in more detail below, may also be involved in the oxidation of **quinoline** to give 2-hydroxyquinoline (Fig. 4.14). The heterocyclic **phthalazine** ring in the drug hydralazine is oxidized by the microsomal enzymes to phthalazinone. The mechanism, which may involve nitrogen oxidation, is possibly involved in the toxicity of this drug (see chap. 7). Again, other enzymes may also be involved (Fig. 4.15).

N-dealkylation. Dealkylation is the removal of alkyl groups from nitrogen, sulfur and oxygen atoms, and is catalyzed by the microsomal enzymes. Alkyl groups attached to ring nitrogen atoms or those in amines, carbamates, or amides are removed oxidatively by conversion to the corresponding aldehyde as indicated in Figure 4.16. The reaction proceeds via a hydroxyalkyl

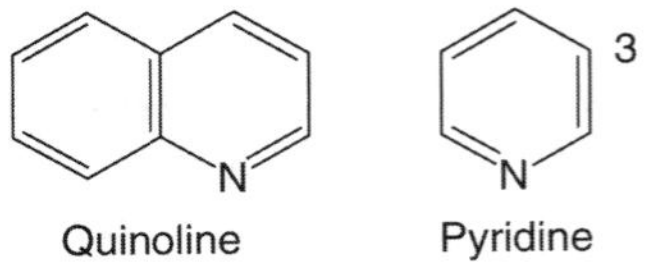

Figure 4.13 Structures of pyridine and quinoline.

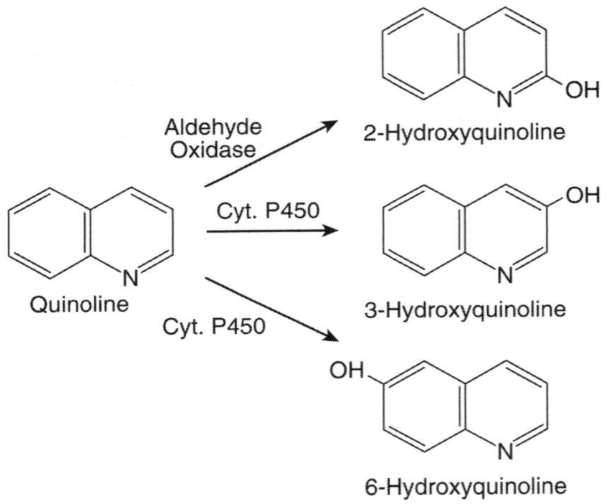

Figure 4.14 Hydroxylation of quinoline.

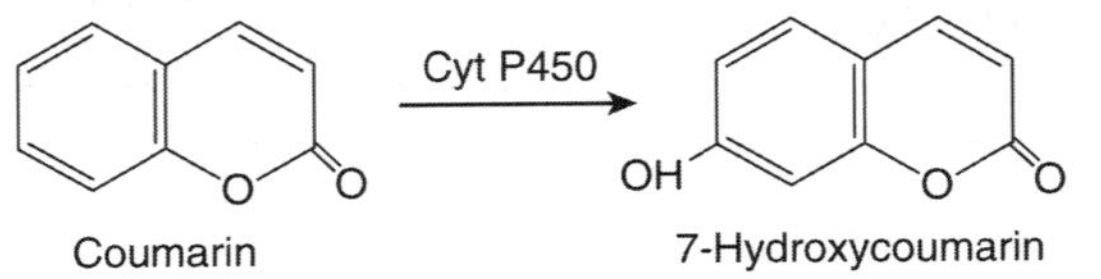

Figure 4.15 Microsomal enzyme-mediated hydroxylation of coumarin.

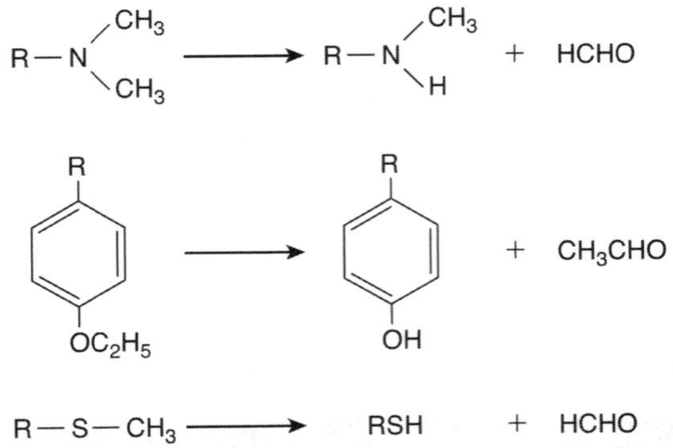

Figure 4.16 Microsomal enzyme-mediated N-, O-, and S-dealkylation.

intermediate, which is usually unstable and spontaneously rearranges with loss of the corresponding aldehyde. However, in some cases, the hydroxyalkyl intermediate is more stable and may be isolated, as for example, with the solvent **dimethylformamide** (Fig. 4.17). N-dealkylation is a commonly encountered metabolic reaction for foreign compounds, which may have important toxicological consequences, as in the metabolism of the carcinogen dimethylnitrosamine (Fig. 7.5).

S-dealkylation. A microsomal enzyme system catalyzes S-dealkylation with oxidative removal of the alkyl group to yield the corresponding aldehyde, as with N-dealkylation (Fig. 4.16). However, as there are certain differences from the N-dealkylation reaction, it has been suggested that another enzyme system, such as the microsomal **FAD-containing monooxygenase** system may be involved. Figure 4.18 shows the S-demethylation of **6-methylthiopurine** to 6-mercaptopurine.

O-dealkylation. Aromatic methyl and ethyl ethers may be metabolized to give the phenol and corresponding aldehyde (Fig. 4.16), as illustrated by the de-ethylation of **phenacetin** (Fig. 4.20). Ethers with longer alkyl chains are less readily O-dealkylated, the preferred route being ω-1-hydroxylation.

N-oxidation. The oxidation of nitrogen in tertiary amines, amides, imines, hydrazines, and heterocyclic rings may be catalyzed by microsomal enzymes or by other enzymes (see below). Thus the oxidation of **trimethylamine** to an *N*-oxide (Fig. 4.19) is catalyzed by the microsomal FAD-containing monooxygenase. The *N*-oxide so formed may undergo enzyme-catalyzed decomposition to a secondary amine and aldehyde. This N to C transoxygenation is mediated by cytochromes P-450. The N-oxidation of 3-methylpyridine, however, is catalyzed by cytochromes P-450. This reaction may be involved in the toxicity of the analogue,

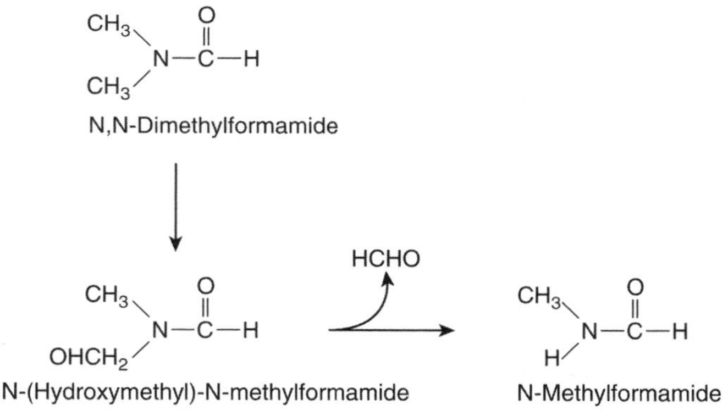

Figure 4.17 Metabolism of dimethylformamide. The hydroxymethyl derivative is stable but may subsequently rearrange to monomethylformamide with release of formaldehyde.

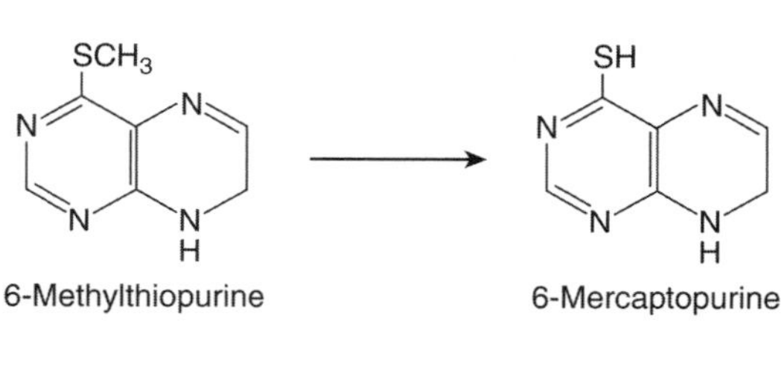

Figure 4.18 Oxidative *S*-demethylation of 6-methylthiopurine.

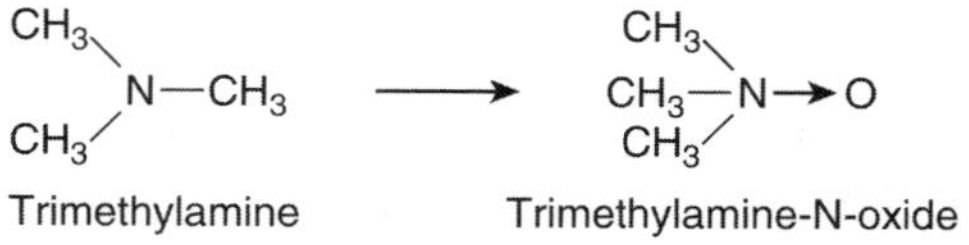

Figure 4.19 N-oxidation of trimethylamine.

trifluoromethyl pyridine, which is toxic to the nasal tissues. It is believed to be metabolized to the *N*-oxide by cytochromes P-450 present in the olfactory epithelium (Fig. 4.20).

N-hydroxylation N-hydroxylation of primary arylamines, arylamides, and hydrazines is also catalyzed by a microsomal mixed-function oxidase, involving cytochromes P-450 and requiring NADPH and molecular oxygen. Thus, the N-hydroxylation of **aniline** is as shown in Figure 4.21. The N-hydroxylated product, phenylhydroxylamine, is thought to be responsible for the production of **methemoglobinemia** after aniline administration to experimental animals. This may occur by further oxidation of phenylhydroxylamine to nitrosobenzene, which may then be reduced back to phenylhydroxylamine. This reaction lowers the reduced glutathione concentration in the red blood cell, removing the protection of hemoglobin against oxidative damage.

 N-hydroxylated products may be chemically unstable and dehydrate, as does phenylhydroxylamine, thereby producing a reactive electrophile such as an imine or iminoquinone (chap. 7, Fig. 7.19).

 An important example toxicologically is the N-hydroxylation of **2-acetylaminofluorene** (Fig. 4.22). N-hydroxylation is one of the reactions responsible for converting the compound into a potent carcinogen. A second example is the N-hydroxylation of **isopropylhydrazine**, thought to be involved in the production of a hepatotoxic intermediate (chap. 7, Fig. 7.25).

S-oxidation. Aromatic and aliphatic sulfides, thioethers, thiols, thioamides, and thiocarbamates may undergo oxidation to form sulfoxides and then, after further oxidation, sulfones (Fig. 4.23).

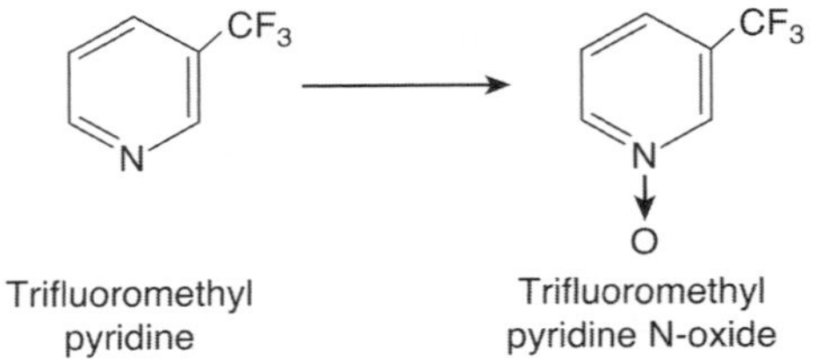

Trifluoromethyl pyridine

Trifluoromethyl pyridine N-oxide

Figure 4.20   N-oxidation of trifluoromethylpyridine.

Aniline

Phenylhydroxylamine

Nitrosobenzene

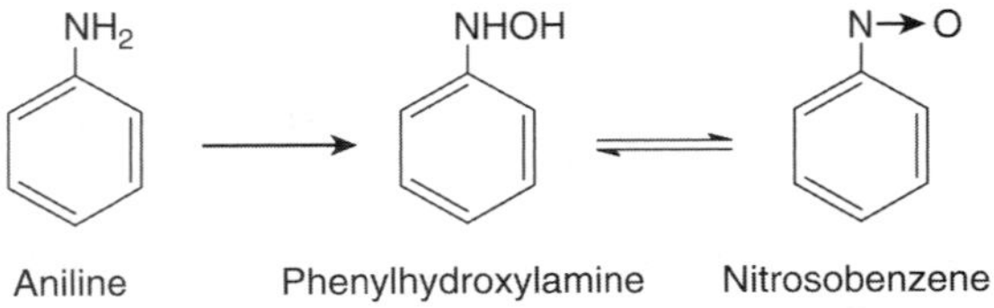

Figure 4.21   N-hydroxylation of aniline.

2-Acetylaminofluorene

N-Hydroxy-2-acetylaminofluorene

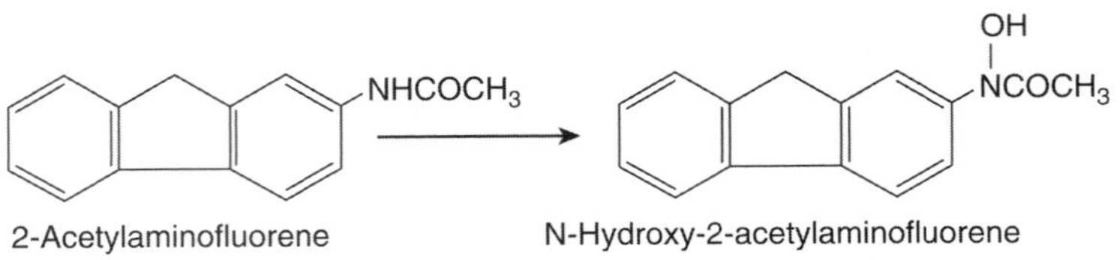

Figure 4.22   N-hydroxylation of 2-acetylaminofluorene.

Sulphoxide

Sulphone

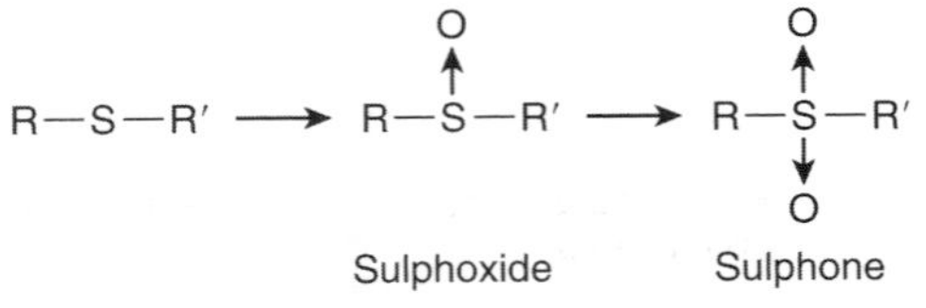

Figure 4.23   S-oxidation to form a sulfoxide and sulfone.

This is catalyzed by a microsomal monooxygenase requiring NADPH and cytochromes P-450. The FAD-containing monooxygenases will also catalyze S-oxidation reactions. A number of foreign compounds, for example, drugs like chlorpromazine and various pesticides such as temik undergo this reaction. An important toxicological example is the oxidation of the hepatotoxin **thioacetamide** (Fig. 4.24).

P-oxidation. In an analogous manner to nitrogen and sulfur, phosphorus may also be oxidized to an oxide, as in the compound **diphenylmethylphosphine**. This is catalyzed by microsomal monooxygenases, both cytochromes P-450 and the FAD-requiring enzyme.

Desulfuration. Replacement of sulfur by oxygen is known to occur in a number of cases, and the oxygenation of the insecticide parathion to give the more toxic paraoxon is a good example of this (Fig. 4.25). This reaction is also important for other phosphorothionate insecticides.

The toxicity depends upon inhibition of cholinesterases, and the oxidized product is much more potent in this respect. The reaction appears to be catalyzed by either cytochromes P-450 or the FAD-containing monooxygenases and therefore requires NADPH and oxygen. The mechanism of desulfuration seems to involve formation of a phospho-oxithirane ring, which rearranges with loss of "active atomic sulfur." This is highly reactive and is believed to bind to the enzyme and also to be involved in toxicity. Oxidative desulfuration at the C-S bond may also occur, such as in the barbiturate **thiopental** (Fig. 3.5), which is metabolized to pentobarbital. The solvent **carbon disulfide**, which is also hepatotoxic, undergoes oxidative metabolism catalyzed by cytochromes P-450 and giving rise to carbonyl sulfide, thiocarbonate, and again atomic sulfur, which is thought to be involved in the toxicity (Fig. 4.26).

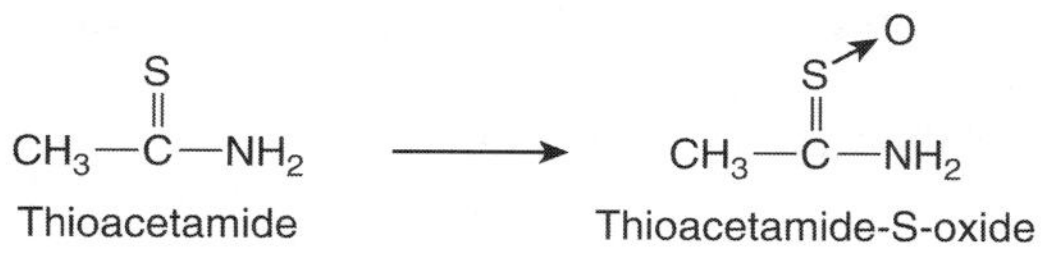

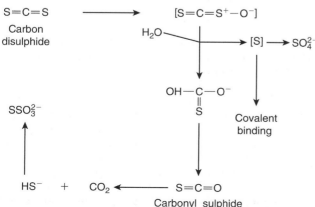

Figure 4.24 S-oxidation of thioacetamide.

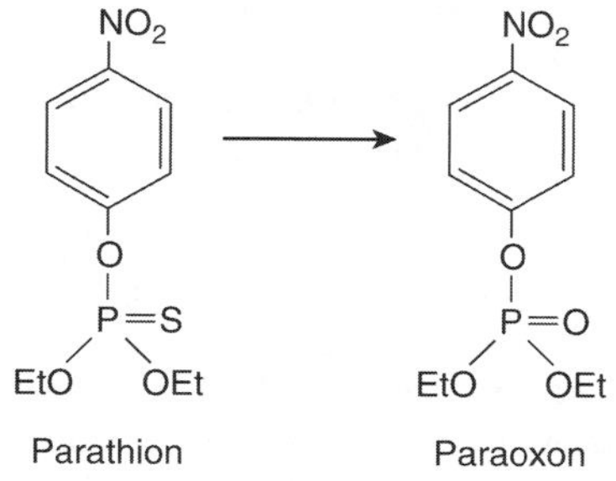

Figure 4.25 Oxidative desulfuration of parathion.

Figure 4.26 Oxidative metabolism of carbon disulfide.

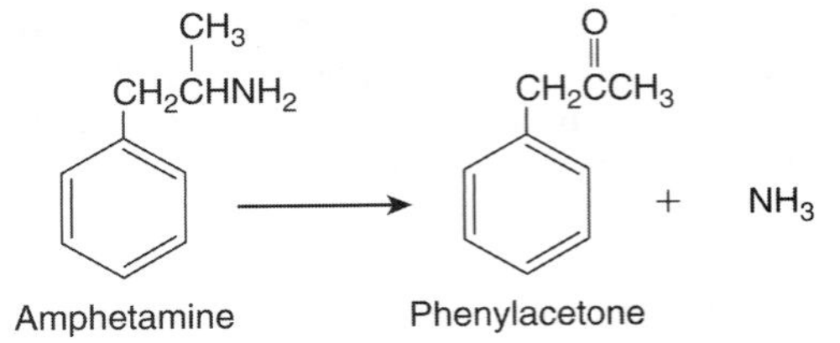

Figure 4.27 Oxidative examination of amphetamine.

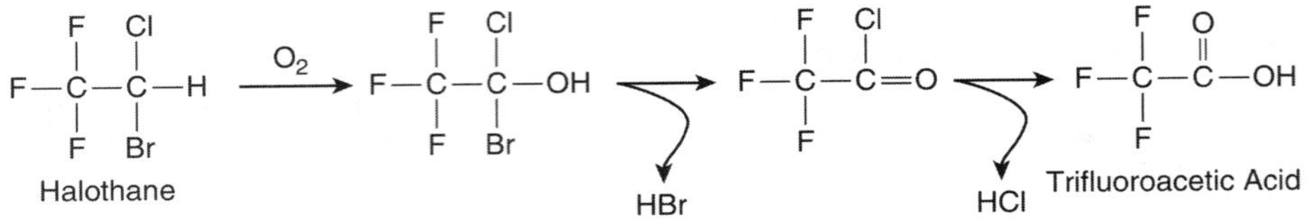

Figure 4.28 Oxidative metabolism of halothane.

Deamination. Amine groups can be removed oxidatively via a deamination reaction, which may be catalyzed by cytochromes P-450. Other enzymes, such as monoamine oxidases, may also be involved in deamination reactions (see below). The product of deamination of a primary amine is the corresponding ketone. For example, **amphetamine** is metabolized in the rabbit to phenylacetone (Fig. 4.27). The mechanism probably involves oxidation of the carbon atom to yield a carbinolamine, which can rearrange to the ketone with loss of ammonia. Alternatively, the reaction may proceed via phenylacetoneoxime, which has been isolated as a metabolite and for which there are several possible routes of formation. The phenyl-acetoneoxime is hydrolyzed to phenylacetone. Also N-hydroxylation of amphetamine may take place and give rise to phenylacetone as a metabolite. This illustrates that there may be several routes to a particular metabolite.

Oxidative dehalogenation. Halogen atoms may be removed from xenobiotics in an oxidative reaction catalyzed by cytochromes P-450. For example, the anesthetic **halothane** is metabolized to trifluoroacetic acid via several steps, which involves the insertion of an oxygen atom and the loss of chlorine and bromine (Fig. 4.28). This is the major metabolic pathway in man and is believed to be involved in the hepatotoxicity of the drug. Trifluoroacetyl chloride is thought to be the reactive intermediate (see chap. 7).

Oxidation of ethanol. Although the major metabolic pathway for alcohols such as ethanol is oxidation catalyzed by alcohol dehydrogenase (see below), ethanol can also be metabolized by cytochrome P-450. The product, ethanal, is the same as produced by alcohol dehydrogenase. The isoform of cytochrome P-450 is CYP2E1. The mechanism may involve a hydroxylation to an unstable intermediate, which loses water to yield ethanal. Alternatively, a radical mechanism could be responsible. The importance of this route of metabolism for ethanol is that it is inducible (see chap. 5), assuming more importance after repeated exposure to ethanol such as in alcoholics and regular drinkers.

Desaturation of alkyl groups. This novel reaction, which converts a saturated alkyl compound into a substituted alkene and is catalyzed by cytochromes P-450, has been described for the antiepileptic drug, **valproic acid** (VPA) (2-n-propyl-4-pentanoic acid) (Fig. 4.29). The mechanism proposed involves formation of a carbon-centered free radical, which may form either a hydroxylated product (alcohol) or dehydrogenate to the unsaturated compound. The cytochrome P-450-mediated metabolism yields 4-ene-VPA (2-n-propyl-4pentenoic acid), which is oxidized by the mitochondrial β-oxidation enzymes to 2,4-diene-VPA (2-n-propyl-2, 4-pentadienoic acid). This metabolite or its CoA ester irreversibly inhibits enzymes of the β-oxidation system, destroys cytochrome P-450, and may be involved in the hepatotoxicity of the drug. Further metabolism may occur to give 3-keto-4-ene-VPA (2-n-propyl-3-oxo-4-pentenoic acid), which inhibits the enzyme 3-ketoacyl-CoA thiolase, the terminal enzyme of the fatty acid oxidation system.

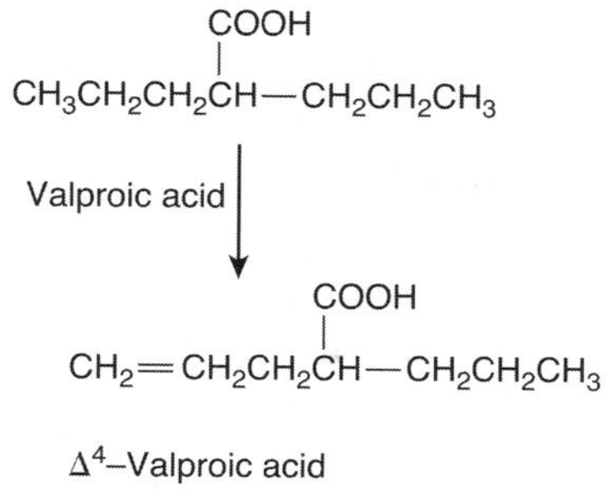

COOH
|
$CH_3CH_2CH_2CH-CH_2CH_2CH_3$

Valproic acid ↓

COOH
|
$CH_2=CH_2CH_2CH-CH_2CH_2CH_3$

Δ^4–Valproic acid

Figure 4.29 Aliphatic desaturation of VPA. *Abbreviation*: VPA, valproic acid.

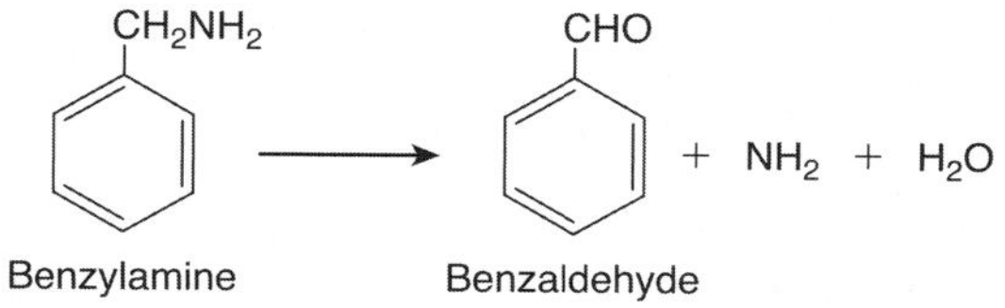

Benzylamine → Benzaldehyde + NH_2 + H_2O

Figure 4.30 Oxidative deamination of benzylamine.

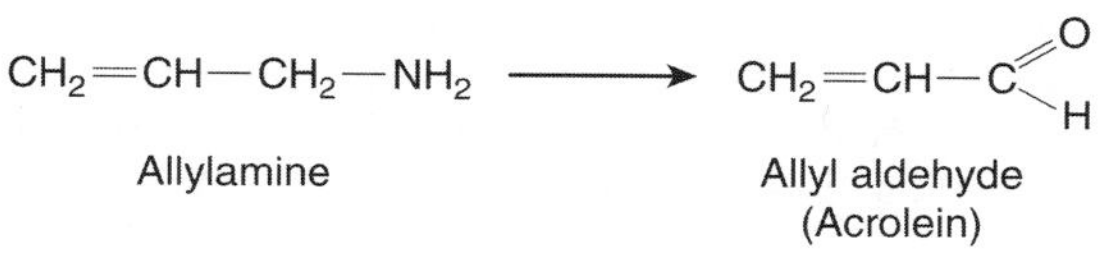

$CH_2=CH-CH_2-NH_2$ → $CH_2=CH-C\overset{O}{\underset{H}{\diagdown}}$

Allylamine → Allyl aldehyde (Acrolein)

Figure 4.31 Oxidative deamination of allylamine.

Non-Microsomal Oxidation

Amine oxidation. As well as the microsomal enzymes involved in the oxidation of amines, there are a number of other amine oxidase enzymes, which have a different subcellular distribution. The most important are the monoamine oxidases and the diamine oxidases. The monoamine oxidases are located in the mitochondria within the cell and are found in the liver and also other organs such as the heart and central nervous system and in vascular tissue. They are a group of flavoprotein enzymes with overlapping substrate specificities. Although primarily of importance in the metabolism of endogenous compounds such as 5-hydroxytryptamine, they may be involved in the metabolism of foreign compounds.

The enzyme found in the liver will deaminate secondary and tertiary aliphatic amines as well as primary amines, although the latter are the preferred substrates and are deaminated faster. Secondary and tertiary amines are preferentially dealkylated to primary amines. For aromatic amines, such as **benzylamine**, electron-withdrawing substituents on the ring will *increase* the reaction rate. The product of the reaction is an aldehyde (Fig. 4.30). Amines such as amphetamine are not substrates, seemingly due to the presence of a methyl group on the α-carbon atom (Fig. 4.27). Monoamine oxidase is important in the metabolic *activation* and subsequent *toxicity* of **allylamine** (Fig. 4.31), which is highly toxic to the heart. The presence of the amine oxidase in heart tissue allows metabolism to the toxic metabolite, allyl aldehyde (Fig. 4.31). Another example is the metabolism of MPTP to a toxic metabolite by monoamine oxidase in the central nervous system, which is discussed in more detail in chapter 7.

Diamine oxidase, a soluble enzyme found in liver and other tissues, is mainly involved in the metabolism of endogenous compounds such as the aliphatic diamine putrescine (chap. 7, Fig. 40).

This enzyme, which requires pyridoxal phosphate, does not metabolize secondary or tertiary amines or those with more than nine carbon atoms. The products of the reaction are aldehydes.

Alcohol and aldehyde oxidation. Although a microsomal enzyme system has been demonstrated, which oxidizes ethanol (see above), probably the more important enzyme *in vivo* is alcohol dehydrogenase, which is a cytosolic enzyme (soluble fraction) and is found in the liver and also in the kidney and the lung.

The coenzyme is normally NAD, and although NADP may be used, the rate of the reaction is slower. The enzyme is relatively nonspecific and so accepts a wide variety of substrates including exogenous primary and secondary alcohols. Secondary alcohols are metabolized at a slower rate than primary alcohols, and tertiary alcohols are not readily oxidized at all.

The product of the oxidation is the corresponding aldehyde if the substrate is a primary alcohol (Fig. 4.32) or a ketone if a secondary alcohol is oxidized. However, secondary alcohols are oxidized much more slowly than primary alcohols. The aldehyde produced by this oxidation may be further oxidized by aldehyde dehydrogenase to the corresponding acid. This enzyme also requires NAD and is found in the soluble fraction. Alcohol dehydrogenase may have a role in the hepatotoxicity of allyl alcohol. This alcohol causes periportal necrosis (see Chap. 6) in experimental animals; this is thought to be due to metabolism to allyl aldehyde (acrolein) (Fig. 4.33) in an analogous manner to allylamine (see above). The use of deuterium-labeled allyl alcohol showed that oxidation was necessary for toxicity: the replacement of the hydrogen atoms of the CH$_2$ group on the C1 with deuterium reduced the toxicity. This was proposed to be due to an isotope effect as breakage of a carbon-deuterium bond is *more difficult* than breakage of a carbon-hydrogen bond. If this bond breakage is a *rate-limiting step* in the oxidation to allyl aldehyde, metabolism will be reduced.

Other enzymes may also be involved in the oxidation of aldehydes, particularly aldehyde oxidase and **xanthine oxidase**, which belong to the **molybdenum hydroxylases**. These enzymes are primarily cytosolic, although microsomal aldehyde oxidase activity has been detected. They are flavoproteins, containing FAD and also molybdenum, and the oxygen incorporated is derived from water *rather than* molecular oxygen. Aldehyde oxidase and xanthine oxidase in fact oxidize a wide variety of substrates, both aldehydes and nitrogen-containing heterocycles such as caffeine and purines (see below). Aldehyde oxidase is found in highest concentrations in the liver, but xanthine oxidase is found in the small intestine, milk, and the mammary gland.

Purine oxidation. The oxidation of purines and purine derivatives is catalyzed by xanthine oxidase. For example, the enzyme oxidizes hypoxanthine to xanthine and thence uric acid (Fig. 4.34). Xanthine oxidase also catalyzes the oxidation of foreign compounds, such as the nitrogen heterocycle phthalazine (Fig. 4.35). This compound is also a substrate for aldehyde oxidase, giving the same product.

Figure 4.32 Oxidation of ethanol by alcohol dehydrogenase.

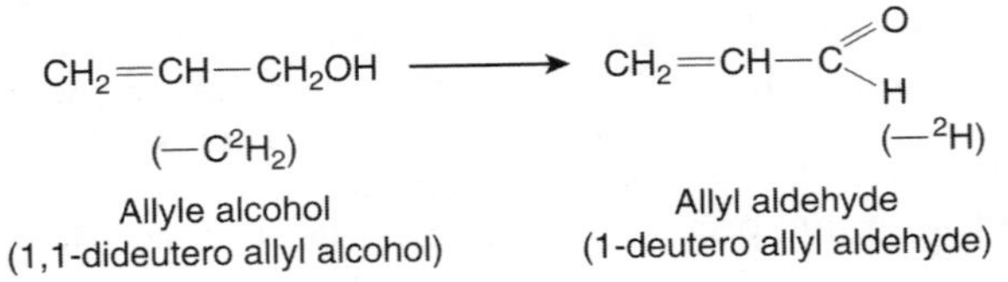

Figure 4.33 Oxidation of allyl alcohol by alcohol dehydrogenase. The figure also shows in brackets the position of deuterium labeling as discussed in the text.

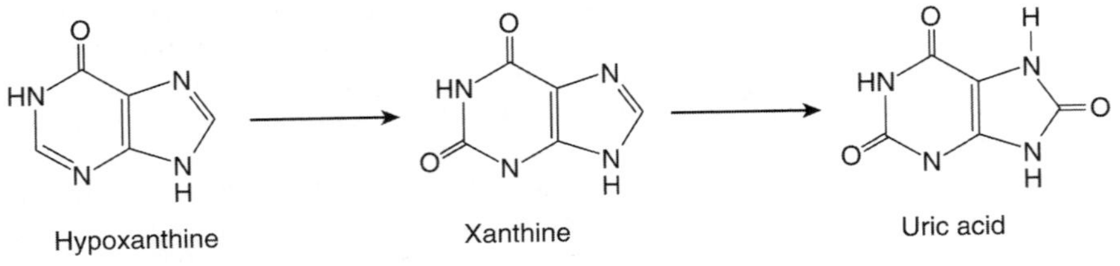

Figure 4.34 Oxidation of hypoxanthine by xanthine oxidase.

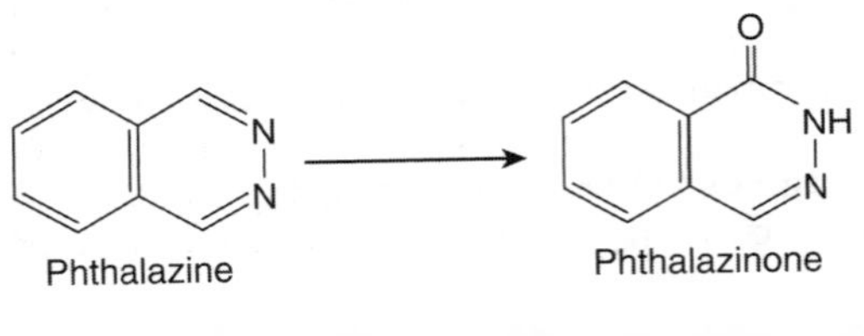

Figure 4.35 Oxidation of phthalazine by xanthine oxidase.

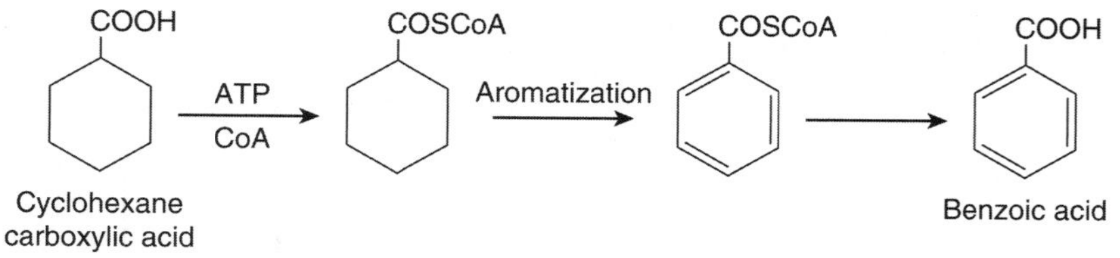

Figure 4.36 Aromatization of cyclohexane carboxylic acid.

Aromatization of alicyclic compounds. Cyclohexane carboxylic acids may be metabolized by a mitochondrial enzyme system to an aromatic acid such as benzoic acid. This enzyme system requires CoA, ATP, and oxygen and is thought to involve three sequential dehydrogenation steps after the initial formation of the cyclohexanoyl CoA (Fig. 4.36). The **aromatase** enzyme also requires the cofactor FAD.

Peroxidases. Another group of enzymes, which is involved in the oxidation of xenobiotics, is the peroxidase. There are a number of these enzymes in mammalian tissues: **prostaglandin synthase** found in many tissues, but especially seminal vesicles and also the kidney, the lung, the intestine spleen, and blood vessels; **lactoperoxidase** found in mammary glands; **myeloperoxidase** found in neutrophils, macrophages, liver Kupffer cells, and bone marrow cells.

The overall peroxidase-catalyzed reaction may be summarized as follows:

$$\text{Peroxidase} + H_2O_2 \rightarrow \text{Compound I}$$
$$\text{Compound I} + RH_2 \rightarrow \text{Compound II} + \bullet RH_2^+$$
$$\text{Compound II} + RH_2 \rightarrow \text{Peroxidase} + \bullet RH_2^+$$

The heme iron in the peroxidase is oxidized by the peroxide from III^+ to V^+ in compound I. The compound I is reduced by two sequential one-electron transfer processes giving rise to the original enzyme. A substrate-free radical is in turn generated. This may have toxicological implications. Thus the myeloperoxidase in the bone marrow may catalyze the metabolic activation of phenol or other metabolites of benzene. This may underlie the toxicity of **benzene** to the bone marrow, which causes aplastic anemia (see below and chap. 6). The myeloperoxidase found in neutrophils and monocytes may be involved in the metabolism and activation of a number of drugs such as isoniazid, clozapine, procainamide, and **hydralazine** (see below). In *in vitro* systems, the products of the activation were found to be cytotoxic *in vitro*.

Similarly, uterine peroxidase has been suggested as being involved in the metabolic activation and toxicity of **diethylstilboestrol** (see page 247, Fig. 6.28). Probably the most important peroxidase enzyme system is prostaglandin synthase. This enzyme is found in most species and is located in many tissues, including the kidney and seminal vesicles. It is a glycoprotein, which is located in the endoplasmic reticulum. This enzyme system is involved in the oxygenation of polyunsaturated fatty acids and the biosynthesis of prostaglandins. The oxidation of arachidonic acid to prostaglandin H2 is an important step in the latter (Fig. 4.37). The enzyme catalyzes two steps, first the formation of a hydroperoxy endoperoxide, prostaglandin G2, then metabolism to a hydroxy metabolite, prostaglandin H2. In the second step, xenobiotics may be co-oxidized. There are a number of examples of xenobiotic metabolism catalyzed by this system, such as the oxidation of **p-phenetidine**, a metabolite of the drug phenacetin (see chap. 5) (chap. 5, Fig. 24), a process, which may be involved in the *nephrotoxicity* of the drug. The prostaglandin synthase–catalyzed oxidation of this compound gives rise to free radicals, which may be responsible for binding to DNA. Horseradish peroxidase will also catalyze the oxidation of *p*-phenetidine. **Paracetamol** can also be oxidized by prostaglandin synthase to a free radical intermediate, a semiquinone, which may be involved in the toxicity and yields a glutathione conjugate, which is the same as that produced via the cytochromes P-450-mediated pathway (for more details see chap. 7). Other examples of metabolic pathways catalyzed by this pathway are N-demethylation of aminopyrine, formation of an aromatic epoxide of 7,8-dihydroxy, 7,8-dihydrobenzo[a]pyrene, and sulfoxidation of methyl phenyl sulfide. The exact mechanism of the co-oxidation is unclear

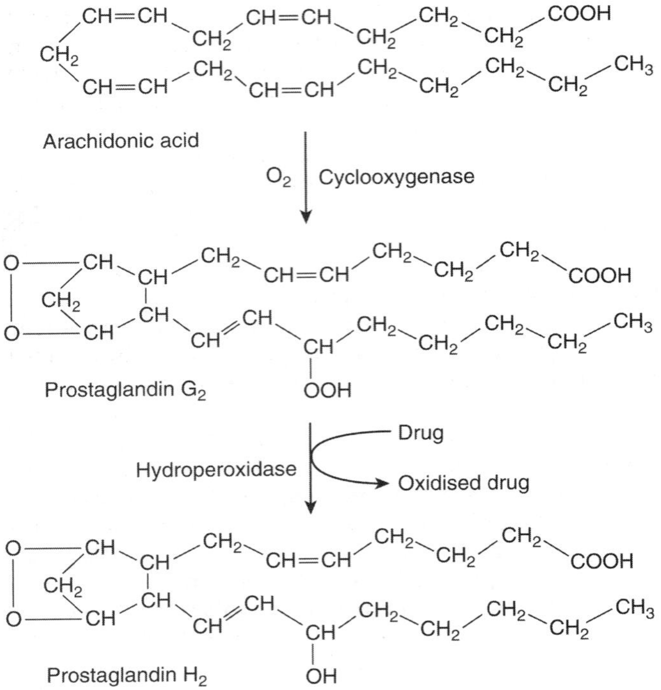

Arachidonic acid

O_2 | Cyclooxygenase

Prostaglandin G_2

Hydroperoxidase — Drug
→ Oxidised drug

Prostaglandin H_2

Figure 4.37 Co-oxidation of a drug by the prostaglandin synthetase enzyme system.

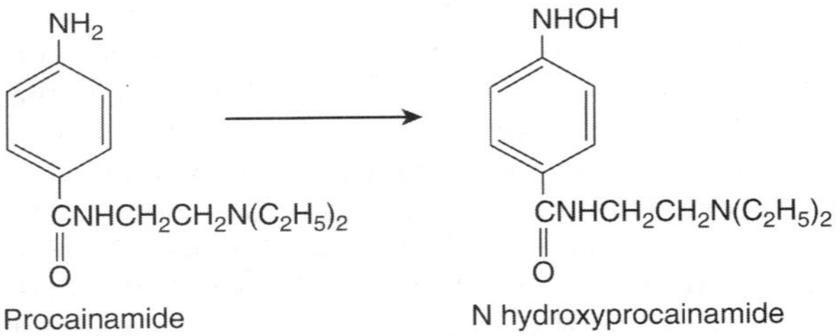

Procainamide

N hydroxyprocainamide

Figure 4.38 N-hydroxylation of procainamide.

at present, but may involve formation of free radicals via one-electron oxidation/hydrogen abstraction as with paracetamol (chap. 7, Fig. 20) or the direct insertion of oxygen as with benzo[a]pyrene. It is not yet clear how significant prostaglandin synthase–catalyzed routes of metabolism are in comparison with microsomal monooxygenase-mediated pathways, but they may be very important in tissues where the latter enzymes are not abundant.

The possible role of peroxidases in metabolic activation and cytotoxicity has only relatively recently attracted attention, but may prove to be of particular importance in underlying mechanisms of toxicity.

For example, it has been suggested that the adverse reactions caused by a number of drugs such as isoniazid, **procainamide**, hydralazine could be due to metabolic activation by myeloperoxidase in neutrophils. Thus neutrophils will metabolize procainamide (Fig. 4.38) to a hydroxylamine metabolite. In the presence of chloride ion, myeloperoxidase will produce hypochlorous acid, a strong oxidizing agent, which may be responsible for metabolic activation and toxicity. One of the products is *N*-chloroprocainamide (see also sect. "Hydralazine," chap. 7).

For more details, the reader is directed to the review by Uetrecht in the Bibliography.

4.3.2 Reduction

The enzymes responsible for reduction may be located in both the microsomal fraction and the soluble cell fraction. **Reductases** in the microflora present in the gastrointestinal tract may also have an important role in the reduction of xenobiotics. There are a number of different reductases, which can catalyze the reduction of azo and nitro compounds. Thus, as well as gut bacterial azo- and nitro-reductase enzymes, cytochrome P-450 and cytochrome P-450 NADPH reductase in the liver and other tissues can reduce nitro and azo groups in conditions of low oxygen concentration. However, oxygen will inhibit these reactions.

FAD alone may also catalyze reduction by acting as an electron donor.

Reduction of the azo dye **prontosil** to produce the antibacterial drug sulfanilamide (Fig. 4.38) is a well-known example of azo reduction. This reaction is catalyzed by cytochromes P-450 and is also carried out by the reductases in the gut bacteria. The reduction of azo groups in food coloring dyes such as **amaranth** is catalyzed by several enzymes, including cytochromes P-450, NADPH cytochrome P-450 reductase, and **DT-diaphorase**, a cytosolic enzyme.

The reduction of nitro groups may also be catalyzed by microsomal reductases and gut bacterial enzymes. The reduction passes through several stages to yield the fully reduced primary amine, as illustrated for **nitrobenzene** (Fig. 4.39). The intermediates are nitrosobenzene and phenylhydroxylamine, which are also reduced in the microsomal system. These intermediates, which may also be produced by the oxidation of aromatic amines (Fig. 4.21), are involved in the toxicity of nitrobenzene to red blood cells after oral administration to rats. The importance of the gut bacterial reductases in this process is illustrated by the drastic reduction in nitrobenzene toxicity in animals devoid of gut bacteria, or when nitrobenzene is given by the intraperitoneal route.

Arylhydroxylamines, whether derived from nitro compounds by reduction or amines by N-hydroxylation, have been shown to be involved in the toxicity of a number of compounds. The possibility of reduction followed by oxidation in a cyclical fashion with the continual production of toxic metabolites has important toxicological implications.

Tertiary amine oxides and hydroxylamines are also reduced by cytochromes P-450. Hydroxylamines, as well as being reduced by cytochromes P-450, are also reduced by a flavoprotein, which is part of a system, which requires NADH and includes NADH cytochrome b_5 reductase and cytochrome b_5. Quinones, such as the anticancer drug adriamycin (doxorubicin) and menadione, can undergo one-electron reduction catalyzed by NADPH cytochrome P-450 reductase. The semiquinone product may be oxidized back to the quinone with the concomitant production of superoxide anion radical, giving rise to redox cycling and potential cytotoxicity. This underlies the cardiac toxicity of **adriamycin** (see chap. 6).

An alternative route of reduction is catalyzed by the enzyme DT diaphorase. This is a two-electron reduction, which produces a hydroquinone. These tend to be less toxic than semiquinones, as they do not tend to undergo **redox cycling**.

Thus, one-electron reduction catalyzed by cytochrome P-450 NADPH reductase tends to cause the toxicity of quinones via oxidative stress, whereas two-electron reduction produces the less toxic hydroquinones and so is a detoxication.

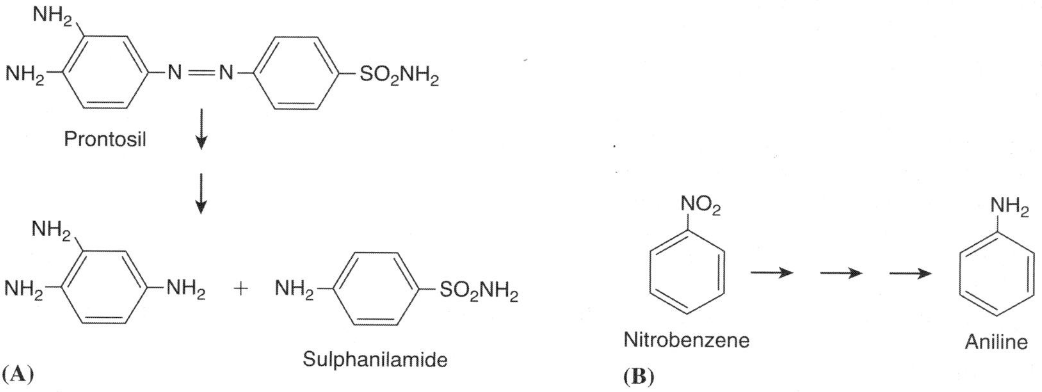

Figure 4.39 (**A**) Reduction of the azo group in prontosil. (**B**) Reduction of the nitro group in nitrobenzene.

Other chemicals, which are toxic as a result of one-electron reduction, are paraquat, 6-hydroxydopamine (see chap. 7) and benzene (see chap. 6).

The one-electron reduction of a chemical to cytotoxic products is the basis of certain anticancer drugs. This is because tumor cells tend to be anaerobic, and so reduction is favored. However, anaerobic conditions can induce DT diaphorase, which carries out two-electron reduction reactions. For example, the anticancer drug tirapazamine is activated by a one-electron reduction catalyzed by NADPH cytochrome P-450 reductase in anaerobic conditions to a nitroxide radical, which is toxic to tumor cells. It is detoxified however, by DT diaphorase, to another product.

An important example is the reduction of **nitroquinoline N-oxide**. This proceeds via the hydroxylamine, which is an extremely carcinogenic metabolite, probably the ultimate carcinogen (Fig. 4.40).

A similar example is 2,6 dinitrotoluene where reduction in rats of a nitro group to a hydroxylamine occurs, which yields a liver carcinogen (see chap. 5).

Other types of reduction catalyzed by non-microsomal enzymes have also been described for xenobiotics. Thus, reduction of aldehydes and ketones may be carried out either by alcohol dehydrogenase or NADPH-dependent cytosolic reductases present in the liver. Sulfoxides and sulfides may be reduced by cytosolic enzymes, in the latter case involving glutathione and glutathione reductase. Double bonds in unsaturated compounds and epoxides may also be reduced. Metals, such as pentavalent arsenic, can also be reduced.

Reductive Dehalogenation

The microsomal enzyme-mediated removal of a halogen atom from a foreign compound may be either reductive or oxidative. The latter has already been discussed with regard to the volatile anesthetic **halothane**, which undergoes both oxidative and reductive dehalogenation and which may be involved in the hepatotoxicity (Fig. 4.28) (chap. 7, Fig. 77). Reductive dehalogenation of halothane is catalyzed by cytochromes P-450 under anaerobic conditions and may lead to reactive radical metabolites (Fig. 4.41) (chap. 7, Fig. 77). Another example of reductive dehalogenation of toxicological importance, also catalyzed by cytochromes P-450, is the metabolic activation of **carbon tetrachloride** by dechlorination to yield a free radical (chap. 7, Fig. 77). In both these examples, the substrate binds to the cytochromes P-450 and then receives an electron from NADPH cytochrome P-450 reductase. The enzyme-substrate complex then loses a halogen ion, and a free radical intermediate is generated. Alternatively, the enzyme-substrate complex may be further reduced by another electron from NADH via

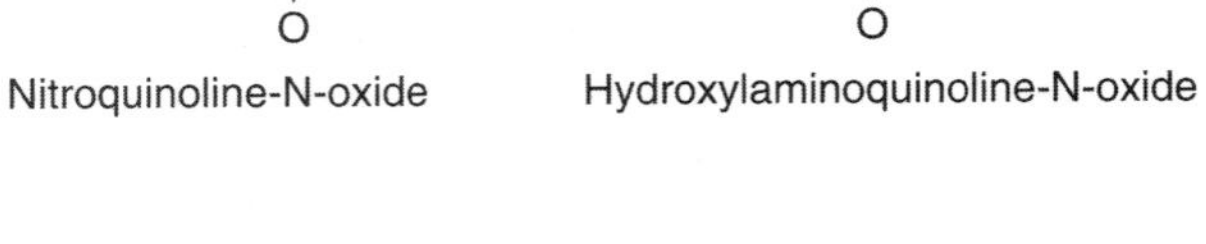

Nitroquinoline-N-oxide   Hydroxylaminoquinoline-N-oxide

Figure 4.40 Reduction of nitroquinoline *N*-oxide.

Halothane   2-Chloro-1,1,1-trifluoroethane   2-Chloro-1,1-difluoroethylene

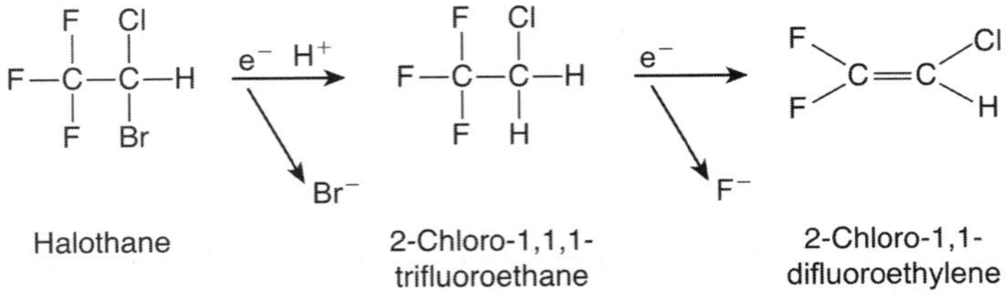

Figure 4.41 Reductive dehalogenation of halothane.

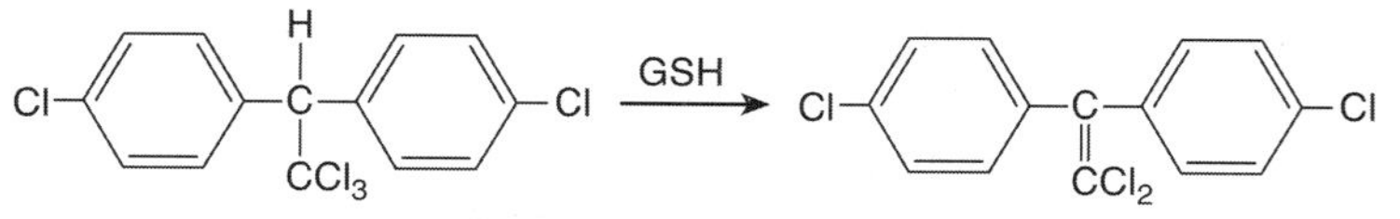

Figure 4.42 Dehydrohalogenation of DDT.

NADH cytochrome b_5 reductase and cytochrome b_5. The carbanion or carbene intermediates may rearrange with loss of a halogen ion (chap. 7, Fig. 77).

Dehalogenation of the insecticide **DDT** is catalyzed by a soluble enzyme and requires glutathione. The overall reaction is a dehydrohalogenation and yields DDE (Fig. 4.42).

4.3.3 Hydrolysis

Esters, amides, hydrazides, and carbamates can all be metabolized by hydrolysis. The enzymes, which catalyze these hydrolytic reactions, carboxylesterases and amidases, are usually found in the cytosol, but microsomal esterases and amidases have been described and some are also found in the plasma. The various enzymes have different substrate specificities, but carboxylesterases have amidase activity and amidases have esterase activity. The two apparently different activities may therefore be part of the same overall activity.

Peptidases such as trypsin are also hydrolytic enzymes and are important considerations for the new generation of peptide and protein drugs (see below).

Hydrolysis of Esters

Although esterase activity is found in blood, this is nonspecific; esterase activity, which is specific for a particular type of chemical structure, tends to be located in tissues, such as the liver.

Various **esterases** exist in mammalian tissues, hydrolyzing different types of esters. They have been classified as type A, B, or C on the basis of activity toward phosphate triesters. A-esterases, which include arylesterases, are not inhibited by phosphotriesters and will metabolize them by hydrolysis. Paraoxonase is a type A esterase (an organophosphatase). B-esterases are inhibited by paraoxon and have a serine group in the active site (see chap. 7). Within this group are carboxylesterases, cholinesterases, and arylamidases. C-esterases are also not inhibited by paraoxon, and the preferred substrates are acetyl esters, hence these are acetylesterases. Carboxythioesters are also hydrolyzed by esterases. Other enzymes such as trypsin and chymotrypsin may also hydrolyze certain carboxyl esters.

Metabolism of the local anesthetic procaine provides an example of esterase action, as shown in Figure 4.43. This hydrolysis may be carried out by both a plasma esterase and a microsomal enzyme.

Esterase activity is important in both the detoxication of organophosphates and the toxicity caused by them. Thus brain acetylcholinesterase is inhibited by organophosphates such as paraoxon and malaoxon, their oxidized metabolites (see above). This leads to toxic effects. Malathion, a widely used insecticide, is metabolized mostly by carboxylesterase in mammals, and this is a route of detoxication. However, an isomer, isomalathion, formed from malathion when solutions are inappropriately stored, is a potent inhibitor of the carboxylesterase. The consequence is that such contaminated malathion becomes highly toxic to humans because detoxication is inhibited and oxidation becomes important. This led to the poisoning of 2800 workers in Pakistan and the death of 5 (see chap. 5 for metabolism and chap. 7 for more details).

Hydrolysis of Amides

The hydrolysis of amides can also be catalyzed by nonspecific plasma esterases but is slower than that of esters. However, hydrolysis of amides is more likely to be catalyzed by the amidases in the liver.

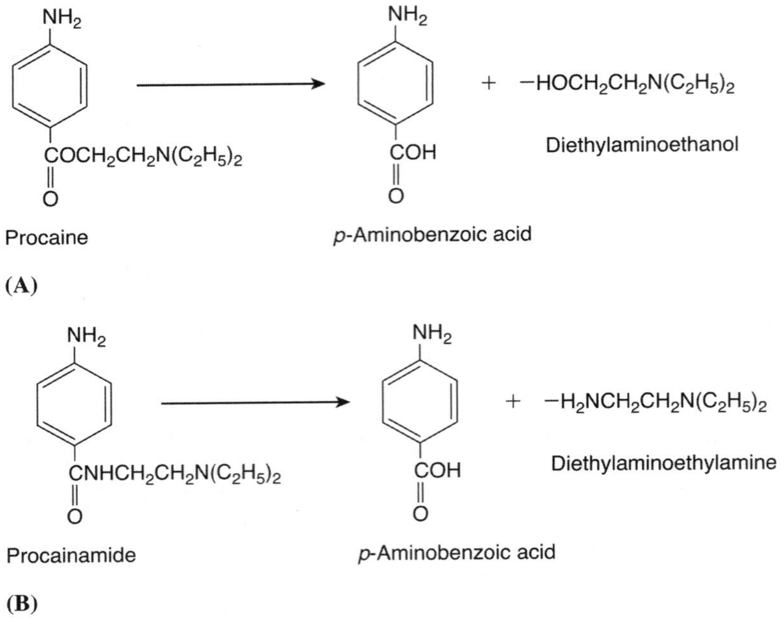

Procaine → p-Aminobenzoic acid + Diethylaminoethanol

(A)

Procainamide → p-Aminobenzoic acid + Diethylaminoethylamine

(B)

Figure 4.43 (**A**) Hydrolysis of the ester procaine. (**B**) Hydrolysis of the amide procainamide.

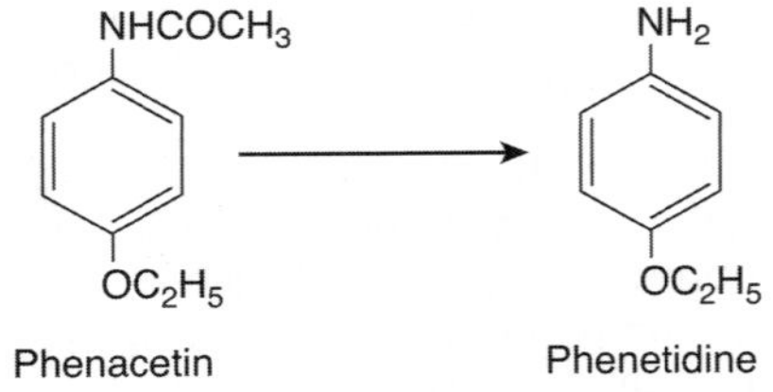

Phenacetin → Phenetidine   **Figure 4.44**   Deacetylation of phenacetin.

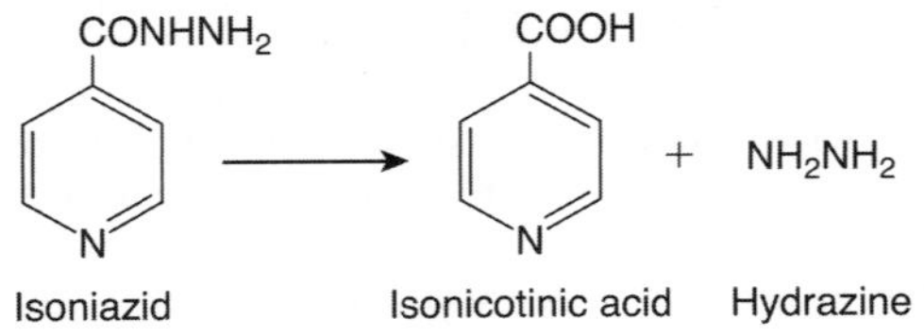

Isoniazid → Isonicotinic acid + Hydrazine   **Figure 4.45**   Hydrolysis of isoniazid.

Thus, unlike procaine, the analogue **procainamide** is not hydrolyzed in the plasma at all, the hydrolysis *in vivo* being carried out by enzymes in other tissues (Fig. 4.43).

The hydrolysis of some amides may be catalyzed by a liver microsomal carboxyl esterase, as is the case with phenacetin (Fig. 4.44). Hydrolysis of the acetylamino group, resulting in deacetylation, is known to be important in the toxicity of a number of compounds. For example, the deacetylated metabolites of **phenacetin** are thought to be responsible for its toxicity, the oxidation of hemoglobin to methemoglobin. This toxic effect occasionally occurs in subjects taking therapeutic doses of the drug and who have a deficiency in the normal pathway of metabolism of phenacetin to paracetamol. Consequently, more phenacetin is metabolized by deacetylation and subsequent oxidation to toxic metabolites (chap. 5, Fig. 24).

Hydrolysis of Hydrazides
The drug **isoniazid** (isonicotinic acid hydrazide) is hydrolyzed *in vivo* to the corresponding acid and hydrazine, as shown in Figure 4.45. Hydrazine is toxic and could be responsible for

some of the adverse effects of isoniazid (see chap. 7). However, in man, *in vivo*, hydrolysis of the acetylated metabolite acetylisoniazid is quantitatively more important and *may* be toxicologically more significant. (Fig. 4.46) (see chap. 7). This hydrolysis reaction accounts for about 45% of the acetylisoniazid produced. These hydrolysis reactions are probably catalyzed by amidases and are inhibited by organophosphorus compounds such as **bis-pnitrophenyl phosphate** (chap. 5, Fig. 38).

Hydrolysis of Carbamates
The insecticide **carbaryl** is hydrolyzed by liver enzymes to 1-naphthol (Fig. 4.47). This compound also undergoes extensive metabolism by other routes.

Hydrolysis of Peptides
As some of the newer drugs such as hormones, growth factors, and cytokines now being produced are peptides and certain toxins are also peptides or proteins, the role of peptidases may be important. Peptidases are especially active in the lumen of the gut, and consequently many such drugs are administered intravenously. Also some natural protein toxins may bypass the gut by via bites or stings into tissue. However, peptidase activity is also found in blood and other tissues. Peptidases are also important in the further metabolism of glutathione conjugates (see below).

Hydration of Epoxides
Epoxides, three-membered oxirane rings containing an oxygen atom, may be metabolized by the enzyme **epoxide hydrolase**. This occurs by a similar mechanism to that involved with the hydrolysis of esters and amides. This enzyme adds water to the epoxide, probably by nucleophilic attack by OH on one of the carbon atoms of the oxirane ring, which may be regarded as electron deficient, to yield a dihydrodiol, which is predominantly trans (Fig. 4.48), although the degree of stereospecificity is variable. Epoxides are often intermediates produced by the oxidation of unsaturated double bonds, aromatic, aliphatic, or heterocyclic, as, for example, takes place during the hydroxylation of bromobenzene, the hepatotoxic solvent (chap. 7, Fig. 22).

There are five forms of the enzyme, but only two, the microsomal and soluble forms are important in xenobiotic metabolism. These two forms have different substrate specificities. The microsomal form is located in the endoplasmic reticulum in close proximity to cytochromes P-450, and like the latter is also present in greater amounts in the centrilobular

Acetylisoniazid   Isonicotinic acid   Acetylhydrazine   **Figure 4.46**   Hydrolysis of acetylisoniazid.

Carbaryl   1-Naphthol   **Figure 4.47**   Hydrolysis of carbaryl.

Benzene-1,2-oxide   Benzene trans-1,2-dihydrodiol   **Figure 4.48**   Hydration of benzene-1, 2-oxide by epoxide hydrolase.

areas of the liver. Epoxide hydrolase is therefore well placed to carry out its important role in detoxifying the chemically unstable and often-toxic epoxide intermediates produced by cytochromes P-450-mediated hydroxylation. Soluble epoxide hydrolases have also been described, and the enzyme has been detected in the nuclear membrane.

The epoxide of bromobenzene is one such toxic intermediate, and this example is discussed in more detail in chapter 7. In the case of some carcinogenic poly cyclic hydrocarbons such as **benzo[a]pyrene**, however, it seems that the dihydrodiol products are in turn further metabolized to epoxide-diols, the ultimate carcinogens (see chap. 7, Figs. 7.2 and 7.3).

4.4 PHASE 2 REACTIONS

4.4.1 Conjugation

Conjugation reactions involve the addition to foreign compounds of endogenous groups, which are generally polar and readily available *in vivo*. These groups are added to a suitable functional group present on the foreign molecule or introduced by phase 1 metabolism. With the exception of acetylation and methylation, conjugation renders the whole molecule more polar and hydrophilic (less lipid soluble). This facilitates excretion and reduces the likelihood of toxicity. Furthermore, phase 2 reactions are generally faster than phase 1 reactions.

The endogenous groups donated in conjugation reactions include carbohydrate derivatives, amino acids, glutathione, and sulfate. The mechanism commonly involves formation of a high-energy intermediate, where either the endogenous metabolite or the foreign compound is activated (type 1 and type 2, respectively). The groups donated in conjugation reactions are often involved in intermediary metabolism.

Glucuronide Formation

This is a major, type 1, conjugation reaction occurring in most species with a wide variety of substrates, including endogenous substances. It involves the transfer of glucuronic acid in an activated form **as uridine diphosphate glucuronic acid (UDPGA)** to hydroxyl, carboxyl, nitrogen sulfur, and occasionally carbon atoms. The UDPGA is formed in the cytosol from glucose-1-phosphate in a two-step reaction (Fig. 4.49). The first step, addition of the UDP, is catalyzed by **UDP glucose pyrophosphorylase,** the second step by UDP glucose dehydrogenase:

The enzyme catalyzing the conjugation reaction is UDP-glucuronosyl transferase (UGT). This is known to exist in multiple forms each with different preferred substrate specificities, although these may overlap. For example, simple phenols may be substrates for UGT1A8 but also UGT1A1 and UGT1A6. Carboxylic acids are substrates for UGT1A8 and UGT1A3 and primary amines for UGT1A6, UGT1A8, and, UGT1A4.

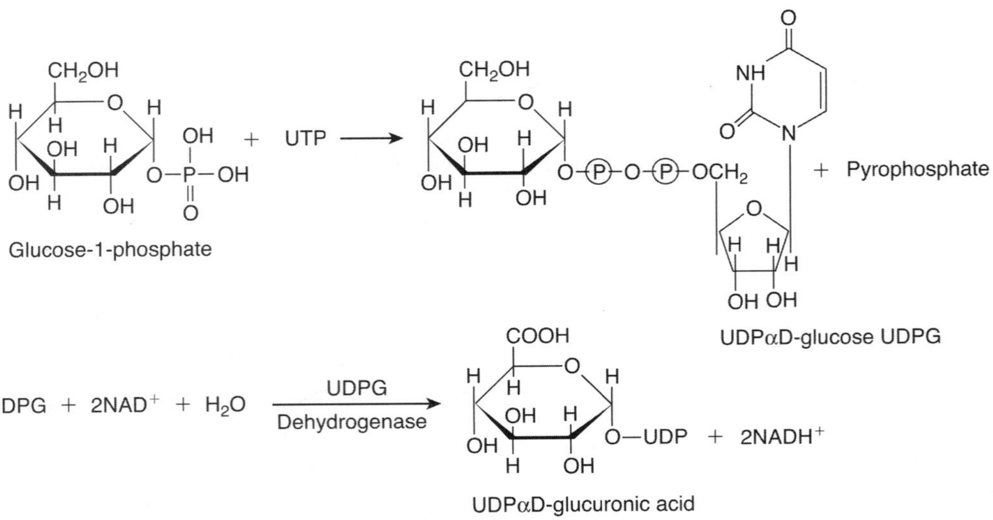

Figure 4.49 Formation of UDPGA. *Abbreviation:* UDPGA, uridine diphosphate glucuronic acid.

There are at least 50 different mammalian functional isoforms with at least 16 in humans and two main families, UGT1 and UGT2.

For further information, the reader is recommended to consult the texts in the Bibliography.

The enzymes are located in the endoplasmic reticulum and are found in many tissues including the liver, kidney, and intestine.

Conjugation with glucuronic acid involves nucleophilic attack by the electron-rich oxygen, sulfur or nitrogen atom in the xenobiotic at the C-1 carbon atom of the glucuronic acid moiety. Glucuronides are therefore generally β in configuration. Conjugation with hydroxyl groups gives ether glucuronides and with carboxylic acids, ester glucuronides (Fig. 4.50). Amino groups may be conjugated directly, as in the case of aniline (Fig. 4.51), or through an oxygen atom as in the case of *N*-hydroxy compounds such as the carcinogen *N*-hydroxyacetylaminofluorene (Fig. 4.52).

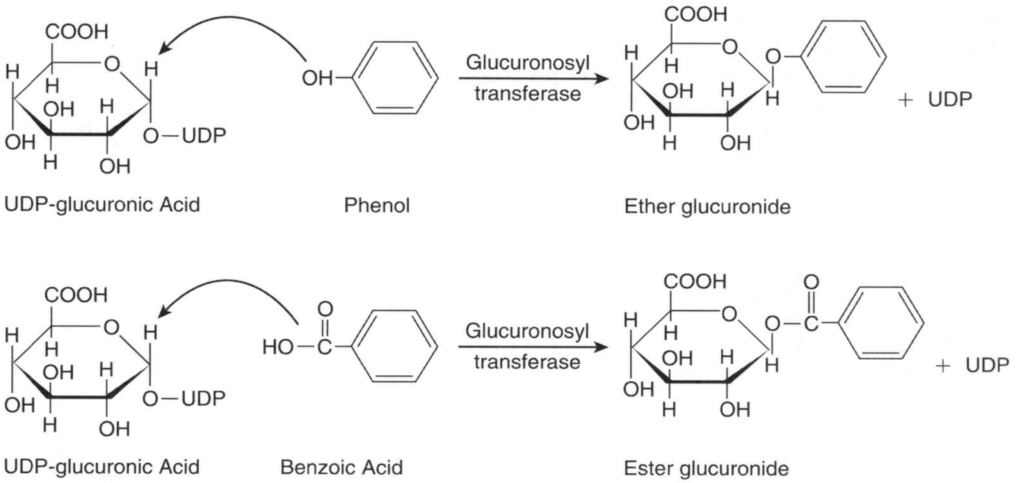

Figure 4.50　Formation of ether and ester glucuronides of phenol and benzoic acid, respectively.

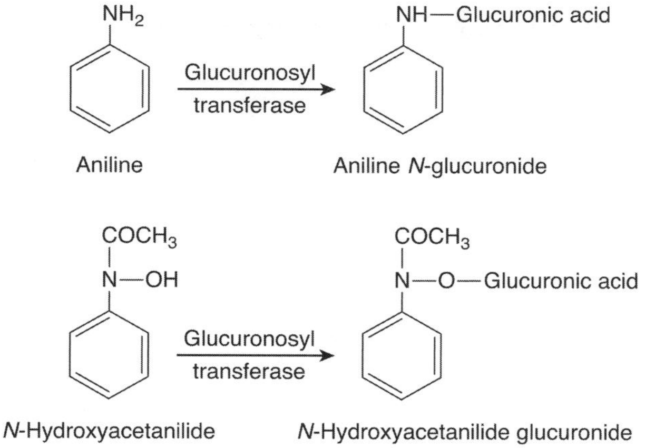

Figure 4.51　Glucuronidation of aniline and *N*-hydroxyacetanilide.

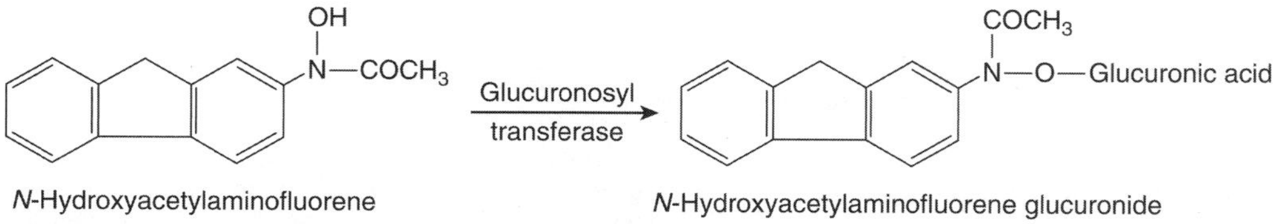

Figure 4.52　Formation of *N*-hydroxyacetylaminofluorene glucuronide.

However, although normally beneficial, glucuronide conjugation can be the cause of toxicity. For example, N-glucuronide conjugation of the aromatic amine N-hydroxy-2-naphthylamine is followed by excretion into the urine. However, once in the bladder, the acidity in the urine cleaves the conjugate, resulting in the formation of a reactive nitrenium ion (Fig. 4.53). This reacts with DNA, which subsequently leads to bladder tumors. A similar mechanism may explain the colon tumors caused by the chemical, resulting from biliary excretion and cleavage of the conjugate in the gut by bacteria.

Some other aromatic amines such as benzidine and 4-aminobiphenyl are also carcinogenic to the bladder by the same mechanism. These amines have industrial uses and have been implicated in bladder cancer in exposed humans.

Another example of glucuronidation being an important step in the toxic process results in the production of protein conjugates. This is of more general importance and is relevant for the toxicity of a number of drugs, especially non-steroidal anti-inflammatory drugs (NSAIDS) containing a carboxylic acid group. As indicated above, carboxylic acid groups can be conjugated with glucuronic acid to give acyl glucuronides. However, in the case of some NSAID drugs, such as diclofenac, the acyl glucuronide is reactive and binds covalently to protein. There are two possible mechanisms for this as shown in Figure 4.54. The resulting protein conjugates might be responsible for rare but serious immune reactions such as immune

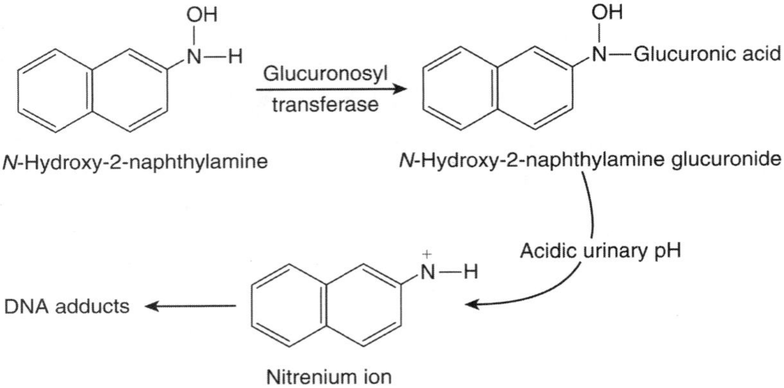

Figure 4.53 Formation of an N-hydroxy glucuronide conjugation of N-hydroxy-2-napthylamine and formation of a reactive nitrenium ion.

Diclofenac

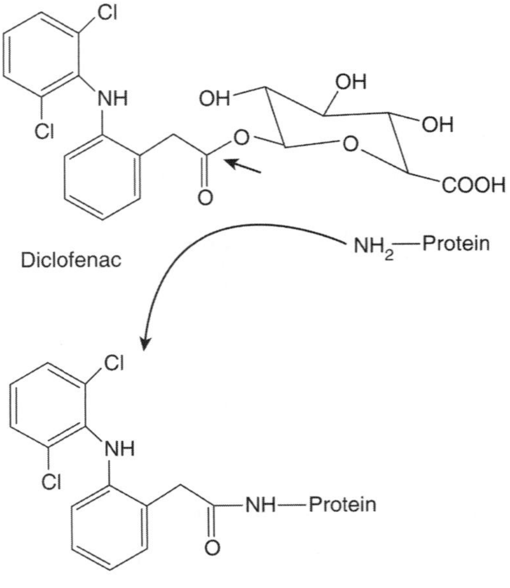

Figure 4.54 Interaction of the acyl glucuronide of diclofenac with protein. The small arrow indicates the reactive carbonyl carbon atom.

hepatitis, which can occur with diclofenac, for example. Because the acyl glucuronide is a reasonably large molecule, a proportion is excreted from the liver where it is synthesized into the bile via the organic anion transporter. This leads to relatively high concentrations in the bile, in fact several orders of magnitude higher than blood. Thus the chances of reaction of the acyl glucuronide with proteins in the bile canaliculus are increased. Indeed in rats, protein adducts with canalicular membrane proteins have been detected in liver. Biliary excretion also leads to enterohepatic recirculation following cleavage by gut bacteria and reabsorption. This will also prolong exposure of the animal or patient to the diclofenac.

Certain thiols may be conjugated directly through the sulfur atom (Fig. 4.55). Glucuronic acid conjugated directly to carbon atoms has been reported such as with the drug phenylbutazone.

Glucuronide conjugates are ionized at urinary pH and being water soluble can be readily eliminated. They are also substrates for the organic anion transporter.

As glucuronide conjugation significantly increases the molecular weight of a chemical, the conjugate may be excreted into the bile via the organic anion transporter in the liver. When presented to the small intestine via the bile, such conjugates can be cleaved by the action of the enzyme β-glucuronidase in gut bacteria. The product (usually the original phase 1 metabolite, know as the aglycone) can then be reabsorbed and transported back into the liver where it may undergo further metabolism. This can possibly lead to toxicity, as is the case with 2,4-dinitro-toluene (see below and chap. 5).

Glucose conjugates (glucosides) may sometimes be formed, especially in insects, and the mechanism is analogous to that involved in the formation of glucuronides. Xylose and ribose are also sometimes used, for example, 2-hydroxynicotinic acid has been shown to form an *N*-ribose conjugate. Analogues of purines and pyrimidines may be conjugated with ribose or ribose phosphates to give ribonucleotides and ribonucleosides.

Sulfate Conjugation
The formation of sulfate esters is a major route of conjugation for various types of hydroxyl group, and may also occur with amino groups. Thus, substrates include aliphatic alcohols, phenols, aromatic amines, and also endogenous compounds such as steroids and carbohydrates (Figs. 4.56 and 4.57).

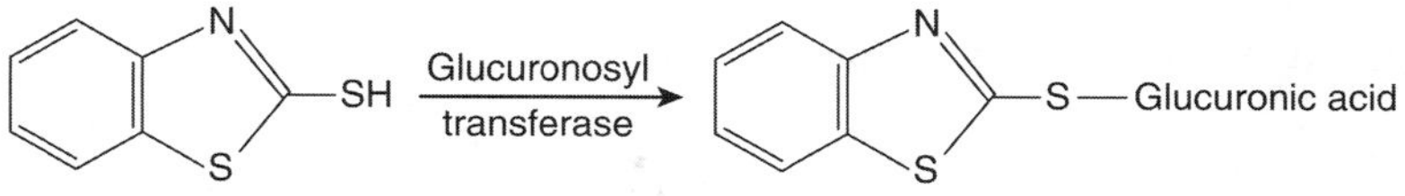

2-Mercaptobenzothiazole → (Glucuronosyl transferase) → 2-Mercaptobenzothiazole-*S*-glucuronide

Figure 4.55 Formation of 2-mercaptobenzothiazole-S-glucuronide.

Phenol + PAPS → (Sulphotransferase) → Phenyl sulphate + 3′-Phosphoadenosine-5′-phosphate

$$C_2H_5OH \xrightarrow{PAPS} C_2H_5OSO_3H + 3'\text{-Phosphoadenosyl-5'-phosphosulphate}$$

Ethyl alcohol → Ethyl sulphate

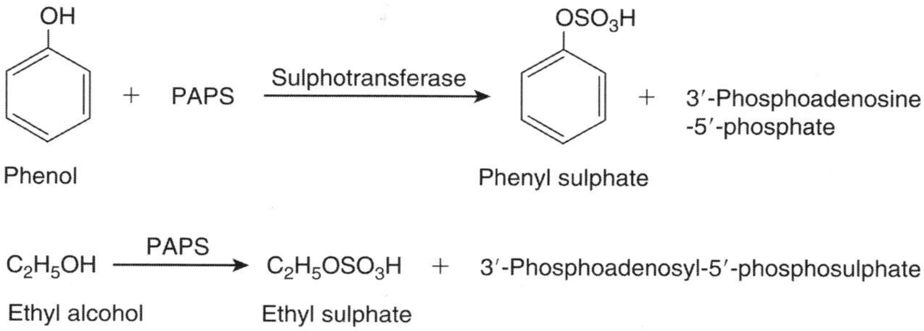

Figure 4.56 Formation of ethereal sulfates of phenol and ethyl alcohol.

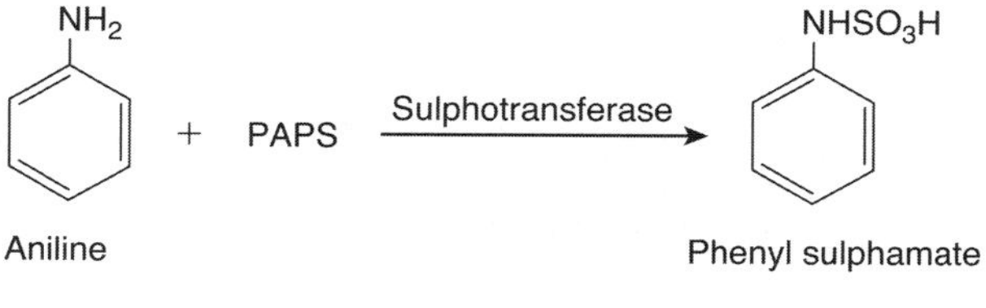

Aniline + PAPS → (Sulphotransferase) → Phenyl sulphamate

Figure 4.57 Sulfate conjugation of aniline.

The sulfate donor for this type 1 reaction is in an activated form, as **3'-phosphoadenosyl-5'-phosphosulfate (PAPS)** (Fig. 4.58), formed from inorganic sulfate and ATP:

The conjugation is catalyzed by a **sulfotransferase** enzyme, which is located in the cytosol and is found particularly in the liver, gastrointestinal mucosa, and kidney. There are at least five gene families for sulfotransferases, these being divided into subfamilies resulting in almost 50 isoforms for the enzyme. Different isoforms catalyze the sulfation of different substrates, for example, phenols (SULT1As), alcohols (SULT2As/SULT2Bs), arylamines (SULT3A1). There are other isoforms for bile salts and steroids. For further information, the reader is recommended to consult the texts in the Bibliography.

The source of sulfate may be dietary or generated by oxidative metabolism of cysteine. PAPS can become depleted when large amounts of a foreign compound conjugated with sulfate, such as paracetamol, are administered.

Sulfate conjugation can increase the toxicity of chemicals in some cases [e.g., acetylaminofluorene, (see chap. 7), safrole, and the drug tamoxifen (see chap. 5)]. This is due to the fact that the sulfate group is described in chemical terms as "a good leaving group." Thus with some conjugates, cleavage of the sulfate ester yields a chemically reactive nitrenium ion or carbocation (chap. 7, Fig. 1), which is electrophilic and can react with biological macromolecules such as DNA.

Glutathione Conjugation

Glutathione is one of the most important molecules in the *cellular defense* against toxic compounds. This *protective function* is due in part to its involvement in conjugation reactions, and a number of toxicological examples of this such as bromobenzene and paracetamol hepatotoxicity are discussed later (see chap. 7). The other protective functions of glutathione are discussed in chapter 6. Glutathione is a tripeptide (Fig. 4.59), composed of glutamic acid, cysteine, and glycine (glu-cys-gly).

The glutathione molecule is significant for several reasons:

1. It has a reactive SH group.
2. It is protected from protease digestion because of the nature of the bond between the cysteine and glutamate.
3. It is present in relatively high concentrations.

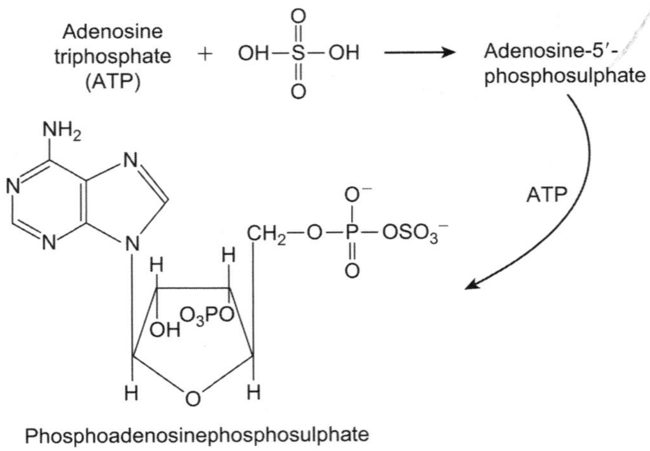

Phosphoadenosinephosphosulphate

Figure 4.58 Formation of PAPS. *Abbreviation*: 3'-phosphoadenosyl-5'-phosphosulfate.

Glutathione (γ-glutamylcysteinylglycine)

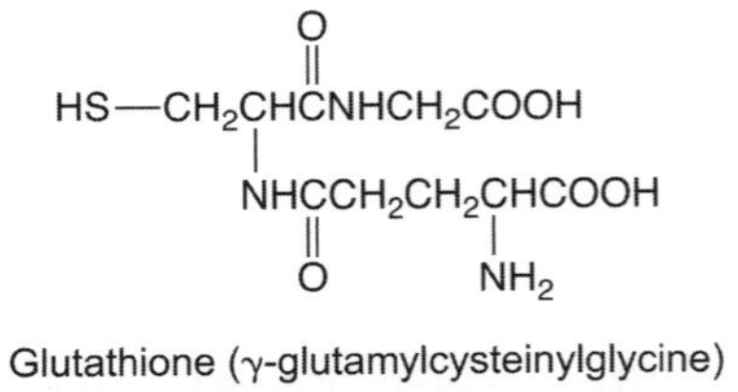

Figure 4.59 Structure of glutathione.

There are three pools of glutathione: in the cytosol, in the mitochondria, and also in the nucleus. It is found in most cells, but is especially abundant in the liver where it reaches a concentration of 5 mM or more in mammals. The presence of cysteine provides a sulfydryl group, which is nucleophilic and so glutathione will react, probably as the thiolate ion, GS^-, with electrophiles. These electrophiles may be chemically reactive, metabolic products of a phase 1 reaction, or they may be more stable foreign compounds, which have been ingested. Thus, glutathione protects cells by removing reactive metabolites. Unlike glucuronic acid or sulfate conjugation, however (type 1 conjugation reactions), the conjugating moiety (glutathione) is not activated in some high-energy form. Rather the substrate is often in an activated form. Glutathione conjugation may be an enzyme-catalyzed reaction or simply a chemical reaction. The glutathione conjugate produced by the reaction may then either be excreted, usually into the bile rather than the urine, or the conjugate may be further metabolized. This involves several steps: removal of the glutamyl and glycinyl groups and acetylation of the cysteine amino group to yield a mercapturic acid or nacetylcysteine conjugate. This is illustrated for the compound **naphthalene,** which is metabolized by cytochromes P-450 to a reactive epoxide intermediate, then conjugated with glutathione, and eventually excreted as an *N*-acetylcysteine conjugate (Fig. 4.60). This sequence of further catabolic steps has been termed "**phase 3** metabolism." Naphthalene is toxic to the lung, and these metabolic pathways are important in this toxicity (see below).

There are many types of substrates for glutathione conjugation, including aromatic, aliphatic, heterocyclic, and alicyclic epoxides, halogenated aliphatic and aromatic compounds, aromatic nitro compounds, unsaturated hydrocarbon benzo[a]pyrene (see chap. 7), and aliphatic compounds and alkyl halides (Figs. 4.61–4.63). In each case, the glutathione reacts with an electrophilic carbon atom in an addition or substitution reaction. With the reactive epoxides, the two carbon atoms of the oxirane ring will be electrophilic and suitable for reaction with glutathione; such that reaction occurs with bromombenzene and the polycyclic hydrocarbon benzo[a]pyrene (see chap. 7). With aromatic and aliphatic halogen compounds and aromatic nitro compounds, the nucleophilic sulfydryl group of the glutathione attacks the electrophilic carbon atom to which the electron-withdrawing halogen or group is attached, and the latter is replaced by glutathione (Fig. 4.62). With unsaturated compounds such as **diethylmaleate,** the electron-withdrawing substituents allow nucleophilic attack on one of the

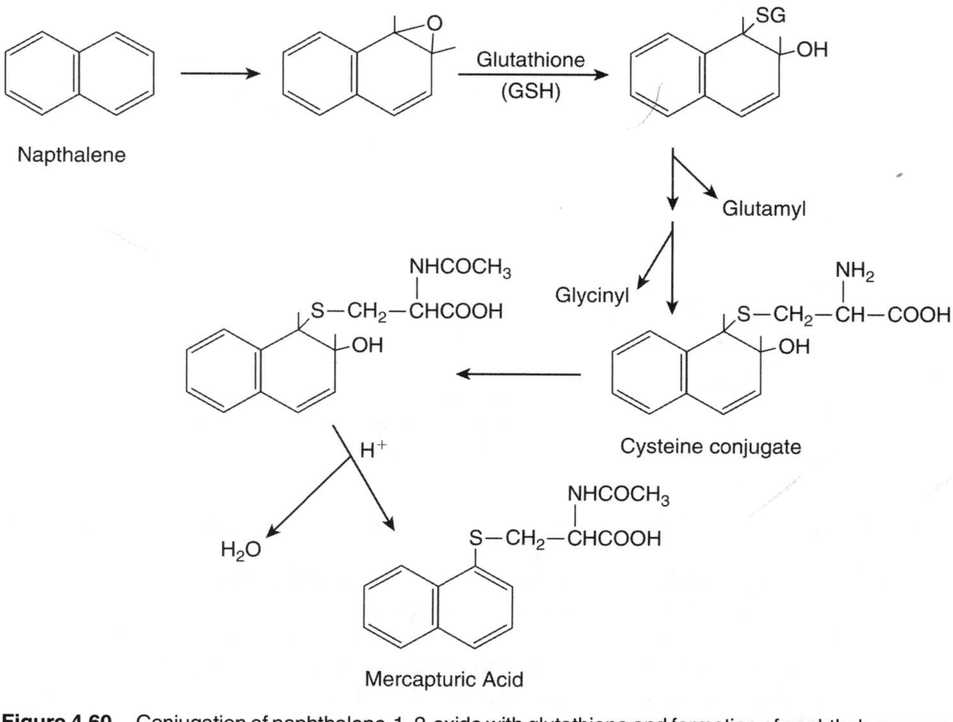

Figure 4.60 Conjugation of naphthalene-1, 2-oxide with glutathione and formation of naphthalene mercapturic acid.

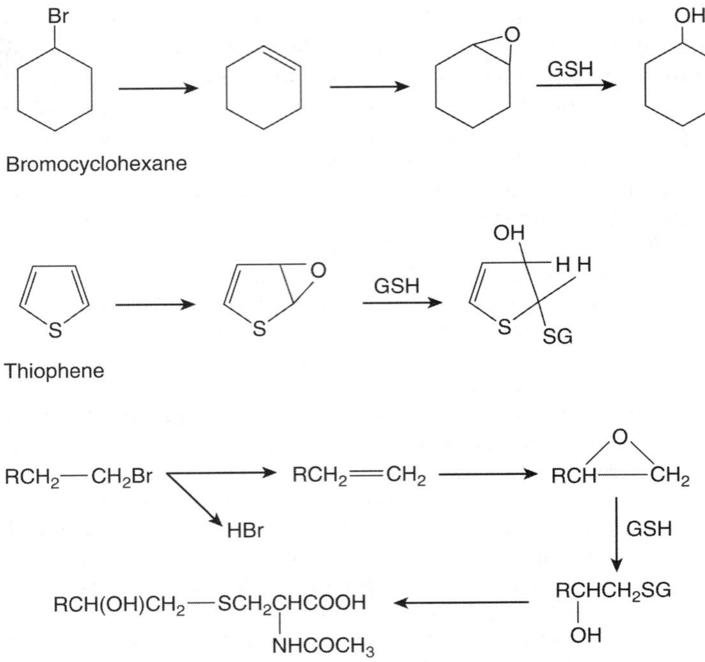

Br

Bromocyclohexane

GSH

OH
SG

Thiophene

GSH

OH
H H
S
SG

RCH_2—CH_2Br → RCH_2=CH_2 → RCH——CH_2 (O)

HBr

GSH

$RCH(OH)CH_2$—$SCH_2CHCOOH$ ← $RCHCH_2SG$
| |
$NHCOCH_3$ OH

Figure 4.61   Conjugation of various epoxides with glutathione.

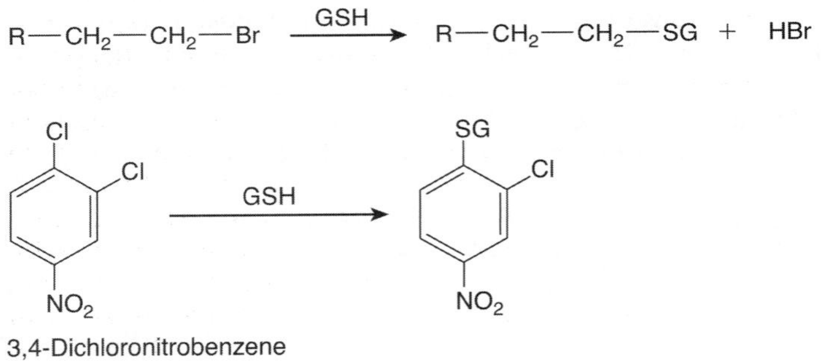

R—CH_2—CH_2—Br $\xrightarrow{GSH}$ R—CH_2—CH_2—SG + HBr

Cl
Cl

GSH

SG
Cl

NO_2

NO_2

3,4-Dichloronitrobenzene

Figure 4.62   Displacement of aliphatic and aromatic halogens by glutathione.

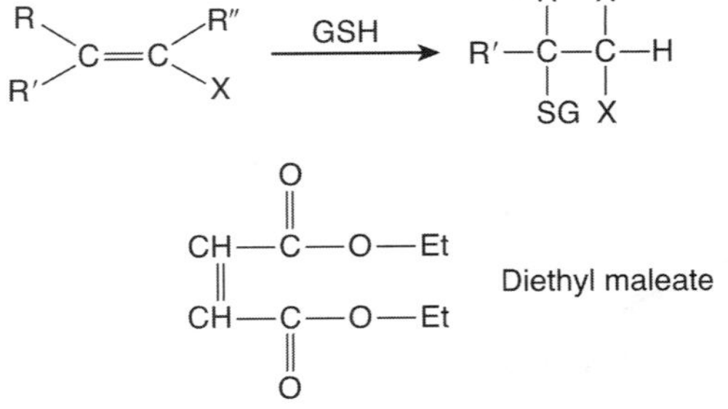

R   R''
\   /
C=C $\xrightarrow{GSH}$ R'—C—C—H
/   \ | |
R'   X   SG X

with R R'' on the carbons

$$\begin{array}{c} CH—C—O—Et \\ \| \quad \| \\ CH—C—O—Et \end{array}$$ Diethyl maleate

Figure 4.63   Conjugation of an unsaturated aliphatic compound with glutathione and structure of diethylmaleate, a typical example.

unsaturated carbon atoms and addition of proton to the other, leading to an addition reaction (Fig. 4.63).

Diethylmaleate reacts readily with glutathione *in vivo* and has been used to deplete tissue levels of the tripeptide. The reaction may be catalyzed by one of the **glutathione-S-transferases (GSTs)**. These are cytosolic enzymes although some distinct isoforms are detectable in the endoplasmic reticulum and are found in many tissues, especially the liver, kidney, gut, testis, and adrenal gland. There are eight families of GSTs, α (A), μ (M), π (P), σ (S), θ (T), ω (O), ς, and κ. The enzyme exists as hetero- or homodimers of subunits arranged to give rise to various isoforms with particular preferred substrates. Some chemicals, such as 1-chloro-2,4-dinitrobenzene, are substrates for conjugation with a number of the isoforms from different families, whereas other chemicals and reaction types are more specific for a particular isoform. More than 20 different isoforms are expressed in humans representing the A, M, P, T, and O families.

Although a wide variety of substrates may be accepted, there is absolute specificity for glutathione. However, the substrates have certain characteristics; namely, they are hydrophobic, contain an electrophilic carbon atom and react nonenzymatically with glutathione to some extent. It appears that as well as catalyzing conjugation reactions, some of the transferases have a binding, transport, or storage function. Thus GSTA1 and GSTA2 are also known as **ligandin**. This binding facility is associated with one particular subunit but is not directly related to catalytic activity.

Thus some of the compounds bound to ligandin are not substrates for the transferase activity (GSTA1, GSTA2).

Substances bound include endogenous compounds—bilirubin, estradiol, cortisol, and also drugs such as tetracycline, penicillin, and ethacrynic acid.

Thus, the enzyme also has a transport or storage function.

There are polymorphisms in the GSTs, including complete absence of some isoforms, which vary with ethnic groups, and the GSTs can also be induced by a variety of chemicals and physiological conditions. These will be considered in the next chapter.

After the conjugation reaction, the first catabolic step, removal of the glutamyl residue, is catalyzed by the enzyme γ-**glutamyltranspeptidase (glutamyltransferase)**.

This is a membrane-bound enzyme found in high concentrations in the kidney. In the second step, the glycine moiety is removed by the action of a peptidase, **cysteinyl glycinase**. The final step is acetylation of the amino group of cysteine by an *N*-acetyl transferase, which uses acetyl CoA and is a microsomal enzyme found in the liver and kidney, but is different from the cytosolic enzyme described below. The resultant *N*-acetylcysteine conjugate, also known as a mercapturic acid, is then excreted. With aromatic epoxides, as in the example of naphthalene shown in Fig. 4.60, the *N*-acetylcysteine conjugate may lose water and regain the aromatic ring structure. This will generally not occur in other types of glutathione conjugation reaction. Also the intermediate such as the cysteinyl-glycine and cysteine conjugates may be excreted as well as being metabolized to the *N*-acetylcysteine derivative. It should be noted, however, that there are now examples of glutathione conjugates being involved in toxicity. For example, the vicinal dihaloalkane **1,2-dibromoethane** forms a glutathione conjugate by displacement of one bromine atom and catalyzed by GST. The conjugate then rearranges with loss of the second bromine to form a highly reactive episulfonium ion, which can interact with DNA (Fig. 4.64). This gives rise to adducts, in particular with guanine. These are believed to be

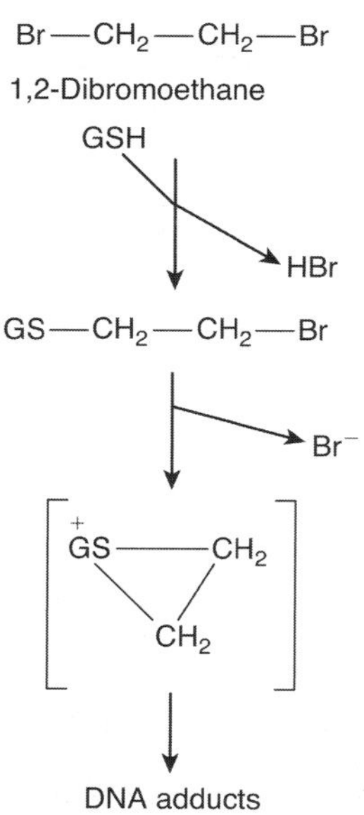

Figure 4.64 Glutathione-mediated activation of 1, 2-dibromoethane. The addition of glutathione is catalyzed by glutathione transferase. Loss of bromide from the glutathione conjugate gives rise to an episulfonium ion. This can react with bases such as guanine in DNA.

responsible for the mutagenicity and tumourigenicity of the compound. This compound is used in industry for synthesis of dyes and as a fumigant.

Another example of a glutathione conjugate responsible for toxicity is the industrial chemical hexachlorobutadiene discussed in chapter 7. The diglutathione conjugate of **bromobenzene** is believed to be involved in the nephrotoxicity after further metabolic activation (chap. 7, Fig. 7.31).

Mono-, di-, tri- and tetra-glutathione conjugates of benzohydroquinone are all formed when it is oxidized to benzoquinone. The di- and especially the tri-glutathione conjugates [2,3,5-(triglutathionyl-S-yl) hydroquinone] are believed to be responsible for the nephrotoxicity of that compound. The conjugates have a high-enough molecular weight to be excreted into bile, and then reabsorption occurs, which presents them to the kidney tubular epithelial cells. Here the γ-glutamyltransferase and cysteinylglycinase cleave the glutamate and glycine residues, leaving a conjugate with one or more cysteine groups. These conjugates are taken up into the proximal tubule, and the hydroquinone moiety is oxidized to benzoquinone. Despite the presence of the cysteine groups, this is a reactive electrophilic compound, which causes oxidative stress and can covalently bind to protein. (Fig. 4.65). The cysteine conjugation lowers the redox potential of the hydroquinone and facilitates the oxidation to a quinone, which is toxic.

The nephrotoxicity of bromobenzene (see chap. 7) and possibly 4-aminophenol is also believed to be due to the cysteine conjugates of the quinone or quinoneimine, respectively. Biliary excretion and reabsorption from the gut and delivery to the kidney is a crucial part of this process as it is with hexachlorobutadiene (chap. 7).

Cysteine Conjugate β-Lyase
Cysteine conjugates resulting from initial glutathione conjugation as described above may undergo further catabolism to give the thiol compound, pyruvate and ammonia (Fig. 4.66). The enzyme, which catalyzes this reaction, cysteine conjugate β-lyase (or C-S lyase), requires pyridoxal phosphate, is cytosolic, and will not accept the acetylated cysteine derivative. The thiol conjugate produced as a result of the action of β-lyase may be further metabolized by

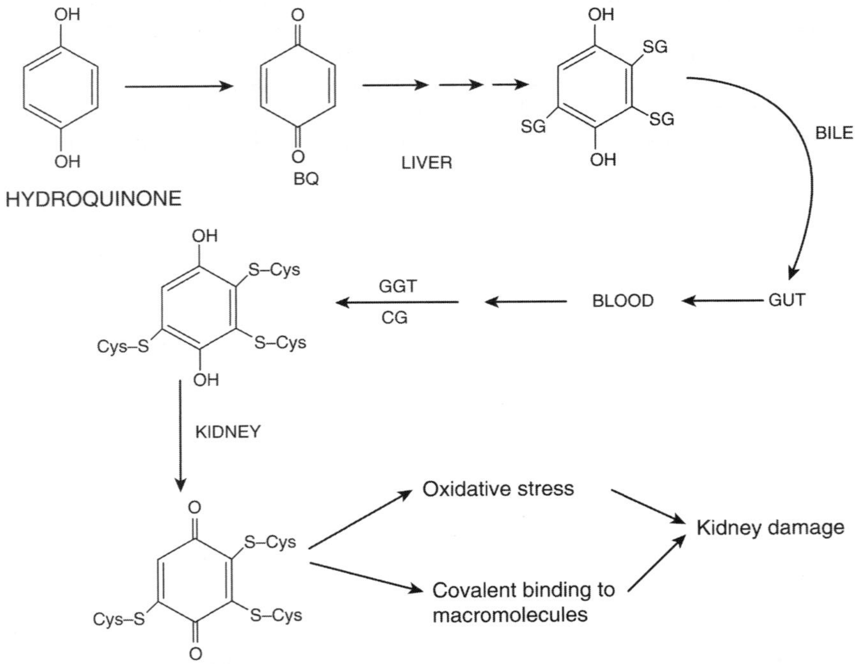

Figure 4.65 Oxidation of hydroquinone to quinone and multiple conjugation with glutathione. Biliary excretion of the conjugate and reabsorption allow further metabolism (phase 3) in the gut and kidney to the cysteine conjugate, which is nephrotoxic.

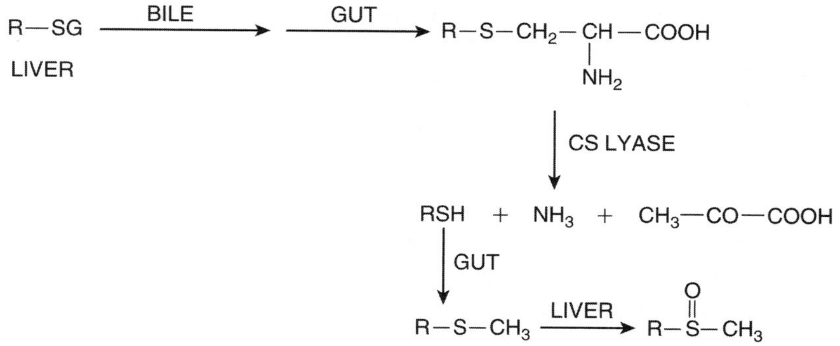

Figure 4.66 Metabolism of a cysteine conjugate by CS lyase (β-lyase). The cysteine conjugate is shown arising from a glutathione conjugate after biliary excretion. The thiol product, which may be toxic (see text) can also be methylated and further oxidized as shown.

methylation and then S-oxidation (see below). The cleavage of cysteine conjugates can occur in the gut, catalyzed by bacteria or the kidney, where the enzyme is found in all three segments of the proximal tubules. The thiol products of the enzyme can in some cases be further metabolized by methylation followed by oxidation as shown in the figure (Fig. 4.66).

These subsequent metabolic transformations are known to be involved in the nephrotoxicity of a number of compounds such as **S-(1,2-dichlorovinyl)-L cysteine** and **hexachlorobutadiene** (see chap. 7). Thus, although the initial glutathione conjugation may be a detoxication step, the final product of this phase 3 reaction may prove to be toxic.

Acetylation

Acetylation is an important route of metabolism for aromatic amines, sulfonamides, hydrazines, and hydrazides, and there is a wide variety of substrates. This metabolic reaction is one of two types of acylation reaction and involves an activated conjugating agent, acetyl CoA. It is hence a type 1 reaction. Acetylation is notable in that the product may be less water soluble than the parent compound. This fact gave rise to problems with sulfonamides when these were administered in high doses. The acetylated metabolites, being less soluble in urine, crystallized out in the kidney tubules, causing tubular necrosis (Table 7.1). However, acetylation *does reduce* the reactivity of amines and hydrazines and *reduces* the possibility of oxidation by cytochrome P-450, which can lead to reactive intermediates. The enzymes, which catalyze acetylation reactions are *N*-acetyltransferases, and there are two distinct forms, NAT1 and NAT2, in humans and some other species such as rabbit and hamster. These cytosolic enzymes, with 87% homology between them, have different preferred substrates and different distribution. NAT1 is widely distributed in many organs and tissues, including red blood cells, whereas NAT2 is primarily found in the Kupffer cells in the liver, and there is some activity in the reticuloendothelial cells in the spleen, gut, and lung.

The mechanism of action of *N*-acetyltransferase has been extensively studied and is well understood.

This involves first acetylation of the enzyme by acetyl CoA (Fig. 4.67), followed by addition of substrate, and then transfer of the acetyl group to the substrate. With loss of the acetylated product, the enzyme is regenerated.

Although the two forms of the enzyme have preferred substrates, there is overlap between them such that no substrate seems to be exclusively acetylated by one or the other. Some preferred NAT1 substrates are *p*-aminobenzoic acid and *p*-aminosalicylic acid and sulfanilamide, whereas preferred substrates for NAT2 include isoniazid, hydralazine, procainamide, and dapsone.

Some chemicals, such as 2-aminofluorene, are acetylated to a similar extent by both NAT1 and NAT2.

The acetylation of sulfanilamide is illustrated in Figure 4.68, and it can be seen that either the N^4 amino nitrogen or the N^1 sulfonamido nitrogen can be acetylated, or indeed both.

$$R\text{—}NH_2 \ + \ Acetyl\text{—}S\text{—}CoA \ \longrightarrow \ RNHCOCH_3 \ + \ CoASH$$

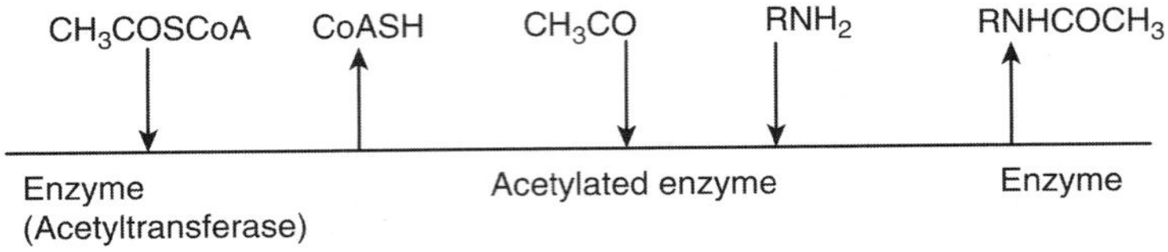

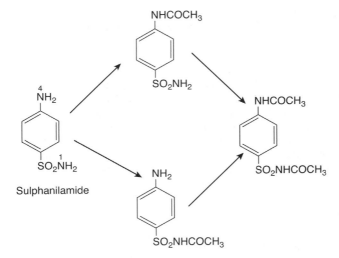

CH$_3$COSCoA   CoASH   CH$_3$CO   RNH$_2$ RNHCOCH$_3$

Enzyme
(Acetyltransferase)   Acetylated enzyme   Enzyme

Figure 4.67   Reaction sequence for *N*-acetyltransferase.

Sulphanilamide

Figure 4.68   Acetylation of sulfanilamide on the N1-sulfonamido or N4-amino nitrogen to give the N1- and N4-acetyl and N1, N4-diacetyl derivatives.

It has been found, however, that the acetylation of certain compounds in man and in the rabbit shows wide interindividual variation. This variation in acetylation is a genetic polymorphism in NAT2 and shows a bimodal distribution with two phenotypes, known as rapid and slow acetylators. The slow acetylator has a mutation, which gives rise to a less active form of the enzyme. This is discussed in more detail in the next chapter.

The acetylation polymorphism has a number of toxicological consequences, which will be discussed more fully in chapters 5 and 7.

Only certain substrates such as **isoniazid, procainamide,** and **hydralazine** (NAT2 substrates) are polymorphically acetylated. Others, **sulfanilamide, p-aminobenzoic acid**, and **p-amino salicylic acid** (NAT1 substrates), do not show the clear bimodal variation in acetylation, leading to the suggestion that NAT1 does not show any genetic polymorphisms. However, there is now evidence that this isoform does also show some genetic polymorphisms, although there is less variation than there is in NAT2.

As with sulfate conjugation, acetylation can lead to toxic metabolites being formed. Thus as well as N-acetylation, O-acetylation can also occur. For example, *N*-hydroxy-2-aminofluorene can undergo acetylation of the hydroxyl group, catalyzed by NAT2, to yield 2-acetoxyaminofluorene. However, this acetyl group is, like sulfate, a good leaving group and can be cleaved off chemically, to leave a reactive nitrenium ion, which can react with DNA (chap. 7, Fig. 1).

Apart from direct O-acetylation, N,O-transacetylation can also occur. Thus 2-acetylaminofluorene can undergo first oxidation to *N*-hydroxyacetylaminofluorene and then N,O-transacetylation takes place, which yields the same product, 2-acetoxyaminofluorene, which then yields the same reactive nitrenium ion. See more details in chapter 7 and Figure 1 in chapter 7. These reactions can also occur with certain other aryl amines and are believed to underlie cancers in various organs following occupational exposure. For example, benzidine and 2-napthylamine (Fig. 4.53) have both been used in industry and are recognized as carcinogens, being associated particularly with bladder cancer in man.

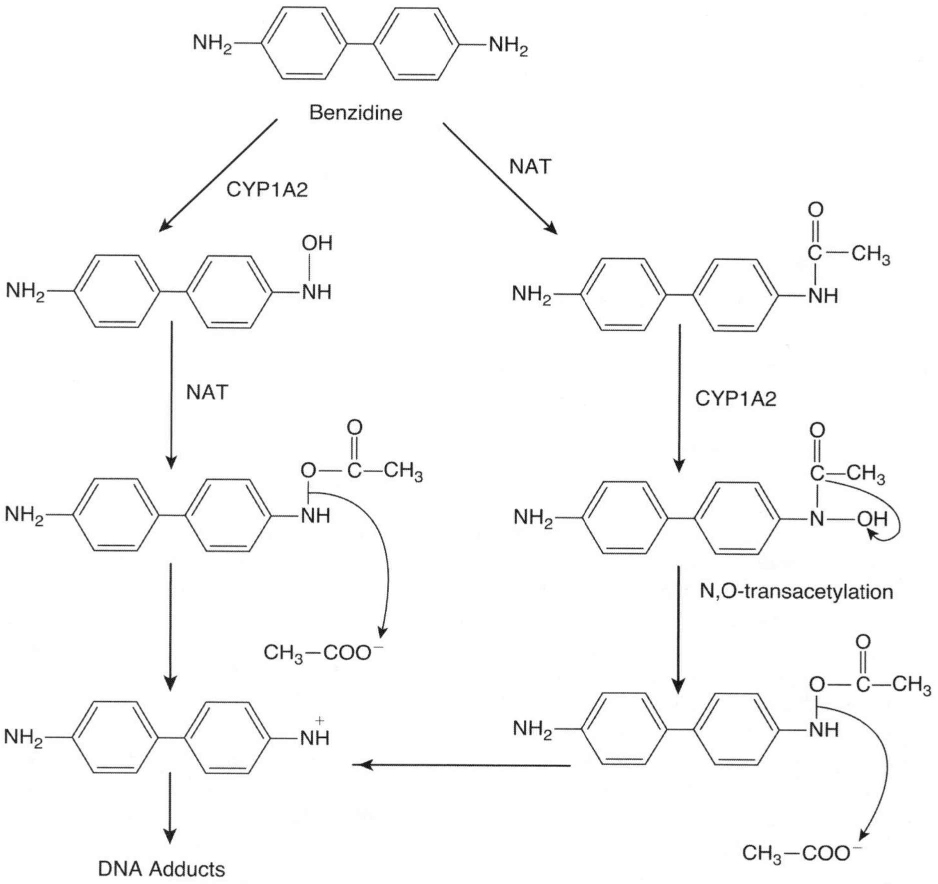

Figure 4.69 Metabolism of the carcinogen benzidine showing oxidation and acetylation. Both routes of acetylation can give rise to a reactive nitrenium ion and DNA adducts. *Abbreviation*: NAT, N-acetyltransferase.

Thus benzidine can be either oxidized by CYP1A2 or N-acetylated and then the hydroxylated product can be O-acetylated. Alternatively the N-acetylated product can be hydroxylated and then undergo an N,O-acetyltransfer to yield the same product. This final product can then lose the *O*-acetyl group in the acidic conditions of the bladder to yield a reactive nitrenium ion, which reacts with DNA (Fig. 4.69).

Both NAT1 and NAT2 N-acetylate benzidine and O-acetylate the *N*-hydroxy metabolite. Because NAT2 and, to a lesser extent, NAT1 both show variation in the human population, this influences susceptibility to the carcinogenic effects of arylamines such as benzidine. With other aromatic amines, such as the heterocyclic amines found as food pyrolysis degradation products, N-acetylation is not favored, N-oxidation being the primary route followed by O-acetylation. This seems to take place in the colon.

The level of NAT therefore influences the cancer risk, with low NAT2 being associated with an increased risk of bladder, liver, breast, and lung cancer, but decreased risk of colon cancer. Conversely, low NAT1 is associated with an increased risk of bladder and colon cancer but decreased lung cancer.

Other examples of the role of the acetylation polymorphism in relation to drug toxicity will be discussed in chapters 5 and 7.

Conjugation with Amino Acids
This is the second type of acylation reaction. However, in this type the xenobiotic itself is activated, and it is therefore a type 2 reaction. Organic acids, either aromatic such as salicylic

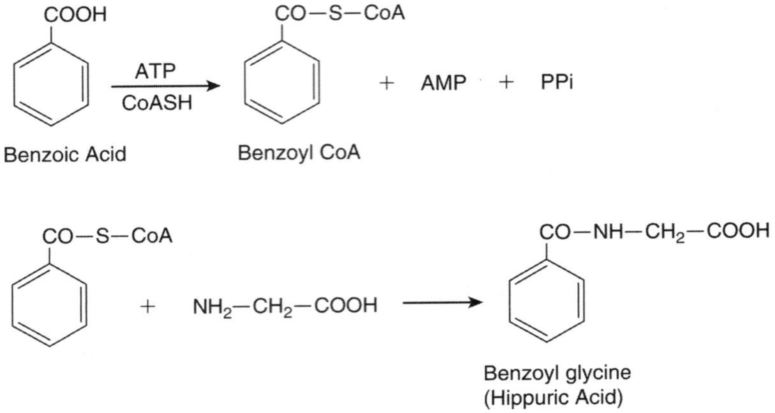

Figure 4.70 Conjugation of benzoic acid with glycine.

acid (see chap. 7) or aliphatic such as 2-methoxyacetic acid, are the usual substrates for this reaction, which involves conjugation with an endogenous amino acid. The particular amino acid depends on the species of animal exposed, although species within a similar evolutionary group tend to use the same amino acid. Glycine is the most commonly used amino acid, but taurine, glutamine, arginine, and ornithine can also be used. The mechanism involves first activation of the xenobiotic carboxylic acid group by reaction with acetyl CoA (Fig. 4.70). This reaction requires ATP and is catalyzed by a ligase or acyl CoA synthetase, which is a mitochondrial enzyme.

The S-CoA derivative then acylates the amino group of the particular amino acid in an analogous way to the acetylation of amine groups described above, yielding a peptide conjugate. This is catalyzed by an amino acid N-acyltransferase, which is located in the mitochondria. Two such enzymes have been purified, each using a different group of CoA derivatives.

Bile acids are also conjugated with amino acids in a similar manner, but different enzymes are involved.

Although bile acid conjugates with amino acids are normally excreted into bile, amino acid conjugates of xenobiotics are usually excreted into urine. Conjugation with endogenous amino acids facilitates urinary excretion because of the organic anion transport systems located in the kidney tubules.

Amino acid conjugation of profen drugs leads to chiral inversion, as discussed in chapter 5.

Generally amino acid conjugation is a detoxication reaction. However, amino acid conjugation with hydroxylamino groups (N-hydroxy) can lead to the formation of reactive nitrenium ions, as already discussed with sulfate conjugation and acetylation. For example, the conjugation of serine with N-hydroxy-4-aminoquinoline-1-oxide (Fig. 4.40 for structure) leads to such a reactive nitrenium ion. This requires the enzyme serine-tRNA synthetase.

Methylation

Amino, hydroxyl, and thiol groups and heterocyclic nitrogen in foreign compounds may undergo methylation *in vivo* (Fig. 4.71) in the same manner as a number of endogenous compounds. The methyl donor is *S*-adenosyl methionine formed from methionine and ATP (chap. 7, Fig. 64), and so again it is a type 1 reaction with an activated donor. The reactions are catalyzed by various **methyltransferase** enzymes, which are mainly cytosolic although some are located in the endoplasmic reticulum. Some of these enzymes are highly specific for endogenous compounds such as the N-methyltransferase, which methylates histamine. However, a nonspecific N-methyltransferase exists in the lung and other tissues, which methylates both endogenous compounds and foreign compounds such as the ring nitrogen in nornicotine and the secondary amine group in desmethylimipramine. Catechol-*O*-methyltransferase is a cytosolic enzyme, which exists in multiple forms and catalyzes the methylation of both endogenous and exogenous catechols and trihydric phenols. A microsomal *O*-methyltransferase

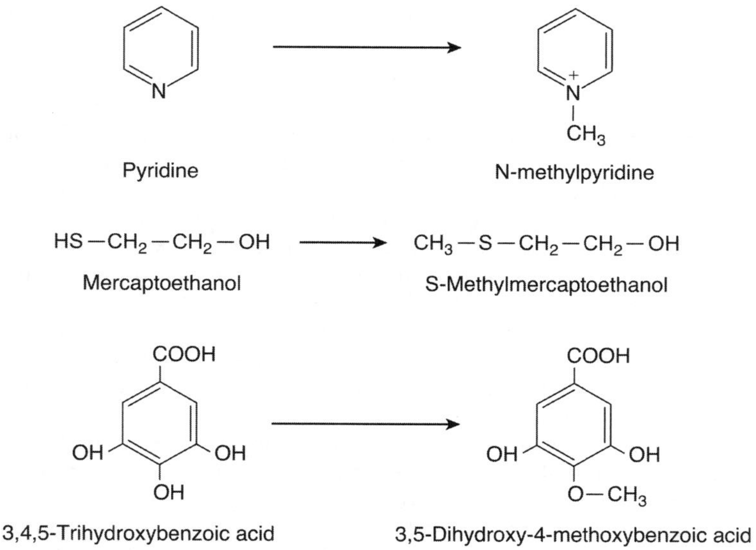

Pyridine → N-methylpyridine

$HS-CH_2-CH_2-OH$ → $CH_3-S-CH_2-CH_2-OH$

Mercaptoethanol → S-Methylmercaptoethanol

3,4,5-Trihydroxybenzoic acid → 3,5-Dihydroxy-4-methoxybenzoic acid

Figure 4.71 N-, O-and S-methylation reactions.

is also found in mammalian liver, which methylates phenols such as paracetamol. Methylation of 4-hydroxyestradiol by catechol-*O*-methyltransferase is believed to be a detoxication reaction.

Thiol *S*-methyltransferase is a microsomal enzyme found in the liver, lung, and kidney, which will catalyze the methylation of a wide variety of foreign compounds. A number of sulfur-containing drugs such as disulfiram, *D*-penicillamine, and 6-mercaptopurine are methylated in humans by thiol *S*-methyltransferase or thiopurine methyltransferase. Methylation of the anticancer drug thiopurine by the latter enzyme is a detoxication reaction, reducing the myelotoxicity of the drug (see Fig. 4.18, which shows the reverse of this reaction).

Hydrogen sulfide, H_2S, is detoxified by methylation first to methanthiol (CH_3SH), which is highly toxic, but is then further methylated to dimethylsulfide (CH_3-S-CH_3). The thiol products of β-lyase may also be methylated by this enzyme.

Metals can also undergo methylation reactions.

For example, **mercury** is methylated by microorganisms:

$$Hg^{2+} \rightarrow CH_3Hg^+ \rightarrow (CH_3)_2Hg$$

In this reaction, the physicochemical properties, toxicity, and environmental behavior are altered. The products methylmercury and dimethylmercury are more lipophilic, neurotoxic, and persistent in the environment than elemental or inorganic mercury (see chap. 7).

Other metals such as **tin**, **lead**, and **thallium** may also be methylated as can the elements **arsenic** and **selenium**. Thus, the methylation reaction, like acetylation, tends to reduce the water solubility of a foreign compound rather than increase it.

4.5 GENERAL SUMMARY POINTS

The foregoing discussion has by no means covered all the possible metabolic transformations that a foreign compound may undergo, and for more information, the reader should consult the Bibliography. However, some general points should be made at this stage.

1. *Generally* the enzymes involved in xenobiotic metabolism are less specific than the enzymes involved in intermediary, endogenous substrate metabolism.
2. However, apart from absolute specificity, foreign compounds may also be substrates for enzymes involved in endogenous pathways, often with toxicological consequences. Thus, for example, with VPA (see above), fluoroacetate, and galactosamine (see chap. 7) involvement in endogenous metabolic pathways is a crucial aspect of

the toxicity. The chemical structure and physicochemical characteristics will determine whether this occurs or not.

3. Foreign compounds do not necessarily only undergo one metabolic transformation. It is obvious from the preceding discussion that there are many possible routes of metabolism for many foreign compounds. What determines which will prevail is not always clear. However, in many cases, several routes will operate at once giving rise to a variety of metabolites each with different biological activity. The balance and competition between these various routes will therefore be important in determining toxicity. This is well illustrated by bromobenzene and isoniazid (see below and chap. 7).

4. Metabolism may involve many sequential steps, not just one phase 1 followed by one phase 2 reaction. Phase 1 reactions can sometimes follow phase 2 reactions, one molecule can undergo several phase 1 reactions, and cyclical or reversible metabolic schemes may operate. Thus, further metabolic transformations, sometimes termed phase 3 reactions, can convert a detoxified metabolite into a toxic product (see hexachlorobutadiene and hydroquinone).

     An example is the metabolism of glutathione conjugates to toxic thiols previously mentioned. Further metabolism may occur as a result of biliary excretion, for example. Thus glucuronic acid conjugates excreted via the bile into the intestine can be catabolized by bacterial β-**glucuronidase** to the aglycone, which may be either reabsorbed or further metabolized by the gut flora and then reabsorbed. Similarly, glutathione conjugates can follow the same type of pathway by the action of intestinal and bacterial enzymes to yield thiol conjugates as already described.

5. The rates of the various reactions will vary. This may be due to the availability of cofactors, concentration of enzyme in a particular tissue, competition with other, possibly endogenous, substrates or to intrinsic factors within the enzymes involved. This variation in rates will clearly affect the concentrations of metabolites in tissues, and the half-life of parent compound and metabolites. It may lead to accumulation of intermediate metabolites.

6. Metabolism of foreign compounds is not necessarily detoxication. This has already been indicated in examples and will become more apparent later in this book. This may involve activation by a phase 1 or phase 2 pathway or transport to a particular site followed by metabolism. Phase 1 reactions, particularly oxidation, can be responsible for the production of reactive intermediates such as epoxides, quinones, hydroxylamines, and free radicals, which lead to toxicity. However, phase 2 reactions can also result in toxicity in some cases.

     Thus, sulfate conjugation and acetylation may be involved in the metabolic activation of *N*-hydroxy aromatic amines, glutathione conjugation may be important in the nephrotoxicity of compounds, methylation in metal toxicity, glucuronidation in the carcinogenicity of β-naphthylamine and 3, 2′-dimethyl-4-aminobiphenyl.

4.6 CONTROL OF METABOLISM

The metabolic pathways, which have been discussed in this chapter are *influenced* by many factors, some of which are discussed in chapter 5. Such factors may have an effect on the toxicity of a compound as indicated below, and these are discussed in more detail in chapter 5.

But it is important to appreciate that the metabolism of foreign compounds is not completely separate from intermediary metabolism, but linked to it. Consequently, this will exert a controlling influence on the metabolism of foreign compounds. Some of the important factors controlling xenobiotic metabolism are

1. the availability of cofactors such as NADPH,
2. the availability of co-substrates such as oxygen and glutathione, and
3. the level of particular enzymes.

The NADPH level is clearly important for phase 1 reactions, yet many biochemical processes, such as fatty acid biosynthesis, use this coenzyme. It is derived from either the pentose phosphate shunt or isocitrate dehydrogenase. Consequently, the *overall metabolic condition* of the organism will have an influence on the NADPH supply as there will be competition for its use. This may be important in paraquat toxicity where NADPH may be depleted (see chap. 7).

Oxygen is normally readily available to all reasonably well-perfused tissues, but deep inside organs such as the liver, especially the centrilobular area (see chap. 6), there will be a reduction in the oxygen concentration. This is clearly important when both oxidative and reductive pathways are available for a particular substrate. Therefore, as conditions in a particular tissue become more anaerobic, reductive pathways will become more important. This is well illustrated by the metabolism of halothane where, in the rat, hypoxia will increase reductive metabolism and hepatotoxicity (see chap. 7). Glutathione is an extremely important cofactor, involved in both protection and conjugation. It may be depleted by both of these processes, or under certain circumstances, such as hereditary glucose-6-phosphate deficiency in man, supply may be reduced (see chap. 5). This will clearly influence toxicity, and there are a number of examples discussed in chapter 7 in which it is important.

Other co-substrates possibly limited in supply are inorganic sulfate and glycine for conjugation; these may be important factors in paracetamol hepatotoxicity and salicylate poisoning, respectively (chap. 7).

The level of a particular enzyme involved in xenobiotic metabolism can obviously affect the extent of metabolism by that enzyme. Again, competition may play a part if endogenous and exogenous substrates are both metabolized by an enzyme, as is the case with some of the forms of cytochromes P-450, which metabolize steroids, or NADPH cytochrome P-450 reductase and cytochrome b_5 reductase, which are also involved in heme catabolism and fatty acid metabolism, respectively.

The synthesis and degradation of enzymes such as cytochromes P-450 are therefore important factors and, as discussed in detail in the next chapter, can be modified by exogenous factors.

4.7   TOXICATION VS. DETOXICATION

Although the biotransformation of foreign compounds is often regarded as a *detoxication* process, this is not always the case. Metabolites or intermediates may sometimes be produced, which are more toxic than the parent compound. These may be the result of phase 1, 2, or 3 reactions, although phase 1 reactions are the most commonly involved. The intermediates or metabolites responsible for the toxicity may be chemically reactive or stable.

When the metabolic process produces a metabolite, which is chemically reactive, this process is known as **metabolic activation** or **bioactivation**. The exact chemical reactivity might indeed be crucial, and there may be an optimum level for this reactivity. Thus, metabolism can underlie the toxicity of a compound, and there will be many examples given throughout this book. However, a discussion of the general principles is appropriate here.

It is often the case that there are several metabolic pathways available for a foreign compound. Some of these pathways could be detoxication pathways, while others might lead to toxicity (Fig. 4.72). When this situation arises, there is the possibility of competition between pathways and, as indicated above, various factors can influence the balance between them. Furthermore, biological systems often have protective mechanisms for the removal of such reactive intermediates. However, these systems may sometimes be overloaded, suffer failure, or be absent in some tissues.

Thus the balance between detoxication and toxication will be affected by many factors such as

1.   relative rates of the toxication and detoxication pathways: these will be influenced by the availability of enzymes and the kinetic parameters of the enzymes involved,
2.   availability of cofactors,
3.   availability of protective agents,
4.   dose and saturability of metabolic pathways,
5.   genetic variation in enzymes catalyzing various pathways,
6.   induction or inhibition of the enzymes involved,
7.   species or strain of animal,

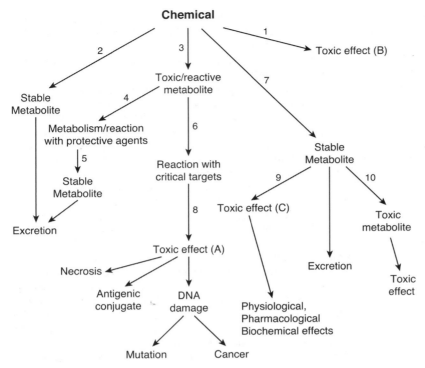

Figure 4.72 The various possible consequences of metabolism of a foreign compound. The compound may undergo detoxication (2); metabolic activation (3), which leads to interaction with critical targets (6) and may cause toxic effects (8; **A**). Alternatively, the parent compound might cause a direct toxic effect (1; **B**). Formation of a stable metabolite (7) could cause a toxic effect (9; **C**). The stable metabolite could be further metabolized to a toxic metabolite (10) responsible for a different toxic effect to C. The reactive metabolite may be detoxified (4/5). The types of toxic effect caused by the metabolites would be one or more of the various types shown.

8. tissue differences in enzyme and isoenzyme patterns,
9. diet,
10. age,
11. disease, and
12. sex.

Because of such factors, differences in enzyme activity or level or availability of cofactors and protective agents will change the balance between toxication and detoxication within an organism between organs or tissues. This leads to particular organs being targets for toxicity, while others are spared. Similarly, some species or individuals will be more or less susceptible or even resistant because of similar variation.

Of course, one particularly important consideration is dose, as this can dramatically change the balance as a result of saturation or depletion of cofactors.

It will become clear from many of the examples in this book that in many cases, especially with drugs, it is not the parent chemical, which is responsible for the toxicity but a metabolic product. This metabolic product may react chemically with an endogenous constituent. Furthermore, toxicity is often a result of failure in detoxication, allowing a chemically reactive substance to wreak damage in the cell.

If detoxication does not remove them, reactive metabolites to which we are exposed can combine covalently with tissue macromolecules of many kinds. Alternatively, a chemical or its metabolite can start a process such as oxidative stress by reacting with oxygen or by changing an endogenous substance such that it in turn will react with oxygen.

Often but by no means always, reactive metabolites are produced by phase 1 reactions, commonly oxidation reactions. Therefore, cytochrome P-450 is regularly a culprit, and we can identify certain reactions, which are often involved. Reactive metabolites are often electrophiles, sometimes free radicals and occasionally nucleophiles.

For example, epoxides, both aliphatic and aromatic, being chemically unstable are usually *potentially* reactive metabolites. Quinones and quinoneimines similarly are known to be able to react chemically with endogenous constituents of the cell.

Other examples of electrophilic toxic chemicals are aldehydes and ketones, especially unsaturated ones, acyl halides and cyanates.

It should be noted, however, that reactive metabolites are not necessarily, *indiscriminately* reactive with all cellular constituents. Indeed to be specifically toxic, they are more likely to be specifically reactive with particular groups on proteins, nucleic acids, or fatty acids.

For example, hexane 2,5-dione, a diketone metabolite of the solvent hexane, is reactive toward particular groups in proteins, the lysine amino groups, but is not especially chemically reactive in general terms (see chap. 7).

Another example is paraquat, which can accept an electron from donors such as NADPH, becoming a *stable* free radical, which is not chemically reactive. However, it will generate reactive oxygen species by donating an electron to available oxygen (see chap. 7).

Highly reactive metabolites, which will react indiscriminately with many molecules in the cell, will be likely to be detoxified either by reaction with abundant molecules such as water or glutathione, possibly the nearest neighbor. They will not be likely to reach a specific target such as DNA.

It should also be appreciated that even though a reactive chemical reacts covalently with a protein such as an enzyme, this may not lead to toxicity. This could be due to the protein not having a critical function or there being significant reserves of activity for an enzyme so that even if it is inhibited, sufficient activity remains (e.g., organophosphate inhibition of cholinesterases, see chap. 7).

Stable metabolites can also sometimes be responsible for toxicity via interaction with critical macromolecules or other mechanisms (see chaps. 6 and 7).

Various enzymes can produce reactive intermediates by metabolism, especially cytochrome P-450, but also peroxidases, FMO, prostaglandin synthetase, MAO.

Such enzyme-mediated reactions can produce different reactive intermediates, namely,

1. an electrophile
2. a nucleophile
3. a free radical
4. a redox reagent

Reactive metabolites can react with important macromolecules including enzymes, DNA, or lipids. As indicated in Figure 4.72, this could lead to necrosis, an immunological reaction, a mutation, or cancer.

4.7.1 Electrophiles

Electrophiles, the most common type of reactive intermediate, are molecules with an electron-deficient atom, so having a full or partial positive charge.

They seek to share electrons with other molecules (e.g., nucleophiles) There are many types of electrophile as indicated in Table 4.5.

A number of these examples will be discussed in detail in chapter 7. Electrophiles are often formed by oxidative metabolism, catalyzed by cytochrome P-450, peroxidases or alcohol dehydrogenase. These can give rise to epoxides, quinones and aldehydes, ketones,

Table 4.5 Formation of Reactive Electrophiles by Metabolic Toxication of Drugs

Electrophile	Parent compound	Enzymes catalyzing toxication	Toxic effect
Acetaldehyde	Ethanol	ADH	Hepatic damage
Acyl glucuronide	Diclofenac	Glucuronosyl transferase	Hepatic toxicity
DES-4-4'quinone	Diethylstilboestrol	Peroxidases	Carcinogenesis
Nitroso sulfamethoxazole	Sulfamethoxazole	Cyt. P-450	Immunotoxicity
N-acetyl-p-benzoquinoneimine	Paracetamol	Cyt. P-450	Hepatic necrosis
Trifluoroacetylchloride	Halothane	Cyt. P-450	Immune mediated hepatic damage
Reactive carbocation	Tamoxifen	Cyt. P-450	Cancer

Abbreviations: ADH, alcohol dehydrogenase; Cyt. P-450, cytochrome P-450.

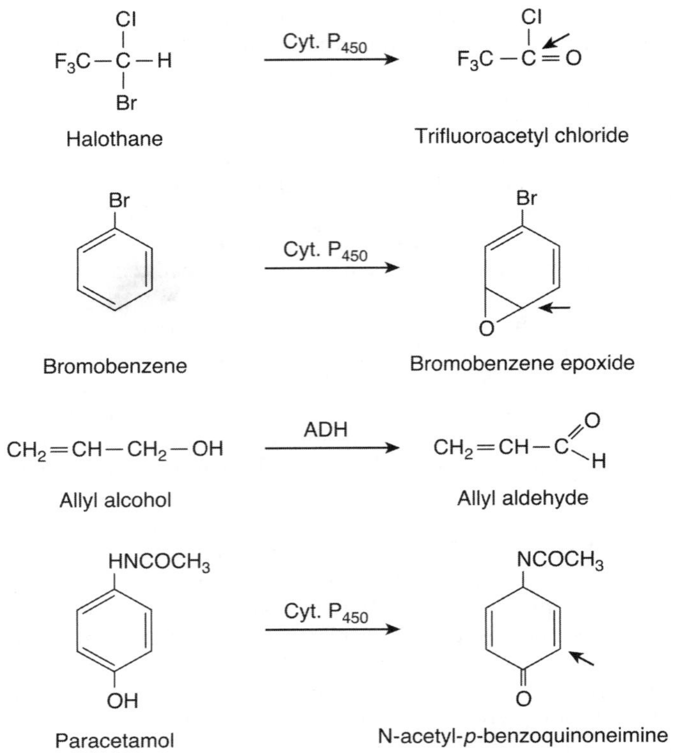

Figure 4.73 Examples of electrophilic metabolites. The small arrow indicates the electrophilic atom.

phosphonates, nitroso compounds, and acyl halides. Oxidation may lead to conjugated double bonds being introduced.

This is due to the fact that the insertion of an oxygen atom, being electron withdrawing, may cause the carbon atom to which it is attached to become positive (Fig. 4.73).

However, phase 2 conjugation can also on occasion produce electrophilic intermediates/metabolites, such as the sulfate conjugation of hydroxytamoxifen (see chap. 7), and the formation of acyl glucuronides with acidic drugs such as diclofenac (Fig. 4.54).

With tamoxifen, loss of the sulfate group cleaves the bond and yields a benzylic carbocation. In the case of diclofenac, the carbonyl carbon is reactive and can react with nucleophiles (Fig. 4.54), but the conjugate can also rearrange and yield a reactive, electrophilic aldehyde.

In other cases, conjugation of an aromatic hydroxylamino group yields a positively charged nitrenium ion. Reactive intermediate metabolites and positively charged electrophilies can also be produced by reduction. For example, metals salts such as chromate can be reduced (to Cr^{3+}) and nitro groups can be reduced to nitroso groups.

Electrophiles can be hard or soft. Thus hard electrophiles react with nucleophilic sites in molecules such as O, N, C in nucleic acids or the S in methionine in proteins. Hard electrophiles are typically genotoxic such as the benzo(a)pyrene diol epoxide (see chap. 7).

In contrast, soft electrophiles react with nucleophilic SH groups in GSH and proteins. Soft electrophiles are typically cytotoxic, such as the metabolite of paracetamol, N-acetyl-p-benzoquinoneimine (Fig. 4.73) (see also chap. 7). So reactivity with GSH depends on the hardness/softness of the electrophile.

4.7.2 Nucleophiles

These are molecules, which have an electron-rich atom. They are therefore electron donating and tend to react with electrophiles.

These are formed much less commonly than electrophiles. An example is cyanide (CN^-), which can be produced by metabolism of acrylonitrile, or of the drug sodium nitroprusside. Carbon monoxide is another example, which can be produced by metabolism of methylene chloride.

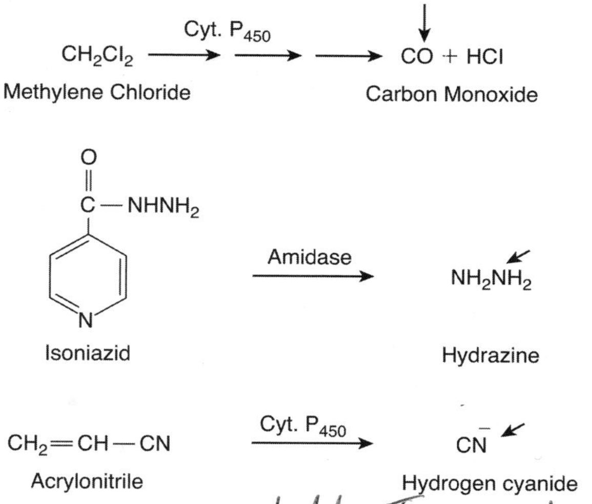

$$CH_2Cl_2 \xrightarrow{\text{Cyt. P}_{450}} \longrightarrow \longrightarrow CO + HCl$$

Methylene Chloride   Carbon Monoxide

Isoniazid $\xrightarrow{\text{Amidase}}$ NH_2NH_2

Isoniazid   Hydrazine

$$CH_2{=}CH{-}CN \xrightarrow{\text{Cyt. P}_{450}} CN^-$$

Acrylonitrile   Hydrogen cyanide

Figure 4.74 Examples of nucleophilic metabolites. The small arrow indicates the nucleophilic atom.

→ highly toxic & dangerously unstable

A target for both these substances (CO and CN) is the ferric cationic form of iron found in cytochromes and hemoglobin (see chap. 7).

Hydrazines may be nucleophiles such as when they interact with aldehyde and keto groups to form hydrazones. This is the basis for the inhibition of enzymes such as transaminases, which rely on pyridoxal phosphate as a coenzyme. Mono-substituted hydrazines can be formed as metabolites when azo groups are reduced, dialkylated hydrazines are dealkylated or hydrazides are hydrolysed.

Thus toxic metabolites of isoniazid are acetylhydrazine and hydrazine. The formation of some of these nucleophiles is shown in Figure 4.74.

It should be noted that acetylhydrazine can also be further metabolized to an electrophilic reactive intermediate by cytochrome P-450. This will be discussed in detail in chapter 7.

4.7.3 Free radicals

These are molecules or fragments, which have one or more unpaired electron. They may be charged or uncharged. Because the electron is unpaired, free radicals are generally, but not exclusively, very reactive. They are formed in one of three ways:

1. By accepting an electron
2. By losing an electron
3. By hemolytic cleavage of a covalent bond

1. The drugs doxorubicin and nitrofurantoin and the herbicide paraquat can acquire electrons from a source such as a reductase enzyme (Fig. 4.75).

   However, even if the free radical is stable and is not especially reactive, as in these cases, if that electron can be donated to another molecule, then a different but possibly very reactive radical can be generated.

   For example, these chemicals all donate electrons to oxygen, producing highly reactive and so potentially toxic superoxide anion radical (see chap. 7).

   With paraquat, doxorubicin, and nitrofurantoin, once the electron is donated to oxygen, another can be acquired, and so the process can continue as long as there is a source of electrons and the chemicals are present. This is called "redox cycling" (see chap. 6).

2. A number of nucleophilic molecules can lose electrons in enzyme-catalyzed reactions (e.g., peroxidase). Thus hydrazines, amines, hydroquinones, phenothiazines, thiols, and aminophenols can form free radicals by this mechanism.

   Hydroquinones, for example, can form semiquinone radicals and then quinones in a two-step oxidation reaction. The semiquinone radical is potentially reactive but so also is the quinone, which can easily form an electrophile. Furthermore, quinones can also take part in redox cycling by accepting electrons and so generate reactive oxygen species.

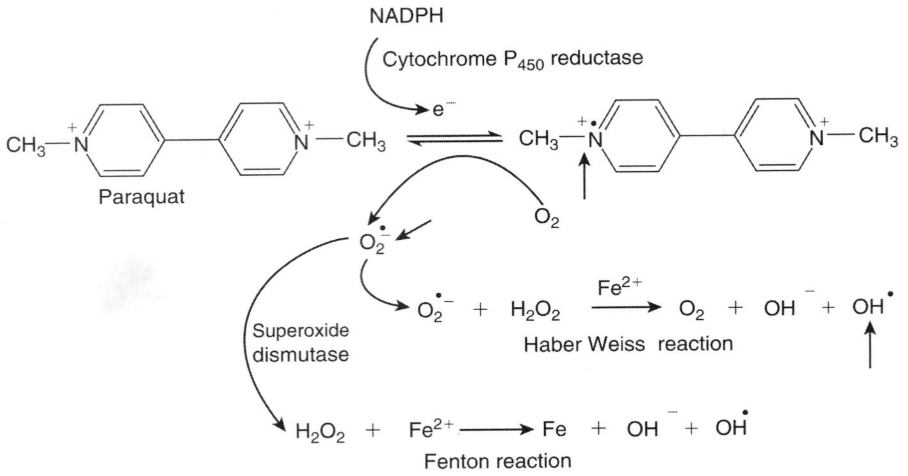

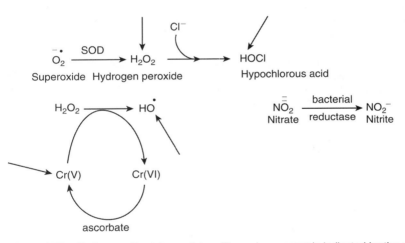

Figure 4.75 Formation of free radicals as reactive intermediates responsible for toxicity. The example shown is paraquat, which acquires an electron and in turn produces superoxide anion radical and hydroxyl radical. The small arrow indicates the free radical intermediate (the small dot represents the extra electron).

Figure 4.76 Redox reactive intermediates. The redox reagent is indicated by the small arrow.

Similarly, aminophenols can be oxidized to semiquinones and quinoneimines. This can be carried out by oxyhemoglobin (Hb-FeII-O_2). However, the reactive products, semiquinones and quinoneimines can oxidize the hemoglobin to ferrihemoglobin (Hb-FeIII), which leads to toxicity as it cannot carry oxygen.

3. Homolytic cleavage of covalent bonds is an alternative means of generating free radicals. This may be assisted by the addition of an electron as in the case of carbon tetrachloride activation. The electron may be donated by cytochrome P-450, allowing the loss of chloride ion and the production of a trichloromethyl radical (Fig. 4.7). This can initiate other radical reactions by reacting with oxygen or unsaturated lipids.

4.7.4 Redox Reagent

These are atoms or molecules able to produce reactive species [such as reactive oxygen species (ROS)] and may be oxidizing agents. Examples are hydrogen peroxide, hypochlorous acid, Cr(V) (Fig. 4.76), the drug metabolites dapsone hydroxylamine and 5-hydroxyprimaquine (see chap. 7). Hypochlorous acid can be produced *in vivo* by neutrophils as part of the oxidative burst. It has been suggested as an important alternative route for the oxidation of drugs such as hydralazine (see chap. 7) and may be involved in the toxicity. Dapsone and primaquine, drugs used for the treatment of infectious diseases, both undergo hydroxylation to products, which can oxidize hemoglobin to methemoglobin and cause hemolytic anemia (Fig. 4.7).

Reactive intermediates are of variable reactivity, but it does not follow that the most reactive are the most toxic.

Ultra Short-Lived Reactive Intermediates
These generally bind to enzyme that produces them and destroy or inhibit it. For example, olefins and acetylenes bind to nitrogen atoms in the pyrrole part of cytochrome P-450. This may lead to the release of the Fe and destruction of the enzyme. The intermediate can be detected as bound to the enzyme protein or active site.

For example, the reactive metabolites of the drug allobarbital (epoxide), the industrial chemical vinyl chloride (epoxide), and the solvent carbon tetrachloride (CCl_3 radical) all damage cytochrome P-450 by this type of mechanism (see chap. 7).

Short-Lived Reactive Metabolites/Intermediates
These can diffuse away from the enzyme where they are produced and interact with other constituents of the cell.

Such intermediates may also travel to adjacent cells. These intermediates can be detected as they are covalently bound to cellular constituents. Examples include the reactive metabolite produced from paracetamol (NAPQUI) and bromobenzene epoxide (see chap. 7).

Longer-Lived Metabolites/Intermediates
These can be transported to neighboring cells and tissues. For example, they may be produced in the liver and transported to the kidney, bladder, or lung. For example, the glutathione conjugates of hexachlorobutadiene (see chap. 7) and methylisocyanate.

4.7.5 Toxicity and Reactivity of Reactive Intermediates
Reactive intermediates can be toxic in a variety of ways:

1. Covalent binding to protein
2. Oxidation of protein sulfydryls
3. Destruction of enzyme-active site
4. Initiation of lipid peroxidation
5. Oxidation of proteins
6. Formation of DNA adducts
7. Formation of immunogenic adducts
8. Depletion of glutathione
9. Depletion or modification of a cofactor
10. Inhibition of enzyme or other protein function

Clearly reactive intermediates are of variable reactivity, but it does not follow that the most reactive ones are the most toxic.

Many of these factors will operate independently, and toxicity will often only result when several conditions apply together. These various factors are discussed in more detail in chapter 5 and are also apparent in some of the examples in chapter 7.

SUMMARY

Metabolism is the (bio)transformation of a molecule into one or more different chemical entities. This is catalyzed by enzymes present in many tissues but predominantly the liver.

The metabolism of a xenobiotic will change its physicochemical properties, usually increasing water solubility, size, and molecular weight and therefore increasing excretion. The consequences of metabolism are therefore a reduction in the biological half-life and hence exposure, a change in the nature and duration of biological activity, and a reduction in accumulation.

Toxicity is often but not always reduced by metabolism.

The biotransformation of a chemical is determined by its structure, physicochemical properties, and enzymes in the tissues exposed. Biotransformation can be divided into phase 1 (oxidation, reduction, and hydrolysis) and phase 2 (conjugation). Further metabolism of conjugates has been termed phase 3.

Phase 1

The most important enzyme involved in biotransformation is cytochrome P-450, which catalyzes many phase 1 reactions. This enzyme is located primarily in the SER (microsomal fraction) of the cell and is especially abundant in liver cells. Cytochrome P-450 primarily catalyzes oxidation reactions and consists of many isoforms (isozymes). These isoenzymes have overlapping substrate specificities. The most important subfamily in humans is CYP3A4, although there is considerable variation in CYP3A4 expression between individuals.

There are other components of the P-450 system, namely, NADPH cytochrome P-450 reductase, NADH cytochrome b_5 reductase, and cytochrome b_5.

Oxidation

Major oxidations are aromatic, aliphatic, alicyclic, heterocyclic, N-oxidation, S-oxidation, dealkylation. Other enzymes also catalyze phase 1 reactions: microsomal flavin monooxygenases, amine oxidases, peroxidases, and alcohol dehydrogenase.

Reduction

Major reduction reactions are azo reduction and nitro reduction. The enzymes (reductases) are found in gut flora but also mammalian tissues. Reduction catalyzed by cytochrome P-450 can occur (e.g., dehalogenation). DT diaphorase carries out two-electron reductions.

Hydrolysis

Major hydrolysis reactions are ester and amide hydrolysis. These are catalyzed by a group of enzymes with overlapping substrate specificity and activity. Hydrazides can also undergo hydrolysis. Some of the newer drugs such as hormones, growth factors, and cytokines now being produced are peptides, and certain toxins are also peptides or proteins, so the role of peptidases may be important.

Hydration

Epoxides may undergo addition of water.

Phase 2

Conjugation involves addition of an endogenous moiety to a foreign molecule, which may be a product of a phase 1 reaction. Major phase 2 routes: conjugation with glucuronic acid, sulfate, glutathione; amino acids; acetylation; methylation. Enzymes involved are transferases except in the case of amino acid conjugation where the first step is catalyzed by an acyl CoA synthetase, then a transferase is involved.

With the exception of acetylation and methylation, conjugation renders the whole molecule more polar and hydrophilic (less lipid soluble). This facilitates excretion and reduces the likelihood of toxicity.

Control of metabolism depends on various factors including the availability of cofactors (e.g., NADPH), co-substrates (e.g., oxygen), and the level and activity of particular enzymes.

Toxication versus Detoxication

Metabolism may be detoxication in many cases, but sometimes it produces a more toxic or reactive metabolite.

Enzymes, especially cytochrome P-450, can produce reactive intermediates by metabolism, namely,

1. An electrophile
2. A nucleophile
3. A free radical
4. A redox reagent

Such reactive metabolites can react with important macromolecules including enzymes, DNA, or lipids.

However, there may be more than one metabolic pathway for a particular chemical; one or more being detoxication pathways, another pathway producing a toxic metabolite. Therefore, the balance between these becomes crucial and the factors, which affect this balance very important.

REVIEW QUESTIONS

1. Write short notes on three of the following:
   a. cytochrome P-450
   b. glutathione
   c. N-acetyltransferase
   d. glucuronic acid
2. List three consequences of metabolism.
3. Where in the cell is cytochrome P-450 mostly located?
4. Give two examples of phase 1 metabolic transformations and two examples of phase 2 transformations.
5. Which of the following are required by cytochrome P-450:
   Oxygen
   NADPH
   Zinc
   Water
   NADH cytochrome P-450 reductase.
6. How many different phase 1 metabolites can benzene form?
7. Is N-demethylation a result of carbon oxidation or nitrogen oxidation?
8. What cofactor(s) is (are) required for sulfate conjugation?
9. Is glutathione conjugation the result of a chemical reaction or an enzyme-mediated reaction?
10. Which amino acid is commonly used for conjugation of aromatic acids?
11. What biotransformation might inorganic mercury undergo in a biological system?
12. Which is the most important isoform of cytochrome P-450 in humans?
13. What are the types of reactive intermediates, which can be produced by enzymes?

REFERENCES

1. Parkinson A. Biotransformation of xenobiotics. In: Klaassen CD, ed. Cassarett and Doull's Toxicology. 6th ed. New York: McGraw Hill, 2001.
2. Weinstein L. In: L.S. Goodman and A. Gilman, eds. The Pharmacological Basis of Therapeutics. New York: Macmillan, 1970.
3. Makris TM, Davydov R, Denisov IG, et al. Mechanistic enzymology of oxygen activation by the cytochromes p450. Drug Metab Rev 2002; 34:691–708.
4. Nebert DW, Gonzales F. The P450 gene superfamily. In: Ruckpaul K, Rein H, ed. Frontiers of Biotransformation. Vol. 2. Principles, Mechanisms and Biological Consequences of Induction. London: Taylor & Francis, 1990.
5. Nebert DW. The P450 gene superfamily: recommended nomenclature and updates. Nelson DR, Koymans L, Kamataki T, et al. The P450 superfamily: update on new sequences, gene mapping, accession numbers, and nomenclature. Pharmacogenetics 1996; 6:1–42. Previous updates(1987/1989/ 1991/1993) in DNA Cell Biol. 6, 1–11; 8, 1–13; 10; 1–14; 12, 1–51.
6. Gibson GG, Skett P. Introduction to Drug Metabolism. 3rd ed. UK: Nelson Thornes, 2001.

BIBLIOGRAPHY

General Drug Metabolism

Alvares AP, Pratt WB. Pathways of drug metabolism. In: Pratt WB, Taylor P, eds. Principles of Drug Action. New York: Churchill Livingstone, 1990.

Armstrong NR. Structure, catalytic mechanism and evolution of the glutathione transferases. Chem Res Toxicol 1997; 10:2.

Beedham C. Molybdenum hydroxylases. In: Gorrod JW, Oelschlager H, Caldwell J, eds. Metabolism of Xenobiotics. London: Taylor & Francis, 1988.

Benedetti MS, Dostert P. Contribution of amine oxidases to the metabolism of xenobiotics. Drug Metab Rev 1994; 26:507.

Bock. UDP-Glucuronosyltransferases. In: Ioannides C, ed. Enzyme Systems that Metabolise Drugs and Other Xenobiotics. Chichester: John Wiley and Sons, 281–318.

Burk O, Wojnowski L. Cytochromes P450 3A and their regulation. Naunyn Schmiedebergs Arch Pharmacol 2004; 369:105–124.

Cashman JR. Flavin-containing monoxygenase. In: Ioannides C, ed. Enzyme Systems that Metabolise Drugs and Other Xenobiotics. Chichester: John Wiley and Sons, 2002:67–93.

Cashman JR, Zhang J. Human flavin-containing monoxygenases. Annu Rev Pharmacol Toxicol 2006; 46:41–64.

Coleman MD. Human Drug Metabolism. Chapter 3. UK: Wiley, 2005.

Coon MJ. Cytochrome P450: natures most versatile catalyst. Annu Rev Pharmacol Toxicol 2005; 45:1–25.

Degen GH, Vogel C, Abel J. Prostaglandin synthetases. In: Ioannides C, ed. Enzyme Systems that Metabolise Drugs and Other Xenobiotics. Chichester: John Wiley and Sons, 2002:188–229.

Ding X, Kaminsky LS. Human extrahepatic cytochromes P450: function in xenobiotic metabolism and tissue-selective chemical toxicity in the respiratory and gastrointestinal tracts. Annu Rev Pharmacol Toxicol 2003; 43:149–173.

Duffel MW. Sulphotransferases. In: Guengerich FP, ed. Biotransformation. Vol. 3. New York: Elsevier, 1997:365–383.

Eling T, Boyd J, Reed G, et al. Xenobiotic metabolism by prostaglandin endoperoxide synthetase. Drug Metab Rev 1983; 14:1023–1053.

Estabrook R, Masters BSS, Prough RA, et al. Cytochromes P450. FASEB J 1996; 10(2):202.

Gonzalez FJ. The molecular biology of cytochrome P-450s. Pharmacol Rev 1988; 40:243.

Gonzalez FJ, Matsunaga T, Nagata K. Structure and regulation of P-450s in the rat P450IIA gene subfamily. Drug Metab Rev 1989; 20:827–837.

Goodwin B, Redinbo, MR, Kliewer SA. Regulation of CYP3A gene transcription by the pregnane X receptor. Annu Rev Pharmacol Toxicol 2002; 42:1–23.

Gram TE, ed. Extrahepatic Metabolism of Drugs and Other Foreign Compounds. Jamaica, New York: Spectrum Publications, 1980.

(This text contains chapters on various aspects of drug metabolism, both phase 1 and phase 2.)

Guengerich FP. Catalytic selectivity of human cytochrome P450 enzymes: relevance to drug metabolism and toxicity. Toxicol Lett 1994; 70:133–138.

Guerengerich JP. Common and uncommon cytochrome P450 reactions related to metabolism and chemical toxicity. Chem Res Toxicol 2001; 14:611–650.

Guengerich FP. Cytochrome P450: what have we learned and what are the future issues? Drug Metab Rev 2004; 36:159–197.

Hathway DE, Brown SS, Chasseaud LF, et al. (reporters) Foreign Compound Metabolism in Mammals. Vols. 1–6. London: The Chemical Society, (1970–1981).

Hawkins DR, ed. Biotransformations. Vols. 1–7. Cambridge: Royal Society of Chemistry, (1988–1996).

Hayes JD, Flanagan JU, Jowsey IR. Glutathione transferases. Annu Rev Pharmacol Toxicol 2005; 45:51–88.

Hodgson E, Smart RC, eds. Introduction to Biochemical Toxicology. 3rd ed. Wiley Interscience, 2001, Chapters 5 and 6.

Houston JB. Utility of *in vitro* drug metabolism data in predicting *in vivo* metabolic clearance. Biochem Pharmacol 1994; 47:1469–1479.

Jakoby WB, Ziegler DM. The enzymes of detoxication. J Biol Chem 1990; 265:20715–20718.

Kaminsky LS, Zhang QY. The small intestine as a xenobiotic-metabolising organ. Drug Metab Dispos 2003; 31:1520–1525.

Kao J, Carver MP. Cutaneous metabolism of xenobiotics. Drug Metab Rev 1990; 22:363.

deBethizy JD, Hayes, JR. Metaolism: A determinant of toxicity. In: Hayes AW, ed. Principles and Methods of Toxicology. 4th ed. Philadelphia: Taylor and Francis, 2001.

Ketterer B, Taylor JB. Glutathione transferases. In: Ruckpaul K, Rein H, eds. Frontiers of Biotransformation. Vol. II. Principles, Mechanisms and Biological Consequences of Induction. London: Taylor & Francis, 1990.

Krishna DR, Klotz U. Extrahepatic metabolism of drugs in humans. Clin Pharmacokinet 1994; 26:144.

Leblanc GA, Dauterman WC. Conjugation and elimination of toxicants. In: Hodgson E, Smart RC, eds. Introduction to Biochemical Toxicology. 3rd ed. New York: Wiley Interscience, 2001).

Mackenzie PI. Structure and regulation of UDP glucuronosyltransferase. In: Ruckpaul K, Rein H, eds. Frontiers of Biotransformation. Vol. II. Principles, Mechanisms and Biological Consequences of Induction. London: Taylor & Francis, 1990.

Morisseau C, Hammock BD. Epoxide hydrolases: mechanisms, inhibitor designs, and biological roles. Annu Rev Pharmacol Toxicol 2005; 45:311–333.

Mukhtar H, Khan WA. Cutaneous cytochrome P-450. Drug Metab Rev 1989; 20:657–673.

Mulder GJ. Drug metabolism: inactivation and bioactivation of xenobiotics. In: Mulder GJ, Dencker L, eds. Pharmaceutical Toxicology. London, UK: Pharmaceutical Press, 2006.

Nebert DW, Gonzalez FJ. P-450 genes: structure, evolution and regulation. Annu Rev Biochem 1987; 56:945–993.

Ortiz De Montellano PR, ed. Cytochrome P-450: Structure, Mechanism, and Biochemistry. New York: Plenum Press, 1986.

Ortiz De Montellano PR, Stearns RA. Radical intermediates in the cytochrome P-450 catalysed oxidation of aliphatic hydrocarbons. Drug Metab Rev 1989; 20:183–191.

Parke DV. The Biochemistry of Foreign Compounds. Oxford: Pergamon, 1968. A classic early text.

Ruckpaul K, Rein H, eds. Frontiers of Biotransformation. Vol. I. Basis and Mechanism of Regulation of Cytochrome P450. London: Taylor & Francis, 1989.

Sies H, Ketterer B, eds. Glutathione Conjugation. Mechanisms and Biological Significance. London: Academic Press, 1989.

Smith DA, Jones BC. Speculations on the substrate-activity relationships (SSAR) of cytochrome P450 enzymes. Biochem Pharmacol 1992; 44:2089–2098.

Tafazoli S, O'Brien PJ. Peroxidases: a role in the metabolism and side effects of drugs. Drug Discov Today 2005; 10:617–625.

Tarloff JB, Goldstein RS, Hook JB. Xenobiotic biotransformation by the kidney: pharmacological and toxicological aspects. In: Gibson GG, ed. Progress in Drug Metabolism. Vol. 12. London: Taylor & Francis, 1990.

Thomas H, Timms CW, Oesch F. Epoxide hydrolases: molecular properties, induction, polymorphisms and function. In: Ruckpaul K, Rein H, eds. Frontiers of Biotransformation. Vol. II. Principles, Mechanisms and Biological Consequences of Induction. London: Taylor & Francis, 1990.

Trager WF. Stereochemistry of cytochrome P-450 reactions. Drug Metab Rev 1989; 20:489–496.

Uetrecht J. Drug metabolism by leukocytes and its role in drug induced lupus and other idiosyncratic drug reactions. Crit Rev Toxicol 1990; 20:213–235.

Williams RT. Detoxication Mechanisms. The Definitive First Complete, Comprehensive Text on Drug Metabolism. London: Chapman & Hall, 1959.

Wrighton SA, Stevens JC. The human hepatic cytochromes P450 involved in drug metabolism. Crit Rev Toxicol 1992; 22:1–21.

Wrighton SA, Vandenbranden M, Stevens JC, et al. *In vitro* methods for assessing human hepatic drug metabolism: their use in drug development. Drug Metab Rev 1993; 25:453–484.

Ziegler DM. Unique properties of the enzymes of detoxification. Drug Metab Dispos 1991; 19:847.

Enzymatic bioactivation

Boerlsterli UA. Xenobiotic acyl glucuronides and acyl CoA thioesters as protein reactive metabolites with he potential to cause idiosyncratic drug reactions. Curr Drug Metab 2002; 3:439–450.

Glatt H. Sulphotransferases in the bioactivation of xenobiotics. Chem Biol Interact 2000; 129:141–170.

Gonzalez FJ. Role of cytochromes P450 in chemical toxicity and oxidative stress: studies with CYP2E1. Mutat Res 2005; 569:101–110.

Kemper R, Hayes JR, Bogdanffy MS. Metabolism: a determinant of toxicity. In: Hayes AW, ed. Principles and Methods of Toxicology. 5th ed. New York: CRC Press, 2007.

Ritter JK. Roles of glucuronidation and UDP-glucuronosyltransferases in xenobiotic bioactivation reacts. Chem Biol Interact 2000; 129:171–193.

Van Bladeren PJ. Glutathione conjugation as a bioactivation reaction. Chem Biol Interact 2000; 129:61–76.

5 | Factors Affecting Metabolism and Disposition

5.1 INTRODUCTION

In the preceding two chapters, the disposition and metabolism of foreign compounds, as determinants of their toxic responses, were discussed. In this chapter, the influence of various chemical and biological factors on these determinants will be considered.

It is becoming increasingly apparent that the toxicity of a foreign compound and its mode of expression are dependent on many variables. Apart from large variations in susceptibility between species, within the same species many factors may be involved. The genetic constitution of a particular organism is known to be a major factor in conferring susceptibility to toxicity in some cases. The age of the animal and certain characteristics of its organ systems may also be important internal factors.

External factors such as the dose of the compound or the manner in which it is given, the diet of the animal, and other foreign compounds to which it is exposed are also important for the eventual toxic response. Although some of these factors may be controlled in experimental animals, in the human population they remain and may be extremely important.

For a logical use of experimental animals as models for man in toxicity testing, these factors must be appreciated and used for the fullest possible exploration of potential toxicity.

The factors affecting the disposition and toxicity of a foreign compound may be divided into chemical and biological factors.

Chemical factors: lipophilicity, size, structure, pK_a, ionization, chirality.
Biological factors: species, strain, sex, genetic factors, disease and pathological conditions, hormonal influences, age, stress, diet, tissue and organ specificity, dose, and enzyme induction and inhibition.

5.2 CHEMICAL FACTORS

The importance of the physicochemical characteristics of compounds has already been alluded to in the previous two chapters. Thus, lipophilicity is a factor of major importance for the absorption, distribution, metabolism, and excretion of foreign compounds. Lipophilic compounds are more readily absorbed, metabolized, and distributed, but more poorly excreted, than hydrophilic compounds.

The distribution of compounds is profoundly affected by lipophilicity, and this may in turn influence the biological activity. Lipophilic compounds are more readily able to distribute into body tissues than hydrophilic compounds and there exert toxic effects or be sequestered and be redistributed later to other tissues. As already discussed (see above, chap. 3, pages 30 and 48), comparison of the two drugs thiopental and pentobarbital illustrates the importance of lipophilicity. Similarly, the influence of size, structure, and ionization have been mentioned. For example, there is a correlation between the nephrotoxicity of the aminoglycoside antibiotics, such as gentamycin, and structure, although other factors are also involved. The more ionizable amino groups on the aminoglycoside molecule, the greater the nephrotoxicity. The underlying reason for this is binding of the drug to anionic phospholipids on the brush border of the proximal tubular cells in the kidney and subsequent accumulation (see chaps. 6 and 7). Some chemicals (including a significant number of drugs) cause the excessive accumulation of phospholipids (phospholipidosis), which is directly related to the physicochemical properties of the drug or metabolite. Thus, the chemical should be *cationic* and *amphipathic* (drugs with this property are called CADs).

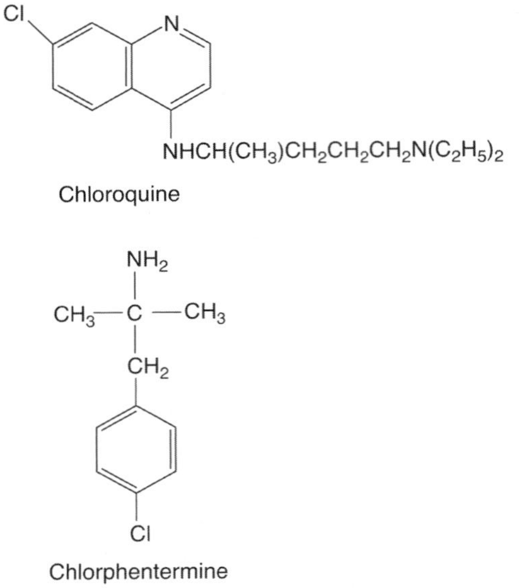

Chloroquine

Chlorphentermine

Figure 5.1 The structures of chloroquine and chlorphentermine, two CADs showing the hydrophilic and hydrophobic regions. *Abbreviation:* CAD, cationic amphipathic drug.

This means that the drug molecule has a hydrophobic section and a cationic hydrophilic section (Fig. 5.1). The hydrophobic section will associate with lipid and the hydrophilic, cationic section will associate with the aqueous phase.

Drugs that cause phospholipidosis are cationic amphiphiles; they contain a hydrophobic ring structure and a hydrophilic side chain with a positively charged amine group. Such chemicals can interact with either the ionic (e.g., chlorophentermine) or hydrophobic (e.g., amiodarone) moieties of phospholipids.

The lipophilicity of the drug can affect cell membrane permeability. The hydrophobicity allows the drug to favorably interact with membrane receptors. However, the cationic nature can affect the movement of cations such as Na^+ and Ca^+ across the cell membrane.

Chemicals such as these bind with phospholipids, inhibiting their hydrolysis by phospholipases. They can also interact with phospholipases, thus limiting the ability of the enzyme to metabolize phospholipid. Drugs, which have this structure, may also influence the synthesis of phospholipids.

The selective uptake of paraquat is another excellent example of the importance of size and shape in the disposition and toxicity of foreign compounds (see chap. 3, pages 40–1, and chap. 7). Similarly, the chemical similarities between carbon monoxide and oxygen are important in the toxic effects caused as are discussed in detail in chapter 7.

Metabolism is also affected by the physicochemical characteristics of a compound as discussed by Hansch (1), for example, see Bibliography. Thus, with the monooxygenases there is a correlation between the lipophilicity of a compound, as measured by the partition coefficient, and metabolism by certain routes, such as N-demethylation, for example. This correlation is not always clear-cut, however, as other factors may be involved. Ionization is another factor, which may inhibit the ability of compounds to be metabolized. Size and molecular structure are clearly important in metabolism. For example, substrates (and inhibitors) for CYP2D6 require an extended hydrophobic region; a positively charged, basic nitrogen; groups with negative potential; and the ability to accept hydrogen bonds positioned 5 to 7 Å(0.5–0.7 nm) from the nitrogen atom. Very large molecules may not be readily metabolized because of their inability to fit into the active site of an enzyme. As will already be apparent from chapter 4, the molecular structure will determine what types of metabolic transformation will take place.

Polychlorinated hydrocarbons, such as polychlorinated dibenzodioxins, dibenzofurans, and biphenyls exist as a number of different congeners. Some of these are geometric isomers. Many cause a range of toxic effects that are believed to be mediated by interaction with the aryl hydrocarbon receptor (AhR) (they are known as pleiotropic effects). However, not all the isomers cause these effects because they do not all interact with the AhR receptor. To interact with this receptor, the molecule needs to be flat (planar). It can be seen in the diagram (Fig. 5.2)

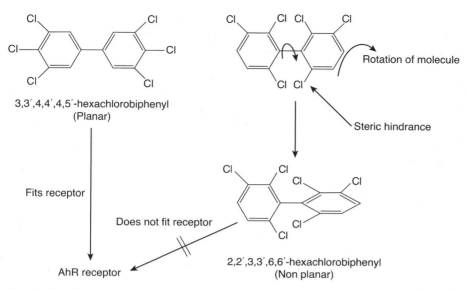

Figure 5.2 The structures of 2,2', 3,3', 6,6'-hexachlorobiphenyl and the geometric isomer 3,3', 4,4', 5,5'-hexachlorobiphenyl. This shows that only one isomer, which is planar, can interact with the AhR receptor.

that of the two isomers shown, only the 3,3',4,4',5,5'-hexachlorobiphenyl is able to be planar, the other hexachlorobiphenyl shown must rotate to avoid the steric interaction between two chlorine atoms. It therefore loses planarity and cannot fit into the receptor.

5.2.1 Chiral Factors

The importance of chiral factors in disposition and toxicity has been fully recognized only relatively recently, although important examples have been known for some time. For instance the S(−) enantiomer of thalidomide is known to have greater embryotoxicity than the R(+) enantiomer (see chap. 7). Another example in which a particular isomer of a metabolite is responsible for a carcinogenic effect is the *exo*-oxide of aflatoxin B$_1$, discussed later in this chapter (Fig 5.14).

Ariens (see Bibliography) has been in the vanguard of those trying to highlight the importance of chirality, and particularly its implications for drugs. The presence of a chiral center in a molecule, giving rise to isomers, may influence the disposition of a compound and therefore its toxicity or other biological activity. It is clear from a consideration of the biochemistry of endogenous compounds, where only one isomer may be metabolized or active, that biological systems are intrinsically chiral. Therefore, it is hardly surprising that these considerations should apply also to foreign compounds.

All four phases of disposition may be influenced by chirality. Absorption is not often directly affected by the presence of a chiral center except when an active transport process is involved. Thus, **L-DOPA** is more readily absorbed from the gastrointestinal tract than the D-isomer. Various aspects of distribution may be affected by chirality such as protein and tissue binding. Thus for the drug **ibuprofen**, the ratio of plasma-protein binding for the (+) and (−) enantiomers is 1.5. It has been shown that the S enantiomer of **propranolol** undergoes selective storage in adrenergic nerve endings in certain tissues such as the heart.

The renal excretion of compounds can also be affected by the presence of a chiral center, probably as a result of active secretion. For example, with the drug **terbutyline**, the ratio for the excretion of the (+) to (−) enantiomers was found to be 1.8. Similarly, biliary excretion of compounds may show **stereoselectivity**.

Probably of more significance are chiral effects in metabolism, which can be divided into

1. substrate stereoselectivity (the effect of a preexisting chiral center in a molecule),
2. product stereoselectivity (the production of metabolic products with chiral centers),
3. inversion of configuration, and
4. loss of chirality as a result of metabolism.

When racemic mixtures of drugs or foreign compounds are administered to animals, as is currently often the case, stereoselective metabolism means that either two or more different isomeric products are formed or only one isomer is metabolized. Alternatively, there may be differences in rates of metabolism for the different isomers. All of these factors may have significant implications with regard to the biological activity of the molecule in question. Thus differences in rates or routes of metabolism for different isomers may underlie species, organ, or genetic differences in metabolism and toxicity. Stereoselectivity in metabolism occurs with cytochromes P-450 and also with other enzymes such as epoxide hydrolase and glutathione transferases (GSTs).

As an example of the first type of chiral effect, metabolism of the drug **bufuralol** may be considered. Hydroxylation in the 1' position only occurs with the (+) isomer, whereas for hydroxylation in positions 4 and 6, the (−) isomer is the preferred substrate (Fig. 5.3). Glucuronidation of the side chain hydroxyl group is specific for the (+) isomer. A further complication in human subjects is that the 1-hydroxylation is under genetic control, being dependent on the **debrisoquine hydroxylator status** (see below). The selectivity for the isomers for the hydroxylations is virtually abolished in poor metabolizers.

There are other examples in which chirality is a factor in determining which particular metabolic route occurs, such as the metabolism of the drugs propranolol, metoprolol, and warfarin.

An example of the second type of chiral effect in metabolism is afforded by **benzo[a]-pyrene**, also discussed in more detail in chapter 7. This carcinogenic polycyclic hydrocarbon is metabolized stereoselectively by a particular cytochrome P-450 isozyme, CYP1A1, to the (+)-7R,8S oxide (chap. 7, Fig. 5.2), which in turn is metabolized by epoxide hydrolase to the (−)-7R,8S dihydrodiol. This metabolite is further metabolized to (+)-benzo[a]pyrene, 7R,8S dihydrodiol, 9S,10R epoxide in which the hydroxyl group and epoxide are trans and which is more mutagenic than other enantiomers. The (−)-7R,8S dihydrodiol of benzo[a]pyrene is 10 times more tumorigenic than the (+)-7R,8S enantiomer. It was reported that in this case the configuration was more important for tumorigenicity than the chemical reactivity.

Another example is the metabolism of **naphthalene**, which may cause lung damage in certain species. Thus, naphthalene is metabolized first to an epoxide as previously discussed (see chap. 4). However, this is stereospecific giving rise predominantly to the 1R,2S-naphthalene oxide (Fig. 5.2) in the susceptible species (mouse) compared with ratios of 1R,2S: 1S,2R of one or less in non-susceptible species (rat and hamster). The 1R,2S enantiomer was found to be a better substrate for epoxide hydrolase than the 1S,2R enantiomer. The relationship of the stereospecificity to the lung toxicity is not yet clear, but differences in cytotoxicity between the two enantiomers were found in isolated hepatocytes. A further complication is the production of two chiral centers and hence diastereoisomers from metabolism of a chiral compound. For example, this may occur when conjugation of isomers occurs with other chiral molecules such as β-D-glucuronic acid or L-glutathione. Thus, using the example of naphthalene, the oxirane ring of the two enantiomeric epoxides may be opened by attack by glutathione at either the 1- or 2-position to give four different diastereoisomeric glutathione conjugates (Fig. 5.4). The formation of the various **diastereoisomers** was found to show considerable species differences *in vitro*.

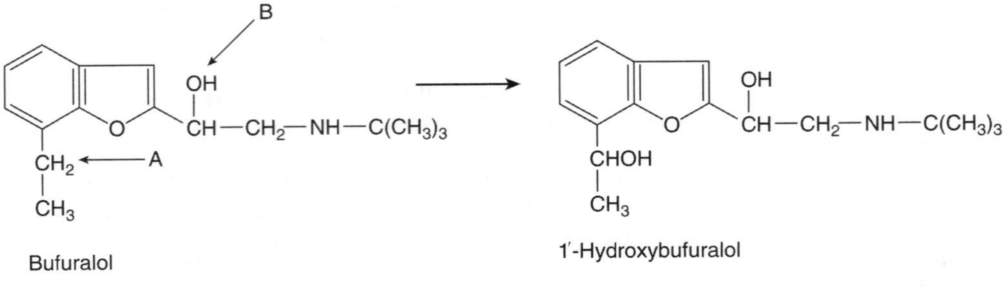

Figure 5.3 The hydroxylation of the drug bufuralol. The arrows show (**A**) the site of 1' hydroxylation and (**B**) the site of glucuronidation.

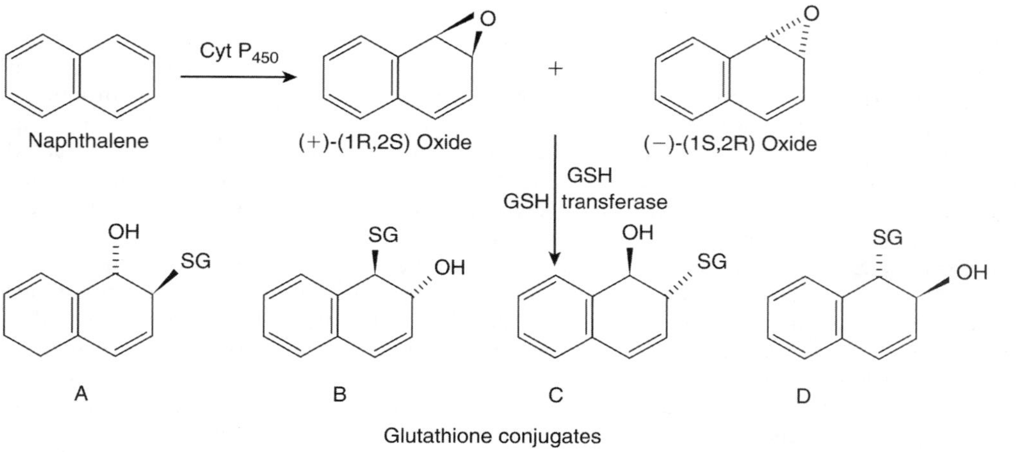

Figure 5.4 The metabolism of naphthalene showing the various possible isomeric glutathione conjugates.

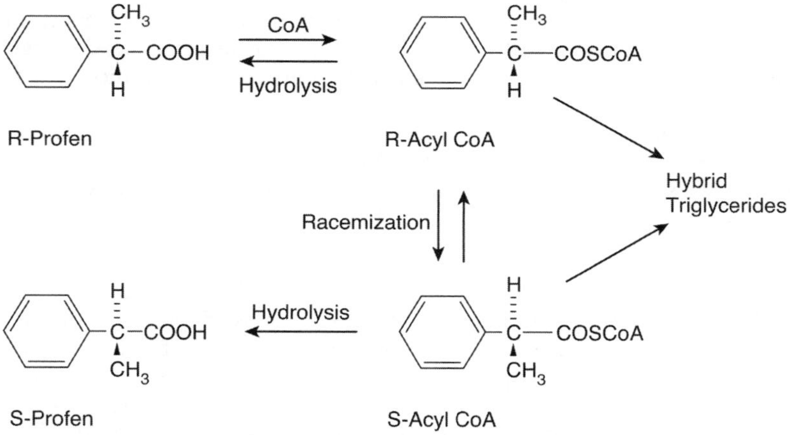

Figure 5.5 The mechanism of chiral inversion of ibuprofen and formation of hydrid triglycerides.

The third type of chiral effect, inversion of configuration, has been shown to occur with a number of compounds. For example, the anti-inflammatory drug ibuprofen, an arylpropionic acid, undergoes inversion from the R- to the pharmacologically active S-isomer. Furthermore, stereoselective uptake of the R-ibuprofen into fat tissue occurs as a result of selective formation of the coenzyme A-thioester of the R-isomer. This thioester may then undergo inversion to the S-thioester. Both thioester isomers are incorporated into triglycerides forming hybrid products (Fig. 5.5). Thus, after S-ibuprofen is administered to rats, only a fraction is found in fat tissue in comparison with the incorporation after R-ibuprofen or the racemate is administered. Although the fate of these hybrid triglycerides is currently unknown, they might potentially interfere in lipid metabolism with possible toxicological consequences. This would be especially likely after chronic administration when accumulation could occur. There are also various factors, which may affect the inversion *in vivo*, such as the species and reduction of renal excretion.

In conclusion, it cannot be stressed too strongly that the physicochemical characteristics of a foreign compound are factors of paramount importance in determining its toxicity.

5.3 BIOLOGICAL FACTORS

5.3.1 Species

There are many different examples of species differences in the toxicity of foreign compounds, some of which are commercially useful to man, as in the case of pesticides and antibiotic drugs where there is exploitation of **selective toxicity**. Species differences in toxicity are often related to differences in the metabolism and disposition of a compound, and an understanding of such differences is extremely important in the safety evaluation of compounds in relation to the extrapolation of toxicity from animals to man and hence risk assessment.

Some species differences are due to differences in the response of the organism to insult or in the repair mechanisms available. There may be very simple differences; for example, rats are susceptible to certain rodenticides, which they ingest by mouth, as unlike most other mammals, they are unable to vomit. There may be differences in receptor sensitivity such as for the organophosphorus compound **paraoxon** (chap. 4, Fig. 25). Thus, the cholinesterase enzyme in the mouse is more sensitive to inhibition than that in the frog ($79\times$), and the frog is correspondingly less sensitive to the toxicity of the compound ($22\times$). Another example is the very big species difference in the acute toxicity of 2,3,7,8-tetrachlorodibenzdioxin (**TCDD or dioxin**) as can be seen from Table 5.1. TCDD causes many toxic effects, some of which are mediated through the AhR. Thus cancer, progressive weight loss, birth defects, effects on hormones, and male and female reproduction are all toxic effects. The guinea pig is the most sensitive species, being at least three orders of magnitude more sensitive than the hamster. However, this does not seem to be due to differences in the AhR as the amounts are similar in guinea pig and hamster liver. Some differences in sensitivity may be due to differences in receptor affinity, but these mostly apply to differences between mouse strains. Although exactly comparable data are not available for humans, the general consensus of opinion is that man is less sensitive than most laboratory animals.

Species differences in the toxicity of the drug digitoxin have been suggested because of differences in the concentration required to inhibit Na^+/K^+-ATPase, which is more sensitive in sensitive species. An alternative suggestion is that differences in metabolism catalyzed by CYP3A4 are responsible.

Species Differences in Disposition

Absorption. Absorption of foreign compounds from various sites is dependent on the physiological and physical conditions at these sites. These, of course, may be subject to species variations. Absorption of compounds through the skin shows considerable species variation. Table 5.2 gives an example of this and shows the species differences in toxicity of an organophosphorus compound absorbed percutaneously. Human skin is generally less permeable to chemicals than that of rabbits, mice, and rats, although there is variation. For some compounds, rat skin has similar permeability to human skin and seems to be less permeable than that of the rabbit.

Table 5.1 Species Differences in the Acute Toxicity of Dioxin[a]

Species	LD$_{50}$ (μg kg^{-1} body weight)
Guinea pig	0.5–2
Rat	22–100
Mouse	114–284
Rabbit	10–115
Chicken	25–50
Rhesus monkey	<70
Dog	>30–300
Hamster	5051[b]

[a]TCDD or dioxin: 2,3,7,8-tetrachlorodibenzodioxin.
[b]Document for PCDDs, Parts 1 and 2, External Reviewers Draft; PB 84-220268.
Source: From Refs. 2 and 3.

Table 5.2 Species Differences in the Relative Percutaneous Toxicity and Skin Penetration of Organophosphorus Compounds

Species	Rate (μg cm^2) min^{-1}	Compound 1[a]	Compound 2[a]	2/1
Pig	0.3	10.0	80.0	8.0
Dog	2.7	1.9	10.8	5.7
Monkey	4.2	4.4	13.0	3.0
Goat	4.4	3.0	4.0	1.3
Cat	4.4	0.9	2.4	2.7
Rabbit	9.3	1.0	5.0	5.0
Rat	9.3	17.0	20.0	1.2
Mouse	—	6.0	9.2	1.5
Guinea pig	6.0	—	—	—

[a]Values are expressed as the ratio of the LD$_{50}$ of that compound to the rabbit LD$_{50}$ of compound 1.
Source: From Refs. 4 and 5.

Table 5.3 Species Differences in pH of Saliva and Gastric Juices

	pH of saliva	pH of gastric juice
Man	6.75	1.2–2.5
Dog	7.5	1.5–2.0[a]; 4.5[b]
Cat	7.5	—
Rat	8.2–8.9	2.0–4.0
Horse	7.3–8.6	4.46
Cattle	8.1–8.8	5.5–6.5
Sheep	8.4.8.7	7.6–8.2
Chicken	—	4.2
Frog	—	2.2–3.7

[a]Fasting.
[b]Fed.
Source: From Refs. 6–10.

Oral absorption depends partially on the pH of the gastrointestinal tract, which is known to vary between species, as shown in Table 5.3. Clearly, therefore, considerable differences in the absorption of weak acids from the stomach may occur between species. Similarly, differences might be seen in compounds, which are susceptible to the acidic conditions of the stomach, such that a foreign compound would be more stable in the gastric juice of a sheep than that of a rat. For example, there is a difference in the acute toxicity in **pyrvinium chloride**, a rodenticide, between rats and mice due to differences in absorption from the gastrointestinal tract. Thus, after intraperitoneal dosing the toxicity is similar in the two species (LD$_{50}$ 3–4 mg kg^{-1}), whereas after an oral dose it is very different, being more toxic in the mouse (LD$_{50}$ 15 mg kg^{-1}) than in the rat (LD$_{50}$ 1550 mg kg^{-1}). Species differences in lung physiology may be important in considerations of absorption of compounds by inhalation. Small animals such as rats and mice and birds have a more rapid respiration rate than larger animals such as humans. Consequently, for compounds with high solubility in the blood where absorption is ventilation limited, exposure will be greater for these small animals when exposed to the same concentration of a compound. This is the basis of the use of canaries in mines to warn of the build up of dangerous gases.

Distribution. The distribution of foreign compounds may vary between species because of differences in a number of factors such as proportion and distribution of body fat, rates of metabolism and excretion and hence elimination, and the presence of specific uptake systems in organs. For instance, differences in localization of **methylglyoxal-bis-guanyl hydrazone** (Fig. 5.6) in the liver account for its greater hepatotoxicity in rats than in mice. The hepatic concentration in mice is only 0.3% to 0.5% of the dose after 48 hours, compared with 2% to 8% in the rat.

The plasma-protein concentration is a species-dependent variable, and the proportions and types of proteins may also vary. The concentration may vary from about 20 g L^{-1} in certain

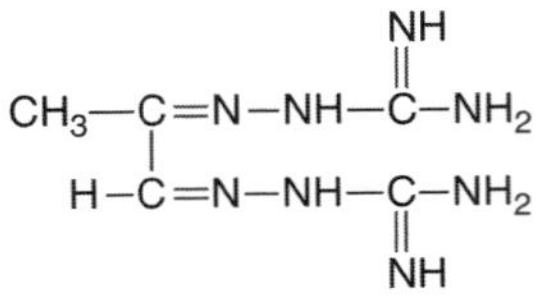

$$CH_3-\underset{\underset{H-\overset{||}{C}=N-NH-\overset{||}{\underset{NH}{C}}-NH_2}{|}}{C}=N-NH-\overset{\overset{NH}{||}}{C}-NH_2$$

Figure 5.6 Structure of methylglyoxal-bis-guanyl hydrazone.

Table 5.4 Binding of Various Sulfonamides to Plasma of Various Species

Percent bound at protein concentration of 100 $\mu g\ mL^{-1}$

Sulfonamide	Human	Monkey	Dog	Cat	Mouse	Chicken	Bovine Plasma	Bovine Albumin
Sulfadiazine	33	35	17	13	7	16	24	24
Sulfamethoxy-pyridazine	83	81	60	49	28	14	66	60
Sulfisoxazole	84	86	68	43	31	5	76	76
Sulfa-ethyl-thiadiazole	95	90	86	76	38	48	87	87

Source: From Ref. 5.

Table 5.5 Variation in Urinary Volume and pH with Species

Species	Urine volume (mL kg.$^{-1}$ day^{-1})	pH
Man	9–29	4.8–7.8
Monkey	70–80	
Dog	20–100	5–7
Cat	10–20	5.0–7.0
Rabbit	50–75	—
Rat	150–300	—
Horse	3–18	7–8
Cattle	17–45	7–8
Sheep	10–40	7–8
Swine	5–30	Acid or alkaline

Source: From Refs. 6,11,12.

fish to 83 g L^{-1} in cattle. Thus, foreign compounds may bind to plasma proteins to very different extents in different species (Table 5.4). Because the extent of binding may be a very important determinant of the free concentration of a compound in the plasma and the tissues, this species difference may be an important determinant of toxicity. The free form of the compound is the important moiety as far as toxicity is concerned.

Excretion

Renal excretion. Although most mammals have similar kidneys, there are functional differences between species and urine pH, and volume and rate of production may vary considerably (Table 5.5). Thus, the rate of urine production in the rat is an order of magnitude greater than the rate in man. Although the pH ranges for the urine of a number of mammals may overlap (Table 5.5), a small change in pH may markedly change the solubility of a foreign compound and therefore its excretion. For instance, some of the sulfonamides and their acetylated metabolites show marked changes in solubility for a pH change of one unit (chap. 4, Table 5.1), and renal toxicity due to crystallization of the drug or its metabolites in the renal tubules has been known to occur when high doses are used. The species differences in renal excretion for an unmetabolized compound, methylglyoxal-bis-guanylhydrazone (Fig. 5.6), are shown in Table 5.6. It can be seen that rats and mice excrete twice the amount excreted by man in 24 hours. This may be at least partially due to the greater rate of urine flow in the rodent. The observation that renal tubular atrophy is caused by certain diuretics such as furosemide (chap. 3, Fig. 33) in the dog, but not in the rat or monkey, has been explained in terms of differences in the vascular system of the dog kidney from that in rats, monkeys, and humans. For compounds, which are not actively secreted in the kidney, species differences in plasma-protein binding may indirectly lead to differences in urinary excretion.

Table 5.6 Urinary Excretion of Methylglyoxal-bis-Guanylhydrazone in Mammalian Species

Species	Dose (mg kg^{-1})	Percent excreted	Time period for excretion (hr)
Mouse	20 (i.v.)	51	24
Rat	20 (i.p.)	65	24
Dog	20 (i.v.)	26	24
		52	48
		66	96
Monkey	25 (i.v.)	47	24
Man	4 (i.v.)	25	24
		42	118
		49	166

Source: From Refs. 5 and 13.

Table 5.7 Biliary Excretion of Compounds of Molecular Weight 300–500 in Various Species

	Percent excreted in bile		
Species	Methylene disalicylic acid MW 288 (10 mg kg^{-1}; i.v. 6 hr)	Stilboestrol glucuronide MW 445 (10 mg kg^{-1}; i.v. 3 hr)	Sulfadimethoxine glucuronide MW 487 (15 mg kg^{-1}; i.v. 3 hr)
Rat	54	95	43
Dog	20	65	43
Rabbit	5	32	10
Guinea pig	4	20	12

Source: From Refs. 14 and 15.

Table 5.8 Biliary Excretion of Indocyanine Green in Various Species

Species	Dose (mg kg^{-1}, i.v.)	% Dose in bile[a]
Rat	0.5	60
Rat	2.5	82
Dog	1–7	97
Rabbit	2.5	94
Man	0.5	High
Man	2.0	High

[a]Excreted unchanged.
Source: From Refs. 16–19. Delaney et al., (1969). Unpublished data.

Biliary excretion. The extent of excretion of foreign compounds via the bile is influenced by a number of factors, the molecular weight of the compound being the major one. However, the molecular weight threshold for biliary excretion may show considerable species differences. Little biliary excretion (5–10% of the dose) occurs for compounds of molecular weight of less than 300. Above this value, however, the bile may become a major route of elimination, and it is probably around this value that species variations are most noticeable. Thus, for methylene disalicylic acid (mol. wt. 288), the dog excretes 65% in the bile, whereas the guinea pig excretes only 40% (Table 5.7). Similarly, the biliary excretion of succinyl sulfathiazole (mol. wt. 355) shows more than a 10-fold variation between the rhesus monkey and the rat (Table 5.7). Thus, the approximate molecular weight thresholds are 500 to 700 in humans, 475 in rabbits, 400 in guinea pig, and 325 in rats.

The species pattern of the rabbit and the guinea pig being poor biliary excretors and the rat being an extensive biliary excretor is maintained with many other compounds. With compounds of higher molecular weight, however, species differences are less, as illustrated by the compound **indocyanine green** (Table 5.8). The metabolism of a compound obviously influences the extent of biliary excretion, and therefore species differences in metabolism may also be a factor.

The rate of bile secretion and the pH of the bile may also be determinants of the extent of biliary excretion of a foreign compound, and these also show species variations. The fate of compounds excreted in the bile may also depend on the species, as differences in intestinal pH and flora occur. A particularly important consequence of biliary excretion is metabolism by the gut flora and reabsorption. This enterohepatic circulation prolongs the length of time the animal is exposed to the foreign compound, and may introduce novel toxic metabolites. This could, therefore, result in marked species differences in toxicity.

Species Differences in Metabolism
Differences in metabolism between species may be either quantitative or qualitative, but quantitative differences are more common. In general, small animals such as mice metabolize foreign compounds at a faster rate than larger animals such as humans, consistent with differences in overall metabolic rate. An extreme example of a difference in rates of metabolism is afforded by the drug **oxyphenbutazone**. In the dog it is rapidly metabolized and has a half-life of around 30 minutes, in several other species such as the rat, rabbit, and monkey, the half-life is between 3 and 6 hours, whereas in humans, metabolism is very slow and therefore the drug has a half-life of about 3 days. Quantitative differences also exist although with a few exceptions, it is generally difficult to discern useful patterns. Even the simplest organisms such as bacteria seem to be able to carry out many different types of reaction. The differences, which are clear and fall within taxonomic groups, are mainly found with phase 2 reactions. Differences in some cases are related to diet, and so herbivores and carnivores may show differences.

Examples of toxicologically important species differences in metabolism will therefore be dealt with by considering the different types of metabolic reactions.

Phase 1 reactions
Oxidation.   Although most of the common mammals used as experimental animals carry out oxidation reactions, there may be large variations in the extent to which some of these are carried out. The most common species differences are in the rate at which a particular compound is oxidized rather than the particular pathway through which it is metabolized. Most species are able to hydroxylate aromatic compounds, but there is no apparent species pattern in the ability to carry out this metabolic transformation.

Fish have a relatively poor ability for oxidative metabolism compared with the commonly used laboratory animals such as rats and mice. Insects such as flies have microsomal enzymes, and these are involved in the metabolism of the insecticide parathion to the more toxic paraoxon as discussed in the previous chapter (chap. 4, Fig. 25).

However, there are known instances of differences in the preferred route of metabolism, which are important in toxicity, as well as simple differences in the route of a particular oxidation. For example, the oxidative metabolism of ethylene glycol gives rise to either carbon dioxide or oxalic acid (Fig. 5.7). The relative importance of these two pathways is reflected in the toxicity. Thus, the production of oxalic acid is in the order: cat>rat> rabbit, and this is also the order of increasing toxicity (Fig. 5.8). The aromatic hydroxylation of aniline (Fig. 5.9) shows marked species differences in the position of substitution, as shown in Table 5.9. Thus carnivores such as the ferret, cat, and dog excrete mainly *o*-aminophenol, whereas herbivores such as the rabbit and guinea pig excrete mainly *p*-aminophenol. The rat, an omnivore, is intermediate.

The preferred route of hydroxylation also correlates with the toxicity, such that those species to which aniline is particularly toxic, such as the cat and dog, produce mainly *o*-aminophenol, whereas those producing *p*-aminophenol, such as the rat and hamster, seem

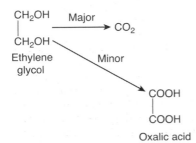

Figure 5.7   Metabolism of ethylene glycol showing the production of the toxic metabolite.

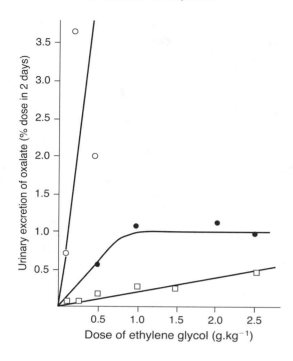

Figure 5.8 Species differences in the metabolism of ethylene glycol to oxalate after increasing doses. Species used were cats (○), rats (●), and rabbits (□). *Source*: From Refs. 20 and 21.

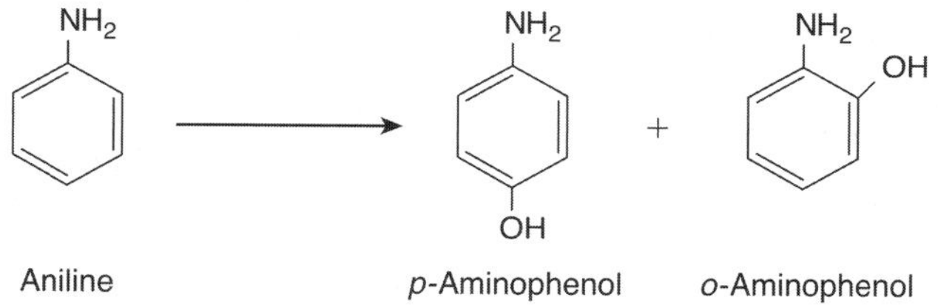

Aniline   *p*-Aminophenol   *o*-Aminophenol

Figure 5.9 Aromatic hydroxylation of aniline.

Table 5.9 Species Differences in the Hydroxylation of Aniline

Species	% Dose excreted	
	o-Aminophenol	*p*-Aminophenol
Gerbil	3	48
Guinea pig	4	46
Golden hamster	6	53
Chicken	11	44
Rat	19	48
Ferret	26	28
Dog	18	9
Cat	32	14

less susceptible. Conversely, the hydroxylation of coumarin at position 7 (chap. 4, Fig. 15) is an important pathway in the rabbit and also the hamster and cat, but not in the rat or mouse. It is clear that, even with aromatic hydroxylation, species cannot be readily grouped.

The N-hydroxylations of acetylaminofluorene and paracetamol are two toxicologically important examples illustrating species differences (see chap. 7).

The carcinogenic heterocyclic amine MeIQX formed in meat when it is cooked has been shown to exhibit species differences in activation, with humans possibly the most sensitive. MeIQX undergoes N-hydroxylation catalyzed by CYP1A2, followed by O-acetylation to give a reactive nitrenium ion, which covalently binds to DNA. *In vitro* studies using liver microsomes from man, rat, and cynomologous monkeys reveal that human liver generates greater adduct formation than the other species. MeIQX is carcinogenic in rats, and adducts are detected both *in vivo* and *in vitro*. However, there is little adduct formation in the monkeys, consistent with absence of tumors when the animals are treated *in vivo* for five years. It seems that unlike man and rat, the cynomologous monkey does not have constitutively expressed CYP1A2, and so metabolic activation does not take place.

Another example is the metabolism of **amphetamine**, which reveals marked species differences in the preferred route, as shown in Figure 5.10.

Species differences in the rate of metabolism of **hexobarbital** *in vitro* correlate with the plasma half-life and duration of action *in vivo* as shown in Table 5.10. This data show that the marked differences in enzyme activity between species is the major determinant of the biological activity in this case.

A recent example of a species difference in metabolism causing a difference in toxicity is afforded by the alicyclic hydroxylation of the oral antiallergy drug, **proxicromil** (Fig. 5.11). After chronic administration, this compound was found to be hepatotoxic in dogs but not in rats. It was found that dogs did not significantly metabolize the compound by alicyclic oxidation, whereas rats, hamsters, rabbits, and man excreted substantial proportions of metabolites in the urine. In the dog, biliary excretion was the route of elimination of the unchanged compound,

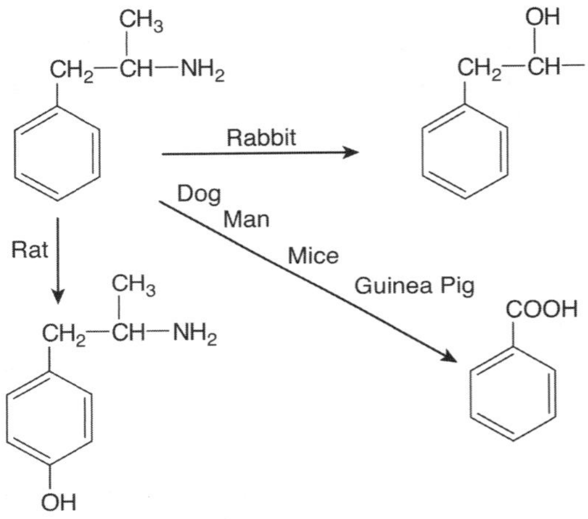

Figure 5.10 Species differences in the metabolism of amphetamine.

Table 5.10 Species Differences in the Duraction of Action and Metabolism of Hexobarbital (Dose of Barbiturate 100 mg kg^{-1}; 50 mg kg^{-1} in Dogs)

Species	Duration of action (min)	Plasma half-life (min)	Relative enzyme activity (μg g^{-1} hr^{-1})	Plasma level on awakening (μg mL^{-1})
Mouse	12	19	598	89
Rabbit	49	60	196	57
Rat	90	140	135	64
Dog	315	260	36	19

Source: From Ref. 22.

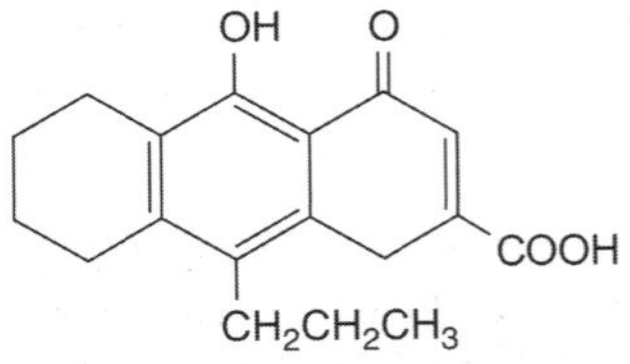

Figure 5.11 The structure of proxicromil.

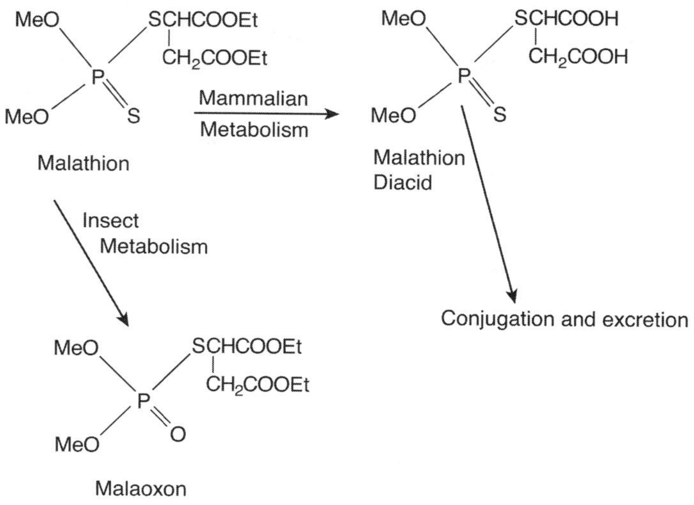

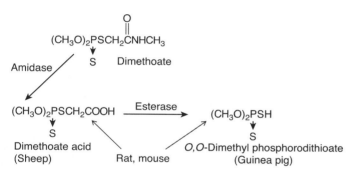

Figure 5.12 Metabolism of malathion.

Figure 5.13 Metabolism of dimethoate.

and after toxic doses were administered, this route was saturated. Hence, the toxicity was probably due to the accumulation of high levels of the unchanged compound.

Hydrolytic reactions. There are numerous different esterases responsible for the hydrolysis of esters and amides, and they occur in most species. However, the activity may vary considerably between species. For example, the insecticide **malathion** owes its selective toxicity to this difference. In mammals, the major route of metabolism is hydrolysis to the dicarboxylic acid, whereas in insects it is oxidation to malaoxon (Fig. 5.12). Malaoxon is a very potent cholinesterase inhibitor, and its insecticidal action is probably due to this property. The hydrolysis product has a low mammalian toxicity (see chap. 7).

Another example is **dimethoate**, the toxicity of which is related to its rate of hydrolysis. Those species, which are capable of metabolizing the insecticide, are less susceptible than those species, which are poor metabolizers. The metabolism of dimethoate is shown in Figure 5.13. Studies on the metabolism *in vitro* of dimethoate have shown that sheep liver produces only the first metabolite, whereas guinea pigs produce only the final product (Fig. 5.13). Rats and mice metabolize dimethoate to both products. The toxicity is in the descending order: sheep>dog>rat>cattle>guinea pig>mouse.

Reduction. The activity of azo- and nitroreductase varies between different species, as shown by the *in vitro* data in Table 5.11. Thus, azoreductase activity is particularly high in the guinea pig, relative to the other species studied, whereas nitroreductase activity is greatest in the mouse liver.

Phase 2 reactions. Species vary considerably in the extent to which they conjugate foreign compounds, but this is generally a quantitative rather than a qualitative difference. Thus, most species have a preferred route of conjugation, but other routes are still available and used.

Table 5.11 Hepatic Azoreductase and Nitroreductase Activities of Various Species

Species	Azoreductase (μmol sulfanilamide formed in liver g^{-1} hr^{-1})	Nitroreductase (μmol p-aminobenzoic acid formed in liver g^{-1} hr^{-1})
Mouse	6.7–9.6[a]	2.1–3.2[a]
Rat	5.9	2.1
Guinea pig	9.0	2.0
Pigeon	7.1	1.1
Turtle	1.4 (0.5)[b]	0.15 (2.5)[b]
Frog	1.2 (0.6)[b]	0 (0)[b]

Substrates used were neoprontosil for the azoreductase and p-nitrobenzoic acid for nitroreductase.
[a]According to strain.
[b]Temperature of incubation 21°C (temperature elsewhere 37°C).
Source: From Ref. 23.

Table 5.12 Conjugation of Phenol with Glucuronic Acid and Sulfate

Species	Phenol conjugation % total	
	Glucuronide	Sulfate
Cat	0	87
Gerbil	15	69
Man	23	71
Rat	25	68
Rhesus monkey	35	65
Ferret	41	32
Rabbit	46	45
Hamster	50	25
Squirrel monkey	70	10
Guinea pig	78	17
Indian fruit bat	90	10
Pig	100	0

Source: From Ref. 24.

Glucuronide conjugation. Conjugation of foreign compounds with glucuronic acid is an important route of metabolism in most animals, namely mammals (Table 5.12) birds, amphibians, and reptiles, but not fish. In insects, glucoside conjugates using glucose rather than glucuronic acid are formed. The major exception with regard to glycoside conjugation is the cat, which is virtually unable to form glucuronic acid conjugates with certain foreign compounds, in particular phenols. However, bilirubin, thyroxine, and certain steroids are conjugated with glucuronic acid in the cat. This may be explained by the presence of multiple forms of the enzyme UDP-glucuronosyl transferase shown in the rat, for example, which catalyze conjugation with different types of substrate (see chap. 4). Presumably, the cat lacks the isoenzyme, which catalyzes the glucuronide conjugation of phenols. The cat is therefore more susceptible to the toxic effects of phenols than species able to detoxify them by glucuronide conjugation.

Sulfate conjugation. Conjugation of foreign compounds with sulfate occurs in most mammals (Table 5.12), amphibians, birds, reptiles, and insects, but, as with glucuronidation, not in fish. The pig, however, has a reduced ability to form certain ethereal sulfate conjugates, such as with phenol, whereas 1-naphthol is excreted as a sulfate conjugate. As there are several forms of sulfotransferase, specific for different substrates, the inability of the pig to form a sulfate conjugate with phenol may be due to the lack of one particular form of the enzyme. The inability of the **guinea pig** to form a sulfate conjugate of N-hydroxyacetylaminofluorene helps confer resistance to the tumorigenicity of acetylaminofluorene on this species (see chap. 7).

It can be seen from Table 5.12 that the relative proportions of glucuronide and sulfate conjugates vary between the species, with the cat and pig being at the opposite extremes.

Interestingly, humans and rats excrete similar proportions of conjugates, but the squirrel monkey, also a primate, is quite different. However, despite the similarity between humans and rats in relation to phenol conjugation, in the case of the anti-estrogenic drug tamoxifen, rat and man are fortunately very different. This drug, which is used to treat breast cancer in humans, is in fact a liver carcinogen in rats. DNA adducts can be detected both *in vivo* in rats and *in vitro* in rat hepatocytes and to a lesser extent in mice, although it is not carcinogenic in this species, whereas there is no indication that it is hepatocarcinogenic in humans, and no DNA adducts have been detected. In rats the drug undergoes CYP-mediated hydroxylation on the α-carbon atom, and this is then conjugated with sulfate, catalyzed by sulfotransferase. As already discussed, the sulfate group is a good leaving group and can be easily lost to yield a reactive positively charged reactive intermediate (a carbocation). This reacts with DNA to form adducts, ultimately leading to liver cancer. Humans, in contrast, produce much less of the α-hydroxy tamoxifen, which is then conjugated with glucuronic acid. This is a detoxication reaction leading to safe elimination. The reason for the species difference therefore is that rats are more active at the hydroxylation (3×) than humans and have much higher sulfotransferase activity but no glucuronosyl transferase activity. Humans have little sulfotransferase activity (5× less than rats) but much more glucuronosyl transferase activity (100× more than rats). This kind of information allows a rationale scientific approach to risk assessment.

Conjugation with amino acids. Considerable species differences exist in the conjugation of aromatic carboxylic acids with amino acids. A number of amino acids may be used, although conjugation with glycine is the most common route (chap. 4, Fig. 70) and occurs in most species except some birds, where ornithine is the preferred amino acid. Humans and Old World monkeys use glutamine for conjugation of arylacetic acids, and in the pigeon and ferret, taurine is used. Reptiles may excrete ornithine conjugates as well as glycine conjugates, and some insects use mainly arginine.

Aromatic acids may also be excreted as glucuronic acid conjugates, and the relative importance of glucuronic acid conjugation versus amino acid conjugation depends on the particular species and the structure of the compound. Herbivores generally favor amino acid conjugation, carnivores favor glucuronide formation, and omnivores, such as man, use both routes of metabolism.

There are also species differences in the site of conjugation; this usually occurs in both the liver and kidney, but dogs and chickens carry out this conjugation only in the kidney.

Glutathione conjugation. Conjugation with glutathione, which results in the urinary excretion of *N*-acetylcysteine or cysteine derivatives (chap. 4, Fig. 60) occurs in man, rats, hamsters, mice, dogs, cats, rabbits, and guinea pigs. Guinea pigs are unusual, however, in generally not excreting *N*-acetylcysteine conjugates, as the enzyme responsible for the acetylation of cysteine is lacking.

Insects are also capable of forming glutathione conjugates, this being probably involved in the dehydrochlorination of the insecticide DDT, a reaction at least some insects, such as flies, are able to carry out (chap. 4, Fig. 42).

Methylation. Methylation of oxygen, sulfur, and nitrogen atoms seems to occur in most species of mammal and in those birds, amphibia, and insects, which have been studied.

Acetylation. Most mammalian species are able to acetylate aromatic amino compounds, the major exception being the dog. Thus, for a number of amino compounds such as procainamide (chap. 4, Fig. 43), sulfadimethoxine, sulfamethomidine, sulfasomizole, and the N4 amino group of sulfanilamide (chap. 4, Fig. 68), the dog does not excrete the acetylated product. However, the dog does have a high level of deacetylase in the liver and also seems to have an acetyltransferase inhibitor in the liver and kidney. Consequently, acetylation may not be absent in the dog, but rather the products may be hydrolyzed or the reaction effectively inhibited.

The dog does, however, acetylate the N1, sulfonamido group of sulfanilamide (chap. 4, Fig. 68), and also acetylates aliphatic amino groups. The guinea pig is unable to acetylate aliphatic amino groups such as that in cysteine. Consequently, it excretes cysteine rather than *N*-acetylcysteine conjugates or mercapturic acids. Birds, some amphibia, and insects are also able to acetylate aromatic amines, but reptiles do not use this reaction.

In some cases, species differences in toxicity are due to more than one metabolic difference. For example, research on the fungal toxin aflatoxin B1 indicates that humans are particularly susceptible, more so than rodents, with rats being more susceptible than mice. Interestingly, cynomologous monkeys are also relatively insensitive probably due to the lack of constitutive CYP1A2.

The toxicity (hepatocarcinogenicity) measured as the TD_{50} was greater than 70 μg AFB_1 kg^{-1} day^{-1} in male mice compared with 1.3 μg AFB_1 kg^{-1} day^{-1} for male rats. Hamsters appeared more sensitive than rats.

Measurement of aflatoxin DNA adducts gave a similar result with mice having the lowest level and rats the greatest, with hamsters in between, giving values for rats 3 times those in hamsters but 40 to 600 times higher than those in mice. Using albumin adducts from peripheral blood as a surrogate biomarker, humans exposed to aflatoxin had relative values of 1.56 compared with 0.3 to 0.5 for rats and 0.025 for mice. However, if adjustment is made for surface area between rats and humans, both species give similar values. A similar result was obtained using hepatocytes with mice showing markedly less DNA binding than human or rat hepatocytes. However, metabolic activation of AFB_1 by cytochrome P-450 seems to be greater in mice than either rats or humans. Furthermore, the *major* metabolic product is the reactive metabolite, AFB_1-8,9-epoxide in mice and rats but not in man, where the major metabolite is AFQ_1 (Fig. 5.14). One enantiomer of this metabolite, AFB_1-*exo*-8,9-oxide, binds extensively to DNA, whereas the *endo*-8,9-oxide, which is also formed, does not.

The aflatoxin epoxide, which is responsible for the carcinogenicity, can be detoxified by conjugation with glutathione, catalyzed by GSTs. However, it seems that the reason for the species differences in susceptibility are as follows: although the mouse produces greater amounts of the epoxide than rats or humans, this species also has much greater levels of α-GST activity compared to the rat. Therefore, the rat does not adequately detoxify the reactive epoxide metabolite. With humans, there is α-GST present, but it has little activity toward the carcinogenic aflatoxin epoxide and human μ-GST preferentially conjugates the *endo*-peroxide. It can be seen therefore that lack of adequate detoxication can be at least as important as the presence of activating systems.

Concluding remarks.   Thus most species differences in metabolism are quantitative rather than qualitative; only occasionally does a particular single species show an inability to carry out a particular reaction, or to be its sole exponent. The more common quantitative differences depend on species differences in the enzyme concentration or its kinetic parameters, the availability of cofactors, the presence of reversing enzymes or inhibitors, and the concentration of substrate in the tissue.

These quantitative differences may often mean, however, that different metabolic routes are favored in different species, with a consequent difference in pharmacological or toxicological activity.

In general, man is able to carry out all the metabolic transformations found in other mammals and does not show any particular differences in the presence or absence of an enzymatic pathway.

However, it would be difficult at this time to pick a single species as the best model for man on metabolic grounds alone. It may be possible to do this after a consideration of the structure of the foreign compound in question, however.

5.3.2 Strain
Differences in the disposition of foreign compounds between different strains of the same species have been well documented. For example, mice and rats of various strains show marked differences in the duration of action of **hexobarbital**, whereas within any one strain, the response is uniform (Table 5.13). It is noteworthy that the variation as indicated by the standard deviation is greatest for the outbred group.

The metabolism of antipyrine in rats varies widely between different strains. A well-known example of a strain difference is that of the Gunn rat, which is unable to form *o*-glucuronides of bilirubin and most foreign compounds. This defect is due to a deficiency in glucuronyl transferase. N-acetyltransferase activity has been found to vary with the strain of rats and mice. The hydrolysis of acetylisoniazid to acetylhydrazine (chap. 4, Fig. 46) was found to be significantly different between two strains of rats, as was the hepatotoxicity of the acetylisoniazid, being greater in the strain with the greater acylamidase activity (see chap. 7).

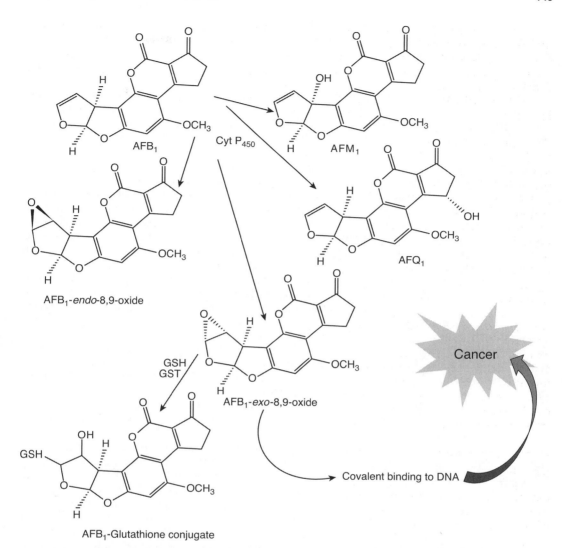

Figure 5.14 Routes of oxidative metabolism of aflatoxin B_1 showing activation to an epoxide catalyzed by CYP1A2 and CYP3A4 and conjugation with glutathione catalyzed by α-GST. The 8,9-*exo*-epoxide is the enantiomer, which binds extensively to DNA and is carcinogenic. The *endo*-epoxide binds less readily. *Abbreviation*: α-GST, α-glutathione transferase.

Table 5.13 Strain Differences in the Duration of Action of Hexobarbital in Mice (Dose of Barbiturate 125 mg kg^{-1} Body Weight)

Species	Numbers of animals	Mean sleeping time (min) ± S.D.
A/NL	25	48 ± 4
BALB/cAnN	63	41 ± 2
C57L/HeN	29	33 ± 3
C3HfB/HeN	30	22 ± 3
SWR/HeN	38	18 ± 4
Swiss (non-inbred)	47	43 ± 15

Source: From Ref. 25.

Humans also show such genetically based variations, and this is discussed later in this chapter. Strain or genetic differences in responsiveness may also be important, such as in the response to enzyme inducers discussed below.

Different strains of mice exhibit different susceptibilities to TCDD (dioxin) and other polycyclic hydrocarbons. Thus the C57Bl 6 mouse is more responsive to substances such as TCCD and 3-methylcholanthrene, which interact with the AhR receptor and induce enzymes such as CYP1A1, whereas the DBA/2 mouse is less responsive. This has been found to be inherited as an autosomal dominant trait. The nonresponsive allele Ah^d expresses a protein with diminished affinity for binding ligands such as TCDD.

5.3.3 Sex

There are a number of documented differences in the disposition and toxicity of foreign compounds, which are related to the sex of the animal. For instance, the organophosphorus compound, **parathion**, is twice as toxic to female as compared with male rats. Many such gender differences in toxicity have been noted in rodents. For example, in rats, females are more susceptible to the toxicity of the drugs phenacetin, salicylic acid, and warfarin and also to nicotine and picrotoxin. One of the first sex differences to be noted was the fact that hexobarbital-induced sleeping time is longer in female than in male rats. This is in accord with the view that, in general, male rats metabolize foreign compounds more rapidly than females. Thus, the biological half-life of **hexobarbital** is considerably longer in female than in male rats, and *in vitro*, the liver microsomal fraction metabolizes both hexobarbital and aminopyrine more rapidly when derived from male rather than female rats. In contrast, the male rat is more susceptible to the toxic effects of monocrotaline, epinephrine, and ergot. As well as phase 1 reactions, phase 2 reactions also show sex differences. Thus, glucuronic acid conjugation of **1-naphthol** (chap. 4, Fig. 4) is greater in male than female rats, and this difference is also found in microsomes *in vitro*. The acetylation of **sulfanilamide** (chap. 4, Fig. 68) is also greater in male than female rats. Sex differences in metabolism depend on the substrate, however. For example, the hydroxylation of aniline or zoxazolamine shows little difference between the sexes, in contrast to the threefold greater metabolism of hexobarbital or aminopyrine in male compared with female rats.

Sex differences in metabolism are less pronounced in species other than the rat. In humans, differences, when noted, are similar to those in the rat, whereas in the mouse, they are often the reverse of those in the rat. The differences in metabolism between males and female animals are due to the influence of hormones and genetic factors. Many of the gender differences in both phase 1 and phase 2 enzymes in rodents develop during puberty in the animal and are influenced by sex hormone concentrations or indirect effects of growth hormone. The administration of androgens to female animals abolishes the differences from the males. The sex hormones appear to have a powerful influence on the metabolism of foreign compounds. For example, study of the metabolism of aminopyrine and hexobarbital in male rats showed that castration markedly reduced metabolism of both drugs, but this was restored by the administration of androgens (Fig. 5.15).

Thus in some cases, enzyme expression is directly influenced in a particular organ by testosterone or estrogens. For example, in the mouse kidney, testosterone directly regulates the expression of cytochrome P-450 isozymes, and this leads to the particular sensitivity of the female kidney to the nephrotoxicity of paracetamol.

Many rat liver cytochromes P-450 show gender differences, some such as CYP2A2 and CYP3A2, which are higher in males, and CYP2A1 and CYP2C7, which are higher in females. Other enzymes also show gender differences such as epoxide hydrolase, glutathione transferases, and some glucuronosyl transferases and sulfotransferases, which are higher in male rat liver. Also levels of many FMO enzymes are higher in females.

Although gender differences are much less apparent in other species including humans, there are some, such as the higher levels of CYP3A4 in human females compared with males.

However, the control of metabolism appears to be more complicated than this as it also involves the **hypothalamus** and **pituitary gland**. It seems that the male hypothalamus produces a factor, which inhibits the release of a hormone and which therefore leaves the liver in a particular male state. In the female, the hypothalamus is inactive, and therefore produces no factor, and hence the pituitary releases a feminizing factor (possibly growth hormone), which changes the liver to the female state.

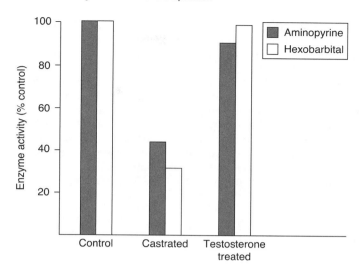

Figure 5.15 The effect of castration and androgen replacement on the metabolism of aminopyrine and hexobarbital in the rat. *Source*: From Ref. 26.

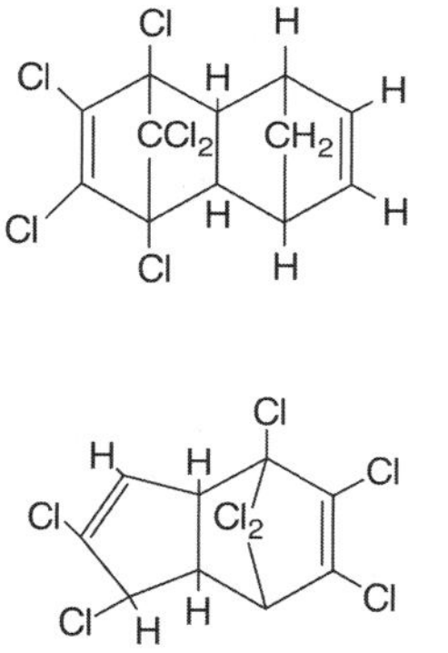

Figure 5.16 Structures of the insecticides aldrin and heptachlor.

This activity of the hypothalamus does not depend on the genotypic sex of the animal, but on events during the perinatal period. For example, male rats have very low activity for the hydroxylation of steroid sulfate conjugates (15 β-hydroxylase). This applies to both normal and castrated male animals. However, if male animals are castrated immediately after birth, they show a greater female level of activity. This effect can be reversed by treatment of castrated males with testosterone during the period immediately after birth, but not by treatment of the adult animal. This has been interpreted as indicating that it is the absence or presence of androgen in the period immediately after birth, which determines the maleness or femaleness of the animal as regard the enzymes involved in the metabolism of foreign compounds. The phenomenon is known as imprinting.

As already mentioned, the pharmacological activity of certain drugs is lengthened in the female rat, and also the toxicity of certain compounds may be increased. **Procaine** is hydrolyzed more in male rats (chap. 4, Fig. 43), with consequently a lower toxicity in the sex, whereas the insecticides **aldrin** and **heptachlor** are metabolized more rapidly to the more toxic epoxides in males and are therefore less toxic to females (Fig. 5.16).

Probably the best example of a sex difference in toxicity is that of the renal toxicity of **chloroform** in mice. The males are markedly more sensitive than the females, and this difference can be removed by castration of the male animals and subsequently restored by administration of androgens.

In vitro, chloroform is metabolized 10 times faster by kidney microsomes from male compared with female mice. In contrast to rats, male mice given hexobarbital metabolize it more slowly, and the pharmacological effect is more prolonged than in females. The excretion of the food additive butylated hydroxytoluene (Fig. 5.17) shows an interesting sex difference in rats, being mainly via the urine for males but predominantly via fecal excretion in females, which probably reflects differences in the rates of production of glucuronide and mercapturate conjugates between the sexes.

A similar but particularly important example, which results in a difference in toxicity, is afforded by 2,6- and 2,4-**dinitrotoluenes** (see Fig. 5.18 for illustration using 2,6-dinitrotoluene). These compounds (to which there may be industrial exposure) have been found to be hepatocarcinogenic, especially in male rats. The compounds are first metabolized by hydroxylation of the methyl group followed by conjugation with glucuronic acid. The greater susceptibility to hepatocarcinogenicity of the male animals has been shown to be due to the greater biliary excretion of the glucuronide conjugate in the males. This is followed by breakdown in the gut by β-glucuronidase in intestinal bacteria to give the aglycone

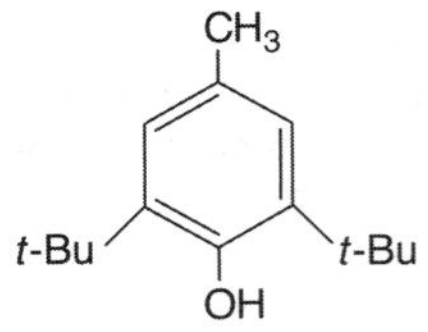

Figure 5.17 Structure of the food additive butylated hydroxytoluene.

R = SO₃H or COCH₃

Nitrenium ion

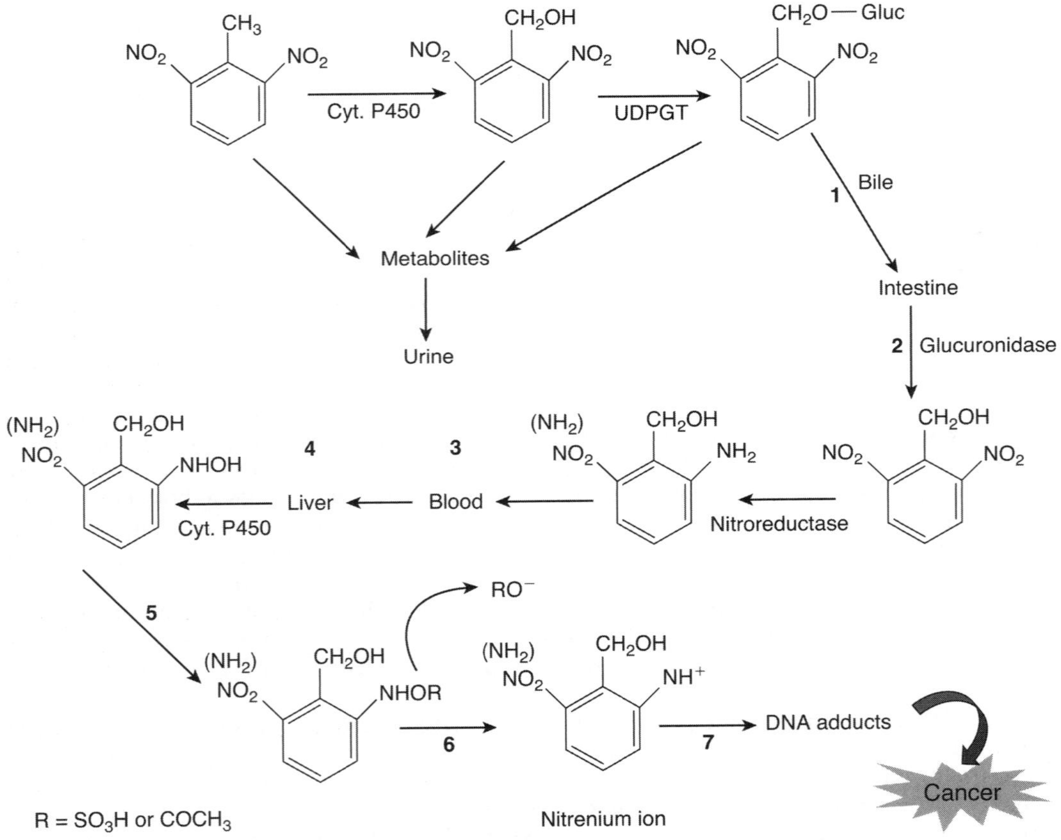

Figure 5.18 The role of metabolism and enterohepatic recirculation in the sex difference in toxicity of 2,6-dinitrotoluene in rats. In the female animal, the glucuronide conjugate is excreted mainly in the urine, whereas in the male, it undergoes enterohepatic recirculation following biliary excretion (1). Metabolism by the gut flora (β-glucuronidase) (2) releases the aglycone, which is further metabolized by gut flora (nitroreductase) and reabsorbed via the blood into the liver (3). In the liver, it is metabolized to a hydroxylamine (cytochrome P-450) (4), which is conjugated either with sulfate or is acetylated (sulfotransferase/acetyltransferase) (5). These are good leaving groups, and a reactive nitrenium ion is formed (6), which interacts with liver macromolecules such as DNA (7) and causes liver tumors.

(1-hydroxymethyl-2,6,-dinitrotoluene) and reduction by nitroreductase in the bacteria. The reduced aglycone is then reabsorbed into the liver where it undergoes metabolic activation by N-oxidation and conjugation with either sulfate or acetyl, which can yield a reactive intermediate nitrenium ion, which binds to liver DNA. This is believed to cause the liver tumors. This example illustrates the importance of the multiple steps (in this case of metabolism and excretion), which may be necessary to cause toxicity.

Kidney tumors caused by several different compounds, including 1,4-dichlorobenzene, isophorone, and unleaded petrol, have been found to be both sex dependent and species dependent. Thus, only male rats suffer from α_2-μ-globulin nephropathy and renal tubular adenocarcinoma as a result of the accumulation of a compound-protein complex in the epithelial cells of renal proximal tubules (see chap. 6). The synthesis of the protein involved, α_2-μ-globulin, is under androgenic control in the male rat.

5.3.4 Genetic Factors

Inherited differences in the metabolism and disposition of foreign compounds, which may be seen as strain differences in animals and as racial and interindividual variability in man, are of great importance in toxicology.

There are many examples of human subjects showing idiosyncratic reactions to the pharmacological or toxicological actions of drugs. In some cases, the genetic basis of such reactions has been established, and these cases underline the need for an understanding and appreciation of genetic factors and their role in the causation of toxicity.

Genetic factors may affect the toxicity of a foreign compound in one of two ways:

1. By influencing the response to the compound
2. By affecting the deposition of the compound

Genetic Factors Affecting the Response to the Compound
Glucose-6-phosphate dehydrogenase deficiency.   A well-known example of the first type in humans is deficiency of the enzyme glucose-6-phosphate dehydrogenase, which is associated with susceptibility to drug-induced **hemolytic anemia**. This is a genetically determined trait, carried on the X chromosome and so it is sex-linked, but the inheritance is not simple. Overall 5% to 10% of Negro males suffer the deficiency, but it is particularly common in the Mediterranean area and in some ethnic groups, such as male Sephardic Jews from Kurdistan; its incidence may be as high as 53%. Worldwide as many as 100 million people are affected by this genetic deficiency. It is of interest to note that the deficiency confers resistance to the malarial parasite. The biochemical basis for this increased sensitivity or response is the result of variants in the glucose-6-phosphate dehydrogenase enzyme rather than a complete absence. There are 34 gene variants due to missense mutants or small in-frame deletions, which result in various degrees of loss of enzyme activity, the most severe having about 10% of normal activity. A complete loss of activity, as can occur in mice, is lethal. The enzyme deficiency gives rise to a deficiency in the concentration of reduced glutathione (GSH) in the red blood cell, as shown in Figure 5.19. The reaction, catalyzed by glucose-6-phosphate dehydrogenase and carried out in the red blood cell, is the first step in the pentose-phosphate shunt, or hexose monophosphate pathway.

The enzyme maintains the level of NADPH in the red cell, which in turn maintains the level of GSH. This functions to protect the hemoglobin from damage caused by oxidizing agents, which might cause hemolysis. Thus, if the enzyme glucose-6-phosphate dehydrogenase has reduced activity, the levels of NADPH, and therefore of GSH, are low. This situation leaves the hemoglobin unprotected, and consequently when the afflicted patient takes a drug such as **primaquine** or is exposed to a chemical such as **phenylhydrazine**, which may generate reactive superoxide and hydrogen peroxide, which are not removed by GSH, hemoglobin can be oxidized and hemolytic anemia ensues.

As illustrated with the drug (Fig 7.46), the metabolites may undergo redox GSH, cycling to yield reactive oxygen species, which can be detoxified by reduced glutathione. In the absence of sufficient GSH, the reactive oxygen species can react with and damage hemoglobin in the red cell. Also reactive metabolites of the drug may be reduced by GSH and form more oxidized glutathione (GSSG). Both the N-hydroxy metabolite and the 5-hydroxy metabolite

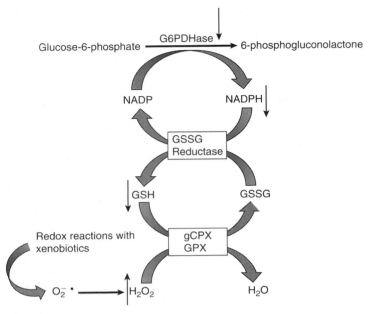

Figure 5.19 The interrelationship between glucose-6-phosphate dehydrogenase, NADPH, GSH, and the reduction of oxidized metabolites. The arrows indicate the consequences of a genetic deficiency in G6PDHase. *Abbreviations*: G6PDHase, glucose-6-phosphate dehydrogenase; GSH, reduced glutathione; GSSG, oxidised glutathione; GSSG reductase, glutathione reductase; GPX, glutathione peroxidase.

Figure 5.20 The structure of the β-glycoside vicine, found in Fava beans, which in the gut is cleaved, as shown, to the toxic product divicine.

will cause damage to red cells *in vitro* and damage to the cytoskeleton, including the formation of hemoglobin-protein adducts. The depletion of red cell glutathione markedly increases the cytotoxicity of the *N*-hydroxy metabolite *in vitro*. Removal of damaged red cells by the spleen prematurely will occur *in vivo*.

Another cause of this type of hemolytic anemia is eating Fava beans, which contain the substance vicine, a β-glycoside, which is cleaved in the gut to divicine (Fig. 5.20). Divicine has been shown to deplete GSH, cause oxidative stress, and oxidative damage to hemoglobin from reactive oxygen species, in red cells *in vitro* and to red cells in rats *in vivo*. Cells exposed *in vitro* become spikey in appearance and hemoglobin becomes covalently bound the cytoskeleton of the red cell. *In vivo* exposure of rats to divicine causes hemoglobinuria and the enlargement of the spleen along with loss of red cells and depletion of glutathione in the red cells.

Oxidized, denatured hemoglobin forms aggregates, which can become attached to the inner surface of the red cell, known as Heinz bodies. This leads to damage to the red cell, which may result in direct destruction of the cell, which can be shown *in vitro*, or removal from the circulation by the spleen *in vivo*. When caused by Fava beans, the syndrome is known as **Favism**. As the deficient enzyme (glucose-6-phosphate dehydrogenase) is intrinsic to the red cell, exposure of such cells *in vitro* to suitable drugs will lead to cell damage and death.

Other similar genetic factors, which lead to a heightened response are those, which affect hemoglobin more directly. Thus, there are genetic traits in which the hemoglobin itself is sensitive to drugs such as **sulfonamides**. Exposure to certain drugs such as primaquine and **dapsone** can oxidize hemoglobin to methemoglobin, which is unable to carry oxygen. Individuals who have a genetically determined deficiency in methemoglobin reductase are unable to remove this methemoglobin and hence suffer cyanosis.

Heightened sensitivity to **alcohol** in individuals of Oriental origin is another example of a genetically determined increased response.

Genetic Factors Affecting the Disposition of the Compound

The second type of genetic factor, where the disposition is affected, is probably more important in terms of the toxicity of foreign compounds. Variability in human populations makes drug therapy unpredictable and risk assessment for chemicals difficult. An indication of the scale of variability is afforded by the drug **paracetamol**, which is discussed in more detail in chapter 7. Thus, rates of metabolic oxidation for this drug were found to vary over a 10-fold range in human volunteers. The highest rate of oxidation, occurring in about 5% of the population studied, was comparable to that in the hamster, the species most susceptible to the hepatotoxicity of paracetamol. However, those human individuals with the lowest rates would probably be relatively resistant to the hepatotoxicity, similar to the rat. This example highlights the problems of using a single, inbred animal model to try to predict toxicity and risk to heterogeneous human populations. The variation seen may be only partly due to genetic influences, there being many factors in the human environment, which may also affect drug disposition, as discussed later in this chapter. Genetically determined enzyme deficiencies, which affect the disposition of foreign compounds, occur in both man and animals, and the example of the Gunn rat, which is deficient in glucuronosyl transferase, has already been mentioned.

Similar defects have been described in man; for example, **Gilbert's syndrome** and the **Crigler–Najjar syndrome** are both associated with reduced glucuronyl transferase and the consequent reduced ability to conjugate bilirubin. The administration of drugs also conjugated as glucuronides and, therefore in competition with the enzyme, may lead to blood levels of unconjugated bilirubin sufficient to result in brain damage.

Acetylation Polymorphism: Acetylator Phenotype

Perhaps one of the best known and fully described genetic factors in drug disposition and metabolism is the acetylator phenotype. It has been known for more than 30 years that for certain drugs which are acetylated, [**isoniazid** (Fig. 7.15) was the prototype] in human populations there is variation in the amount of this acetylation. This variation was found to have a genetic basis and did not show a normal Gaussian distribution, but one interpreted as bimodal, suggesting a *genetic polymorphism* (Fig. 5.21). The two groups of individuals, were

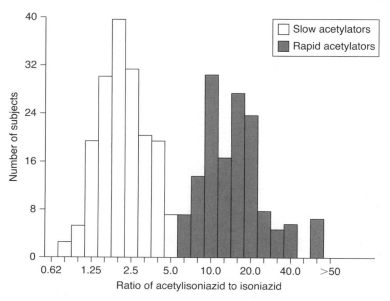

Figure 5.21 Frequency distribution for isoniazid acetylation. The acetylated metabolite (acetylisoniazid) and total isoniazid (acid labile isoniazid) were measured in the urine. *Source*: From Ref. 27.

Table 5.14 Some Toxicities and Diseases Proposed to be Associated with Acetylator Phenotype

Foreign compound	Adverse effect	Higher incidence
Isoniazid	Peripheral neuropathy	Slow acetylators
Isoniazid	Hepatic damage	Slow acetylators
Hydralazine	Lupus erythematosus	Slow acetylators
Procainamide	Lupus erythematosus	Slow acetylators
Isoniazid plus phenytoin	Central nervous system toxicity	Slow acetylators
Sulfasalazine	Hemolytic anemia	Slow acetylators
Aromatic amines	Bladder cancer	Slow acetylators
Aromatic amines	Mutagenesis/carcinogenesis	Rapid acetylators
Food pyrolysis products[a]	Colorectal cancer	Rapid acetylators
Amonafide	Myelotoxicity	Rapid acetylators

[a]Has yet to be fully substantiated.

termed "rapid acetylators" and "slow acetylators" because of the difference in both the amount and rate of acetylation. There are now a number of drugs, which show this polymorphism in humans, including isoniazid, **hydralazine, procainamide, dapsone, phenelzine, aminoglutethimide, sulfamethazine, caffeine** metabolites and probably aromatic amines such as the carcinogen **benzidine** and heterocyclic amines such as **PhIP** (see chap. 4). The acetylation polymorphism is an important genetic factor in a number of toxic reactions (Table 5.14), and both isoniazid and hydralazine are discussed in more detail in chapter 7. The situation with regard to aromatic amine and heterocyclic amine-induced cancer is somewhat complex.

With aromatic amines such as benzidine, 4-aminobiphenyl and 2-aminonaphthalene, which cause bladder cancer, epidemiological evidence suggests that those with the slow acetylator phenotype who have occupational exposure are more at risk. In contrast, with the heterocyclic amines produced in food by cooking, such as PhIP, which cause colon cancer, it seems from similar evidence that fast acetylators are more at risk.

The reasons for these differences have been already discussed in chapter 4. The trait has now been observed in other species, including the rabbit, mouse, hamster, and rat. Studies in human populations suggest that the acetylator phenotype is a single gene trait, with two alleles at a single autosomal genetic locus with slow acetylation a simple Mendelian recessive trait. The dominant-recessive relationship is unclear, however. Genetic studies in the rabbit, mouse, and hamster indicate that there is codominant expression of these alleles. Thus there are three possible genotypes with the following arrangement of the slow (r) and rapid (R) alleles: rr, rR, and RR.

These would be manifested as slow, intermediate, and rapid acetylators. However, the ability to distinguish the three phenotypes in human populations is dependent on the method used and the particular drug administered. The heterozygous group, rR, may therefore be indistinguishable from the homozygous rapid group. The genetic trait especially affects the activity or stability of the N-acetyltransferase enzyme NAT2. The mutations in the gene give rise to many allelic variants (26), which results in an enzyme with decreased activity or stability, and the activity can vary by as much as 100 times. Thus slow acetylators have reduced tissue enzyme activity.

Substrates for NAT2 therefore have been termed "polymorphic" and substrates for NAT1 "monomorphic." However, it is *now* recognized that NAT1 can also show genetic variation, indeed there are 22 allelic variants, but these are less prevalent than those affecting NAT2, and consequently there is less genetic variation with NAT1. Thus *p*-aminosalicylic acid is metabolized by NAT1, yet there is evidence that suggests a bimodal distribution in humans.

However, with substrates for NAT1, the picture is not entirely clear. For example, *p*-aminobenzoic acid is another NAT1 substrate, and there is evidence from one study suggesting that at least in individuals who are phenotpyically fast or slow acetylators for NAT2, there is no clear phenotype distinction for NAT1. This is illustrated in Figures 5.22 and 5.23, where individuals given procainamide, a NAT2 substrate, separated into fast and slow acetylators. However, the metabolite of procainamide, *p*-aminobenzoic acid, which is also acetylated, does not show a bimodal distribution in those same individuals. However, because

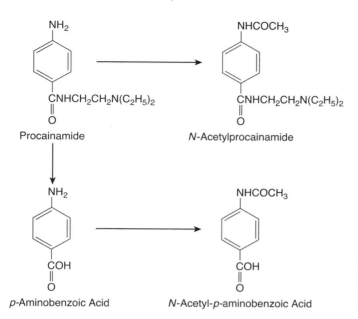

Figure 5.22 Metabolism of procainamide. Procainamide and the hydrolysis product *p*-aminobenzoic acid are both acetylated.

(A)

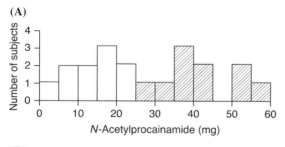

(B)

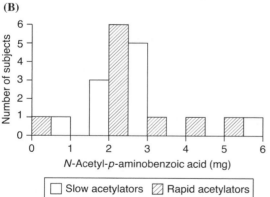

☐ Slow acetylators ▨ Rapid acetylators

Figure 5.23 Frequency distribution for the acetylation of (**A**) procainamide (**B**) and *p*-aminobenzoic acid in human subjects. Data represents excretion of acetylated product in the urine 6 hours after dosing. *Source*: From Ref. 28.

there are different allelic variants for each NAT and presuming they are independent, perhaps this datum is not surprising. Indeed the two forms of the enzyme, NAT1 and NAT2, are both independently regulated. A further complication is that some chemicals are substrates for both NAT1 and NAT2, and so in a slow acetylator for NAT2, deficiency in activity could be made up by NAT1. Thus, because NAT1 may also metabolize a particular substrate, the overall effect of phenotypic variation in NAT2 may be modulated.

Studies using liver from rapid and slow acetylator phenotype rabbits *in vitro* showed that there were indeed small differences in acetylation (2–2.5×) for "monomorphic" substrates such as *p*-aminobenzoic acid, although much smaller than for "polymorphic" substrates (10–100×).

Table 5.15 Characteristics of Rabbit N-Acetyltransferase with Various Substrates

Substrate	Apparent K_m (µM)		Apparent V_{max} (nmol min^{-1} mg^{-1})	
	Rr	RR	rr	RR
PABA	<5	105	0.24	9.3
PAS	<5	74	0.31	5.0
PA	200	67	0.35	4.4
SMZ	160	90	0.38	4.8

Abbreviations: PABA, p-aminobenzoic acid; PAS, p-aminosalicylic acid; PA, procainamide; SMZ, sulfamethazine.
Source: From Ref. 29.

However, using "polymorphic" substrates with the isolated enzyme, no correlation could be found between K_m and acetylator status; only when "monomorphic" substrates (p-aminobenzoic acid and p-aminosalicylic acid) were used was this apparent (Table 5.15). It is notable that both these substrates are negatively charged at physiological pH.

In rabbits, the activity of liver N-acetyltransferase was studied *in vitro* and found to show trimodal or bimodal distributions.

Thus, the K_m and V_{max} data obtained for the isolated rabbit N-acetyltransferase enzyme using p-aminobenzoic acid and p-aminosalicylic acid show striking differences between genotypes and support the hypothesis that the enzymes from slow and rapid acetylators are structurally different (Table 5.5.15). The enzyme from heterozygous animals was intermediate between the homozygous values. The slow acetylator enzyme thus has a much greater affinity for these particular substrates, but a lower capacity than the rapid acetylator enzyme. It seems that in the rabbit the N-acetyltransferase from the slow phenotype is probably saturated *in vivo* at normal concentrations of drug and is therefore operating at maximum capacity, whereas the rapid enzyme has such a high K_m that it would remain unsaturated and would be below maximum capacity. For monomorphic substrates therefore, the overall rates of metabolism *in vivo* would be similar. For polymorphic substrates (e.g., sulfamethazine and procainamide), the K_m values are much greater and more similar in the two phenotypes, but the V_{max} values are markedly different, and therefore the metabolism of such compounds *in vivo* is different in the two phenotypes (Table 5.15).

It should be noted that in the hamster the situation is the reverse, with p-aminobenzoic acid and p-aminosalicylic acid being polymorphically acetylated and sulfamethazine and procainamide monomorphically acetylated.

Thus in man and a number of other species, the basis for the acetylator polymorphism is qualitative differences in NAT2 and to some extent NAT1, resulting from mutations. There may also be posttranslational modifications of the primary gene product. In mice, there is evidence that other factors such as modifier genes may also affect the expression of acetyltransferase activity.

As can be seen from Table 5.14, the acetylator phenotype is a factor in a number of toxic effects due to foreign compounds, including the carcinogenicity of aromatic amines. As already mentioned in chapter 4, generally the slow acetylators are more at risk, probably because acetylation protects the amino or hydrazine group from metabolic activation. Thus in the case of carcinogenic aromatic amines such as benzidine (chap. 4, Fig. 69), it has been suggested that slow acetylators are more susceptible to bladder cancer. (Thus one study showed a relative risk of slow to rapid of 1.36 and for individuals exposed to aromatic amines in industry a relative risk of slow to rapid of 1.7.) However, this is by no means certain, as one recent study of the role of the acetylator phenotype in bladder cancer in Chinese workers exposed to benzidine found no association of increased risk with the slow acetylator phenotype, which, it was suggested, may even have a protective effect. Contrary to this, another very recent study in Taiwanese individuals did find an association between the slow acetylator (NAT2) genotype and bladder cancer. Conversely, evidence is accumulating suggesting rapid acetylators may be more at risk from colorectal cancer, possibly as a result of exposure to aromatic amines. Pyrolysis of food during cooking produces various mutagenic and carcinogenic amines, which are known to be acetylated. However, as explained in chapter 4, as well as N-acetylation, which detoxifies, O-acetylation can occur and yield toxic and potentially carcinogenic products (see chapter 4 for details and chap. 4, Fig. 69). Studies have shown that rapid acetylator mice have greater DNA adduct formation with 2-aminofluorene than slow acetylator mice. As some aromatic amines (such as 2-aminofluorene) are metabolized equally by NAT1 and NAT2 in some species, clearly the impact of the acetylator phenotype on cancer risk could be minimal depending on which combination of mutant alleles were present. The role of acetylation in carcinogenicity is discussed in more detail in chapter 7 with regard to aminofluorene derivatives.

There are examples where several genetic factors, including the acetylator phenotype, operate together. Hydralazine toxicity is one such example, which is discussed in detail in chapter 7. Another is the hemolytic anemia caused by the drug **thiozalsulfone** (Promizole), which occurs particularly in those individuals who are *both* glucose-6-phosphate dehydrogenase deficient and slow acetylators. Promizole is acetylated, and studies in rapid and slow acetylator mice confirmed that acetylation was a factor as well as an extent of hydroxylation. The latter may also be another factor in humans as is discussed below.

The acetylator polymorphism also exhibits an interesting **ethnic distribution** in humans, as shown in Table 5.16, which may have important implications for the use of drugs in different parts of the world. As well as being a factor in the toxicity of several drugs, the acetylation polymorphism may influence the efficacy of treatment. For example, the plasma half-life of isoniazid is two to three times longer, and the concentration higher in slow acetylators compared with rapid acetylators. Therefore, the therapeutic effect will tend to be greater in the slow acetylators as the target microorganism is exposed to higher concentrations of drug.

Polymorphisms in cytochrome P-450

CYP2D6 polymorphisms (hydroxylator status). More recently than the acetylator phenotype was discovered, it has become apparent that genetic factors also affect phase 1 oxidation pathways. Early reports of the defective metabolism of diphenylhydantoin in three families and of the defective de-ethylation of phenacetin in certain members of one family indicated a possible genetic component in microsomal enzyme-mediated reactions. Both these cases resulted in enhanced toxicity. Thus, diphenylhydantoin, a commonly used anticonvulsant, normally undergoes aromatic hydroxylation and the corresponding phenolic metabolite is excreted as a glucuronide (fig. 7.54). Deficient hydroxylation results in prolonged high blood levels of diphenylhydantoin and the development of toxic effects, such as nystagmus, ataxia, and dysarthria. The deficiency in the ability to hydroxylate diphenylhydantoin is inherited with dominant transmission.

The defective de-ethylation of phenacetin was discovered in a patient suffering methemoglobinemia after a reasonably small dose of the drug. This toxic effect was observed in a sister of the patient but not in other members of the family. The metabolism of phenacetin in the patient and in the sister was found to involve the production of large amounts of the normally minor metabolites 2-hydroxyphenacetin and 2-hydroxyphenetidine, with a concomitant reduction in the excretion of paracetamol, the major metabolic product of

Table 5.16 Acetylator Phenotype Distribution in Various Ethnic Groups

Ethnic group	Rapid acetylators (%)
Eskimos	95–100[a]
Japanese	88[a]
Latin Americans	70[a]
Black Americans	52[a]
White Americans	48[a]
Africans	43[b]
South Indians	39[a]
Britons	38[b]
Egyptians	18[a]

[a]Phenotype determined using isoniazid.
[b]Phenotype determined using sulfamethazine.
Source: From Ref. 30.

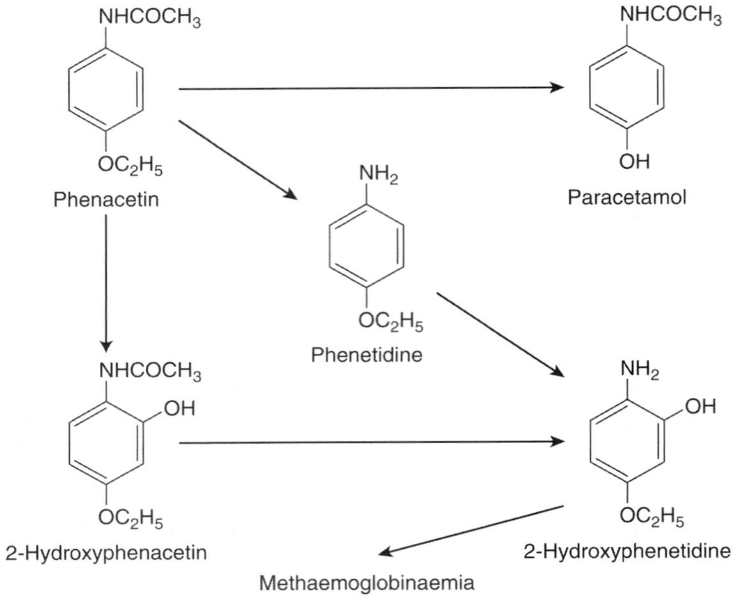

Figure 5.24   Metabolism of phenacetin showing the metabolite believed responsible for methemoglobinemia.

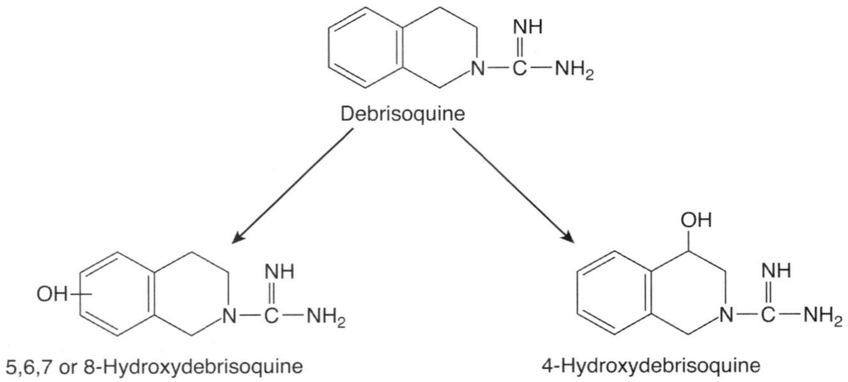

Figure 5.25   Metabolism of debrisoquine in man. CYP2D6 catalyzes the production of 4-hydroxydebrisoquine, the major metabolite.

de-ethylation in normal individuals (Fig. 5.24). It was suggested that autosomal recessive inheritance was involved, with the 2-hydroxylated metabolites probably responsible for the methemoglobinemia.

These early observations suggesting that genetic factors affect the oxidation of foreign compounds were confirmed by studies on the metabolism of the antihypertensive drug **debrisoquine**. The benzylic oxidation in position 4 of the alicyclic ring (Fig. 5.25) has been found to be defective in 5% to 10% of the white population of Europe and North America. This is detected as a bimodal distribution when the metabolic ratio, urinary 4-hydroxydebrosoquine to debrisoquine, is plotted against frequency of occurrence in the population (Fig. 5.26). The two phenotypes detectable are known as poor metabolizer (PM) and extensive metabolizer (EM), and the poor metabolizer phenotype behaves as an autosomal recessive trait. Thus the extensive metabolizers are either homozygous (DD) or heterozygous (Dd), and the poor metabolizers are homozygous (dd). Poor metabolizers suffer an exaggerated pharmacological effect after a therapeutic dose of the drug as a result of a higher plasma level of the unchanged drug (Fig. 5.27). Extensive metabolizers excrete 10 to 200 times more 4-hydroxydebrisoquine than poor

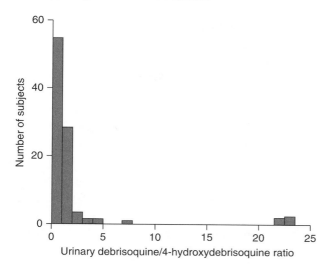

Figure 5.26 Frequency distribution for the ratio of urinary debrisoquine to 4-hydroxyde-brisoquine in human subjects. *Source*: From Ref. 31.

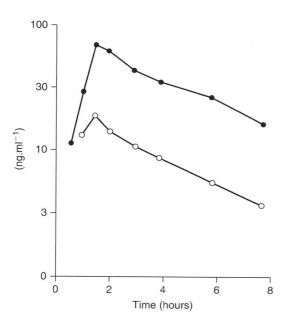

Figure 5.27 The plasma concentration of debriso-quine after a single oral dose (10 mg) in human subjects of the extensive (○) and poor (●) metabolizer phenotypes. *Source*: From Ref. 32.

metabolizers. The deficiency extends to more than 20 different drugs, and various types of metabolic oxidation reaction. For example, the aromatic hydroxylation of **guanoxon**, the O-demethylation of **codeine**, oxidation of **sparteine** (Fig. 5.28), and the hydroxylation of **bufuralol** (Fig. 5.3) have all been shown to be polymorphic. There are now a number of adverse drug reactions, which are associated with the poor metabolizer status. For example, perhexiline may cause hepatic damage and peripheral neuropathy in poor metabolizers in whom the half-life is significantly extended; lactic acidosis may be associated with the use of phenformin in poor metabolizers. Poor metabolizers have a reduced response to pain relief from codeine as the metabolite produced by CYP2D6 is morphine, which is responsible for the analgesia.

The biochemical basis for the trait is an almost complete absence of one form of cytochrome P-450, CYP2D6. It seems that there are several mutations, which give rise to the poor metabolizer phenotype. These mutations produce incorrectly spliced, variant mRNAs in the liver from poor metabolizers. At least three mutant alleles have been described for the CYP2D6 gene and in those with "poor metabolizer" status. A large number of mutations have also been described some of which, but not all, affect enzyme activity. However, for some of

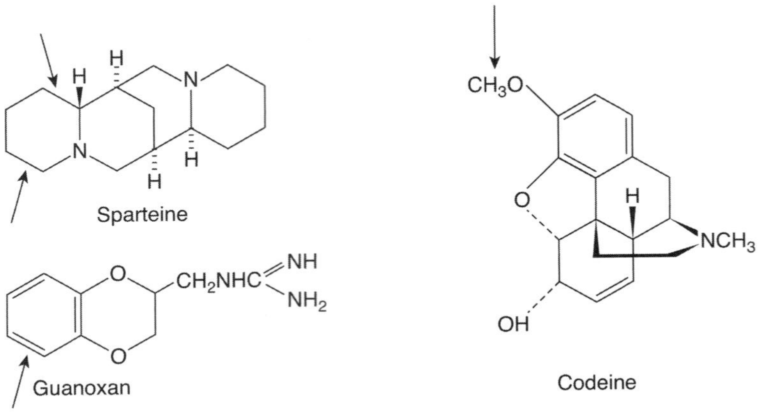

Figure 5.28 Compounds known to show the same oxidation polymorphism in humans as debrisoquine. The arrows indicate the site of oxidation.

the substrates of the polymorphism, such as **bufuralol**, biphasic kinetics has been demonstrated *in vitro*. This datum and data from studies using inhibitors such as quinidine, which is specific for the debrisoquine hydroxylase form of cytochromes P-450, have indicated that two functionally different forms of cytochrome P-450 isoenzymes are involved. These two forms differ in that one has low affinity whereas the other has high affinity. Only one of these may be under polymorphic control. The metabolism of bufuralol is further complicated by the fact that it is stereospecific. The aliphatic hydroxylation (1-hydroxylation) is selective for the (+) isomer, whereas the aromatic 4- and 6-hydroxylations are selective for the (–) isomer. Both aliphatic and aromatic hydroxylation of bufuralol are under the same genetic control, and the selectivity is virtually abolished in the poor metabolizer. Only the high-affinity polymorphic enzyme is stereospecific; the low-affinity enzyme is not.

Studies with sparteine have also indicated that there are two isoenzymes involved in metabolism to 2- and 5-dehydrosparteine. Thus, there is a high-affinity, quinidine-sensitive form and a low-affinity, quinidine-insensitive form. The formation of the metabolites by the two isoenzymes is, however, quantitatively and qualitatively different. Both isoenzymes exhibit large interindividual differences, but it is believed that the low-affinity enzyme is not controlled by the debrisoquine polymorphism. The variation in the high-affinity isoenzyme is suggested as being due to allozymes of cytochrome P-450; that is, different enzymes from the alleles comprising one gene.

The importance of such polymorphisms in human susceptibility to diseases, such as cancer, is now increasingly being recognized. For example, studies in humans with amino biphenyl, a liver and bladder carcinogen, have shown that both the acetylator phenotype and hydroxylator status are important in the formation of adducts (see chap. 6). **Another similar, but distinct, genetic polymorphism concerns the aromatic 4-hydroxylation** of the drug mephenytoin. This hydroxylation deficiency occurs in 2% to 5% of Caucasians and in 20% of Japanese. Again, it is an autosomal recessive trait. The enzyme, cytochrome P-450, form CYP2C19, from poor metabolizers has a high K_m and low V_{max}. The other route of metabolism, N-demethylation, is increased in poor metabolizers.

As with the hydroxylation of bufuralol, the hydroxylation is stereo-selective. Thus, only S-mephenytoin undergoes aromatic 4-hydroxylation, and only this route is affected by the polymorphism. The R isomer undergoes N-demethylation. Poor metabolizers may suffer an exaggerated central response when given therapeutic doses (Fig. 5.29).

There is considerable variation in CYP3A4 expression in individual humans, as much as one to two orders of magnitude. This may be due to exposure to dietary or industrial chemicals or drugs. Also genetic variation occurs, as exemplified by studies in Finnish, Taiwanese, and black populations. Studies found that 2.7% of Finns had an allelic variant, designated CYP3A4*2. This base change in the coding gene lead to a change in one amino acid in the enzyme. The effect of that was to decrease the intrinsic activity of the enzyme when determined *in vitro* using the enzymes generated from the cDNAs for the gene and nifedipine

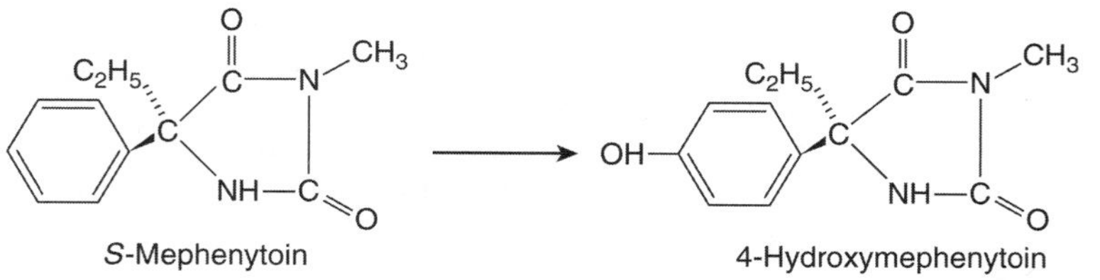

Figure 5.29 Aromatic hydroxylation of *S*-mephenytoin.

Table 5.17 Metabolism of Nifedipine by a Mutant Form of CYP3A4 *In Vitro* in Comparison with the Wild-Type Enzyme

Kinetic parameters	CYP3A4 wild-type enzyme	CYP3A42 mutant enzyme
V_{max} (min^{-1})	36	23
K_m (μM)	2	11
Intrinsic clearance V_{max}/K_m	18	2

Source: From Ref. 33.

as substrate. Thus the V_{max} was considerably reduced and the K_m more than five times higher (Table 5.17).

Clearly this would mean that substrates for this enzyme could be more poorly metabolized in individuals with the mutant enzyme. The mutant was not found in the Taiwanese or black populations, which were also studied.

Other mutants of CYP3A4 have also been found in a Chinese population, which were associated with a reduction in the ratio of 6-β hydroxyl cortisol to cortisol, an indicator of metabolic activity *in vivo*.

Alcohol Metabolism
Another oxidation reaction, which shows variation in human populations, is the oxidation of ethanol. This has been shown to be significantly lower in Canadian Indians compared with Caucasians, and thus the Indians are more susceptible to the effects of alcoholic drinks. The rate of metabolism *in vivo* in Indians is 0.101 g kg^{-1} hr^{-1} compared with 0.145 g kg^{-1} hr^{-1} in Caucasians. This seems to be due to variants in alcohol dehydrogenase, although differences in aldehyde dehydrogenase may also be involved. Variants of alcohol dehydrogenase resulting in increased metabolism have also been described within Caucasian and Japanese populations.

Esterase Activity
Hydrolysis reactions can also exhibit genetic influences such as the plasma enzyme, a **pseudocholinesterase**, which is responsible for the metabolism and inactivation of the drug **succinylcholine** (fig. 7.55). Certain individuals may be defective in the ability to hydrolyze, and therefore inactivate, this drug. Such individuals have a form of the enzyme with a decreased hydrolytic activity, and the half-life of the drug is dramatically increased. Consequently, such individuals may be affected by the drug for two to three hours instead of the more normal two to three minutes. In extreme cases, apnea may result from the prolonged neuromuscular blockade and muscular relaxation caused by the drug. Family studies are consistent with this deficiency being a recessive trait governed by an autosomal autonomous gene with two alleles. The three genotypes are manifested as two phenotypes: rapid and slow hydrolysis of succinyl choline, although careful analysis may reveal three. Thus, the atypical pseudocholinesterase occurs in the individuals homozygous for the abnormal gene, the individuals homozygous for the normal gene have the normal enzyme and the heterozygotes produce a mixture of enzymes. The trait has a frequency of about 2% in some populations studied (British, Greek, Portuguese, North African), but it is absent from other populations such as Japanese and Eskimo, for example. There are in fact a number of variants of the atypical enzyme, which can be distinguished by using specific inhibitors.

5.3.5 Environmental Factors

There are many factors in the environment, which may influence drug disposition, metabolism, and toxicity to a greater or lesser extent. However, as the influence of certain foreign compounds, both drugs and those in the environment, on microsomal enzymes has been well studied, this will constitute a separate section "Enzyme induction and Inhibition".

5.3.6 Stress and Diurnal Variation

Adverse environmental conditions or stimuli, which create stress in an animal, may influence drug metabolism and disposition. Cold stress, for instance, increases aromatic hydroxylation, as does stress due to excessive noise. It should be noted that the microsomal monooxygenases show a diurnal rhythm in both rats and mice, with the greatest activity at the beginning of the dark phase.

As well as microsomal enzymes showing diurnal variation, other factors important in toxicology also show differences over time. For example, the level of glutathione varies significantly. Studies have shown that the hepatic GSH level in mice is significantly lower at 8 p.m. versus 8 a.m. Correspondingly, the susceptibility to paracetamol toxicity is much greater when administered at 8 p.m. versus 8 a.m. (Table 5.18).

5.3.7 Diet

The influence of diet on drug metabolism, disposition, and toxicity consists of many constituent factors. Food additives and naturally occurring contaminants in food may influence the activities of various enzymes by induction or inhibition. However, these factors are discussed in a later section Enzyme induction and Inhibition. The factors with which this section will be concerned are the nutritional aspects of diet.

The multitude of factors contained within the environment, which may influence drug disposition and metabolism, are difficult to separate.

The finding that race and diet may affect the clearance of drugs such as antipyrine is such an example. Thus, meat-eating Caucasians have a significantly greater clearance and a shorter plasma half-life for antipyrine than Asian vegetarians. However, the relative importance of race versus diet as contributions to those differences is not clear.

The nutritional status of an animal is well recognized as having an important influence on drug metabolism, disposition, and toxicity. The lack of various nutrients may affect drug metabolism, though not always causing a depression of metabolic activity. Lack of protein has been particularly well studied in this respect and shows a marked influence on drug metabolism.

Thus, rats fed on low-protein diets (5%) show a marked loss of microsomal enzyme activity when compared with those animals fed a 20% protein diet (Table 5.19). The decline in

Table 5.18 Variation in Liver Glutathione Levels with Time and Effect on Paracetamol Toxicity

Time of assay and dosing	Plasma ALT (U/L)	Hepatic GSH (μmol g^{-1})	
		Before dosing	After dosing
8 am	27 $\pm$ 3	8.7 $\pm$ 0.2	4.2 $\pm$ 0.6
2 pm	70 $\pm$ 28	6.8 $\pm$ 0.4	2.4 $\pm$ 0.5
8 pm	3451 $\pm$ 1036	6.1 $\pm$ 0.4	1.5 $\pm$ 0.2

Mice were dosed with paracetamol (400 mg kg^{-1}) and the transaminase level (ALT) was measured 24 hr later.
Source: From Ref. 34.

Table 5.19 Effect of a Reduced Protein Diet on Hepatic Cytochrome P-450 Enzyme Activity in Rats

	Ethylmorphine N-demethylation nmol HCHO 100g^{-1} body weight 10 min^{-1}
20% protein diet	10.5
5% protein diet for 4 days	6
5% protein diet for 8 days	<1

Source: From Ref. 35.

activity is accompanied by a decline in the level of liver microsomal protein. Both cytochrome P-450 content and cytochrome P-450 reductase activity are reduced by a 5% protein diet. Similarly in humans, lowering the protein in the diet from 20% to 10% decreased theophylline clearance by 30% . However, the different forms of cytochrome P-450 are differently affected. Thus, in studies in rats, CYP3A fell dramatically with reduction of the dietary protein from 18% to 1%, whereas CYP1A2 only dropped after a 0.5% protein diet.

Measurements *in vivo*, such as barbiturate sleeping times, are in agreement with these findings, sleeping times being longer in protein-deficient animals. Toxicity may also be influenced by such factors as a low-protein diet. For example, the hepatotoxicity of carbon tetrachloride is markedly less in protein-deficient rats than in normal animals, and this correlates with the reduced ability to metabolize the hepatotoxin in the protein-deficient animals.

However, the reverse is the case with the hepatotoxicity of paracetamol, which is increased after a low-protein diet. This may be due to the reduced levels of glutathione in rats fed low-protein diets, which offsets the reduced amount of cytochromes P-450 caused by protein deficiency.

The carcinogenicity of aflatoxin is reduced by protein deficiency, presumably because of reduced metabolic activation to the epoxide intermediate, which may be the ultimate carcinogen, which binds to DNA (Fig. 5.14). A deficiency in dietary fatty acids also decreases the activity of the microsomal enzymes. Thus, ethylmorphine, hexobarbital, and aniline metabolism are decreased, possibly because lipid is required for cytochromes P-450. Thus, a deficiency of essential fatty acids leads to a decline in both cytochromes P-450 levels and activity *in vivo*.

Mineral and **vitamin** deficiencies also tend to reduce the metabolism of foreign compounds. Carbohydrates, however, do not seem to have major direct effects on the metabolism of foreign compounds. **Starvation** or changes in diet may also reduce supplies of essential cofactors such as sulfate, which is required for phase 2 conjugation reactions and may be readily depleted (see chap. 7). Overnight fasting of animals may also have significant effects on metabolism and toxicity. Thus, the dealkylation of **dimethylnitrosamine** (chap. 7, Fig. 5) is increased by this short-term food deprivation, and the hepatotoxicity is consequently increased. The hepatotoxicity of **paracetamol** and **bromobenzene** are also both increased by overnight fasting of animals, probably because this results in the depletion of glutathione to 50% of normal levels. There is thus less glutathione available for the detoxication of these compounds (see chap. 7).

The nutritional status of an animal may affect the disposition of a foreign compound *in vivo* as well as the metabolism. Many drugs are protein-bound in the plasma, and alteration of the extent of binding for compounds extensively bound may have important toxicological implications. Thus, the decreased plasma levels of **albumin** after low-protein diets, such as occur in the human deficiency disease Kwashiorkor, might lead to significantly increased plasma levels of the free drug and therefore the possibility of increased toxicity.

5.3.8 Age Effects

It is well known that the drug-metabolizing capacity and various other metabolic and physiological functions in man and other animals are influenced by age. Furthermore, sensitivity to the toxic and pharmacological effects of drugs and other foreign compounds is often different in young and geriatric animals. As might be expected, at the extremes of age, drug-metabolizing activity is often impaired, plasma-protein binding capacity is altered, and the clearance of foreign compounds from the body may be less efficient. These differences generally lead to exposure to a higher level of unchanged drug for longer periods in the young and geriatric animals than in adults, with all the implications for toxicity that this has. The case of the fetus is rather special, because of the influence of the maternal organism, and will therefore be considered separately at the end of this section.

Although the toxicological significance has yet to be studied, age-related differences in the absorption of foreign compounds are demonstrable. Neonatal and geriatric human subjects have low gastric acid secretion, and consequently the absorption of some foreign compounds may be altered. Thus in the neonate, **penicillin** absorption is enhanced, whereas paracetamol absorption is decreased.

Intestinal motility may also be influenced by age, with various effects on the absorption of foreign compounds dependent upon the site of such absorption. The absence of gut flora in the neonate may have as yet unknown influences on the disposition of foreign compounds in the gastrointestinal tract.

Once absorbed, foreign compounds may react with plasma proteins and distribute into various body compartments. In both neonates and elderly human subjects, both total plasma-protein and plasma-albumin levels are decreased. In the neonate, the plasma proteins may also show certain differences, which decrease the binding of foreign compounds, as will the reduced level of protein. For example, the drug lidocaine is only 20% bound to plasma proteins in the newborn compared with 70% in adult humans. The reduced plasma pH seen in neonates will also affect protein binding of some compounds as well as the distribution and excretion. Distribution of compounds into particular compartments may vary with age, resulting in differences in toxicity. For example, morphine is between 3 and 10 times more toxic to newborn rats than adults because of increased permeability of the brain in the newborn. Similarly, this difference in the **blood-brain** barrier underlies the increased neurotoxicity of lead in newborn rats.

Total body water, particularly extracellular water, has been found to be greater in neonates than in adults and to decrease with age. The distribution of water-soluble drugs could clearly be influenced by this; lower plasma levels being one possible result. Both glomerular filtration and renal tubular secretion are lower in neonates and geriatrics, and human infants achieve adult levels of glomerular filtration by one year of age at the earliest. The consequences of this are reduced excretion, and hence reduced body clearance of foreign compounds. Consequently, toxicity may be increased by prolongation of exposure or accumulation, if chronic dosing is involved.

This was an important factor in the development of adverse reactions to the antiarthritis drug benoxaprofen **(Opren)**, which caused serious toxicity in some elderly patients, necessitating withdrawal of the drug from the market.

Many drug-metabolizing enzyme systems show marked changes around the time of birth, being generally reduced in the fetus and neonate. In some cases, adult activity may be achieved within a few days, whereas for some enzymes, such as cytochromes P-450, several weeks may be necessary to obtain optimum levels. The development of these enzyme systems does, however, depend on the species. For example, in the rat, microsomal monooxygenase activity is negligible at birth whereas in humans at six months of gestation, enzyme activity is not only detectable, but may be between 20% and 50% of adult levels.

In the rat, development to adult levels of activity takes about 30 days after which levels decline toward old age. In humans, however, hydroxylase activity increases up to the age of 6 years, reaching levels greater than those in the adult, which only decrease after sexual maturation. Thus the elimination of antipyrine and theophylline was found to be greater in children than in adults. It should be noted, however, that proportions of isoenzymes may be very different in neonates from the adult animal, and the development of the isoenzymes may be different. Thus, in the rat there seem to be four types of development for phase 1 metabolizing enzymes: linear increase from birth to adulthood, type A (aniline 4-hydroxylation); low levels until weaning, then an increase to adult levels, type B (*N*-demethylation); rapid development after birth followed by rapid decline to low levels in adulthood, type C (hydroxylation of 4-methylcoumarin); and rapid increase after birth to a maximum and then decline to adult levels, type D. Patterns of development may be different between sexes as well as between species. For example, in the rat, steroid 16-α-hydroxylase activity toward androst-4-ene-3,17-dione develops in type B fashion in both males and females, but in females, activity starts to disappear at 30 days of age and is undetectable by 40 days. It seems that the monooxygenase system develops largely as a unit, with the rate dependent on species and sex of the animal and the particular substrate.

With some of the phase 2 metabolizing enzymes, there may be strict ontogenetic patterns of expression. Sulfate conjugation ability occurs early in rats, whereas glucuronidation (of xenobiotics), and conjugation with glutathione and amino acids, only develop over about 30 days from birth.

These deficiencies in metabolic capacity in neonates may have significance for the toxicity of foreign compounds.

When the parent compound is responsible for the pharmacological or toxicological effect, reduced metabolic activity may lead to prolonged and exaggerated responses. For example, in mice there is very little oxidation of the side chain of hexobarbital, and the sleeping time is

Table 5.20 Effect of Age on Metabolism and Duration of Action of Hexobarbital

Age (days)	Percent hexobarbital metabolized *in vitro*[a] in 1 hr (guinea pigs)	Percent hexobarbital metabolized *in vivo* in 3 hr (mice)	Sleeping time in mice (min)		
			(10 mg kg^{-1})	(50 mg kg^{-1})	(100 mg kg^{-1})
1	0	0	>360	Died	Died
7	25–35	11–24	107 ± 26	243 ± 30	>360
21	13–21	21–33	27 ± 11	64 ± 17	94 ± 27
Adult	28–39		<5	17 ± 5	47 ± 11

[a]Guinea pig liver microsomes.
Source: From Ref. 36.

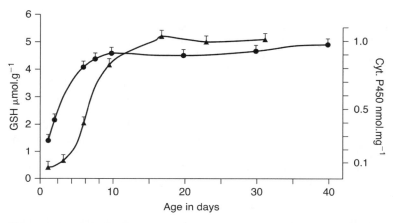

Figure 5.30 Development of hepatic reduced glutathione and cytochrome P-450 levels with age in mice. *Abbreviations*: GSH, glutathione (•); cyt.P450, cytochrome P-450 (▲). *Source*: From Ref. 37.

excessively prolonged (by 70×), as shown in Table 5.20. Doses giving sleeping times of less than one hour in adult mice are fatally toxic to neonates. This activity correlates with the inability to metabolize the drug.

The converse is true of drugs requiring metabolic activation for toxicity. For example, **paracetamol** is less hepatotoxic to newborn than to adult mice, as less is metabolically activated in the neonate. This is due to the lower levels of cytochromes P-450 in neonatal liver (Fig. 5.30). Also involved in this is the hepatic level of glutathione, which is required for detoxication. Although levels of this tripeptide are reduced at birth, development is sufficiently in advance of cytochrome P-450 levels to ensure adequate detoxication (Fig. 5.30). The same effect has been observed with the hepatotoxin bromobenzene. (For further details of paracetamol and bromobenzene see chap. 7.) Similarly, **carbon tetrachloride** is not hepatotoxic in newborn rats as metabolic activation is required for this toxic effect, and the metabolic capability is low in the neonatal rat.

The deficiency in phase 2 metabolizing enzymes may also influence the toxicity of foreign compounds. For example, glucuronosyl transferase activity in human neonates is often low, and hence conjugation is impaired, which may be important for the detoxication of a drug. A relative deficiency of UDP-glucuronic acid may also contribute to the lower level of conjugation. Thus, with **chloramphenicol**, which is 90% conjugated with glucuronc acid, neonatal deficiency of glucuronide conjugation can lead to severe cyanosis and death in some human infants due to the high plasma levels of unchanged drug. However, in the case of paracetamol metabolism in neonates, as sulphate conjugation is at normal levels, this will compensate for any reduction in glucuronidation. When conjugation of bilirubin with glucuronic acid is impaired, **neonatal jaundice** results leading to brain damage due to the elevated levels of free bilirubin. This may be exacerbated by the administration of drugs, which displace bilirubin from its binding sites on plasma albumin.

As with neonates, drug-metabolizing capacity in elderly subjects is also reduced in comparison with young adults for some compounds, such as **benoxaprofen, propranolol**, and

lignocaine. Other drugs such as **isonazid** and **warfarin** show a similar plasma half-life in elderly as in young adults. It seems that phase 1 metabolism is more likely to be affected in the elderly than phase 2 reactions. However, it should be noted that clearance of some drugs (propranolol and lignocaine) may be decreased in the elderly, partly as a result of reduced hepatic blood flow and reduced hepatic extraction as well as reduced metabolism. Thus, as in neonates, reduced metabolism of foreign compounds in the elderly animal may lead to differences in toxicity in comparison with the adult animal.

The situation with regard to the fetus is a rather special case. Although many of the same comments apply to the fetus as to neonatal animals, the relationship with the maternal circulation exerts a modifying effect. Thus, clearance of compounds from fetal plasma may be more efficient, although if polar compounds are produced in the fetal liver, these may be unable to leave the embryo because they are unable to cross the placenta. Alternatively, drugs ingested by the mother may be metabolized maternally and then be unable to cross the placenta. However, most drugs, particularly on chronic administration, attain the same steady-state levels in the fetal as in the maternal plasma, unless metabolism and/or excretion by the fetus is significant. The fetal kidney, however, excretes certain compounds into the allantoic fluid. Plasma-protein binding capacity is less in the fetus due to lower levels of albumin. The ability of the fetus to metabolize foreign compounds is generally less than that of the adult animal, as it is with the neonate. However, this may depend on the metabolic pathway and enzyme system involved.

Cytochromes P-450 show negligible activity in the fetus until birth, although activity is measurable in embryonic tissue. This is the case in most mammals, except man, where the fetus six weeks after gestation shows measurable cytochromes P-450 activity, which is between 20% and 40% of adult values by mid-term, this level being maintained until birth. Thus in rat, mouse, rabbit, and hamster fetuses, cytochrome P-450 levels are between 0% and 5% of the adult level. In guinea pigs and macaque monkeys, levels are higher, but the human fetus seems to have the highest levels at 20% to 40% of the adult level.

Glucuronyl transferase activity has been shown to be low in the fetus, using substrates such as p-nitrophenol and 1-naphthol (chap. 4, Figs. 8 and 4). **Diazepam** is metabolized to N-demethyldiazepam, which is conjugated with glucuronic acid in adults and children; however, no glucuronide is detectable in premature infants. It is clear that if cytochromes P-450 activity is present in the fetus but conjugating ability is impaired, toxic metabolites rather than stable, less toxic conjugates could accumulate. The activity of cytosolic enzymes such as the GSTs is also lower in the fetus, being only 5% to 10% of the adult value in rabbits and guinea pigs. The glutathione content of fetal mouse liver was found to be approximately 10% of that of the adult, but studies indicate that the mouse fetus is still protected from the hepatotoxic effects of compounds such as paracetamol by this level of glutathione.

It is therefore apparent that the fetus may not always be at greater risk from the toxic effects of foreign compounds than the neonate. Its susceptibility depends very much on the particular compound. The human neonate, however, may be unusual in having a comparatively high level of microsomal enzyme activity, which may have important toxicological consequences for the fetus exposed to compounds metabolized by such enzymes to more toxic products. The rapidity of clearance of the compound from the maternal circulation is of paramount importance to the fetal toxicity of a foreign compound. Teratogenicity is a special case of fetal toxicity in its dependence on the stage of development, and will be considered in a later chapter.

5.3.9 Hormonal Effects

As well as the sex hormones already mentioned (see above), many other hormones seem to affect the metabolism of foreign compounds and therefore may have effects on toxicity. A number of **pituitary** hormones may directly, as well as indirectly, affect metabolism, for example, growth hormone, follicle-stimulating hormone, adrenocorticotrophic hormone, luteinizing hormone, and prolactin. Thus, **hypophysectomy** in male rats results in a general decrease in metabolism, but the effects of some of the individual hormones may depend on the sex of the animal and the particular enzyme or metabolic pathway. For example, **adrenocorticotrophic hormone (ACTH)** administration decreases oxidative metabolism in males but increases N-demethylation in female rats. Removal of the **adrenal gland** reduces metabolism, such as ethylmorphine demethylation and aniline 4-hydroxylation, and this can be partly restored by the administration of corticosteroids. However, there are probably many factors involved as well as the loss of a particular hormone. Similarly, **thyroidectomy** causes changes in metabolism, but these vary depending on the sex of the animal and on the particular metabolic reaction studied. **Diabetes** induced by **streptozotocin**, which specifically damages the pancreas (see chap. 6), seems to cause a general decrease in monooxygenase metabolic capacity, and the activity may be restored by **insulin** treatment. In fact, although there is no change in the total cytochromes P-450 level, diabetes may induce one particular form of the enzyme (see below), and so the diabetic condition may cause a change in the isoenzyme proportions. Thus overall, the thyroid, adrenal, and pituitary hormones and insulin mainly act directly on the liver to affect metabolism, whereas the sex hormones probably act indirectly via the pituitary and hypothalamus. Clearly, however, the potential interactions are complex.

5.3.10 Effects of Pathological Conditions
Although most experimental studies using foreign compounds are carried out in healthy animals or healthy human volunteers, animals and humans with diseases may be exposed to foreign compounds. This is clearly often the case with drugs designed to treat specific diseases, or to be used in particular conditions. The influence of pathological conditions on the disposition and metabolism of such compounds is therefore a very important consideration. Furthermore, chronic exposure to foreign compounds may result in pathological damage, which may influence the disposition of the compound on subsequent exposure.

The absorption of foreign compounds from the gastrointestinal tract may be altered in certain malabsorption syndromes and may be increased after subcutaneous or intramuscular injection if vasodilation accompanies the particular disease. The disposition of foreign compounds, once absorbed, can be influenced by changes in plasma proteins, which are sometimes reduced in certain disease states. Consequently, for foreign compounds, which are highly protein bound, the plasma concentration of the free compound may be significantly increased in such circumstances. This may alter renal excretion, increasing it in some cases, but could also increase the toxicity of a drug with a narrow therapeutic ratio if the compound dissociated from the protein was responsible for the toxicity. Thus, thiopental anesthesia is prolonged when plasma albumin is reduced by chronic liver disease, and more unbound diphenylhydantoin and sulfonamides result from changes in plasma proteins in chronic liver disease.

Hepatic disease and damage clearly have the potential to be major factors in the metabolism of foreign compounds. Thus, in patients with liver necrosis due to paracetamol, the half-life of the latter was increased from 2 to 8 hours and that of **antipyrine** from 12 to 24 hours. Acute hepatic necrosis in animals caused by the administration of hepatotoxins resulted in the plasma half-lives of barbiturates, **diphenylhydantoin**, and antipyrine being approximately doubled. However, there may be several factors operating such as displacement of a drug from plasma-protein-binding sites by bilirubin. In liver damage, plasma-bilirubin levels may be high due to lack of conjugation with glucuronic acid. This may alter elimination of some drugs such as tolbutamide. The level of plasma albumin may often be reduced by liver disease as this is the site of synthesis of albumin, hence the binding to plasma albumin will tend to be reduced. However, the effects of liver disease can be somewhat unpredictable. For example, in patients with cirrhotic liver, the glucuronidation of **chloramphenicol** and the acetylation of **isoniazid** are both reduced, and hence the half-lives are prolonged. However, chronic liver disease did not affect the hepatic clearance of **lorazepam** or **oxazepam** despite the fact that glucuronidation is the route of metabolism for these compounds. In general, the formation of glucuronic acid and sulfate conjugates tends to be impaired in liver diseases such as hepatitis, obstructive jaundice, and cirrhosis.

Hepatic damage may also affect the disposition of some drugs by altering **hepatic blood flow**. For example, **Indocyanine** *Green* is not metabolized, and the clearance is related to hepatic blood flow. In cholestatic liver damage, clearance is reduced to 70% of the control. Both **antipyrine** and **aminopyrine** are cleared by metabolism in the liver, yet only the clearance of antipyrine is affected by the cholestatic damage. Although it seems that hepatocellular disease will generally affect drug clearance, it is still difficult to extrapolate from the known effects on the disposition of a particular drug to the likely effects on another. It is clearly important to know the relevant factors in the disposition of a particular drug and also the severity and type of pathological damage. Thus, liver cirrhosis will alter blood flow but will not necessarily affect enzyme levels. Mild to moderate hepatitis in humans was found to have no influence on hepatic cytochromes P-450 content, yet severe hepatitis and cirrhosis reduced the content by 50%. The liver synthesizes cholinesterases, and therefore drugs hydrolyzed by these enzymes, such as **aspirin, procaine**, and **succinylcholine**, show reduced metabolism.

Thus, liver damage and disease may have a number of effects on the disposition of a compound, which may be due to the following:

1. Alteration in enzyme levels or activities
2. Alteration in blood flow
3. Alteration in plasma albumin and hence binding

Decreased ability to form glucuronides may also occur in Gilbert's disease and the Crigler–Najjar syndrome. In these genetic disorders, glucuronyl transferase is reduced, and consequently bilirubin conjugation may be affected when drugs, which are also conjugated with glucuronic acid, are administered.

Renal disease is another important factor, particularly if renal excretion is the major route of elimination for the pharmacologically or toxicologically active compound. This may be particularly important with drugs showing a low therapeutic ratio, such as digoxin and the aminoglycoside antibiotics. As already mentioned, plasma-protein binding may be affected in renal disease, such binding being reduced particularly with regard to organic acids. Increased toxicity of drugs undergoing significant metabolism, such as chloramphenicol, has been found in uraemic patients. The half-lives of a number of drugs are prolonged in renal failure, although this effect is variable and by no means the general rule, different drugs being differently affected.

Chronic renal disease may also affect metabolism, not necessarily because of impaired metabolism in the kidney, but because of an indirect effect of renal failure on liver metabolism. For example, in animals with renal failure, it was observed that there was a decrease in hepatic cytochromes P-450 content, and consequently, zoxazolamine paralysis time and ketamine narcosis time were prolonged.

Cardiac failure may also affect metabolism by altering hepatic blood flow. However, even after heart attack without hypotension or cardiac failure, metabolism may be affected. For example, the plasma clearance of lidocaine is reduced in this situation. Other diseases such as those, which affect hormone levels: **hyper-or hypothyroidism**, lack of or excess growth hormone, and **diabetes** can alter the metabolism of foreign compounds.

Infectious diseases, viral, bacterial, and protozoal, and inflammation may all affect metabolism, some partially by increasing levels of the endogenous compound interferon, which will inhibit some metabolic pathways. Levels of cytochrome P-450 are known to fall in certain infectious and other diseases. Cytokines appear to be involved in the effect of a number of diseases on drug metabolism.

5.3.11 Tissue and Organ Specificity

The disposition or localization, and in some cases metabolism, of foreign compounds may be dependent upon the characteristics of a particular tissue or organ, which may in turn affect the toxicity. There are many examples of organotropy in toxicology, but the mechanisms underlying such organ-specific toxic effects are often unknown.

It is clear that a foreign compound, which is chemically similar to, or at least has certain structural similarities with, an endogenous compound, may become localized in a particular tissue(s). An example of this is **6-hydroxydopamine**, a single dose of which is selectively toxic to sympathetic nerve endings. Because of its similarity to noradrenaline, 6-hydroxydopamine is distributed specifically to the sympathetic nerve endings. There it is oxidized to a quinone (fig. 7.43), which binds covalently to the nerve endings and permanently inactivates them. In other cases, however, the mechanism of specificity is less clear. Certain carcinogens show organ or tissue specificity, such as *trans*-4-dimethylaminostilbene, which causes earduct tumors after repeated administration, or hydrazine, which causes lung tumors (Fig. 5.31). This may reflect the ability of the tissue to carry out repair of damaged macromolecules, such as DNA. Thus, if DNA repair is poor in a particular tissue, such as nervous tissue, that tissue may be particularly susceptible to certain carcinogens.

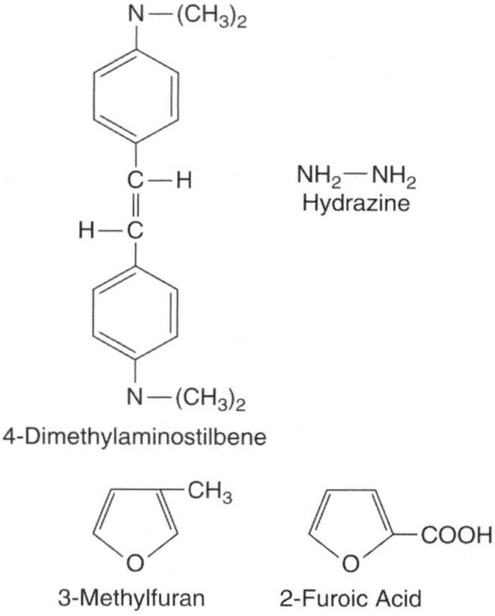

Figure 5.31 Structure of 4-dimethylaminostilbene, hydrazine, 2-methylfuran, and 2-furoic acid.

4-Dimethylaminostilbene

3-Methylfuran   2-Furoic Acid

The carcinogen diethylstilboestrol induces tumors in those female organs particularly exposed to estrogens, namely, the mammary glands, uterus, and vagina (fig. 6.28). In male hamsters, however, this compound causes kidney tumors.

Although certain organs are prime targets for toxic effects because of their anatomical position and function, such as the gut, liver, and kidney, they may not necessarily be the most susceptible. Such organs may have a particular ability to cope with a toxic insult, and this has been termed "reserve functional capacity." Examples of organ-specific toxicity are the lung toxicity of the natural furan ipomeanol (fig. 7.37), 3-methylfuran (Fig. 5.31), the herbicide paraquat (Fig. 7.40) and the kidney toxicity of chloroform (Fig. 7.27), 2-furoic acid (Fig. 5.31), and *p*-aminophenol (Fig. 5.9). However, some of these compounds, such as 3-methylfuran and paraquat, are rarely, if ever, hepatotoxic, even after oral administration. The lung toxin ipomeanol is considered in greater detail in the final chapter 7.

The phospholipidoses caused by certain amphiphilic drugs typified by chlorphentermine (chap. 3, Fig. 5.19) tend to be organ specific and tissue specific, occurring primarily in tissues with high cell membrane turnover such as macrophages in lung or the immune system or in organs with high lipid or phospholipid biosynthesis such as the adrenals and retina, for example. In the case of chlorphentermine, this correlates with the accumulation of the drug, which is localized in fatty tissue, particularly that associated with the adrenals and lungs (see chap. 6). However, the organs affected can be different for the same compound in different species. For example, the drug tafenoquine causes phospholipid accumulation in the lung in rat and dog, but cornea in man. Phospholipidosis is discussed in greater detail in chapters 6 and 7.

5.3.12 Dose

It is clear from chapter 2 that the dose of a toxic compound is a major factor in its toxicity. This may be a simple relationship resulting from the increasing concentration of toxin at its site of action, with a proportional increase in response with increasing dose. However, the size of the dose may also influence the disposition or metabolism of the compound.

Thus, a large dose may be ineffectively distributed and remain at the site of administration as a depot. A large dose of a compound given orally, for instance, may not be all absorbed, depending on the rates of absorption and transit time within the gut. Saturable active absorption processes would be particularly prone to dose effects, which could result in unexpected dose-response relationships.

Once a toxic compound has been absorbed, the disposition of it *in vivo* may also be affected by the dose. Thus, saturation of plasma-protein binding sites may lead to a significant rise in the plasma concentration of free compound, with possible toxic effects. This, of course, depends on the fraction of the compound that is bound and is only of significance with highly protein-bound substances. The result of this is a disproportionate rise in toxicity for a small increase in dose. As already discussed in chapter 3, transporter systems (e.g., MDR proteins) can pump some foreign chemicals out of exposed tissues or organs back into the blood or lumen of the organ (e.g., the gut). As these are active transport systems, they can be saturated. So a large dose of a chemical could reduce the ability of the organ to protect itself and the body by removal of the drug.

Saturation of the processes involved in the elimination of a foreign compound from the plasma, such as metabolism and excretion, may also have toxicological consequences. Thus, ethanol exhibits zero-order elimination kinetics at readily attainable plasma concentrations, because the metabolism is readily saturated. Therefore, once these plasma concentrations have been attained, the rate of elimination from the plasma is constant. Increasing the dosage of ethanol leads to accumulation and the well-known toxic effects.

Inappropriate dosing was almost certainly one of the factors in benoxaprofen- (Opren) induced cholestatic liver damage, which occurred in some elderly patients (see above). With repeated dosing the drug accumulated, probably as a result of the reduced kidney function, which is often seen in old age. The elimination of the drug was therefore less efficient than in the healthy volunteers used in phase 1 clinical trials, resulting in longer elimination half-lives in susceptible patients when similar doses were given. Thus the drug accumulated in the body. It now seems that the benoxaprofen acylglucuronide metabolite, which has low solubility, is also eliminated into the bile by an active transport system (Mrp2), so exposing the canalicular membranes to relatively high concentrations. The glucuronide can also be hydrolyzed and acylate nearby proteins in the biliary system. The accumulated doses of the drug and its metabolites in the hepatocytes of such patients may also have saturated the biliary transporter (s) leading to increased exposure of these cells.

The biliary excretion of **furosemide** is also an active process and saturable, and after high doses of furosemide, there is a disproportionate increase in the plasma level of the drug. This appears to be responsible for the toxic dose threshold, above which hepatic necrosis ensues (figs 3.33, 3.34). Similarly, saturation of the biliary excretion of proxicromil in dogs is believed to be the cause of the hepatotoxicity (Fig. 5.11).

The metabolism of toxic compounds may also be influenced by dose, with possible toxicological consequences. Saturation of metabolic routes may increase toxicity if the parent compound is more toxic than its metabolites, or if minor but toxic routes become more important. Conversely, toxicity may not increase proportionately with dose if the pathway responsible becomes saturated. For example, paracetamol hepatotoxicity is dose dependent, but only above a threshold dose. This threshold is the result of saturation of the glutathione conjugation pathway. Another example is isoniazid hepatotoxicity, which results from an acetylated metabolite. Giving large doses of isoniazid to experimental animals does not cause hepatic necrosis, whereas giving several smaller doses does, probably because acetylation is saturated at high doses, and the drug is metabolized by other routes. These and other examples such as salicylic acid and vinyl chloride are discussed in greater detail in the final chapter. The problem of dose-dependent metabolism is also important in considerations of the extrapolation from animals to man in the prediction of risk and safety assessment. For example, the food additive estragole undergoes side chain, aliphatic oxidation followed by conjugation to form a carcinogenic metabolite (Fig. 5.32). Estragole is carcinogenic in mice at a

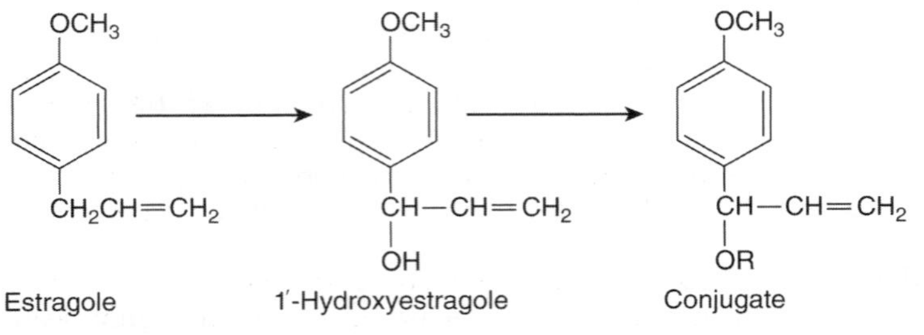

Figure 5.32 Metabolic activation of estragole. Conjugation (e.g., with sulfate) can yield a reactive intermediate.

dose of around 511 mg kg^{-1}, but humans are only exposed to around 1 µg kg^{-1} day^{-1}. When the metabolism of estragole in mice and rats is examined, it is found that it is dose dependent, with the proportion of the dose metabolized to the toxic metabolite increasing by 10-fold from about 1% to 10%. In humans, the pathway only represents 0.3% of the dose at the normal daily intake. Thus, in this case the ratio of the amount of toxic metabolite required to produce tumors in mice to the amount that man is normally exposed to is about 15 million to 1. These findings clearly have implications with regard to setting safe limits for daily intake of such substances. The data also underline the crucial importance of metabolic data gathered at different doses for the sensible interpretation and use of toxicity data. Another example of this is the case of saccharin, the artificial sweetener. This compound caused bladder cancer in experimental animals when these were exposed to levels of 5% to 7% of the diet but not if exposed to levels up to 5%. However, pharmacokinetic studies revealed that at these high exposure levels the plasma clearance of saccharin was saturated, and therefore tissue levels would be higher than expected on the basis of a linear extrapolation from lower doses. Consequently, prediction of the incidence of bladder tumors at the lower exposure levels to which humans would be exposed could not reasonably be based on a simple linear extrapolation from the data obtained at the high exposure levels.

5.3.13 Enzyme Induction and Inhibition

Most biological systems, and especially humans, are exposed to a large number of different chemicals in the environment. Thus, pesticides, natural contaminants in food, industrial chemicals, and agricultural pollutants may all contaminate the environment and thereby affect various biological systems in that environment. Humans and certain other animals may also be exposed to drugs and food additives, and humans are exposed to substances in the workplace. These chemicals may modify their own disposition and that of other chemicals in several ways. One way this may occur is by an effect on the enzymes involved in the metabolism of foreign compounds; these enzymes may be induced or inhibited. By altering the routes or rates of metabolism of a foreign compound, either induction or inhibition clearly can have profound effects on the biological activity of the compound in question. The biological importance is probably an adaptation to chemical exposure and hence potential stress, by increasing the removal of the offending substance.

Enzyme Induction

Although first reported with the cytochrome(s) P-450 mixed function oxidases, it is now known that a number of the enzymes involved in the metabolism of foreign compounds are inducible. Thus, as well as the CYPs, NADPH cytochrome P-450 reductase, cytochrome b$_5$, glucuronosyl transferases, epoxide hydrolases, and GSTs are also induced to various degrees. However, this discussion concentrates on the induction of the CYPs with mention of other enzymes where appropriate.

The induction of the CYPs has been demonstrated in many different species including humans, and in various different tissues as well as the liver. Induction usually results from repeated or chronic exposure, although the extent of exposure is variable. The result of induction is an increase in the amount of an enzyme; induction requires de novo protein synthesis, and therefore an increase in the apparent metabolic activity of a tissue *in vitro* or animal *in vivo*. Consequently, inhibitors of protein synthesis, such as cycloheximide, inhibit induction. It is a reversible cellular response to exposure to a substance. Thus, it can be shown in isolated cells, such as hamster fetal cells in culture, that exposure to benzo[a]anthracene induces aryl hydrocarbon hydroxylase (AHH) activity (CYP1A1).

A large variety of substances have been shown to be inducers, probably numbering several 100 and including many drugs. Table 5.21 shows some inducers of different CYPs. Some of these inducers will be specific form one CYP isoform, but others will induce a number of different isoforms, complicating the effect on metabolism. Also, although enzyme induction occurs most commonly in the liver, enzymes can be induced in other tissues also, sometimes to a greater extent.

Apart from the fact that many are lipophilic and organic, there are no common factors, and many chemical classes of compound are included (Table 5.21). However, there are several types of induction, which can be differentiated, and within some of these types, inducers show certain structural similarities. For example, inducers of the **polycyclic** hydrocarbon type (CYP1A1) tend to be planar molecules.

Table 5.21   Examples of Different Inducers of Cytochrome P-450

Enzyme induced	Inducer
CYP1A1	Dioxin
CYP1A2	Omeprazole, cigarette smoke, 3-methylcholanthrene
CYP2B1/2B2	Phenobarbital
CYP2B6[a]	Phenobarbital
CYP2C9	Rifampicin
CYP3A1	Dexamethasone
CYP3A4[a]	Carbamazepine, phenobarbital, phenytoin, rifampicin
CYP2E1	Isoniazid, ethanol
CYP4A6	Clofibrate

[a]Form induced in humans.

Planar and nonplanar **polychlorinated biphenyls** (Fig. 5.2) differ in the type of induction they will cause. Thus, 3, 3′, 4, 4′, 5, 5′-hexachlorobiphenyl is a planar molecule, which is an inducer of the polycyclic hydrocarbon type. 2, 2′, 4, 4′, 6, 6′-Hexachlorobiphenyl is a nonplanar molecule due to the steric hindrance between the chlorine atoms in positions 2 and 6 and is a **phenobarbital type** of inducer. The variety and type of inducing agents are shown in Table 5.21. Some compounds may indeed be mixed types of inducers, and thus mixtures of planar and nonplanar polychlorinated biphenyls are found to act as inducers of both the polycyclic and phenobarbital type.

As already discussed in chapter 4, cytochrome P-450 has many forms or isoenzymes, which differ in their ability to catalyze particular reactions. Some of these forms of cytochrome P-450 are found in normal liver tissue and are "constitutive," whereas others are only apparent after induction. Constitutive as well as non-constitutive forms of cytochrome P-450 are inducible. Some of the major forms of cytochrome P-450, which are induced, are shown in Table 5.21. It should be noted, however, that this is not an exhaustive list, and there are species and tissue differences in the constitutive and induced forms of cytochrome P-450.

Thus, induction can change the proportions of isoenzymes in a particular tissue and may increase the activity of a normally insignificant form by many times. Although phenobarbital induction increases the overall concentration of cytochromes P-450 in the liver by about threefold, specific isoenzymes may be increased up to 70-fold. Treatment with **3-methylcholanthrene** can increase a specific form of the enzyme by a similar order.

Induction of the microsomal enzymes may also have other effects as well as the increased production of particular enzymes and isozymes. Here again the different types of inducer vary. Thus, the barbiturate type of inducer differs significantly from the polycyclic hydrocarbon type as can be seen from the list below. Some of the characteristic changes caused by the barbiturate type of inducer are

1. increase in smooth endoplasmic reticulum,
2. increase in liver blood flow,
3. increase in bile flow,
4. increase in protein synthesis,
5. liver enlargement,
6. increase in phospholipid synthesis,
7. increase in cytochromes P-450 content ($3\times$),
8. increase in NADPH cytochrome P-450 reductase ($3\times$),
9. increase in glucuronosyl transferases,
10. increase in GSTs,
11. increase in epoxide hydrolases, and
12. induction of cytochrome P-450, which mostly occurs in the centrilobular area of the liver.

The **polycyclic hydrocarbon type** of inducer does not have such major effects, it only causes slight liver enlargement and has no effect on liver blood or bile flow. The increase in

Table 5.22 Effect of Various Enzyme Inducers on Metabolism of Different Compounds

Compound	Inducer control	Pb	PCN	3MC	Aro
	nmol product min^{-1} nmol cyt P-450^{-1}				
Ethymorphine	13.7	16.8	24.9	6.4	—
Aminopyrine	9.9	13.9	9.7	7.6	9.5
Benzphetamine	12.5	45.7	6.6	5.7	13.7
Caffeine	0.48	0.65	—	0.52	0.64
Benzo[a]pyrene	0.14	0.14	0.14	0.33	—

Abbreviations: Pb, phenobarbital; PCN, pregnenolone-16 α-carbonitrile; 3MC, 3-methylcholanthrene; Aro, arochlor 1254. *Source*: From Ref. 38.

cytochromes P-450 is not confined to the centrilobular area of the liver, protein synthesis is only slightly increased, and there is no increase in phospholipid synthesis. Other enzymes than cytochromes P-450 are also induced by polycyclic hydrocarbons, although generally to a lesser extent than with barbiturate induction. NADPH cytochrome P-450 reductase is not induced by polycyclic hydrocarbons, however.

Alcohol is an inducer of CYP2E1, which can lead to situations of enhanced drug toxicity in alcoholics or heavy drinkers (e.g., from paracetamol overdose).

With the **clofibrate type** of inducer, other changes are also apparent. Thus, there is a proliferation in the number of **peroxisomes** (an intracellular organelle) as well as induction of a particular form of cytochrome P-450 involved in fatty acid metabolism. A number of other enzymes associated with the role of this organelle in fatty acid metabolism are also increased, such as **carnitine acyltransferase** and **catalase.** This phenomenon is discussed in more detail in chapter 6.

The onset of the inductive response is in the order of a few hours (3–6 hours after polycyclic hydrocarbons, 8–12 hours after barbiturates), is maximal after 3 to 5 days with barbiturates (24–48 hours with polycyclic hydrocarbons), and lasts for at least 5 days (somewhat longer with polycyclic hydrocarbon induction). The magnitude of the inductive effect may depend on the size and duration of dosing with the inducer and will also be influenced by the sex, species, strain of animal, and the tissue exposed.

It is clear from these comments that the biochemical and toxicological effects seen after various inducers may be markedly different. This is illustrated by the effects of different inducers on the metabolism of various substrates examined *in vitro* and shown in Table 5.22. It can be seen that in some cases the inducers cause no change in the metabolism, whereas in other cases, metabolism is increased or even decreased. Thus studies *in vitro* showed that pretreatment with phenobarbital markedly increased benzphetamine metabolism but had no effect on benzo(a)pyrene metabolism. Conversely, 3-methylcholanthrene pretreatment increased benzo(a)pyrene metabolism but markedly decreased benzphetamine metabolism.

These effects are compounded by species and tissue differences in response to inducers and the differences in isoenzymes present in these species and tissues.

Induction therefore may cause (*i*) increased rate of metabolism of a foreign compound through one pathway (*ii*), altered metabolite profiles if the foreign compound is metabolized by several routes and only one is induced, (*iii*) no effect on metabolism if the particular isoenzymes induced are not involved in metabolism of the particular compound and (*iv*) decreased metabolism if induction increases levels of certain isoenzymes at the expense of the one(s) metabolizing the compound in question.

Depending on the role of metabolism in the toxicity of a compound, therefore, enzyme induction may increase, decrease, or cause no change in the toxicity of a particular compound. The effects of induction need to be considered in the light of distribution and excretion as competing processes (chap. 4, Fig. 72 and chap. 6, Fig. 1). However, the consequences can be simply summarized and explained as follows:

1. If a metabolite is responsible for the toxic effect of a compound, then induction of the enzyme responsible may increase that toxicity. However, if there is only one route of metabolism and elimination is dependent on this, then only the rate of metabolism to the single metabolite will be increased rather than the total amount. This may not

increase toxicity if no other factors are involved or are not time dependent. However, as most toxic effects are multistage, involving repair and protection, this is unlikely.

2. If the parent compound is responsible for a toxic effect, then induction of metabolism may decrease that toxicity. However, induction of metabolism may lead to a different toxic effect due to a metabolite.
3. If a foreign compound is metabolized by several routes, then induction may alter the balance of these routes. This may lead to either increases or decreases in toxicity.
4. Induction may change the stereochemistry of a reaction.

Although some of these principles will be illustrated in more detail by the examples in chapter 7, it is worthwhile examining some briefly at this point. A simple example is the pharmacological effect of a barbiturate, measured as sleeping time, which is dramatically reduced by induction of the enzymes of metabolism (Table 5.23). This effect correlates with the plasma half-life, whereas the plasma level of pentobarbital on awakening is similar in the control and the induced groups (Table 5.23). A toxicological example is afforded by **paracetamol**, which is metabolized by several routes. The hepatotoxicity of paracetamol in the rat is increased by induction with phenobarbital due to an increase in the cytochrome P-450 isoenzyme, which activates the drug. However, in the hamster hepatotoxicity is decreased due to an increase in glucuronidation, a detoxication pathway induced by phenobarbital (see chap. 7). With the hepatotoxin **bromobenzene**, 3-methylcholanthrene induction decreases the toxicity. This is due to an increase in an alternative, nontoxic pathway, 2,3-epoxidation and an increase in the detoxication pathway catalyzed by epoxide hydrolase (chap. 7, Table 7.5). With the lung toxic compound **ipomeanol**, phenobarbital induction decreases the toxicity, metabolic activation, and the LD_{50}, whereas pretreatment with 3-methylcholanthrene changes the target organ from the lung to the liver (see chap. 7).

The pharmacological action of **codeine** is increased by induction as this increases demethylation to morphine. Induction by phenobarbital decreases the toxicity of **organophosphates**, but increases that of **phosphorothionates**. Studies with the drug **warfarin** have shown that induction by both phenobarbital and 3-methylcholanthrene will change the stereochemistry of the product, as can be seen in Table 5.24. Thus, hydroxylation in the 8-position in the R-isomer is increased 12 times compared with only 4 times with the S-isomer following 3-methylcholanthrene induction.

Table 5.23 Effect of Pentobarbital Pretreatment on the Duration of the Pharmacological Effect and Disposition of Pentobarbital in Rabbits

Pretreatment	Sleeping time after pentobarbital (30 mg kg^{-1})	Plasma level of pentobarbital on waking (µg mL^{-1})	Pentobarbital half-life (min)
None	67 ± 4	9.9 ± 1.4	79 ± 3
Pentobarbital 3×60 mg kg^{-1}	30 ± 7	7.9 ± 0.6	26 ± 2

Source: From Ref. 39.

Table 5.24 Influence of Cytochrome P-450 Induction on the *In Vitro* Metabolism of R- and S-Warfarin

	Hydroxylated warfarin metabolites			
	R-isomer		S-isomer	
	7-OH	8-OH	7-OH	8-OH
Inducer	(nmol warfarin metabolite nmol P-450^{-1})			
Uninduced	0.22	0.04	0.04	0.01
Phenobarbitone	0.36	0.07	0.09	0.02
3-Methylcholanthrene	0.08	0.50	0.04	0.04

Source: From Ref. 40.

Thus, the importance of enzyme induction is that it may alter the toxicity of a foreign compound. This can have important clinical consequences and underlie drug interactions. Thus, the antitubercular drug rifampicin is thought to increase the hepatotoxicity of the drug isoniazid, and **alcohol** may increase susceptibility to the hepatotoxicity of paracetamol. However, it should also be noted that induction can alter the metabolism of endogenous compounds. For example, the antitubercular drug, rifampicin, is a microsomal enzyme inducer in human subjects. As well as increasing the toxicity of drugs such as isoniazid, this compound also alters steroid metabolism and may lead to reduced efficacy of the contraceptive pill.

It has only been discovered relatively recently that a natural remedy, **St. John's Wort**, used by millions for the treatment of depression, is a potent inducer of the major human drug–metabolizing enzyme, CYP3A4. This induction can lead to the therapeutic failure of a drug when the remedy is taken at the same time.

That the influence of environmental agents is sufficient to cause significant changes in xenobiotic metabolism in humans has been shown in a number of studies. This is illustrated by studies of the effects of cigarette smoking and cooking meat over charcoal on the metabolism of the drug **phenacetin** in human volunteers. Both these activities produce polycyclic hydrocarbons such as benzo[a]pyrene, which is a potent microsomal enzyme inducer. Eating charcoal-grilled steak was shown to cause a significant increase in the rate of metabolism of phenacetin by de-ethylation to paracetamol (Fig. 5.24). This was indicated by the plasma level of phenacetin, which was significantly lower (20–25%) in human volunteers after eating meat exposed to charcoal compared with foil-wrapped meat. There was no decrease in half-life as phenacetin undergoes a significant first-pass effect, and enzyme induction in the gastrointestinal tract may have been a factor in this study responsible for a significant proportion of the increased metabolism. Cigarette smoking similarly increased the rate of metabolism of phenacetin. A study comparing antipyrine metabolism in Caucasians with that in Asians revealed that there were significant differences in the rate of metabolism between the two ethnic groups.

The greater rate of metabolism (shorter half-life, increased clearance) in the Caucasians was ascribed at least in part to the influence of dietary factors such as eating meat and exposure to coffee, cigarette smoke, and alcohol in the Caucasians.

Mechanisms of Induction and Gene Regulation and Expression
The mechanisms of induction of the cytochromes P-450 will be considered in terms of the different types of inducer. However, some general principles can be considered first.

Induction involves increased synthesis of enzyme protein, which may be detected as an increase in total enzyme level as with phenobarbital induction or increase in a particular isoenzyme. Protein synthesis is increased, and this usually seems to be necessary as inhibition of protein synthesis results in inhibition of induction. The increased protein synthesis may involve increased mRNA synthesis and inhibitors of this, such as **actinomycin D**, block induction. For a simple diagram explaining the relationship of protein synthesis to DNA see Figure 6.38.

Some of the possible mechanisms of induction are

1. increased synthesis of mRNA or precursor rRNA,
2. increased stability of mRNA or rRNA,
3. decreased heme degradation,
4. decreased apoprotein degradation,
5. increased transport of RNA, and
6. effects on DNA-dependent RNA polymerase.

The polycyclic type of inducer has been the most successfully studied, and this work led to the discovery that a cytosolic receptor was involved in induction by polycyclic hydrocarbons. The receptor, AhR, is found in many different cell types.

Indeed, it is now known that there are several receptor-mediated mechanisms responsible for the action of the various different inducers. These are centered on the nucleus and involve xenobiotic responsive elements (XREs) or glucocorticoid responsive elements (GREs). The process is driven by ligand-activated transcription factors. Thus induction of CYP1A1, CYP1A2, CYP2B6, CYP2C9, and CYP3A4 and CYP4A involve similar systems. Induction of CYP2E1 is different and CYP2D6 is not inducible.

Polycyclic Hydrocarbons-Induction of CYP1A1

This type of induction is caused by a large group of environmental chemicals, both natural such as plant indoles and man made such as polycyclic hydrocarbons. Examples include benzo [a]pyrene, benzo[a]anthracene, 3-methylcholanthrene, β-napthoflavone, polychlorinated biphenyls and benzo-*p*-dioxins, and dibenzofurans. Many of these compounds are of interest as they are carcinogenic.

The discovery of strains of mice, which are nonresponsive to induction by polycyclic hydrocarbons (see below), has been a major factor in the study of mechanisms of induction. Thus, strains of mice (DBA/2) were found in which AHH could not be induced by 3-methylcholanthrene in comparison with responsive mice (C57BL/6). It was found that this nonresponsiveness was dependent on a recessive gene at a single locus. The alleles were known as Ah^b for the responsive trait and Ah^d for the nonresponsive, recessive trait. It is significant that the responsive mouse strain is often more susceptible to the various carcinogenic effects of polycyclic hydrocarbons. Other toxic effects are also associated with the possession of responsiveness. For example, mice responsive to AHH induction are susceptible to corneal damage caused by paracetamol, whereas nonresponsive mice are not. Variations in responsiveness to induction have also been observed in strains of rats, and there may be a genetic factor operating in humans. Polycyclic hydrocarbons such as 3-methylcholanthrene and benz[a]anthracene induce AHH, which is known to be cytochrome CYP1A1. However, by far the most potent inducer was found to be **TCDD**, which could in fact cause induction in the nonresponsive mice. TCDD will cause an increase in the rate of transcription of the CYP1A1 gene in cells in culture within minutes.

It was subsequently discovered that the nonresponsive mice (Ah^d) possessed a defective cytosolic *receptor* rather than being deficient for the cytochrome P-450 gene. Thus, this receptor had reduced binding affinity for polycyclic hydrocarbons, which could be overcome with larger amounts of a potent inducer such as TCDD. The nonresponsive trait was due to a defective regulatory gene. The AhR is cytosolic protein, about 88 kDa, localized to the cytosol bound to a heat shock protein (HSP) and aryl hydrocarbon receptor–interacting protein (AIP). It may have several functional domains: ligand binding, DNA binding, a domain for transactivation, for binding HSP, for dimerization, and for nuclear import/export.

The associated proteins act as chaperones, and stop the receptor binding to DNA, keeping it in the cytosol. This complex binds the ligand (hydrocarbon e.g., TCDD), whereupon the HSP and AIP dissociate away and the liganded receptor (hydrocarbon-receptor complex) is translocated to the nucleus (Fig. 5.33). The liganded receptor heterodimerizes with another protein, the AhR nuclear translocator (ARNT), which allows it to bind to a number of genes. The receptor-ARNT complex binds to DNA sequences upstream of the CYP1A1 gene. Target genes are CYP1A1, CYP1A2, and CYP1B1. Genes for some phase 2 enzymes are also induced by ligands for this receptor so the response is pleiotropic.

The binding is with regulatory elements or enhancer/promoter regions of the gene and involves XREs or drug responsive elements (DRES). These are positive control elements or transcriptional enhancers, which function like switches. They are about 1 kilobase upstream of the CYP1A1 gene, which is part of the polycyclic hydrocarbon responsive transcriptional enhancer.

An enhancer is a segment of DNA, which can increase the transcription rate. This causes an alteration in chromatin structure leading to increased accessibility of regulatory DNA and increased transcription of the CYP1A1 gene. There is thus an increase in mRNA, which results in synthesis of the protein portion of cytochrome P-450. CYP1A1 and CYP1A2 are only normally expressed at a low level and mostly in extrahepatic tissues such as the lung. Here they have an important role in the metabolism and detoxication of inhaled environmental pollutants. The potency of a hydrocarbon as an inducer relates to binding ability. Regulation of the cytochrome P-450 genes by inducers such as TCDD is clearly very complex and may also involve posttranscriptional stabilization of mRNA, especially for CYP1A2.

Phenobarbital-Induction of CYP2B2

The overall mechanism for this type of induction is still unclear but is different from that for CYP1A1.

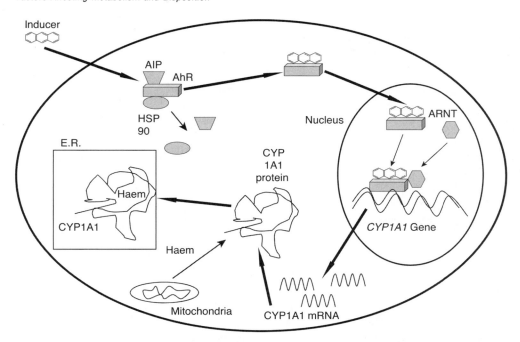

Figure 5.33 Mechanism of the receptor-mediated induction of CYP1A1 by a polycyclic hydrocarbon such as TCDD. The inducer-receptor (AhR) complex enters the nucleus, binds with ARNT, and the complex binds to the CYP1A1 gene in the DNA. This induces the production of CYP1A1 mRNA, which leads to the production of CYP1A1 protein and functional enzyme. *Abbreviations*: AIP, AhR interacting protein; AhR, aryl hydrocarbon receptor; HSP, heat shock protein 90; ARNT, AhR nuclear translocator; E.R., endoplasmic reticulum.

Pretreatment with phenobarbital leads to an increase in mRNA in the liver within four hours. The mRNA, which is translatable, polysomal, and is coded for a cytochrome P-450, is increased 14-fold. The major effect seems to be an increase in gene transcription rate. Phenobarbital causes induction of cytochrome P-450 via transcriptional activation of genes encoding two members of the rat CYP2B family. The result is an increase in CYP2B1 and CYP2B2, the latter being a constitutive form of cytochrome P-450. CYP2B2 is inactive in untreated rat liver, but is a constitutive form in lung and testis. In humans, the CYP2B isoform induced is CYP2B6. Phenobarbital induces a number of other CYPs such as CYP2C8, CYP2C9, and isoforms of the CYP3A family, notably CYP3A4 in man. The induction is mediated by a nuclear receptor, constitutive androstane receptor or CAR. The activity of the gene is controlled by the binding of a complex consisting of the CAR, the retinoic X receptor (RXR) and a co-activator, SCR-1. It seems there is a binding site for the complex in the CYP2B2 gene [phenobarbital-responsive enhancer module (PBREM)] and similarly in the CYP3A4 gene (responsive element ER6). The control of the gene can in turn be modulated by substances such as drugs, for example, phenobarbital or steroids (Fig. 5.34). Some ligands such as certain steroids reduce the activity by splitting SCR-1 off the complex, so that the binding does not occur and the gene is switched off. Hence the production of the CYP2B2 is decreased or stopped. Conversely, inducers may affect the binding of the CAR-SCR-1 RXR complex, perhaps making it bind tighter, and so increase the transcription rate. It has also been suggested that phenobarbital can affect the phosphorylation state of CAR, which may influence its interaction with the binding site. However, although the details are not entirely clear transcriptional activation of the CYP2B gene and message stabilization occur and treatment with phenobarbital clearly leads to an increase in mRNA and hence enzyme protein.

Phenobarbital Pretreatment Also Induces CYP2C and CYP3A1 and CYP3A2
Induction of CYP2E1. This isoform is induced by a variety of small, generally hydrophilic molecules such as ethanol, acetone, pyrazole and drugs such as isoniazid. Diabetes and

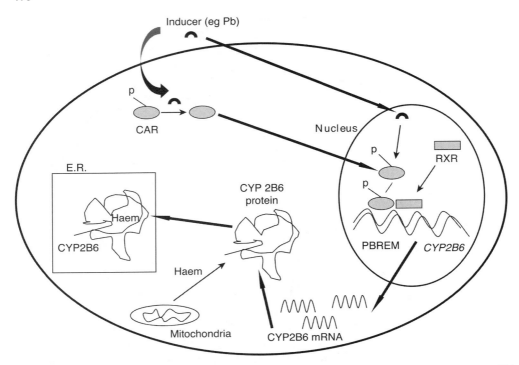

Figure 5.34 Mechanism of induction of CYP2B6 by a chemical such as the drug phenobarbital. This drug activates a nuclear receptor (CAR).This combines with the retinoid X receptor and binds to PBREM, as specific section of the CYP gene, which stimulates the production of CYP2B6 mRNA leading to the production of CYP2B6 protein and enzyme. *Abbreviations*: CAR, constitutive androstane receptor; RXR: retinoid X receptor; PBREM, phenobarbital-responsive enhancer module.

starvation will also lead to induction. Although not the major form of the enzyme in humans (10% in human liver), it is particularly important toxicologically. This is because it is involved in the metabolic activation of a number of chemicals, including carbon tetrachloride, paracetamol, thioacetamide, and vinyl chloride.

In contrast to the other CYPs, with induction of CYP2E1, there is no increase in mRNA, and more than one mechanism may be involved. There is, however, an increase in enzyme protein, up to eightfold in animals exposed to inducers. Thus, it seems that there is posttranscriptional regulation. The suggested mechanisms are

1. the substrate stabilizes the enzyme protein making it more functional,
2. more protein is synthesized more efficiently, and
3. degradation may be inhibited.

In diabetes, it appears that stabilization of mRNA occurs, leading to an increased amount rather than an increase in synthesis. The induction caused by this disease may reflect the necessity to metabolize the ketone bodies produced.

Induction of CYP3A family. CYP3A4 is the most abundant isoform, and in human liver, it represents 30% to 50% total cytochrome P-450, metabolizing a wide variety of chemical types, including many drugs (possibly 60% of those in current use), natural toxicants, and environmental chemicals. This isoform also has a role in the metabolism of endogenous steroids such as testosterone. It is induced by a similarly wide variety of chemicals including phenobarbital as already indicated, dexamethazone, pregnenolone 16α-carbonitrile, rifampicin, imidazole, antifungal drugs, synthetic steroids, organochlorine and organophosphate insecticides, and natural chemicals such as hyperforin found in St. John's Wort.

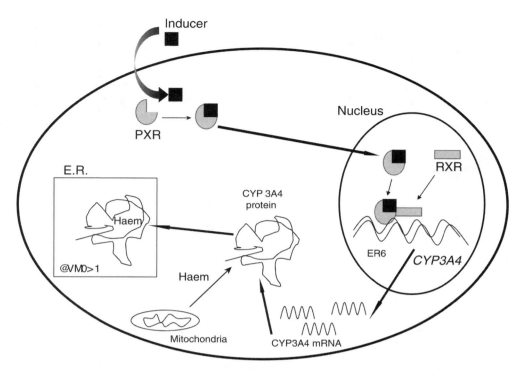

Figure 5.35 Mechanism of the receptor-mediated induction of CYP3A4 by a chemical. The inducer-receptor (PXR) complex enters the nucleus, binds with RXR, and the complex binds to the ER6 region of the CYP gene. This induces the production of CYP3A4 mRNA, which leads to the production of CYP3A4 protein and functional enzyme. *Abbreviations*: RXR, retinoid X receptor-α; PXR, pregnane X receptor; ER6, everted repeat with a 6-kilobase spacer.

The mechanism of induction has many similarities with the induction of CYP2B6, with the exception that a receptor detects the presence of the inducer. This nuclear receptor, found in tissues such as liver and intestine is the pregnane X receptor (PXR), and it binds various steroid hormones. It also binds enzyme inducers, such as the chemicals given above, and this binding eventually leads to activation of the gene. There is a large variety of chemical types, which can bind, suggesting that there is a relatively large binding cavity, which seems to be hydrophobic, and this can accept a variety of molecular orientations. This allows great flexibility and low specificity of binding. Hyperforin is the most potent ligand yet discovered. PXR also has a DNA-binding domain.

Once the inducer (ligand) has bound to PXR, it then translocates to the nucleus where it forms a heterodimer complex with the retinoid X receptor–α (RXR-α). The complex in turn binds to the response elements, ER6 (and maybe another), upstream of the CYP3A4 gene. The result is an upregulation of gene expression, increased transcription, and after translation, more CYP3A4 enzyme protein. Other isoforms of the CYP3A group may also be induced (Fig. 5.35).

Other nuclear receptors, notably the glucocorticoid receptor and CAR, are also thought to be involved, but the details are currently unclear. Thus it seems that activation of the glucocorticoid receptor induces PXR, and hence CYP3A4. Conversely, PXR can be down-regulated by some cytokines, which underlies the decreased level of some CYPs in certain pathological conditions.

Induction of CYP4A. This gene family can be induced by xenobiotics, such as the hypolipidemic drugs (e.g., clofibrate) and other peroxisome proliferators. The enzymes in this family tend to metabolize lipids, in particular fatty acids, rather than xenobiotics. As discussed in chapter 7, the effects of these peroxisome proliferators are mediated by the peroxisome proliferator–activated receptors (PPAR), which are receptors of the steroid hormone family. The receptor involved with induction of CYP4A is PPARα, which has a ligand-binding domain and a DNA-binding domain. The latter binds to the regulatory sections of, for example, the

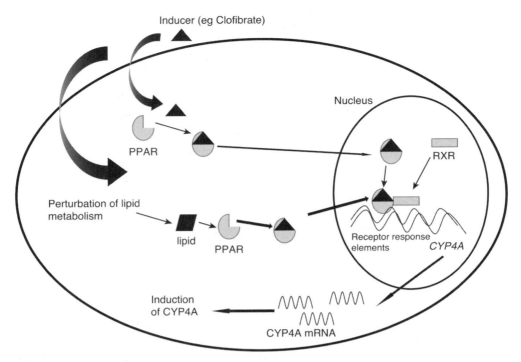

Figure 5.36 Mechanism of the receptor-mediated induction of CYP4A by a chemical such as the drug clofibrate. The inducer-receptor (PPAR) complex enters the nucleus, binds with RXR, and the complex binds to the receptor response elements in the CYP gene. This induces the production of CYP4A mRNA, which leads to the production of CYP4A protein and functional enzyme. Alternatively, the drug may perturb lipid metabolism leading to increases in a lipid(s), which will bind to the receptor and cause the same response. *Abbreviations*: PPAR, peroxisome proliferator–activated receptor; RXR, retinoid X receptor.

CYP4A6 gene. Again RXR is involved in the mechanism. With clofibrate, there is transcriptional activation within 1 hour of dosing (Fig. 5.36).

As endogenous molecules, in particular fatty acids, can bind to the PPAR, the mechanism of induction by xenobiotics could also involve disturbance of lipid metabolism, leading to induction by fatty acids.

The levels of CYP4A1 are very low in liver and kidney but are markedly increased by inducers.

As well as cytochrome P-450, other enzymes are induced by peroxisome proliferators such as clofibrate. Thus enzymes involved in the β-oxidation of fatty acids are increased and other structural proteins are also increased (see chap. 7). This is, therefore, another example of a pleiotropic effect, similar to the induction via the AhR discussed above.

With the hypolipidemic drug-inducible cytochromes P-450 (CYP4A), there is transcriptional activation within 1 hour of administration of clofibrate. The levels of CYP4A1 are very low in the liver and kidney, but are markedly increased by the inducers. As well as cytochromes P-450, a number of other enzymes involved in peroxisomal β-oxidation reactions are also induced.

As well as induction of the synthesis of the apoprotein portion of cytochrome P-450, there is also induction of the synthesis of the heme portion. Clearly, it is also necessary to have an increased amount of heme if there is an increase in the amount of the enzyme apoprotein being synthesized. Thus, the rate-limiting step in heme synthesis, the enzyme δ-**aminolaevulinate synthetase**, is inducible by both phenobarbital and TCDD. This is the result of transcriptional activation of the gene, which codes for the δ-aminolaevulinate synthetase. It may be that the decrease in the heme pool, which results from incorporation of heme into the newly synthesized apoprotein, leads to derepression of the gene and hence increased mRNA synthesis. The gene repression could be heme-mediated, or heme may modulate P-450 genes.

Table 5.25 Induction of Enzymes Other than Cytochrome-P-450

Enzyme	Inducing agent
Glucuronosyl tranferases	Pb, 3-MC, TCDD, PCBs
NADPH cytochrome P-450 reductase	Pb, PCBs, isosafrole
Epoxide hydrolases	Pb, 3-MC, arochlor, isosafrole
GSTs	Pb, 3-MC, TCDD
Cytochome b_5	2-Acetylaminofluorene, BHT
Catalase	Clofibrate, phthalates
Carnitine acyl transferase	Clofibrate

Abbreviation: GSTs, glutathione transferases; Pb, penobarbital; 3-MC, 3-methylcholanthrene; TCDD, 2,3,7,8-tetrachlorodibenzo-*p*-dioxin; PCB's, polychlorinated biphenyls; BHT, butylated hydroxytoluene.

Induction of Non-Cytochrome P-450 Enzymes

As already indicated above, other enzymes as well as cytochromes P-450 may be induced by a variety of compounds. Some of the important enzymes known to be induced and their inducers are shown in Table 5.25. These other enzymes may have a crucial bearing on the overall toxicity of a compound as seen in examples in chapter 7.

Detection of Enzyme Induction

Enzyme induction may be detected in a variety of ways:

1. *In vitro*. The activity or absolute level of enzymes such as cytochrome P-450 and glucuronosyl transferase can be measured in cells, tissue fractions, or subcellular fractions (e.g., microsomes) and compared with those from control animals. The activity is measured by using a particular substrate for each of the isoforms of the enzyme (e.g., cytochrome P-450 or UDPGT) of interest. The total level of cytochrome P-450 could be determined by spectrophotometry using standard methods (e.g., carbon monoxide binding and difference spectra). Alternatively, the level of protein can be determined by gel electrophoresis and Western blotting, and this would allow the separation of different isoforms.
2. *In vivo*
    a. a decrease in half-life or plasma level
    b. a change in proportions of metabolites
    c. a change in a pharmacological effect such as sleeping time after a barbiturate or paralysis time after zoxazolamine
    d. a change in a toxic effect
    e. an increase in plasma γ-glutamyl transferase
    f. an increase in the urinary excretion of D-glucaric acid
    g. an increase in urinary 6- β-hydroxycortisol excretion in humans

Enzyme Inhibition

In contrast to enzyme induction, inhibition usually only requires a single dose or exposure. Although environmentally it may be of less consequence than induction, with drug interactions it is probably of greater importance. Many of the different enzymes involved in the metabolism of foreign compounds may be inhibited (Table 5.26) with consequences for the biological activity of those compounds. As with enzyme induction, the consequences of inhibition will depend on the role of metabolism in the mechanism of toxicity or other biological activity. Thus, enzyme inhibition may increase or decrease toxicity or cause no change.

Enzyme inhibition will be considered in the broadest sense as a decrease in the activity of an enzyme *in vivo* or *in vitro*, which is caused by a foreign compound. Inhibition may be divided into several types involving

1. interaction of the compound with the active site of the enzyme to give a complex; this may involve reversible or irreversible binding;
2. competitive inhibition by two different compounds at the same enzyme-active site;
3. destruction of the enzyme;

Table 5.26 Examples of Inhibitors of Enzymes Involved in Drug Metabolism

Enzyme	Inhibitor
CYP1A2	Furafylline, a-napthoflavone
CYP2C9	Sulfaphenazole, sulfinpyrazone
CYP2D6	Ajmalicine, quinidine, trifluperidol, yohimbine
CYP3A4	Clotrimazole, ketaconazole, naringenin (grapefruit juice)
Esterases	Parathion,
Aldehyde dehydrogenase	Disulfiram (antabuse)
Xanthine oxidase	Allopurinol
Monoamine oxidases	Phenelzine, iproniazid

4. reduced synthesis;
5. allosteric effects; and
6. lack of cofactors.

Each of these types is examined in the following sections.

Complexes.   There are a number of well-known inhibitors of cytochromes P-450, which form stable complexes with the enzyme. Thus SKF 525A, piperonyl butoxide, triacetyloleandomycin, **amphetamine, isoniazid**, and **cimetidine** are a few of those known to form complexes with cytochromes P-450. In the case of several of these inhibitors, metabolism catalyzed by cytochromes P-450 takes place, and the metabolite so produced inhibits the enzyme. Thus, amphetamine is believed to be metabolized to a nitroso intermediate, which complexes with the enzyme. **Piperonyl butoxide** is a commonly used microsomal enzyme inhibitor. It is active both *in vivo* and *in vitro* and is metabolized to a reactive carbene, which binds to the enzyme. It may also inhibit δ-aminolaevulinic acid synthetase, the rate-limiting step in heme biosynthesis. It is used as an insecticide **synergist** (see chap. 2) to block the microsomal enzyme-mediated metabolism of insecticides. For example, when a structurally similar synergist, 2,3-methylenedioxynaphthalene, is mixed with the insecticide carbaryl, the toxicity of **carbaryl** to houseflies occurs at around 5 mg g^{-1}, whereas no toxicity is observed in mice exposed to a dose of carbaryl of 750 mg kg^{-1}. This impressive and selective difference in toxicity is partly due to the fact that mice are able to metabolize and therefore remove the synergist, 2,3-methylenedioxynaphthalene, and so metabolism of the carbaryl is not inhibited. In contrast, the housefly does not metabolize the synergist, and so microsomal metabolism of carbaryl is inhibited, giving rise to greater toxicity. This is an example of the creative use of inhibition in selective toxicity.

Triacetyloleandomycin, an antibiotic, is associated with a number of drug interactions, which may be the result of inhibition of CYP3A4. Thus, when co-administered with oral contraceptives, triacetyloleandomycin may cause problems of cholestasis, and with carbamazepine and theophylline, signs of neurologic intoxication can occur. Again it is metabolized by cytochromes P-450, first by demethylation and then by oxidation, and the resulting metabolite forms a stable complex with the enzyme. The inhibition has a marked effect on metabolism and biological activity of compounds such as hexobarbital (Table 5.27). Although one dose is sufficient, the inhibition increases with time and clearly repeated dosing markedly increases the inhibition.

The type of inhibition in which a complex is formed will tend to be noncompetitive, although if initial metabolism is required, then there may be competition at this stage. Thus, piperonyl butoxide itself is a competitive inhibitor, but its metabolite is a noncompetitive inhibitor.

Metyrapone, another compound widely used as an inhibitor of cytochromes P-450, binds to the reduced form of the enzyme and acts as a noncompetitive inhibitor.

A clinically important example of microsomal enzyme inhibition is the interaction between the drugs **isoniazid** and **diphenylhydantoin**. Termination of the pharmacological activity of diphenylhydantoin depends on microsomal metabolism (figure 7.54). This may be inhibited

Table 5.27 Effect of the Enzyme Inhibitor Triacetyloleandomycin on Metabolism of Hexobarbital

Treatment	Hexobarbital metabolism (nmol min^{-1} mg^{-1})	Sleeping time (min)
Control	1.8 ± 0.7	22 ± 8
TAO 1 hr	1.7 ± 0.7	27 ± 9
TAO 24 hr	1.2 ± 0.7	40 ± 18
TAO × 4	0.3 ±0.1	168 ± 58

TAO, triacetyloleandomycin; 1 hr, 1 hr after the dose; 24 hr, 24 hr after the dose; ×4, 4 daily doses. All the doses were 1 mmol. kg^{-1}.
Source: From Ref. 41.

by isoniazid when the two drugs are administered together. When this occurs, the plasma half-life of diphenylhydantoin is significantly increased, and signs of toxic effects to the central nervous system are apparent. This interaction is seen to a greater extent in slow acetylators, which have higher plasma levels of isoniazid. **SKF 525A** is another inhibitor of this type, which is active both *in vivo* and *in vitro*, but seemingly may sometimes act as a competitive inhibitor and at other times as a noncompetitive inhibitor. This compound not only inhibits cytochromes P-450-mediated reactions, but also inhibits glucuronosyl transferases and esterases.

Inhibitors of cytochromes P-450 may be specific for only one form and SKF 525A, for example, is specific for the phenobarbital inducible form(s), whereas the inhibitor **7, 8-benzoflavone** is specific for the polycyclic hydrocarbon-inducible form.

Other important enzyme inhibitors of this type are the organophosphorus compounds. Thus, after metabolism to the oxygen analogues, the insecticides parathion and malathion (chap. 4, Fig. 25) (Fig. 5.12) form complexes with the enzyme acetylcholinesterase as described in more detail in chapter 7.

Carbamates such as carbaryl (chap. 4, Fig. 47) will also inhibit esterases, although usually less potently than the organophosphorus compounds. The toxicity of compounds metabolized by hydrolysis may thus be altered by such inhibitors. For example, **bis-*p*-nitrophenylphosphate (BNPP)** is an organophosphate type of inhibitor (Fig. 5.37), which blocks the hydrolysis of compounds such as phenacetin and acetylisoniazid (Fig. 5.24) (chap. 4, Fig. 46). Thus, **phenacetin** will cause methemoglobinemia in both experimental animals and man due to the metabolite 2-hydroxyphenetidine (Fig. 5.24). This metabolite is a product of deacetylation and hydroxylation, and the deacetylation step, catalyzed by an acyl amidase, is inhibited by BNPP. Consequently, methemoglobinemia can be prevented in experimental animals by treatment with the organophosphate. BNPP has also been used as an experimental tool to study isoniazid toxicity as it inhibits the hydrolysis of acetylisoniazid (see fig. 7.24). Organophosphorus compounds may cause prolonged and cumulative inhibition of iproniazid, phenelzine, and isocarboxazid. cholinesterases in humans occupationally exposed, which can have important toxicological consequences as discussed in chapter 7.

Another enzyme involved in the metabolism of foreign compounds, which may be purposefully inhibited, is aldehyde dehydrogenase. This is irreversibly and noncompetitively inhibited by the drug **disulfiram** (Antabuse). This inhibition, which lasts for about 24 hours, is used in the treatment of alcoholism. Intake of alcohol after disulfiram leads to unpleasant effects due to the accumulation of acetaldehyde. This is not removed by metabolism as the enzyme aldehyde dehydrogenase is inhibited by disulfiram. Another example of the clinical use of enzyme inhibition is the use of **allopurinol**, which inhibits xanthine oxidase *in vivo* and consequently inhibits the metabolism of compounds such as the anticancer drug, 6-mercaptopurine (chap. 4, Fig. 18), thereby increasing its efficacy. Disulfiram also inhibits cytochromes P-450, and consequently other drug interactions may occur, such as with diphenylhydantoin, in a similar manner to that described for isoniazid. Disulfiram reduces the toxicity of **1,2-dimethylhydrazine**, a colon carcinogen (Fig. 5.38). The N-oxidation of the intermediate metabolite, azomethane, essential for the carcinogenicity, was found to be inhibited by disulfiram.

Monoamine oxidase inhibitors such as phenelzine and isocarboxazid (Fig. 5.37) are used clinically for the treatment of depression. These compounds also form complexes with

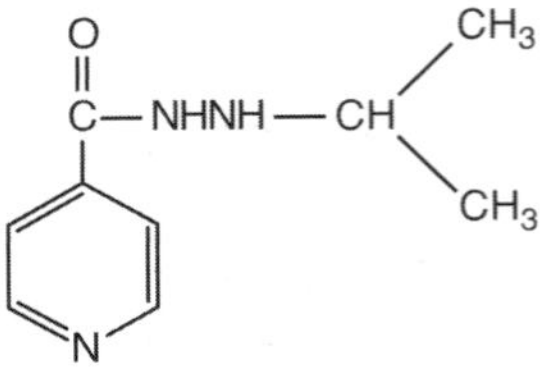

Iproniazid

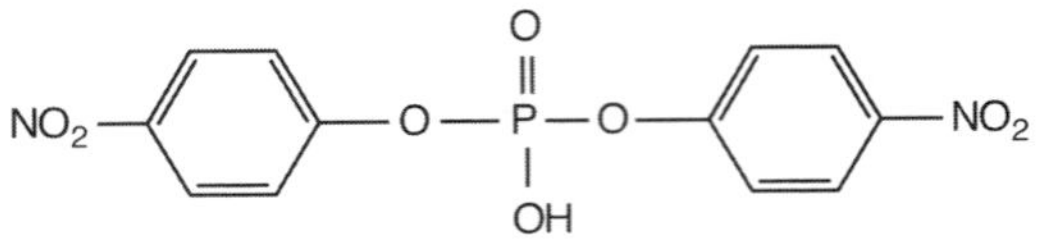

bis-p-Nitrophenyl phosphate

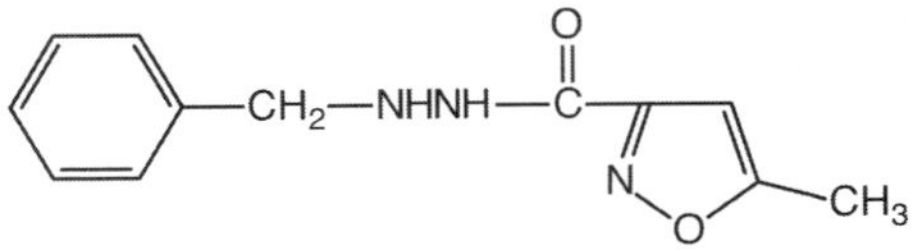

Isocarboxazid

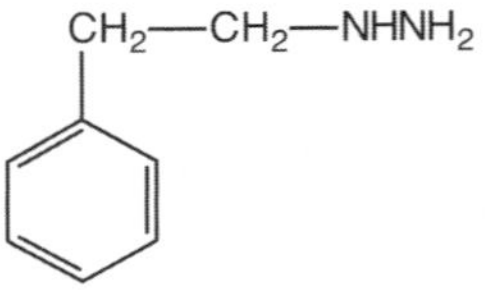

Phenelzine

Figure 5.37 Structures of the enzyme inhibitors iproniazid, BNPP, isocarboxazid, and phenelzine. *Abbreviation*: BNPP, bis-*p*-nitrophenyl phosphate.

$$CH_3NHNHCH_3 \rightarrow CH_3N{=}NCH_3 \rightarrow CH_3N{=}NCH_3$$

1,2-Dimethylhydrazine   Azomethane   Azoxymethane

$$CH_3^+ \leftarrow CH_3N{=}NCH_3$$

Azoxymethanol

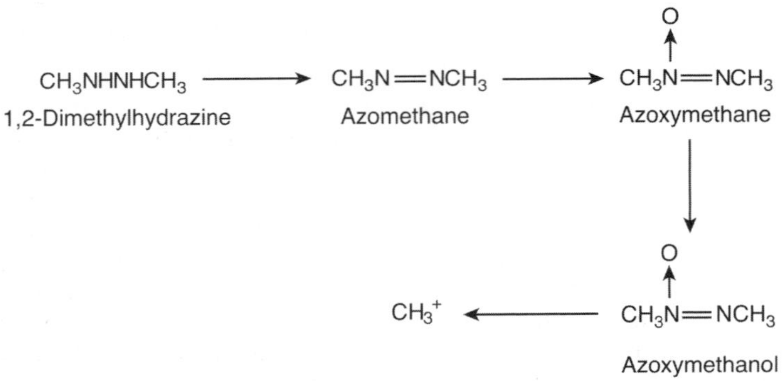

Figure 5.38 Metabolism of 1, 2-dimethylhydrazine.

the enzyme and irreversibly block its action. With iproniazid, complete inhibition of the enzyme lasts for 24 hours, and normal activity is not regained for 5 days. This inhibition may have important toxicological consequences because endogenous compounds and amines in food are also metabolized by monoamine oxidase. Consequently, intake of significant quantities of such amines as tyramine found in cheese may lead to life-threatening hypertension as the amine accumulates in patients who have taken monoamine-inhibiting drugs.

Competitive inhibition. Any two compounds, which are metabolized by the same enzyme, may competitively inhibit the metabolism of the other. The extent of this will depend on the affinity each compound has for the enzyme. One example where this is important toxicologically is in the treatment of **ethylene glycol** and **methanol** poisoning. Both of these

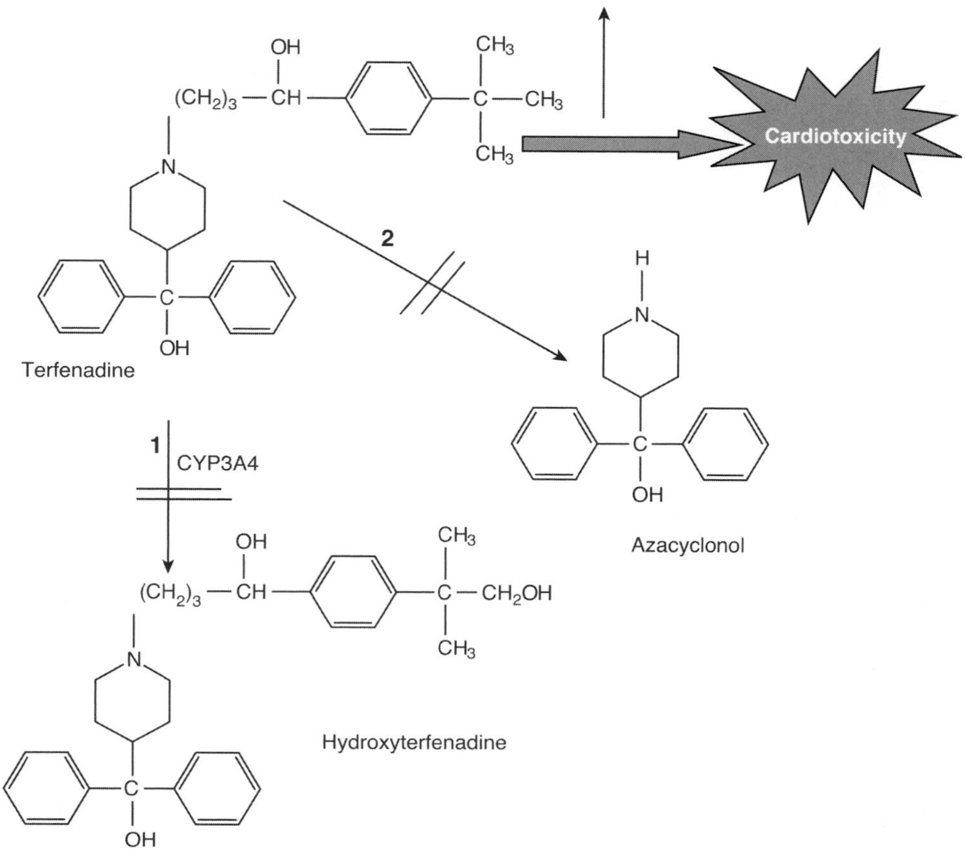

Figure 5.39 The metabolism of the drug terfenadine. The double lines indicate the inhibition of metabolism by other drugs such as ketoconazole or the natural product found in grapefruit juice. This leads to a rise in the blood level of the drug and toxicity.

compounds are toxic as a result of metabolism by the enzyme alcohol dehydrogenase (see chap. 7). Consequently, one method of treatment is to reduce this by administration of ethanol, which has a greater affinity for the enzyme and so reduces metabolism and toxicity. Cytochromes P-450 type I ligands, which bind to the enzyme as substrates, but not to the iron, act as competitive inhibitors. Thus, dichlorobiphenyl, a high-affinity type I ligand for P-450, is a competitive inhibitor of the O-demethylation of p-nitroanisole. Azole antifungal drugs, such as ketoconazole, are well known competitive inhibitors of CYP3A4 as well as other isoforms (Fig. 5.39). As this is the major isoform in man, such inhibition affects the metabolism of many chemicals, and also the metabolism of endogenous steroids can be inhibited. The inhibition of drug metabolism can have serious ramifications, such as when ketoconazole is given at the same time as terfenadine, for example. Given the popularity of the herbal remedy St. John's Wort, the ability of the active ingredient, hyperforin, to inhibit CYP2D6 *in vitro*, has significant potential. With competitive inhibition, the K_m of the enzyme is found to change, but the V_{max} remains the same, whereas with noncompetitive types of inhibition, the reverse is the case. With uncompetitive inhibition, where the inhibitor interacts with the enzyme substrate complex, both V_{max} and K_m change. Epoxide hydrolase can be inhibited by 1,1,1-trichloropropene oxide and a number of other chemicals such as the drug valpromide and progabide. The latter two drugs can potentiate the neurotoxicity of carbemazepine by inhibiting the breakdown of its epoxide metabolite(s). An example of inhibition of metabolism leading to toxicity is afforded by the following drug combination.

In 1993, fifteen Japanese patients died when they took a combination of sorivudine, a new antiviral drug and tegafur, an anticancer drug. The reason was that sorivudine was metabolized by the gut flora in the patients to a product, which was metabolized by the

Table 5.28 The Effect of Allylisopropylacetamide (AIA) on Cytochrome P-450 and Drug-Metabolizing Activity *In Vitro* and Pharmacological Activity *In Vivo*

Parameter	Non-AIA-treated animal control[a] (%)		
	Noninduced	Pb induced	3-MC induced
Cyt P-450[a]	84	33	74
Hexobarbital hydroxylase[a]	80	22	62
Ethylmorphine N-demethylase[a]	62	8	35
p-Chloro-N-methylaniline N-demethylase[a]	75	51	75
Hexobarbital sleeping time (min)[b]	38	236	
Zoxazolamine paralysis time (min)[b]	258	478	

Results for *in vitro* enzyme activity are expressed as percentage of the corresponding non-AIA-treated control. Results for *in vivo* sleeping and paralysis time are expressed as absolute values in minutes.
[a]*In vitro* results.
[b]*In vivo* results.
Source: From Refs. 42 and 43.

enzyme dihydropyrimidine dehydrogenase. Tegafur was converted to 5-fluorouracil, the active drug, which was also metabolized by the same enzyme. The sorivudine bacterial metabolite, however, was a suicide substrate, which inhibited the enzyme and so the 5-fluorouracil was not metabolized and detoxified and caused lethal toxicity after a therapeutic dose.

Destruction. A number of foreign compounds will destroy enzymes such as cytochromes P-450. Thus, halogenated alkanes, hydrazines, compounds containing carbon-carbon double and triple bonds may all interact with and destroy cytochrome P-450. In many cases, this is **suicide inhibition**, whereby the substrate is metabolized by the enzyme but the product reacts with and destroys the enzyme. Thus drugs containing the alkene group such as **secobarbital, allobarbital, fluoroxene** and compounds such as **allylisopropylacetamide** and **vinyl chloride** (chap. 7, Fig. 6 and see chap. 7) all destroy hepatic cytochrome P-450. Similarly, drugs such as ethinyloestradiol and norethindrone, which contain the alkyne group, will also destroy the enzyme. Carbon tetrachloride is an example of a halogenated alkane, which destroys cytochrome P-450, as will several aliphatic and aromatic hydrazines such as ethylhydrazine and phenylhydrazine.

The cytochrome P-450 destroyed may be a specific isoenzyme, as is the case with **carbon tetrachloride** and allylisopropylacetamide (Table 5.28). Indeed, with carbon tetrachloride the isoenzyme destroyed is the one, which is responsible for the metabolic activation (CYP1A2). With allylisopropylacetamide, it is the phenobarbital-inducible form of the enzyme, which is preferentially destroyed as can be seen from Table 5.25. It seems that it is the heme moiety, which is destroyed by the formation of covalent adducts between the reactive metabolite, such as the trichloromethyl radical formed from carbon tetrachloride (see chap. 7), and the porphyrin ring.

A natural product, which is a potent of inhibitor, particularly of CYP3A4, is found in **grapefruit juice**. This has caused serious and indeed fatal interactions with some drugs. The active ingredient responsible is a furanocoumarin, possibly **bergamottin**, 6,7-dihydroxyberga-mottin or bergamottin dimers. This is believed to be a suicide inhibitor, which blocks gut CYP3A4 and leads to inhibition, which lasts several days, even after a single glass (about 200–300 mL). It thus affects drugs with high first-pass metabolism, such as terfenadine, nifedipine, simvastatin, and amiodarone. Although it will also inhibit liver CYP3A4, this only normally occurs after large amounts of grapefruit juice. So the half life of the drug is not normally affected , only the plasma level. In the case of nifedipine, it increases the AUC by fivefold.

The effect is exacerbated as the inhibitor(s) will also inhibit *P*-glycoprotein in the gut cells and so inhibits the efflux mediated by this transporter. Bergamottin will also inhibit CYP1A2, CYP2C9, CYP2C19, and CYP2D6.

Another suicide inhibitor, but of CYP2D6, is the illegal drug 3,4-methylene-dioxy-metamphetamine (MDMA).

Reduced synthesis.   The synthesis of enzymes may be decreased, resulting in a decrease in the *in vivo* activity. With cytochrome P-450 there are a number of ways in which this occurs. Thus, administration of the metal **cobalt** to animals will decrease levels of cytochromes P-450 by inhibiting both the synthesis and increasing the degradation of the enzyme. Thus, cobalt inhibits δ-aminolaevulinic acid synthetase, the enzyme involved in heme synthesis. Cobalt will also increase the activity of heme oxygenase, which breaks down the heme portion to biliverdin. The compound **3-amino, 1, 2, 3-triazole** decreases cytochromes P-450 levels by inhibiting porphyrin synthesis.

Clearly, any agent, which inhibits protein synthesis, will lead to a general decrease in the levels of enzymes.

Allosteric effects.   When **carbon monoxide** interacts with hemoglobin as well as competing with oxygen, it also causes an allosteric change, which affects the binding of oxygen. Although hemoglobin is, strictly speaking, not an enzyme, this allosteric effect has important toxicological implications as discussed in more detail in chapter 7.

Lack of cofactors.   Clearly, where cofactors are involved in a metabolic pathway, a lack of these will result in a decrease in metabolic activity. Thus depletion of hepatic glutathione by compounds such as **diethyl maleate** or inhibition of its synthesis by compounds such as **buthionine sulfoximine** will decrease the ability of the animal to conjugate foreign compounds with glutathione. **Salicylamide** and **borneol** will deplete animals of UDP-glucuronic acid by forming glucuronide conjugates. Galactosamine (fig. 7.62) will inhibit the synthesis of UDP-glucuronic acid. Sulfate required for conjugation is easily depleted by giving large doses of compounds, which are conjugated with it, such as paracetamol. Salicylamide also inhibits sulfate conjugation, and the combined effect of inhibition of glucuronidation and sulfate conjugation, the two major pathways for elimination of paracetamol, markedly increases the hepatotoxicity of the latter.

SUMMARY

There are many factors, both chemical and biological, which affect the disposition of xenobiotics. Chemical factors include size and structure, pK_a, chirality, and lipophilicity. Biological factors include species, sex and strain, genetic factors, hormonal influences, disease and pathological conditions, age, stress, diet, dose, enzyme induction and inhibition, and tissue and organ specificity. All of these factors can affect the toxicity of a chemical by changing its disposition, especially its metabolism.

Chemical factors. Lipophilicity, polarity, and size of a molecule are all important in determining absorption, distribution, metabolism, and excretion of the chemical and hence its toxicity. The chirality of a molecule may affect its absorption, distribution, excretion, and the route of metabolism or extent to which it is metabolized and hence its toxicity. Planarity may affect the interaction with a receptor.

Biological factors. The species of animal is a very important factor especially affecting metabolism but also influencing the other phases of disposition. The pH of the gastrointestinal tract, the nature of the skin, and breathing rate all may affect absorption. Plasma proteins and fat can affect distribution. Biliary excretion is affected by species in relation to the molecular weight threshold.

Species differences in metabolism are mainly found in phase 2 reactions.

Strain differences in route and rate of metabolism and hence toxicity have been documented in inbred strains of mice. Species and strain differences in receptors occur. *Sex* differences in metabolism and toxicity are mostly confined to rodents. Generally male rats metabolize chemicals faster than female rats. These differences can be removed by castration of males and returned by treatment with androgens.

Genetic factors are particularly important in humans and can influence the response to the compound or the disposition of the compound and hence its toxicity. Several genetic factors affecting metabolism are known in which a nonfunctional, less functional, or unstable form of the enzyme is produced in a particular phenotype, for example, acetylator phenotype

(*N*-acetyltransferase, NAT2); hydroxylation status (CYP2D6; CYPC19); esterase deficiency (pseudocholinesterase).

Age effects. Metabolic capabilities of animals at the extremes of age are generally reduced. Also other functions are decreased by old age such as kidney function. Therefore, toxicity may be increased if elimination or detoxication is slowed in old or young animals. Conversely, toxicity may be decreased if metabolic activation is responsible for toxicity.

Hormonal effects. Many hormones influence metabolism (often increasing), including the metabolism of foreign chemicals. Therefore changes in hormone levels may influence toxicity if metabolism is increased or decreased.

Disease/pathological conditions. Disposition of chemicals is potentially altered by disease and hence toxicity. Generalization, however, is difficult as the effects are unpredictable. Thus liver disease may decrease metabolism, but this depends on type of disease and particular pathway of metabolism. Disease in one organ may affect the response of another, for example, chronic renal disease decreases hepatic cytochrome P-450.

Organ/tissue specificity. Particular tissues and organs may be damaged by chemicals because of uptake mechanisms, other biochemical/physiological characteristics or metabolic capabilities. For example, the lung is susceptible to paraquat because of an uptake mechanism (see chap. 6).

Dose. Disposition and metabolism may be affected by the dose of the compound. Hence, the toxicity may be increased or may be limited by saturation/inhibition of metabolism or other processes (e.g., saccharin and estragole).

Stress. The information available suggests that stress may increase metabolism of foreign compounds.

Diet. The constituents and amount of food (deficiency/starvation) may influence disposition and hence toxicity of chemicals. Food constituents may be enzyme inducers or inhibitors. Lack of food or specific constituents (e.g., protein or vitamins) may decrease metabolic capability, for example, a protein-deficient diet decreases cytochrome P-450 activity. Lack of sulfur amino acids decreases glutathione level. The effect on toxicity will depend on the role of metabolism.

Enzyme induction. Some of the enzymes responsible for biotransformation may be induced by exposure to chemicals and other factors (such as diabetes). Induction requires repeated exposure. If only one isoform of the enzyme (e.g., of cytochrome P-450) is induced, the route of metabolism/proportion of metabolites may be changed as well as the rate. The amount of enzyme is increased (and the overall activity in the particular tissue), but inducers may cause other changes such as increases in bile and blood flow and in the smooth endoplasmic reticulum.

The mechanism of induction for cytochrome P-450 is variable but usually involves increased protein synthesis. For CYP1A1, CYP1A2, CYP2B6, CYP2C9, CYP3A4, and CYP4A induction is via similar system via interaction with a receptor. CYP2E1 is different, and CYP2D6 is not inducible.

Toxicological consequences of induction will depend on whether metabolism is a detoxication process or not, but both the rate and route of metabolism of the compound may be changed.

Enzyme inhibition. The enzymes of biotransformation may be inhibited by a single exposure to chemicals. This occurs by several mechanisms: formation of a complex, competition between substrates, destruction of the enzyme, reduced synthesis of the enzyme, allosteric effects, and lack of cofactors. The consequences will depend on the role of metabolism in toxicity in the same way as induction (see above).

REVIEW QUESTIONS

1. Write short notes on three of the following:
   a. glucose-6-phosphate dehydrogenase deficiency
   b. cytochrome P-450 2D6
   c. diet
   d. age of the animal.

2. List four ways in which chirality may feature in the metabolism of a foreign chemical.
3. List two reasons for species differences in absorption of a chemical.

4. Why does the biliary excretion of chemicals differ between different species?
5. Give an example of a qualitative difference in phase 1 metabolism between species.
6. Give an example of a quantitative difference in phase 2 metabolism between species.
7. A sex difference in the toxicity of chloroform has been reported. In which species does this occur and which is the more sensitive sex and why?
8. In which two ways can genetic factors affect toxicity. Give an example of each.
9. Name an enzyme involved in (a) phase 1 metabolism and (b) phase 2 metabolism, which show genetic variation in human.
10. When would a protein-deficient diet (a) increase or (b) decrease the toxicity of a chemical and why?
11. Explain how a chemical might be less toxic in a neonatal animal.
12. List five different microsomal enzyme (cytochrome P-450) inducers. Name one, which acts through a receptor and one, which increases liver blood flow.
13. Give three ways by which enzyme induction can be detected *in vivo*.
14. Give an example of (a) a phase 1 enzyme (apart from cytochrome P-450); (b) a phase 2 enzyme, which can be induced.
15. Indicate four ways in which chemicals may inhibit the enzymes of drug metabolism with an example of each.
16. Write notes on the toxicological importance of three of the following:
    a. enzyme induction
    b. biliary excretion
    c. blood flow through an organ
    d. the lipophilicity of a xenobiotic

17. Explain the term "amphipathic" and why it has toxicological significance. Use an example to illustrate you answer.
18. Explain the species differences in carcinogenicity of tamoxifen.

REFERENCES

1. Hansch C. Quantitative relationships between lipophilic character and drug metabolism. Drug Metab Rev 1972; 1:1–14.
2. Reggiani G. Medical problems raised by the TCDD contamination in Seveso, Italy. Arch Toxicol 1978; 40:161–188.
3. US EPA. Health Assessment Document for 2,3,7,8-Tetrachlorodibenzo-P-Dioxin (TCDD) and Related Compounds Volume I of III. U.S. Environmental Protection Agency, Washington, D.C., EPA/600/BP-92/001a (NTIS PB94205465), 1994.
4. McCreesh AH. Percutaneous toxicity. Toxicol Appl Pharmacol 1965; 7(suppl 2):20–26.
5. Adamson RH, Davies DS. Comparative aspects of absorption, distribution, metabolism and excretion of drugs. In: International Encyclopaedia of Pharmacology and Therapeutics, Section 85, Chapter 9. Oxford: Pergamon, 1973.
6. Altman PL and Dittmer DS. Blood and Other Body Fluids. Washington D.C.: Federation of American Societies for Experimental Biology, 1961:453–464.
7. Dobson A. Physiological peculiarities of the ruminant relative to drug distribution. Fed Proc 1967; 26:994–1000.
8. Levine RJ. Stimulation by saliva of gastric acid secretion in the rat. Life Sci 1965; 4:959–964.
9. Prosser CL and Brown FA. Comparative Animal Physiology. Philadelphia: W.B. Saunders, 1961.
10. Bishop DW. Comparative Animal Physiology. In: Prosser CL ed. Philadelphia: W.B. Saunders, 1950.
11. Bloom F. The Urine of the Dog and Cat New York: Gamma Publications, 1960.
12. Cornelius CE, Kaneko JJ. Clinical Biochemistry of Domestic Animals. New York: Academic Press, 1963.
13. Oliverio VT, Adamson RH, Henderson ES, et al. The distribution, excretion, and metabolism of methylglyoxal-bis-guanylhydrazone-C14. J Pharmac Exp Ther 1963; 141:149–156.
14. Abou-El-Makarem MM, Millburn P, Smith RL, et al. Biliary excretion in foreign compounds. Species difference in biliary excretion. Biochem J 1967; 105:1289–1293.
15. Davison C, Williams RT. The metabolism of 5,5'-methylenedisalicylic acid in various species. J Pharm Pharmac 1968; 20:12–18.
16. Caesar J, Shaldon S, Chiandussi L, et al. The use of indocyanine green in the measurement of hepatic blood flow and as a test of hepatic function. Clin Sci 1961; 21:43–57.
17. Cherrick GR, Stein SW, Leevy CM, et al. Indocyanine green: observations on its physical properties, plasma decay, and hepatic extraction. J Clin Invest 1960; 39:592–600.

18. Levine WG, Millburn P, Smith RL, et al. The role of the hepatic endoplasmic reticulum in the biliary excretion of foreign compounds by the rat. The effect of phenobarbitone and SKF 525-A (diethylaminoethyl diphenylpropylacetate Biochem Pharmac 1970; 19:235–244.
19. Wheeler HO, Cranston WI, Meltzer JI. Hepatic uptake and biliary excretion of indocyanine green in the dog. Proc Soc Exp Bio Med NY 1958; 99:11–14.
20. Gessner PK, Parke DV, Williams RT, et al. Studies in detoxication. 86. The metabolism of 14C-labelled ethylene glycol. Biochem J 1961; 79:482–489.
21. Parke DV. The Biochemistry of Foreign Compounds. Oxford: Pergamon, 1968.
22. Quinn GP, Axelrod J, Brodie BB, et al. Species, strain and sex differences in metabolism of hexobarbitone, amidopyrine, antipyrine and aniline. Biochem Pharmac 1958; 1:152–159.
23. Adamson RH, Dixon RL, Francis FL, et al. Comparative biochemistry of drug metabolism by azo and nitro reductase. Proc Natl Acad Sci U S A 1965; 54:1386–1391.
24. Hodgson E, Levi PE. A Textbook of Modern Toxicology. New York: Elsevier, 1987.
25. Jay GE Jr. Variation in response of various mouse strains to hexobarbital (Evipal). Proc Soc Exp Biol Med 1955; 90:378–381.
26. Kato R, Onoda K. Studies on the regulation of the activity of drug oxidation in rat liver microsomes by androgen and estrogen. Biochem Pharmacol 1970; 19:1649–1660.
27. Ellard GA, Gammon PT, Titinen H. Determination of the acetylator phenotype using matrix isoniazid Tubercle. 1975; 56:203–209.
28. De Souich P, Erill S. Metabolism of procainamide in patients with chronic heart failure, chronic respiratory failure and chronic renal failure. Eur J Clin Pharm 1976; 10:283–287.
29. Weber WW. The Acetylator Gene and Drug Response. New York: Oxford University Press, 1987.
30. Lunde PK, Frislid K, Hansteen V. Disease and acetylation polymorphism. Clin Pharmacokinet 1977; 2:182–187.
31. Magoub A, Idle JR, Dring LG, et al. Polymorphic hydroxylationof debrisoquine in man. Lancet 1977; 2:584–586.
32. Sloan TP, Lancaster R, Shah RR, et al. Genetically determined oxidation capacity and the disposition of debrisoquine. Br J Clin Pharmacol 1983; 15:443–450.
33. Sata F, Sapone A, Elizondo G, et al. CYP3A4 allelic variants with amino acid substitutions in exons 7 and 12: evidence for an allelic variant with altered catalytic activity. Clin Pharmacol Ther 2000; 67:48–56.
34. Kim YC, Lee SJ. Temporal variation in hepatotoxicity and metabolism of acetaminophen in mice. Toxicology 1998; 128:53–61.
35. Gibson GG, Skett P. Introduction to Drug Metabolism. Cheltenham: Nelson Thornes, 2001.
36. Jondorf WR, Maickel RP, Brodie BB. Inability of newborn. mice and guinea pigs to metabolize drugs. Biochem Pharmac 1958; 1:352.
37. Hart JG, Timbrell JA. The effect of age on paracetamol hepatotoxicity in mice. Biochem Pharmac 1979; 28:3015–3017.
38. Powis G, Talcott RE, Schenkman JB. In: Ullrich V, Roots A, Hildebrandt A, et al., eds. Microsomes and Drug Oxidations. Oxford: Pergamon Press:137.
39. Remmer H. In: Brodie BB, Erdos EG, eds. Metabolic Factors Controlling Duration of Drug Action, Proceedings of First International Pharmacological Meeting, Vol. 6. New York: Macmillan, 1962.
40. Gibson GG, Skett P. Introduction to Drug Metabolism. London: Chapman Hall, 1986.
41. Pessayre D, Konstantinova-Mitcheva M, Descatoire V, et al. Hypoactivity of cytochrome P-450 after triacetyloleandomycin administration. Biochem Pharmacol 1981; 30:559–564.
42. Farrell H, Correia MA. Structural and functional reconstitution of hepatic cytochrome P-450 *in vivo*. Reversal of allylisopropylacetamide-mediated destruction of the hemoprotein by exogenous heme. J Biol Chem 1980; 255:10128–10133.
43. Unseld B, de Matteis F. Destruction of endogenous and exogenous haem by 2-allyl-2-isopropylace-tamide: role of the liver cytochrome P-450 which is inducible by phenobarbitone. Int J Biochem 1978; 9:865–869.

BIBLIOGRAPHY

General

Abdel-Rahman S, Kauffman RE. The integration of pharmacokinetics and pharmacodynamics: under-standing dose response. Ann Rev Pharmacol Toxicol 2004; 44:111–136.
Alvares AP, Pratt WB. Pathways of drug metabolism. In: Pratt WB, Taylor P, eds. Principles of Drug Action, The Basis of Pharmacology. New York: Churchill Livingstone, 1990.
Burk O, Wojnowski L. Cytochromes P450 3A and their regulation. N-S Arch Pharmacol 2004; 369:105–124.
Caldwell J, Jakoby WB, eds. Biological Basis of Detoxication. New York: Academic Press, 1983.
Coleman MD. Human Drug Metabolism, Chapter 3. UK: Wiley, 2005.

Gibson GG, Skett P. Introduction to Drug Metabolism, Chapter 7. 3rd ed. UK: Nelson Thornes, 2001.

Hill MJ, ed. Role of Gut Bacteria in Human Toxicology and Pharmacology. London: Taylor & Francis, 1995.

Hodgson E. Chemical and environmental factors affecting metabolism of xenobiotics. In: Hodgson E, Levi PE, eds. Introduction to Biochemical Toxicology. 2nd ed. Connecticut: Appleton-Lange, 1994.

Jakoby WE, ed. Enzymatic Basis of Detoxication. New York: Academic Press, 1980.

Parke DV. The Biochemistry of Foreign Compounds, Chapter 6. Oxford: Pergamon, 1968.

Ronis MJJ, Cunny HC. Physiological (endogenous) factors affecting Xenobiotic Metabolism In: Hodgson E, Smart RC, eds. Introduction to Biochemical Toxicology. 3rd ed. New York: Wiley Interscience, 2001.

Rose RL, Hodgson E. Adaptation to Toxicants In: Hodgson E, Smart RC, eds. Introduction to Biochemical Toxicology. 3rd ed. New York: Wiley Interscience, 2001.

Chemical Factors

Ariens EJ, Wins, EW, Veringa EJ. Stereoselectivity of bioactive xenobiotics. Biochem Pharmacol 1988; 37:9.

Lewis DFV. MO-QSARs: a review of molecular orbital-generated quantitative structure-activity relationships. In: Gibson GG, ed. Progress in Drug Metabolism. London: Taylor & Francis, 1990.

Trager WF. Jones JP. Stereochemical considerations in drug metabolism. In: Bridges JW, Chasseaud LF, Gibson GG, eds. Progress in Drug Metabolism, Vol. 10. London: Taylor & Francis, 1987.

Tucker GT, Lennard MS. Enantiomer specific pharmacokinetics. Pharmacol Ther 1990; 45:309.

Species, Strain, and Sex Effects

Calabrese EJ. Toxic Susceptibility: Male/Female Differences. New York: John Wiley, 1985.

Caldwell J. The current status of attempts to predict species differences in drug metabolism. Drug Metab Rev 1981; 12:221.

Caldwell J. Conjugation reactions in foreign compound metabolism: definition, consequences and species variations. Drug Metab Rev 1982; 13:745.

Caldwell J. Conjugation mechanisms of xenobiotic metabolism: Mammalian aspects. In: Paulson GD, Caldwell J, Hutton DH, et al., eds. Xenobiotic Conjugation Chemistry. Washington: American Chemical Society, 1986.

Gandhi M, Aweeka F, Greenblatt R, et al. Sex differences in pharmacokinetics and pharmacodynamics. Ann Rev Pharmacol Toxicol 2004; 44:499–523.

Gustafsson JÅ. Sex steroid induced changes in hepatic enzymes. Annu Rev Physiol 1983; 45:51.

Hathway DE, Brown SS, Chasseaud LF, et al. Foreign Compound Metabolism in Mammals, Vols. 1–6, London: The Chemical Society, 1970–1981. (This series contains chapters on species, strain and sex differences in metabolism.)

Hengstler JG, Oesch F. Interspecies differences in xenobiotic metabolizing enzymes and their importance for interspecies extrapolation of toxicity. In: Ballantyne B, Marrs TC, Syversen T, eds. General and Applied Toxicology. Vol. 1. London: Macmillan, 2000.

Jondorf WR. Drug metabolism as evolutionary probes. Drug Metab Rev 1981; 12:379.

Kato R, Yamazoe Y. Sex-dependent regulation of cytochrome P450 expression. In: Ruckpaul K, Rein H, eds. Frontiers of Biotransformation, Vol. II, Principles, Mechanisms and Biological Consequences of Induction. London: Taylor & Francis, 1990.

Walker CH. Species variations in some hepatic microsomal enzymes that metabolise xenobiotics. Prog Drug Metab 1980; 5:113.

Walker CH. Comparative toxicology. In: Hodgson E, Levi PE, eds. Introduction to Biochemical Toxicology. 2nd ed. Connecticut: Appleton-Lange, 1994.

Williams RT. Interspecies variations in the metabolism of xenobiotics. Biochem Soc Trans 1974; 2:359.

Genetic Factors

Alvan G. Clinical consequences of polymorphic drug oxidation. Fund Clin Pharmacol 1991; 5:209.

Alvan G. Genetic polymorphisms in drug metabolism. J Internal Med 1992; 231:571.

Evans DAP. N-Acetyltransferase. In: Kalow W, ed. Pharmacogenetics of Drug Metabolism. New York: Pergamon, 1993:95.

Grant DM, Blum M, Meyer UA. Polymorphisms of N-acetyltransferase genes. Xenobiotica 1992; 22:1073.

Human cytochrome P450s: assessment, regulation and genetic polymorphisms. Xenobiotica, 28, number 12, 1998. A series of relevant articles.

Hahidi H, Guzey G, Idle JR. Pharmacogenetics and toxicological consequences of human drug oxidation and reduction. In: Ballantyne B, Marrs TC, Syversen T, eds. General and Applied Toxicology, Vol. 1. London: Macmillan, 2000.

Idle J. Pharmacogenetics. Enigmatic variations. Nature 1988; 331:391.

Kadlubar FF. Biochemical individuality and its implications for drug and carcinogen metabolism:Recent insights from acetyltransferase and cytochrome PA4501A2 phenotyping and genotyping in humans. Drug Metab Rev 1994; 26:37.

Kalow W, ed. Pharmacogenetics of Drug Metabolism. Oxford: Pergamon Press, 1992.

Meyer UA. The molecular basis of genetic polymorphisms of drug metabolism. J Pharm Pharmacol 1994; 46(suppl 1):409.

Nebert DW. The genetic regulation of drug-metabolising enzymes. Drug Metab Disp 1986; 16:1.

Nebert DW, Weber WW. Pharmacogenetics. In: Pratt WB, Taylor P, eds. Principles of Drug Action. The Basis of Pharmacology. New York: Churchill Livingstone, 1990.

Tucker GT. Clinical implications of genetic polymorphism in drug metabolism. J Pharm Pharmacol 1994; 46(suppl 1):417.

Diet

Cambell TC, Hayes JR, Merrill AH, et al. The influence of dietary factors on drug metabolism in animals. Drug Metab Rev 1979; 9:173.

Dollery CT, Fraser HS, Mucklow JC, et al. Contribution of environmental factors to variability in human drug metabolism. Drug Metab Rev 1979; 9:207.

Krishnaswamy K. Drug metabolism and pharmacokinetics in malnutrition. TIPS 1983; 4:295.

Parke DV, Iaonnides C. The role of nutrition in toxicology. Annu Rev Nutr 1981; 1:207.

Age

Kinirons MT, Crome P. Clinical pharmacokinetic considerations in the elderly. Clin Pharmacokinet 1997; 33(4):302.

Klinger W. Biotransformation of xenobiotics during ontogenetic development. In: Ruckpaul K, Rein H, eds. Frontiers of Biotransformation, Vol. II, Principles, Mechanisms and Biological Consequences of Induction. London: Taylor & Francis, 1990.

Neims AH, Warner M, Loughnan PM, et al. Developmental aspects of the hepatic cytochrome P-450 mono-oxygenase system. Annu Rev Pharmacol Toxicol 1976; 16:427.

Schmucker DL. Age-related changes in drug disposition. Pharmacol Rev 1979; 30:445.

Skett P, Gustafsson J-Å. Imprinting of enzyme systems of xenobiotic and steroid metabolism. Rev Biochem Tox 1979; 1:27.

Pathological Conditions

Azri S, Renton KW. Factors involved in the depression of hepatic mixed function oxidase during infection with Listeria monocytogenes. Int J Immunopharmacol 1991; 13:197.

Hoyumpa AM, Schenker S. Major drug interactions: effect of liver disease, alcohol and malnutrition. Annu Rev Med 1982; 33:113.

Kato R. Drug metabolism under pathological and abnormal physiological states in animals and man. Xenobiotica 1977; 7:25.

Kraul H, Truckenbrodt J, Huster A, et al. Comparison of in vitro and in vivo biotransformation in patients with liver disease of differing severity. Eur J Clin Pharmacol 1991; 41:475.

Prescott LF. Pathological and physiological factors affecting drug absorption, distribution, elimination and response in man. In: Gillette JR, Mitchell JR, eds. Handbook of Experimental Pharmacology, Vol. 28, Part 3, Concepts in Biochemical Pharmacology. Berlin: Springer, 1975.

Wilkinson GR, Schenker S. Drug disposition and liver disease. Drug Metab Rev 1975; 4:139.

Tissue and Organ Specificity and Circadian Effects

Gram TE, ed. Extrahepatic Metabolism of Drugs and Other Foreign Compounds. Jamaica, New York: Spectrum, 1980.

Wood PA, Lincoln DW, Imam I, et al. Circadian toxicology. In: Ballantyne B, Marrs TC, Syversen T, eds. General and Applied Toxicology, Vol. 1. London: Macmillan, 2000.

Enzyme Induction

Alvares AP, Pantuck EJ, Anderson KE, et al. Regulation of drug metabolism in man by environmental factors. Drug Metab Rev 1979; 9:185.

Batt AM, Siest G, Magdalou J, et al. Enzyme induction by drugs and toxins. Clin Chem Acta 1992; 209:109.

Bresnick E, Houser WH. The induction of cytochrome P-450c by cytosolic hydrocarbons proceeds through the interaction of a 4S cytosolic binding protein. In: Ruckpaul K, Rein H, eds. Frontiers of Biotransformation, Vol. II, Principles, Mechanisms and Biological Consequences of Induction. London: Taylor & Francis, 1990.

Handschin C, Meyer UA. Induction of drug metabolism: the role of nuclear receptors. Pharmacol Rev 2003; 55; 649–673.

Nebert DW. Role of the aryl hydrocarbon receptor mediated induction of the CYP1 enzymes in environmental toxicity and cancer. J Biol Chem 2004; 279; 23847–23850.

Okey AB. Enzyme induction in the cytochrome P450 system. Pharmacol Ther 1990; 45:241.

Park BK. Assessment of the drug metabolism capacity of the liver. Br J Clin Pharmacol 1982; 14:631.

Parke DV. Induction of cytochromes P450—General principles and biological consequences. In: Ruckpaul K, Rein H, eds. Frontiers of Biotransformation, Vol. II, Principles, Mechanisms and Biological Consequences of Induction. London: Taylor & Francis, 1990.

Plant NJ, Gibson GG. Evaluation of the toxicological relevance of CYP3A4 induction. Curr Opin Drug Dis Dev 2003; 6:50–56.

Snyder R, Remmer H. Classes of hepatic microsomal mixed function oxidase inducers. Pharmacol Ther 1979; 7:203.

Tukey RH, Johnson EF. Molecular aspects of regulation and structure of the drug-metabolizing enzymes. In: Pratt WB, Taylor P, eds. Principles of Drug Action. New York: Churchill Livingstone, 1990.

Waxman DJ, Azarnoff L. Phenobarbital induction of cytochrome P450 gene expression. Biochem J 1992; 281:577.

Whitlock JP. The control of cytochrome P-450 gene expression by dioxin. TIPS 1989; 10:285.

Whitlock JP. Mechanistic aspects of dioxin action. Chem Res Toxicol 1993; 6:754.

Enzyme Inhibition

Halpert JR, Guengerich FP, Bend JR, et al. Contemporary issues in toxicology: Selective inhibitors of cytochromes P450. Toxicol App Pharmacol 1994; 125:163.

Ortiz de Montellano PR, Correia MA. Suicidal destruction of cytochrome P-450 during oxidative metabolism. Annu Rev Pharmacol Toxicol 1983; 23:481.

Shitara Y, Sato H, Sugiyama Y. Evaluation of drug-drug interaction in the hepatobiliary and renal transport of drugs. Ann Rev Pharmacol Toxicol 2005; 45:689–723.

Stockley IH. Drug Interactions. Wallingford: Pharmaceutical Press, 1996.

Testa B, Jenner P. Inhibitors of cytochrome P-450s and their mechanisms of action. Drug Metab Rev 1981; 12:1.

6 | Toxic Responses to Foreign Compounds

6.1 INTRODUCTION

There are many ways in which an organism may respond to a toxic compound, and the type of response depends upon numerous factors. Although many of the toxic effects of foreign compounds have a biochemical basis, the expression of the effects may be very different. Thus, the development of tumors may be one result of an attack on nucleic acids, another might be the birth of an abnormal offspring. The interaction of a toxic compound with normal metabolic processes may cause a physiological response such as muscle paralysis, or a fall in blood pressure, or it may cause a tissue lesion in one organ. The covalent interaction between a toxic foreign compound and a normal body protein may in some circumstances cause an immunological response, in others a tissue lesion.

Thus, although all these toxic responses may have a biochemical basis, they have been categorized according to the manifestation of the toxic effect. Therefore, although there will be overlap between some of the types of toxic response, for the purposes of this discussion, it is convenient to divide them into the following:

1. Direct toxic action: tissue lesions
2. Pharmacological, physiological, and biochemical effects
3. Teratogenesis
4. Immunotoxicity
5. Mutagenesis
6. Carcinogenesis

Toxic responses may be detected in a variety of ways in animals, and some of these have already been alluded to in previous chapters. Toxic responses may be the **all-or-none** type such as the death of the organism or they may be **graded responses**. Thus, the main means of detection are as follows:

1. Death: The LD_{50} assay was previously used but has been superseded as an indicator of toxicity by other assays.
2. Pathological change: This could be the development of a tumor or destruction of tissue, but it would be detectable by observation either macroscopically or microscopically.
3. Biochemical change: This might involve an effect on an enzyme such as inhibition or alteration in a particular metabolic pathway. Alternatively, the appearance of an enzyme or other substance in body fluids may indicate leakage from tissue due to damage and be indicative of pathological change.
4. Physiological change: This could be measured in the whole conscious animal as, for example, a change in blood pressure, in temperature, or in a response to a particular stimulus.
5. Changes in normal status: There are a number of markers of toxicity, which are simple to determine yet can indicate a toxic response. Thus, changes in body weight, food and water intake, urine output, and organ weight may all be sensitive indicators of either general or specific toxicity. Thus, animals often consume less food and lose weight after exposure to toxic compounds, and increased organ weight may be due to a tumor, fluid, or triglyceride accumulation, hypertrophy, or enzyme induction. These changes may of course be confirmed by chemical, biochemical, or histopathological measurements.

Examples of some of these manifestations of toxicity will be covered in the various sections of this chapter.

6.2 DIRECT TOXIC ACTION: TISSUE LESIONS

6.2.1 Introduction

Some toxic compounds cause direct damage to tissues, leading to the death of some or all the cells in an organ, for instance. The damage may be *reversible* or *irreversible*, and the overall toxic response will depend on many factors, including the importance of the particular tissue to the animal, the degree of specialization and its reserve functional capacity, and ability to repair the damage. It is beyond the scope of this book to examine all the types of tissue lesion, although the major types of toxic damage are discussed. However, it is instructive to discuss first the principles underlying this type of toxicity, the factors, which affect it, and the general mechanisms of cellular damage.

6.2.2 Target Organ Toxicity

Although any organ or tissue may be a target for a toxic compound, such compounds often damage specific organs. Therefore it is instructive first to examine the principles underlying the susceptibility of certain organs to damage by toxic substances.

Thus, there are a number of different reasons why an organ might be a target:

1. Its blood supply
2. The presence of a particular enzyme or biochemical pathway
3. The function or position of the organ
4. The vulnerability to disruption or degree of specialization
5. The ability to repair damage
6. The presence of particular uptake systems
7. The ability to metabolize the compound and the balance of toxication/detoxication systems
8. Binding to particular macromolecules

Thus, those organs well supplied with blood such as the kidneys will be exposed to foreign compounds to a greater extent than poorly perfused tissue such as the bone. Therefore, factors, which affect the absorption, distribution, and excretion of foreign compounds and the physicochemical characteristics of those compounds, can all affect the toxicity to particular target organs. Those organs, which are metabolically active, such as the liver, may be more vulnerable than metabolically less active tissues such as the skin. This is due to the fact that many compounds require metabolic activation to be toxic (chap. 4, Fig. 72) (see chap. 7). However, the toxicity to the organ will depend on the balance of toxication and detoxication pathways and other factors such as the ability to repair toxic damage. Thus, although the liver may have the greatest metabolic activity if this extends to detoxication pathways as well as toxication pathways, overall the toxicity may be less than in an organ, which has lower activity for the toxication pathway, but lacks the detoxication pathway. Organs or tissues, which are at sites of entry, such as the gastrointestinal tract, may be exposed to higher concentrations of foreign compounds prior to dilution by blood and other fluids, and hence may be more susceptible than deep tissues such as muscle. Such tissues as the skin, gastrointestinal tract, and lungs may suffer local irritation, as they tend to be exposed to substances as a result of their position and function. Highly specialized and vital organs such as the central nervous system (CNS) are susceptible to disruption and are not easily repaired in comparison with adipose tissue, for example, which is less vital and less specialized. Uptake into and concentration of foreign compounds by some organs, in some cases, such as the kidney and liver, as a result of excretory function, may impart vulnerability. Binding to specific macromolecules such as melanin in the eye, for example, may lead to target organ toxicity. Some of the interrelationships between metabolism, distribution, and excretion, which may affect toxicity are shown in Figure 6.1.

Thus, the distribution of the parent compound or metabolite(s) into the target tissue(s), metabolism and excretion in such tissues, and the interaction with receptors or other critical

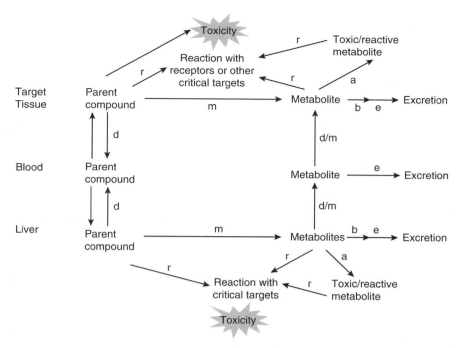

Figure 6.1 The interrelationships between the disposition and toxicity of a foreign compound. The parent compound may undergo distribution out of the blood (d) to other tissues, and there cause toxicity by reaction with receptors (r). Alternatively, metabolism (m) may also occur and give rise to toxic metabolites (a), which react with critical targets (r). Metabolites may also distribute back into the blood (b) and be excreted (e).

cellular macromolecules are all dynamic processes occurring at particular rates. Factors that affect these processes therefore will influence toxicity and the particular target organ, and may even change the target organ.

Thus, it is clear from Figure 6.1 that if the parent compound is toxic, factors such as enzyme induction or inhibition, which change metabolism (m) or change distribution to tissues (d) or excretion (e), will tend to change the toxicity. This is provided that the toxic response is dose related. Specific uptake systems will influence the distribution (d), and specific excretory routes (e) may be saturated. Metabolism may cause the appearance of another, different, type of toxicity. If a metabolite is toxic, then factors that increase metabolism may increase the toxicity, provided that detoxication pathways (b) are not also increased and therefore compensate. However, the presence or absence of such pathways (b) may determine whether a tissue becomes the target. Toxic metabolites or proximate toxic metabolites may be produced in one organ such as the liver and transported to another (d/m). The interaction of toxic or reactive metabolites with receptors (r) may be tissue specific, and there may be protective agents such as glutathione (GSH) in some tissues. These principles are further exemplified in this section with regard to the various target organs considered, and also in the final chapter with specific examples of toxic compounds.

6.2.3 Liver

As it is one of the portals of entry to the tissues of the body, the liver is exposed to many potentially toxic substances via the gastrointestinal tract from the diet, food additives and contaminants, and drugs and is frequently a target in experimental animals. In humans, liver damage is less common, and only around 9% of adverse drug reactions affect the liver. By virtue of its position, structure, function, and biochemistry, the liver is especially vulnerable to damage from toxic compounds. Substances taken into the body from the gastrointestinal tract are absorbed into the hepatic-portal blood system and pass via the portal vein to the liver. Thus, after the gastrointestinal mucosa and blood, the liver is the next tissue to be exposed to a compound, and as it is prior to dilution in the systemic circulation, this exposure will often be at a higher concentration than that of other tissues. The liver, the largest gland in the body,

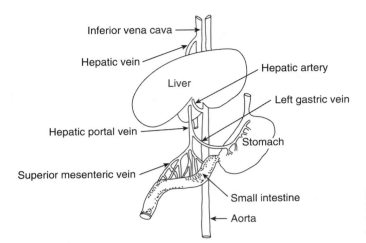

Figure 6.2 The vasculature supply-ing and draining the liver and its relationship to the systemic circulation. *Source*: From Ref. 3.

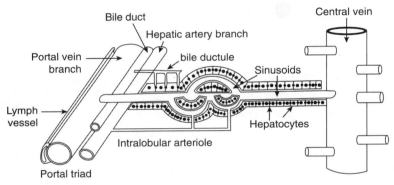

Figure 6.3 Schematic representation of the arrangement and relationship of vessels and sinusoids in the liver. The central vein drains into the hepatic vein. *Source*: Modified from Ref. 4.

represents around 2% to 3% of the body weight in humans and other mammals such as the rat. It is served by two blood supplies, the portal vein, which accounts for 75% of the hepatic blood supply, and the hepatic artery. The portal vein drains the gastrointestinal tract, spleen, and pancreas and therefore supplies nutrients, and the hepatic artery supplies oxygenated blood (Fig. 6.2). The liver receives around 25% of the cardiac output, which flows through the organ at around 1 to 1.3 mL min^{-1} g^{-1} and drains via the hepatic vein into the inferior vena cava. In between the blood entering the liver via hepatic artery and portal vein and leaving via the hepatic vein, the blood flows through sinusoids (Fig. 6.3). Sinusoids are specialized capillaries with discontinuous basement membranes, which are lined with Kupffer cells and endothelial cells. There are large fenestrations in the sinusoids, which allow large molecules to pass through into the interstitial space and into close contact with the hepatocytes (Fig. 6.4). The liver is mainly composed of hepatocytes arranged as plates approximately two-cells thick, each plate bounded by a sinusoid (Fig. 6.4). The membranes of adjacent hepatocytes form the bile canaliculi into which **bile** is secreted.

The bile canaliculi form a network, which feed into ductules, which become bile ducts (Fig. 6.3). The structural and functional unit of the liver is the lobule, which is usually described in terms of the hepatic acinus (Fig. 6.5), based on the microcirculation in the lobule. When the lobule is considered in structural terms, it may be described as either a classical or a portal lobule (see "Glossary"). The acinus comprises a unit bounded by two portal tracts and terminal hepatic or central venules, where a portal tract is composed of a portal venule, bile ductile, and hepatic arteriole (Fig. 6.5). Blood flows from the portal tract toward the central

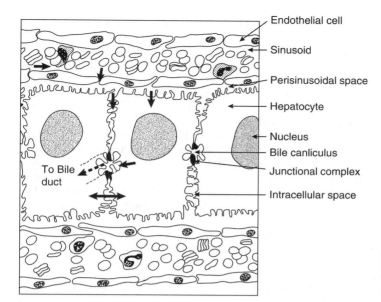

Figure 6.4 Diagrammatic representation of the arrangement of hepatocytes within the liver and the relationship to the sinusoids. From Ref. 4.

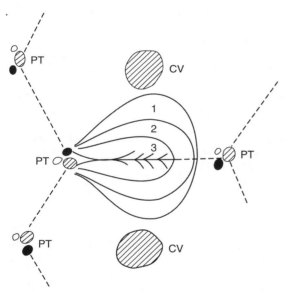

Figure 6.5 Schematic representation of a hepatic acinus. PT represents the portal tract, consisting of branches of the portal vein and hepatic artery and a bile duct. CV represents a branch of the central vein. The areas 1, 2, and 3 represent the various zones draining the terminal afferent vessel. *Source*: Adapted from Ref. 5.

venules, whereas bile flows in the opposite direction. There are three circulatory zones in the acinus, with zone 1 receiving blood from the afferent venules and arterioles first, followed by zone 2, and finally zone 3. Thus, there will be metabolic differences between the zones because of the blood flow. Zone 1 will receive blood, which is still rich in oxygen and nutrients, such as fats and other constituents. The hepatocytes in zone 3, however, will receive blood, which has lost much of the nutrients and oxygen. Zone 1 approximates to the periportal region of the classical lobule and zone 3 to the **centrilobular** region. Zone 3, particularly where several acini meet, is particularly sensitive to damage from toxic compounds. The acinus is also a secretory unit, the bile it produces flowing into the terminal bile ductules in the portal tract.

The close proximity of the blood in the sinusoids with the hepatocytes allows efficient exchange of compounds, both endogenous and exogenous, and consequently foreign compounds are taken up very readily into hepatocytes. For example, the drug **propranolol** is extensively extracted in the "first pass" through the liver.

The liver is a target organ for toxic substances for four main reasons:

1. The large and diverse metabolic capabilities of the liver enable it to metabolize many foreign compounds, but as metabolism does not always result in detoxication, this may make it a target (see sects. 7.2.1 and 7.2.4 chap. 7).
2. The liver also has an extensive role in intermediary metabolism and synthesis, and consequently, interference with endogenous metabolic pathways may lead to toxic effects, as discussed in chapter 7 (see sects. 7.8.2 and 7.8.3).
3. The secretion of bile by the liver may also be a factor. This may be due to the biliary excretion of foreign compounds leading to high concentrations, especially if saturated, as occurs with the hepatotoxic drug **furosemide**. Alternatively, enterohepatic circulation can give rise to prolonged high concentrations in the liver. Interference with bile production and flow as a result of precipitation of a compound in the canalicular lumen or interference with bile flow may lead to damage to the biliary system and surrounding hepatocytes.
4. The blood supply ensures that the liver is exposed to relatively high concentrations of toxic substances absorbed from the gastrointestinal tract.

The hepatocytes, or parenchymal cells, represent about 80% of the liver by volume and are the major source of metabolic activity. However, this metabolic activity varies depending on the location of the hepatocyte. Thus, zone 1 hepatocytes are more aerobic and therefore are particularly equipped for pathways such as the β-oxidation of fats, and they also have more GSH and GSH peroxidase. These hepatocytes also contain alcohol dehydrogenase and are able to metabolize **allyl alcohol** to the toxic metabolite acrolein, which causes necrosis in zone 1. Conversely, zone 3 hepatocytes have a higher level of **cytochromes P-450** and **NADPH cytochrome P-450 reductase**, and lipid synthesis is higher in this area. This may explain why zone 3 is most often damaged, and lipid accumulation is a common response (see "Carbon Tetrachloride," for instance, chap. 7).

The Kupffer cells are known to contain significant **peroxidase** activity and also **acetyltransferase**. The differential distribution of isoenzymes may also be a factor in the localization of damage.

There are various types of toxic response, which the liver sustains and which reflect its structure and function. Viewed simply, liver injury is usually due either to the metabolic capabilities of the hepatocyte or involves the secretion of bile.

The various types of liver damage, which may be caused by toxic compounds, are discussed in the following sections.

Fatty Liver (Steatosis)

This is the accumulation of **triglycerides** in hepatocytes, and there are a number of mechanisms underlying this response as is discussed below (see the sect. "Mechanisms of Toxicity"). The liver has an important role in lipid metabolism, and triglyceride synthesis occurs particularly in zone 3. Consequently, fatty liver is a common response to toxicity, often the result of interference with protein synthesis, and may be the only response as after exposure to **hydrazine, ethionine**, and **tetracycline**, or it may occur in combination with necrosis as with **carbon tetrachloride**. It is normally a reversible response, which does not usually lead to cell death, although it can be very serious as is the case with tetracycline-induced fatty liver in humans. Repeated exposure to compounds, which cause fatty liver, such as alcohol, may lead to cirrhosis.

The specific accumulation of phospholipids (phospholipidosis) can occur but it also occurs in other organs and tissues and will be discussed later in this chapter.

Cytotoxic Damage

Many toxic compounds cause direct damage to the hepatocytes, which leads to cell death and necrosis. This is a general toxic response, not specific for the liver, and there are undoubtedly many mechanisms, which underlie cytotoxicity, but most are still poorly understood. The mechanisms underlying cytotoxicity in general are discussed below, and several examples of hepatotoxins are discussed in more detail in chapter 7.

The zone of the liver damaged may depend on the mechanism, but may also be the result of the microcirculation. Damage may be zonal, diffuse, or massive. For example, **cocaine** and **allyl alcohol** cause zone 1 (periportal) necrosis. With allyl alcohol, this is partly as a result of the presence of alcohol dehydrogenase and partly because this is the first area exposed to the compound in the blood. Conversely, carbon tetrachloride, bromobenzene, and paracetamol cause zone 3 (centrilobular) necrosis as a result of metabolic activation occurring primarily in that region (see chap. 7). Midzonal, zone 2 necrosis is less common, but has been described for the natural product **ngaione** and **beryllium**. Galactosamine causes diffuse hepatic necrosis (see chap. 7), presumably because it interferes with a metabolic pathway, which occurs in all regions of the liver lobule. The explosive **trinitrotoluene** (TNT) can cause massive liver necrosis.

Ischemia may also be a component of cytotoxic damage, and consequently interference with liver blood flow by toxic compounds such as **phalloidin**, which causes swelling of the endothelial cells lining the sinusoids, may cause or contribute toward cytotoxicity.

Other compounds cause liver necrosis because of biliary excretion. Thus, the drug **furosemide** causes a dose-dependent centrilobular necrosis in mice. The liver is a target as a result of its capacity for metabolic activation and because furosemide is excreted into the bile by an active process, which is saturated after high doses. The liver concentration of furosemide therefore rises disproportionately (chap. 3, Fig. 34), and metabolic activation allows the production of a toxic metabolite (Fig. 6.6). The drug **proxicromil** (chap. 5, Fig. 11) caused hepatic damage in dogs as a result of saturation of biliary excretion and a consequent increase in hepatic exposure.

Cholestatic Damage

There are various types of interference with the biliary system, and this can lead to bile stasis or damage to the bile ducts, ductules, or canaliculi. In some cases, such as with **chlorpromazine**, damage to the hepatocytes may ensue. Thus, some foreign compounds, such as the antibiotic **rifampicin**, interfere with bilirubin transport and conjugation giving rise to **hyperbilirubinemia**. Other compounds, **icterogenin**, for example, cause bile stasis and bilirubin deposits in the canaliculi. This canalicular damage may also be accompanied by damage to hepatocytes, such as caused by chlorpromazine. This drug is a surface-active agent, which can reach high

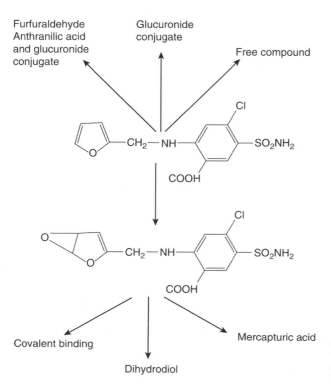

Figure 6.6 The metabolism of furosemide. The epoxide intermediate is the postulated reactive intermediate.

concentrations in the bile and so directly damage the lining cells. It can also cause precipitation of insoluble substances in the lumen of the canaliculi. Accumulation of bile and its constituent bile salts may indeed be the cause of damage, and some are surface-active agents. Consequently, if high concentrations are reached, the cells of the biliary system and hepatocytes exposed can be damaged. The secondary bile acid **lithocholate** will cause direct damage to the canalicular membrane, for example. Some compounds damage the bile ducts and ductules directly such as α-**naphthylisothiocyanate**. The result of the destruction of bile duct-lining cells will be cholestasis as debris from the necrotic cells will block the ductules.

Cirrhosis
This is a chronic lesion resulting from repeated injury and subsequent repair. It may result from either hepatocyte damage or cholestatic damage, each giving rise to a different kind of cirrhosis. Thus, carbon tetrachloride will cause liver cirrhosis after repeated exposure, but also compounds, which do not cause acute necrosis, such as ethionine and alcohol may cause cirrhosis after chronic exposure.

Vascular Lesions
Occasionally toxic compounds can directly damage the hepatic sinusoids and capillaries. One such toxic compound is **monocrotaline**, a naturally occurring pyrrolozidine alkaloid, found in certain plants (*Heliotropium*, *Senecio*, and *Crotolaria* species). Monocrotaline (Fig. 7.7) is metabolized to a reactive metabolite, which is directly cytotoxic to the sinusoidal and endothelial cells, causing damage and occlusion of the lumen. The blood flow in the liver is therefore reduced and ischemic damage to the hepatocytes ensues. Centrilobular necrosis results, and the venous return to the liver is blocked. Hence, this is known as veno-occlusive disease and results in extensive alteration in hepatic vasculature and function. Chronic exposure causes cirrhosis.

Liver Tumors
Both benign and malignant liver tumors may arise from exposure to hepatotoxins and can be derived from various cell types. Thus, adenomas have been associated with the use of **contraceptive steroids** and exposure to **aflatoxin B_1**, and dimethylnitrosamine can produce hepatocellular carcinomas, whereas vinyl chloride causes hemangiosarcomas derived from the vasculature (see chap. 7).

Proliferation of Peroxisomes
A response to exposure to certain foreign compounds, which occurs predominantly in the liver is the phenomenon of peroxisomal proliferation. Peroxisomes (microbodies) are organelles found in many cell types, but especially hepatocytes. Repeated exposure of rodents to certain

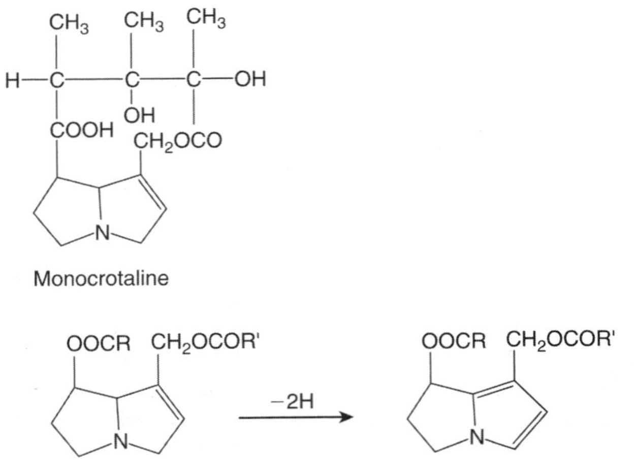

Monocrotaline

Figure 6.7 The structure of the pyrrolizidine alkaloid monocrotaline and the microsomal enzyme-mediated metabolic activation of the pyrrolizidine alkaloid nucleus.

types of foreign compound leads to an increase in the number of these organelles and an increase in the activities of various enzymes. As well as exposure *in vivo*, exposure of isolated hepatocytes *in vitro* will also lead to proliferation of peroxisomes, indicating that it is a cellular response. The function of the organelle is mainly the oxidative metabolism of lipids and certain other oxidative metabolic pathways. Thus, the enzymes for the β-oxidation of fatty acids are found in peroxisomes. The importance of this phenomenon in toxicity and especially carcinogenicity is discussed later in this chapter. The types of compounds, which cause the proliferation, are generally acids or compounds, which can be metabolized to acids. Thus, the hypolipidemic drug **clofibrate**, and a number of similarly acting drugs, will cause the proliferation. **Phthalate esters**, which are commonly used as plasticizers, are another group of compounds, which have been shown to be active. The results of exposure in rodents are the following: a large increase in the numbers of peroxisomes (e.g., 140%); a large increase in liver weight (e.g., 3.9–8.5% body weight); an increase in DNA content (e.g., 1.5–2×); increases in RNA and protein synthesis; large increases in the enzymes of β-oxidation such as palmitoyl CoA oxidase (e.g., 6–15×, but some enzymes maybe increased up to 150×); and increases in the cytochrome P-450 enzymes, which metabolize fatty acids such as lauric acid (CYP4A1, e.g., 5–10×). This type of response is achieved after exposure to a compound such as clofibrate for 28 days.

There is clearly a modulation of gene expression occurring for RNA, DNA, and protein synthesis to be increased, and this seems to be mediated by a receptor interaction. Several possible mechanisms have been proposed to account for the phenomenon, and these are not necessarily exclusive: (*i*) interaction between peroxisome proliferators and a receptor [peroxisome proliferator–activated receptor (PPAR)]; (*ii*) perturbation by peroxisome proliferators of lipid metabolism, leading to substrate overload; (*iii*) action of peroxisomal proliferators as substrates for lipid metabolism. Thus, there is evidence for a receptor, which can be activated by peroxisome proliferators *in vitro*. This interaction seems to activate genes involved with peroxisomal and microsomal fatty acid oxidation. However a perturbation, such as inhibition, of lipid metabolism may also be involved and could lead to increased levels of an endogenous ligand for the receptor. For example, fatty acids such as oleic acid are known to bind to and activate the PPAR *in vitro*. If drugs and other chemicals, which are peroxisomal proliferators, acted as substrates for enzymes, such as those catalysing β-oxidation, they could perturb lipid metabolism, leading to changes in the levels of crucial fatty acids. Therefore, increased peroxisomal enzyme activity could be an adaptive response.

The receptor protein (52 kDa) is a member of the steroid hormone receptor superfamily, which has a DNA-binding as well as ligand-binding domain. Another receptor, the retinoid X receptor is also involved, and after binding of the peroxisome proliferator, the two receptors form a heterodimer. This binds to a regulatory DNA sequence known as the peroxisome proliferator response element. The end result of the interaction between peroxisome proliferators and this system is that genes are switched on, leading to increases in synthesis (induction) of both microsomal and peroxisomal enzymes and possibly hyperplasia.

Structure activity studies *in vitro* have revealed that the relative potency of peroxisome proliferators seems to be determined by a combination of lipophilicity and the calculated binding affinity to the mouse PPARα ligand-binding domain.

The mechanisms underlying some of these types of injury will be discussed in general terms below and in chapter 7.

Detection of Hepatic Damage

Simple quantitative tests can be used such as measurement of the liver weight/body weight ratio. Overt damage to the liver can be detected by light and electron microscopy of liver sections. However, damage can be detected by other noninvasive means such as the urinary excretion of conjugated bilirubin or the amino acid taurine.

Various parameters may be measured in plasma. Thus, determination of the enzymes aspartate transaminase (**AST**) and alanine transaminase (**ALT**) is the most common means of detecting liver damage, the enzymes being raised several fold in the first 24 hours after damage. However, there are a number of other enzymes, which may be used as markers. Plasma bilirubin can also be measured, being increased in liver damage, and plasma albumin is decreased by liver damage (although also by renal damage). Liver function may be determined using the hepatic clearance of a dye such as sulfobromophthalein.

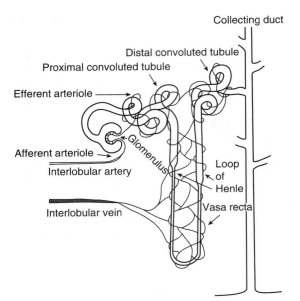

Collecting duct

Distal convoluted tubule

Proximal convoluted tubule

Efferent arteriole

Afferent arteriole

Interlobular artery

Glomerulus

Loop
of
Henle

Vasa recta

Interlobular vein

Figure 6.8   Schematic representation of a mammalian nephron.

6.2.4 Kidney

The function of the kidney is to filter waste products and toxins out of the blood while conserving essential substances such as glucose, amino acids, and ions such as sodium. Anatomically, the kidney is a complex arrangement of vascular endothelial cells and tubular epithelial cells, the blood vessels and tubules being intertwined. The functional unit of the kidney is the nephron (Fig. 6.8). Although all nephrons have their glomeruli and primary vascular elements in the cortex, the position of the nephron within the kidney does vary in that it may lie completely within the cortex (cortical nephron) or the glomerulus may be located at the cortex-medulla boundary (juxtamedullary nephron). The environment both inside and outside the nephron varies along its length, and this influences the types of toxic effect produced by nephrotoxic agents. The kidney is a target organ for toxicity for the following reasons:

1. Renal blood flow. The blood passing through the mammalian kidney represents 25% of the cardiac output, despite the fact that the kidney only represents about 1% of the body mass. Therefore, the exposure of kidney tissue to foreign compounds in the bloodstream, especially the cortex, which receives more blood than the medulla, is relatively high.
2. The concentrating ability of the kidney. After glomerular filtration, many substances are reabsorbed from the tubular fluid. Thus, 98% to 99% of the water and sodium are reabsorbed. Hence, the concentration of foreign substances in the tubular lumen is considerably higher than that in the blood, and the tubular fluid/blood ratio may reach values of 500:1. For example, some **sulfonamides** when given in high doses may cause renal tubular necrosis as a result of the crystallization of less soluble, acetylated metabolites in the lumen of the tubule due to the increased concentration in the tubular fluid and pH-dependent solubility (chap. 4, Table 1). Similarly, **oxalate** crystals may occur in the renal tubules after ingestion of toxic amounts of ethylene glycol (see chap. 7). Foreign substances may also be reabsorbed from the tubular fluid along with water and endogenous compounds. In this case, the tubular cells may contain relatively high concentrations of the foreign compound. The countercurrent exchange of small molecules may lead to very high concentrations of compounds in the interstitial fluid of the renal medulla. An example of a compound, which is nephrotoxic as a result of uptake from the tubular fluid, is the aminoglycoside antibiotic **gentamycin**. This drug is excreted into the tubular fluid by glomerular filtration, but binds to anionic phospholipids on the brush border of the proximal tubular cells. The resulting complex is absorbed into the cell by phagocytosis, is stored in secondary lysosomes, and hence accumulates in the proximal tubular cells. Gentamycin causes a variety of biochemical derangements in the cell, and the

lysosomes are destabilized such that their enzymes are released with resulting degradation of cellular components.

3. Active transport of compounds by the tubular cells. Compounds that are actively transported from the blood into the tubular fluid may accumulate in the proximal tubular cells, especially at concentrations where saturation of the transport system occurs. Again, concentrations to which tubular cells are exposed may be very much higher than in the bloodstream. An example of this is the drug **cephaloridine**, which causes proximal tubular damage as discussed in more detail in chapter 7.

4. Metabolic activation. Although the kidney does not contain as much cytochromes P-450 as the liver, there is sufficient activity to be responsible for metabolic activation, and other oxidative enzymes such as those of the prostaglandin synthetase system are also present. Such metabolic activation may underlie the renal toxicity of chloroform and paracetamol (see chap. 7). Other enzymes such as C-S lyase and GSH transferase may also be involved in the activation of compounds such as hexachlorobutadiene (see chap. 7). In some cases, hepatic metabolism may be involved followed by transport to the kidney and subsequent toxicity.

It is clear that the tissues of the kidney are often exposed to higher concentrations of potentially toxic compounds than most other tissues. The toxic effects caused may be due to a variety of mechanisms ranging from the simple irritant effects of sulfonamides and oxalate crystals to the enzyme inhibition and tubular damage proposed to underlie aspirin-induced medullary lesions. It has been suggested that the latter is due to inhibition of prostaglandin synthesis by aspirin, giving rise to vasoconstriction of the vasa recta (Fig. 6.8), and hence reduction of blood flow and ischemic damage to the kidney. The proximal tubule, and hence the cortex, is the most common site of damage from foreign compounds, whereas the medulla is less commonly damaged and the collecting ducts only rarely. The glomerulus may be damaged by nephrotoxins such as the aminoglycosides, although these compounds also damage the proximal tubule. The proximal convoluted tubule is damaged by chromium, whereas other metals such as mercury (see chap. 7) damage the straight portion (pars recta). The pars recta section of the proximal tubule contains the highest concentration of cytochromes P-450, and this is one reason for its particular susceptibility to toxic injury. Another is the particular capacity for secretion of organic compounds. Hence, cephaloridine damages this section (see chap. 7). The loop of Henle is particularly susceptible to damage from analgesics, such as **aspirin** and **phenacetin**, **fluoride**, and **2-bromoethylamine**, which damage the thin limb and collecting ducts. This is manifested as papillary necrosis, following damage to the renal papillae. The distal convoluted tubule is less commonly a target, although **cisplatin**, the anticancer drug, damages this portion of the kidney. Interestingly, the *trans* isomer does not cause kidney damage at the same doses as the *cis* isomer. **Amphoterecin** causes damage to both proximal and distal tubules. It is a surface-active compound, which is believed to act by binding to membrane phospholipids, causing the cells to become leaky.

The kidney has a marked ability to compensate for tissue damage and loss, and consequently, unless it is evaluated immediately, nephrotoxic effects may not be recognized. Similarly, chronic toxicity may not be detected because of this compensatory ability.

Detection of Kidney Damage

There are a variety of ways in which kidney damage can be detected ranging from simple qualitative tests to more complex biochemical assays.

Simple tests include urine volume, pH, and specific gravity measurement; kidney weight/body weight ratio; and detection of the presence of protein or cells in the urine.

Clearly, overt pathological damage can be detected by light or electron microscopy, but this requires sacrifice of the animal. Damage can also be detected by measurement of a variety of urinary constituents. Thus, damage to tubular cells results in the leakage of enzymes such as **γ-glutamyltransferase** and **N-acetylglucosaminidase** into the tubular fluid, and therefore these can be detected and quantitated in the urine. Also, such damage will be reflected in altered renal tubular function, leading to the excretion of glucose and amino acids. Measurement of urea or creatinine in the plasma will also indicate renal dysfunction if these are raised. The more common measurement is urea or blood urea nitrogen (**BUN**). Measurement of urea or creatinine in the urine and plasma allows the renal clearance of these endogenous compounds to be determined, and this will also indicate renal dysfunction. However, clearance of the polysaccharide inulin is a better indicator of glomerular filtration as it is less affected by other factors such as protein metabolism.

Different types of damage will cause different urinary profiles for endogenous compounds as has been investigated using high-resolution proton nuclear magnetic resonance (NMR) of urine to generate urinary metabolite data patterns (see "Bibliography").

6.2.5 Lung

The lung is a particularly vulnerable organ with regard to toxic substances, since it can be exposed to foreign compounds both in the external environment and also internally from the bloodstream. The lungs receive 100% of the cardiac output and therefore are extensively exposed to substances in the blood. Also, the air breathed in may contain many potentially toxic substances: irritant gases such as sulfur dioxide; particles of dust, metals, and other substances such as **asbestos**; solvents; and other chemicals, which may be local irritants. The lung is well adapted for the efficient exchange of gases between the ambient air and the blood. This also means it is efficient at absorbing potentially toxic substances from the air. The surface area is large, approximately 70 m^2 in humans, and the arrangement of air spaces, the alveoli, and blood capillaries is such that the distance and cellular barrier between air and blood is minimal (chap. 3, Fig. 10). Consequently, foreign compounds may be absorbed very rapidly if they are small enough to pass through pores in the alveolar membrane (8–10 Å radius) or are lipid soluble and able to cross the membrane by diffusion. Solid particles may be taken up by phagocytosis in some cases. The basic functional unit of the lung is the acinus, consisting of a respiratory bronchiole, alveolar ducts and sacs, alveoli, and blood vessels (chap. 3, Fig. 10). There are various cell types within the lung, and these exhibit different susceptibility to toxic damage.

Thus, the lung is a target organ for the following reasons:

1. It is perfused with the whole of the cardiac output of blood and so is exposed to substances in the blood.
2. It is exposed to substances in the air.
3. It has a high oxygen concentration and consequently may be damaged by reactive oxygen species (ROS).

However, the lungs are equipped with defense mechanisms especially with regard to the intake of foreign substances from the air. Thus, the upper airways of the respiratory system are lined with ciliated cells, and mucus is secreted, which also lines the airways. Solid particles are therefore trapped by the mucus and cilia and are transported out of the respiratory system. Other substances may be removed after dissolving in the mucus and then being transported out by the ciliary escalator.

For substances entering the lung tissue from either the blood side or the air, metabolic capability and metabolic activation may be important considerations in the toxicity. The lung is capable of metabolizing many foreign compounds, and the enzymes are present, which catalyze both phase 1 and phase 2 reactions. However, the activity of the monooxygenase enzymes relative to the liver would depend on the particular cell type and species. Although, overall, the activity is generally less, for particular isoenzymes it may indeed be higher. For instance, biphenyl hydroxylase activity is higher in rabbit lung, but lower in rat lung than the corresponding liver. Aniline hydroxylase activity, however, is lower in rabbit lung than liver, as is glucuronyl transferase. The non-ciliated bronchiolar epithelial cell, or **Clara cell** has the highest level of cytochromes P-450, and for some isozymes, the activity is greater than that in the liver. It is this cell, which is responsible for the metabolic activation of ipomeanol (see chap. 7) and other lung toxins such as **naphthalene** and **3-methylfuran**.

There are over 40 different cell types in the respiratory system and those particularly susceptible are the lining cells of the trachea and bronchi, endothelial cells, and the interstitial cells (fibroblasts and fibrocytes).

Some toxic substances such as sulfur dioxide, nitrogen dioxide, and ozone cause acute, direct damage to lung tissue, whereas others, such as nickel carbonyl, may lead to the formation of tumors.

The types of toxic response shown by the lungs are briefly discussed below.

Irritation

Volatile irritants such as ammonia and chlorine initially cause constriction of the bronchioles. These two gases are water soluble, are absorbed in the aqueous secretions of the upper airways of the respiratory system, and may not cause permanent damage. Irritant damage may however lead to changes in permeability and edema, the accumulation of fluid. Some irritants such as arsenic compounds cause bronchitis.

Lining-Cell Damage

Direct damage to cells of the respiratory tract, either by airborne compounds or those in the blood, may cause necrosis and changes in permeability leading to edema. This may be due to the action of the parent compound such as **nitrogen dioxide** or **ozone** or to a metabolite as is the case with ipomeanol (see chap. 7). The former two compounds cause peroxidation of cellular membranes (see below), and phosgene destroys the permeability of the alveolar cell membrane following hydrolysis to carbon dioxide and hydrogen chloride in the aqueous environment of the alveolus. **Phosgene** will also react directly with cellular constituents. The result is a change in permeability and extensive edema. This also occurred with the highly reactive industrial chemical **methyl isocyanate** that caused death and injury to thousands in Bhopal in India when it leaked from a factory. In the case of ipomeanol, metabolic activation in the Clara cell results in necrosis of these cells and edema of the lung tissue. Similarly, the solvent **tetrachloroethylene** undergoes metabolic activation in the lung and causes necrosis and edema. In contrast, **pyrrolizidine alkaloids** such as monocrotaline (Fig. 6.7) are metabolically activated in the liver and then cause damage in the lungs after being transported there by the blood.

In contrast to compounds such as nitrogen dioxide and ozone, compounds, which are very water soluble, will tend to damage the cells higher up in the system such as in the trachea or bronchi.

Fibrosis

This type of response is caused by substances such as silica (silicosis) and **asbestos** (asbestosis). It is thought to involve phagocytic uptake of particles by macrophages. The particles are taken up by lysosomes, which then rupture at some stage and release hydrolytic enzymes, resulting in the digestion of the macrophage and release of the particles. The cyclical process then starts again. The result is aggregation of lymphoid tissue and stimulation of collagen synthesis leading to fibrotic lesions.

Other types of lung damage may also lead to fibrosis, however, such as that caused by paraquat.

Stimulation of an Allergic Response

A variety of particles and chemicals may stimulate an allergic response in the lungs when there is repeated exposure via the lungs. Microorganisms, spores, dusts such as cotton dust, and certain chemicals may produce an allergic-type response with pulmonary symptoms. For instance, **toluene di-isocyanate**, a chemical used in the plastics industry, and **trimellitic anhydride**, another industrial chemical, cause an allergic-type reaction when inhaled. It is thought that these compounds form protein conjugates in the respiratory system, which then act as antigens (see below).

Pulmonary Cancer

Many different types of materials produce lung tumors after inhalation, and also after exposure via the bloodstream. Thus, **asbestos, polycyclic aromatic hydrocarbons, cigarette smoke, nickel carbonyl, nitrosamines, chromates**, and **arsenic** may cause lung tumors after inhalation. Other compounds cause lung tumors after systemic administration, such as **hydrazine** and **1,2-diethylhydrazine**. Although the activity of the drug-metabolizing enzymes is in general less than that in the liver, for certain isoenzymes of cytochrome P-450, this is not the case. So the lung has the ability to metabolize foreign compounds and therefore can metabolically activate compounds to which the organ is exposed. Thus, the polycyclic aromatic hydrocarbon benzo[a] pyrene is metabolically activated by lung tissue as described in more detail in chapter 7, and thus in situ activation may be responsible for the initiation of tumors.

Detection of Lung Damage

Unlike liver and kidney damage, there are currently no general biochemical tests for lung damage. Lung damage may be detected pathologically by light and electron microscopy of lung tissue, and simple measurements such as the lung wet weight/dry weight ratio will detect edema. Bronchoscopy can be employed for the detection of gross changes, and samples may be taken at the same time for histopathology. Lung dysfunction is detected by physiological tests such as forced expiratory volume **(FEV)** over a particular time, and **forced vital capacity**. Measurement of pulmonary mechanics such as **flow resistance** may also be useful for detection of lung irritants or pharmacologically active agents. These measurements are more difficult to perform in conscious animals than in humans, but determination of lung function is possible in experimental animals.

6.2.6 Other Target Organs

Although the liver, kidney, and lung are perhaps the most important and most common target organs, many other organs and tissues in the body may be specifically damaged by toxic chemicals. It is beyond the scope of this book to consider all of these, but a few other examples are briefly highlighted, and the reasons given where known. Some further examples are also considered in the final chapter.

Nervous System

The nervous system, both peripheral and central, is a common target for toxic compounds, and the cells, which make up the system, are particularly susceptible to changes in their environment. Thus, anoxia, lack of glucose and other essential metabolites, restriction of blood flow, and inhibition of intermediary metabolism may all underlie damage to cells of the nervous system as well as direct, cytotoxic damage. The nervous system is a highly complex network of specialized cells, and damage to parts of this system may have permanent and serious effects on the organism as there is little capacity to regenerate and little reserve functional capacity. **Peripheral neuropathy** is a toxic response to a variety of foreign compounds such as organophosphorus compounds, methyl mercury, and isoniazid, for example.

The "designer drug" contaminant 1-methyl-4-phenyl-1, 2, 3, 6-tetrahydropyridine **(MPTP)** causes specific damage to the dopamine-containing cells in the substantia nigra area of the brain. MPTP is highly lipophilic and readily enters the brain. Once there, it is metabolized to a toxic metabolite and is taken up by dopamine neurons and hence damages these particular cells. Thus, the chemical structure of the compound and its physicochemical characteristics determine its toxicity, and the specific uptake system determines the target organ and cell. This example is considered in more detail in the final chapter. A similar example discussed in chapter 7 is 6-hydroxydopamine.

The solvent **hexane** causes a different type of neurotoxicity, involving swelling and degeneration of motor neurons. This leads to paresthesia and sensory loss in the hands and feet and weakness in toes and fingers. Hexane has been widely used in industry as a solvent, and there have been many cases of neuropathy reported from different parts of the world. The toxicity is due to the metabolite 2,5-hexanedione, which arises by first ω-1 oxidation at the 2- and 5- positions to 2,5-hexandiol, then further oxidation to give 2,5-hexanedione (chap. 4, Fig. 9). The 2,5-hexanedione then reacts with protein to form pyrrole adducts. The γ-diketone structure is important as 2,3- and 2,4-hexanedione are not neurotoxic. **Methyl n-butyl ketone** also causes similar neurotoxic effects and is also metabolized to 2,5-hexanedione. The lipophilicity of the molecule allows distribution to many tissues including the nervous system. Thus, chemical structure and metabolism are important prerequisites for this toxicity. Exposure to the solvent **carbon disulfide** in industry causes neuronal damage in the central and peripheral nervous system. The mechanism may involve chelation of metal ions essential for enzyme activity by the oxothiazolidine and dithiocarbamate metabolites of carbon disulfide, which are formed by reaction with GSH and glycine, respectively. However, reactive metabolites may also be formed such as active sulfur (chap. 4, Fig. 26), which might play a role in the various toxic effects.

Gonads

A number of compounds cause specific damage to the male reproductive system. Thus, cadmium, **2-methoxyacetic acid, dibromochloropropane**, and **diethylhexylphthalate** all cause different types of damage to the testis. The spermatocytes (germ cells) may be susceptible to different compounds at different stages of development, such as the pachytene stage, which is specifically damaged by 2-methoxyethanol. Also, the Sertoli and Leydig cells may be damaged such as by **1,3-dinitrobenzene** and **ethanedimethylsulfonate**, respectively. The testis is a target organ for several reasons. It has rapidly growing and dividing tissue, and thus compounds such as anticancer drugs may damage the testis. It has a limited blood supply and hence is susceptible to a reduction in this supply. For example, cadmium is believed to cause ischemic damage to the testis by reducing blood flow. There are particular biochemical pathways, such as the use of lactate as an energy source, which may suffer interference. Thus, the testicular toxicity of **2-methoxyethanol** and **2-methoxyacetic acid** are thought to involve interference with the metabolism of lactate.

The female reproductive system may also be a target for damage by foreign compounds. Thus, the oocytes may be specifically damaged or destroyed by compounds such as **polycyclic aromatic hydrocarbons, bleomycin**, and **cyclophosphamide**. Studies in mice have suggested that the immature oocyte is particularly susceptible. With the polycyclic aromatic hydrocarbons, metabolic activation appears to be necessary. The uterus may be affected by foreign compounds, and in turn this can affect fertility. Thus, **DDT** can cause an estrogenic response in rats, leading to thickening of the endometrium and increase in uterus weight. This is due to competitive inhibition of the binding of estradiol to receptor sites in the uterus. A number of other pesticides may cause similar effects. Exposure to the solvent **carbon disulfide** can cause menstrual disorders and reduced fertility. The compound most well known for its toxic effects to the female reproductive system is **diethylstilbestrol**, which causes vaginal cancer as described later in this chapter.

Heart

The heart is occasionally a target for toxic compounds. Thus, **allyllamine** specifically damages heart tissue. This is partly due to metabolism of this compound by amine oxidases present in cardiac tissue. The product is allyl aldehyde (acrolein), which is highly reactive and toxic (chap. 4, Fig. 31) as already mentioned in the context of the hepatotoxicity of allyl alcohol. Some compounds such as **hydralazine** (see Fig. 7.82) may cause myocardial necrosis as a result of their pharmacological action. Heavy metals such as **cobalt** cause cardiomyopathy when given repeatedly to animals. The lesions seen are similar to those detected in humans who were exposed to cobalt as a result of heavy consumption of beer, which contained cobalt in a stabilizing additive. The toxic effect of cobalt and certain other heavy metals is due to interference with Ca^{2+} in the muscle tissue. The anticancer drug **adriamycin** is cardiotoxic, possibly for several reasons. It has been suggested that it has a high affinity for, and hence binds to, complex lipids such as cardiolipin, present in the membranes of cardiac cells. Also, it has been suggested that the heart has reduced protective mechanisms (see below) such as superoxide dismutase and catalase. Adriamycin is believed to be toxic as a result of cyclical oxidation-reduction processes involving the adriamycin-iron complex, which results in the production of superoxide. Occupational exposure to **carbon disulfide** is known to be a factor in coronary heart disease. Other components of the vascular system may also be specifically damaged.

Pancreas

The compounds **alloxan** and **streptozotocin** both specifically damage the pancreas and are used experimentally to induce diabetes. It is the β cells that are particularly susceptible to these two compounds.

Olfactory Epithelium

This tissue is known to contain a high level of cytochromes P-450 activity. Clearly, it is similar to the lung in exposure. An example is the industrial chemical **trifluoromethylpyridine**, which specifically damages the olfactory epithelium in animals exposed to it in the inspired air. The compound is metabolized by N-oxidation, and the *N*-oxide product (or a nitroxide radical) is believed to be responsible for the toxicity (chap. 4, Fig. 20). The activity for this metabolic pathway is particularly high in the olfactory epithelium.

Blood

As almost all foreign compounds are distributed via the bloodstream, the components of the blood are exposed at least initially to significant concentrations of toxic compounds. Damage to and destruction of the blood cells result in a variety of sequelae such as a reduced ability to carry oxygen to the tissues if red blood cells are destroyed. Aromatic amines such as aniline and the drug **dapsone** (4,4-diaminodiphenyl sulfone) are metabolized to hydroxylamines, and in the latter case, the metabolite is concentrated in red blood cells. Also, nitro compounds such as **nitrobenzene**, which can be reduced to hydroxylamines, are similarly toxic to red blood cells. These hydroxylamines are often unstable and can be further oxidized to reactive products, in the presence of oxygen in the red cell, and thereby damage the hemoglobin. The hemoglobin may be oxidized to methemoglobin, and thus be unable to carry oxygen. Irreversible damage to hemoglobin can occur as a result of oxidation of thiol groups in the protein, with the subsequent denaturation of the protein and hemolysis of the red cell. **Phenylhydrazine** is another compound, which damages red cells both *in vitro* and *in vivo*, causing hemolysis.

The production of blood cells, which takes place in the bone marrow, is also a target for foreign compounds as the cells in the bone marrow are actively dividing. Damage to the bone marrow can result in reduced numbers of red and/or white blood cells, as is the case with the solvent **benzene**. This compound destroys the bone marrow cells, which form both the red and white blood cells, giving rise to the condition of aplastic anemia.

Eye

The eye may be a target organ as a result of its external position in the organism and direct exposure, but also from systemic exposure. Thus, the various components of the eye may be specifically damaged. The presence of pigments in the eye such as melanin has been suggested as the cause of **chloroquine** toxicity to the retina.

Methanol ingestion causes blindness by damage to the optic nerve, as discussed in more detail in chapter 7. The drug practolol caused a serious keratinization of the cornea by an unknown mechanism and had to be withdrawn (see chap. 7). Paracetamol when given in large doses to certain strains of mice causes opacification of the cornea by a similar mechanism to that involved in the hepatic damage (see chap. 7). The lens may be damaged with the formation of cataracts such as by **2,4-dinitrophenol** and **naphthalene**, and the latter will also damage the retina. Carbon disulfide causes various effects on the eye in exposed humans.

Ear

The auditory system may be damaged by systemic exposure to foreign compounds. Perhaps, the most well-known example is the damage to the cochlear system by aminoglycoside antibiotics such as **gentamycin**. The hair cells in this part of the ear seem to be particularly sensitive and may be destroyed by these drugs. The cells can be readily seen by electron microscopy, and the dead or missing cells can be counted, giving a quantifiable indication of ototoxicity. Exposure to carbon disulfide causes hearing loss to high-frequency tones in humans.

Skin

Although the skin is usually the target for external foreign compounds, systemic exposure can also lead to damage such as the photosensitization caused by use of the drug **benoxaprofen** (Opren) and the skin reactions following immune responses (see below). Exposure to **organochlorine compounds** gives rise to the condition of chloracne, which may occur after oral as well as cutaneous exposure. The sebaceous gland ducts become keratinized and are replaced with a keratinous cyst. As with some of the other toxic effects, chloracnegenicity is dependent on the lateral symmetry and position of the chlorine atoms and correlates with the ability to induce the monooxygenase enzymes. Thus, the physicochemical characteristics of the compound are important in this toxic effect.

For further examples of target organs, the reader is referred to the systemic toxicology section of *Casarett and Doull's Toxicology, Target Organ Toxicity*, Vols I and II, and the *Target Organ Toxicity* series (see "Bibliography").

Detection of Organ Damage

There are various ways of detecting damage to organs, in some cases using specific biochemical tests, and in others, function tests are used. Such biomarkers of toxic response have been discussed in chapter 2. Histopathology is generally used at some stage, however, for all types of damage.

Thus, for damage to the nervous system and ear, functional tests and histopathology are used. For damage to the eye and skin, gross observation may be useful before microscopy. For damage to the reproductive system, hormonal changes may be detected. For testicular damage, a urinary marker, creatine, has been proposed, and changes in the plasma levels of **lactate dehydrogenase** isoenzyme C4 is also used diagnostically. Damage to blood cells is readily detected by light microscopy and automated counting techniques. For damage to the heart and pancreas, specific enzymes may be measured in the plasma or serum such as **creatine kinase** and **amylase**, respectively.

6.3   MECHANISM AND RESPONSE IN CELLULAR TOXICITY

There are many different mechanisms underlying toxicity, leading to the different types of responses. However, most toxicity is due to the interaction between the *ultimate toxicant* and a *target molecule*. The ultimate toxicant may be a reactive metabolite, a stable metabolite, or the parent compound. The molecule could be part of a structure, the cell membrane, for example, or an individual macromolecule such as an enzyme.

There are different types of reaction that could occur:

1. Covalent bonding
2. Noncovalent bonding
3. Hydrogen abstraction
4. Electron transfer
5. Enzyme reaction

1. Covalent bonding. As already indicated in chapter 4, the covalent interaction of reactive metabolites with macromolecules is irreversible. As well as electrophiles free radicals, both charged and noncharged can also bind covalently to target molecules, and both proteins and lipids are known to be targets. The trichloromethyl free radical from carbon tetrachloride activation will react with unsaturated lipids, and hydroxyl radical will hydroxylate bases in DNA giving rise to 8-hydroxypurines, 5-hydroxymethylpyrimidines, and thymine and cytosine glycols.

Clearly, nucleophilic toxicants will react with electrophilic sites in target molecules but is not common. Carbon monoxide and cyanide are good examples where the electrophilic site in both cases is the heme group in proteins. Hydrazines will react with the electrophilic carbonyl carbon in keto groups such as in pyridoxal phosphate.

It may have different outcomes ranging from genetic damage to immune responses. This is true of all interactions between chemicals and macromolecules, as will be seen later in this chapter and in chapter 7.

2. Non-covalent bonding includes hydrogen bonding, ionic bonds, or hydrophobic bonds. These types of bonding are involved in binding of chemicals to plasma proteins. They could also underlie the interaction between a chemical and a receptor or enzyme. Thus, the interaction between TCDD and the Ah receptor (AhR) and the intercalation of **doxorubicin** in DNA involve non-covalent bonds.

3. Hydrogen abstraction involves the removal of a proton from an endogenous target molecule by a free radical, typically an unsaturated lipid, but thiols (e.g., GSH) may also be targets. The result will be an endogenous free radical, which also is reactive. Probably the most well known and destructive interaction is between unsaturated fatty acids, such as found in lipids—triglycerides and phospholipids. The result of the interaction is a lipid radical. Because the unsaturated lipid has a number of double bonds, the single unpaired electron will facilitate the movement of the bonds and the formation of conjugated dienes.

4. Electron transfer. Charged species such as nitrite ions can oxidize the iron in hemoglobin. This is a potentially toxic effect as the methemoglobin produced will not carry oxygen.

5. Enzyme reactions. Some toxicants can act as enzymes. A number of natural toxicants operate in this way. For example, snake venoms often contain hydrolytic enzymes, which degrade tissue. Ricin from the castor oil plant is more sophisticated and causes the hydrolytic destruction of ribosomes, so disrupting protein synthesis.

The outcomes of the interaction between a chemical and a target molecule will depend both on the *attributes* and on the *function* of the target molecule. Thus, a covalent adduct could be formed, which might be recognized as a neo-antigen (e.g., see sect. "Halothane Hepatotoxicity," chap. 7), or a mutation could be caused if the target is DNA [see sect. "Benzo(a)pyrene," chap. 7].

The molecule could be destroyed, and therefore its function is lost (see "Carbon Tetrachloride," chap. 7).

The interaction, whether covalent or non-covalent, might cause the dysfunction of the molecule (see "Cyanide" and "Carbon Monoxide," chap. 7).

If the reactivity of the molecule toward the ultimate toxicant is low, no interaction may ensue. Thus, if the ultimate toxicant is an electrophile but the macromolecule has no suitable nucleophilic groups, which are accessible, reaction is unlikely.

The accessibility may be low if it is located in a tissue, which is distant from the site of exposure or with poor vascular perfusion. Alternatively, it may be in a compartment of the cell, which is protected by membranes, which are not readily permeable to the toxicant; mitochondria, for example. The accessibility of DNA will vary with the cell cycle.

The function of the target molecule may be *critical* or *noncritical*. Thus, if the target molecule is an enzyme, this could be involved in a crucial metabolic pathway, such as mitochondrial oxidative phosphorylation. In this case, an adverse interaction with the ultimate toxicant is likely to lead to cell dysfunction and possibly death (e.g., as with cyanide or salicylate). Chemicals such as methimazole and resorcinol, which are activated to free radical intermediates by thyroperoxidase, cause destruction of the enzyme. This then disturbs thyroid hormone synthesis and thyroid function with pathological consequences such as thyroid tumors.

However, the target molecule could be an enzyme involved with a process, which is not critical to the cell, such as cytochrome P-450. Then an interaction with the toxicant, even one leading to damage to or destruction of the enzyme may not have immediately serious consequences for the cell. It may, however, be an important basis for a toxic drug interaction.

When proteins other than enzymes are targets for binding, other processes may be disrupted. Thus, the binding of carbon monoxide to hemoglobin can result in serious physiological consequences and death (see chap. 7).

The consequences of the interaction between the chemical and the target therefore depend on a number of factors, and there may even be several consequences if there is more than one target molecule. In general terms, the consequences could be impaired maintenance or regulation of the cell or of the organism. Hence, changes in activities such as cell division or beating of heart muscle cells or alterations in ATP synthesis or of functions such as the control of clotting.

Thus, toxic compounds can damage cells in target organs in a variety of ways and the cellular injury caused by such compounds leads to a complex sequence of events. The eventual response may be reversible injury or an irreversible change leading to the death of the cell. The processes leading to cell death and the "point of no return" for a cell are not yet all clearly understood and are the subject of considerable research, speculation, and some controversy. However, some of the key elements are known, and a picture of at least part of the sequence of cellular changes is emerging. It is convenient for this discussion to divide the stages following exposure of a cell to a toxic compound into **primary, secondary**, and **tertiary events**. The primary events are those that result in the initial damage, the secondary events are cellular changes, which follow from that initial damage, and the tertiary events are the observable and final changes.

1. Primary events. As already mentioned, many compounds are toxic following metabolism to reactive metabolites. These reactive metabolites may then initiate one or more primary events. For example, inhibition of mitochondrial function and paracetamol- and

bromobenzene-induced liver damage result from metabolic activation, this is
more detail in chapter 7. In other cases metabolic activation is not necessary, ar
compound or a stable metabolite initiates the primary event. For example, cyanid
as a result of inhibition of crucial enzymes, and carbon monoxide deprives the ce
(see chap. 7 for more details).

The major primary events are

lipid peroxidation,
covalent binding to macromolecules,
changes in thiol status,
enzyme inhibition, and
ischemia.

In some cases, several of these primary events may occur and may be interrelated; and in
others perhaps only one occurs. Each of these are discussed in more detail below and will also
be considered in the final chapter as part of detailed discussions of specific examples.

2. Secondary events. These are the changes that may occur in cells exposed to toxic
compounds following the primary events.

When a cell is damaged, a number of changes may be detected both biochemically and
structurally, and some of these are interrelated. However, some of these changes are
consequences of damage rather than causal and result from the loss of cellular control and the
inability of the cell to compensate.

The major secondary events are changes in membrane structure and permeability,
changes in the cytoskeleton, mitochondrial damage, depletion of ATP and other cofactors,
changes in Ca^{2+} concentration, DNA damage and poly ADP-ribosylation, lysosomal
destabilization, stimulation of apoptosis, and damage to the endoplasmic reticulum.

In some cases, these may be primary events; in other situations, they may follow primary
events or follow sequentially from another secondary event.

3. Tertiary events. These are the final, observable manifestations of exposure to a toxic
compound. Several may occur together, or they may occur sequentially.

The major tertiary events are as follows:

steatosis/fatty change
hydropic degeneration
blebbing
apoptosis
necrosis

All of these except the last two are potentially reversible.

6.3.1 Primary Events

Lipid Peroxidation

As already discussed in chapter 4 and mentioned above, many foreign compounds are
metabolically activated to reactive intermediates, which are responsible for initiating toxic
effects. One particular type of reactive intermediate is the free radical, and these are implicated
in the hepatotoxicity of carbon tetrachloride and **white phosphorus**, the pulmonary damage
caused by paraquat, the destruction of pancreatic β-cells by **alloxan** and the destruction of
nerve terminals by **6-hydroxydopamine**. The role of free radicals in carbon tetrachloride
hepatotoxicity is discussed in more detail in chapter 7. Free radicals arise by either homolytic
cleavage of a covalent bond in a molecule, the addition of an electron as in the case of carbon
tetrachloride, or by the abstraction of a hydrogen atom by another radical. They have an
unpaired electron, but are usually uncharged and centered on a carbon, nitrogen, sulfur, or
oxygen atom. Consequently, free radicals are extremely reactive, electrophilic species, which
can react with cellular components. Alternatively, molecular oxygen may accept electrons from
a variety of sources to produce the oxygen-free radical superoxide. This can then produce
further reactive molecules such as singlet oxygen and hydroxyl radical. The process of lipid
peroxidation is initiated by the attack of a free radical on unsaturated lipids, and the resulting

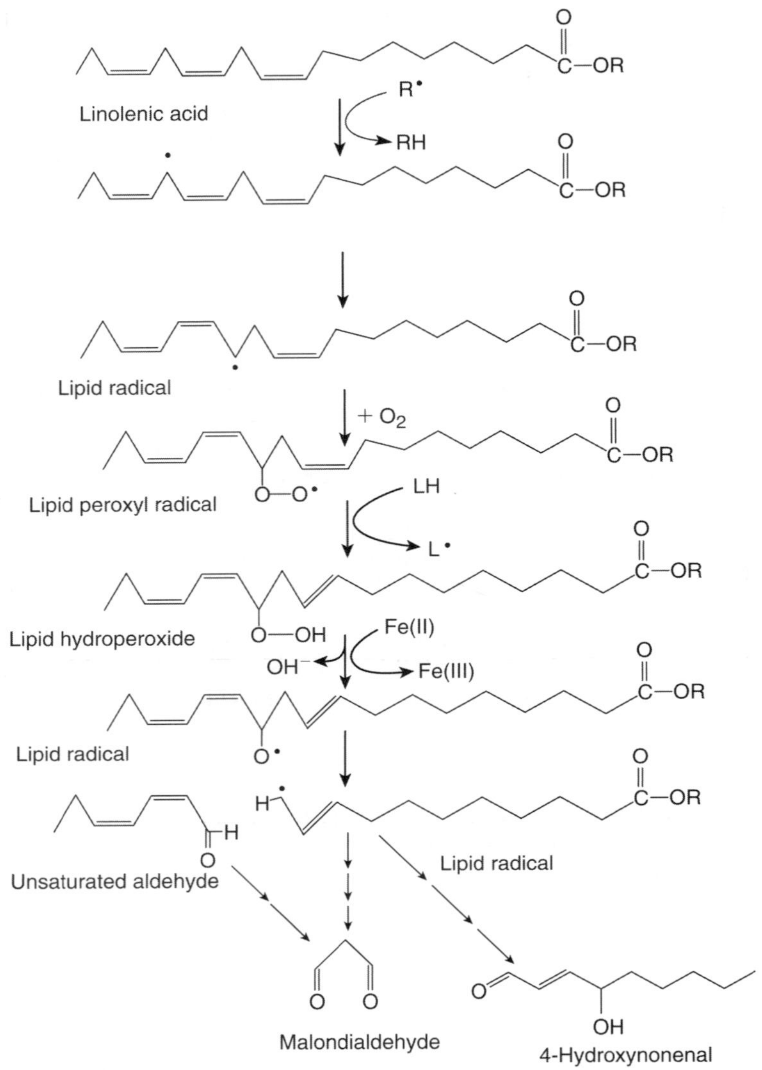

Figure 6.9 Peroxidative destruction of a polyunsaturated lipid initiated by a free radical attack such as by the trichloromethyl radical (CCl$_3$•).

chain reaction is terminated by the production of lipid breakdown products: lipid alcohols, aldehydes, or smaller fragments such as **malondialdehyde** (Fig. 6.9). Thus, there is a cascade of peroxidative reactions, which ultimately leads to the destruction of the lipid (Fig. 6.9), and possibly the structure in which it is located:

$$L\bullet + O_2\bullet \rightarrow LO_2\bullet$$

$$\text{Propagation} \qquad LO_2\bullet + LH \rightarrow LOOH + L\bullet$$

where L is a lipid

This cascade however may be propagated throughout the cell unless terminated by a protective mechanism (see below) or a chemical reaction such as disproportionation, which gives rise to a non-radical product. Polyunsaturated fatty acids, found particularly in membranes, are especially susceptible to free radical attack. The effects of lipid peroxidation are many and various. Clearly, the structural integrity of membrane lipids will be adversely affected. In the lipid radical produced, the sites of unsaturation may change, thereby altering the fluidity of the membrane (see chap. 3). Lipid radicals may interact with other lipids and macromolecules such as proteins to cause again potentially altering membrane structure and function. These membrane effects could lead to increased permeability of either the plasma membrane or the membranes of organelles. Thus, the lysosomal membrane may be destabilized leading to further damage to the cell from the loss of the hydrolytic contents. Some of the mitochondrial enzymes and those of the endoplasmic reticulum are membrane bound, and consequently membrane damage will compromise the function of these organelles leading to changes in cellular homeostasis (see below). Sulfydryl-containing enzymes are particularly susceptible to inhibition by lipid peroxidation. As well as the destruction of lipids and possibly of membranes, lipid peroxidation gives rise to breakdown products, which themselves have biological activity (Fig. 6.9). One cytotoxic breakdown product in particular, which is known to be formed, is **4-hydroxynonenal**. This causes a number of the effects seen with toxic compounds, which are responsible for lipid peroxidation. Such reactive carbonyl compounds may have a variety of cellular effects such as changing membrane permeability, inhibiting enzymes, and hence causing the disruption of the cell via such effects as changes in ion levels and energy production. Thus, 4-hydroxynonenal is known to be a potent inhibitor of Ca^{2+} sequestration by the endoplasmic reticulum. However, after low doses of carbon tetrachloride, Ca^{2+} sequestration is inhibited in the absence of lipid peroxidation, perhaps due to a direct effect of carbon tetrachloride on the endoplasmic reticulum. There is evidence that lipid peroxidation, at least in some cases, may be a consequence rather than a cause of cellular injury. Thus, the compound **diquat** causes lipid peroxidation via active oxygen species, yet cell death requires depletion of cellular GSH. However, depletion of cellular GSH does not occur with all cytotoxic compounds. In other cases, protection against lipid peroxidation does not prevent cell death. Certainly there are many toxic compounds, which do not cause lipid peroxidation, such as the hepatotoxins **hydrazine** and **thioacetamide**. Therefore, although lipid peroxidation may be an important part of cellular injury for some compounds, it is clearly not a general mechanism underlying all toxic responses.

Oxidative Stress

The production of active oxygen species may lead to a cycle of oxidation and reduction (redox cycle) with electrons being donated to oxygen to yield superoxide. This is the case with paraquat and also a number of cytotoxic quinones. Thus, there is a cyclic process, which produces superoxide by adding electrons from paraquat (see chap. 7) or a semiquinone, for example, to oxygen (Fig. 6.10). The superoxide produced may then be metabolized to hydrogen peroxide by superoxide dismutase, which is further metabolized to water by catalase (Fig. 6.10).

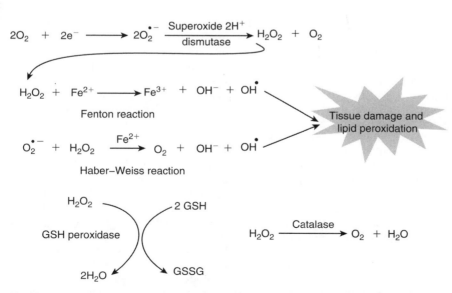

Figure 6.10 Production of various reactive oxygen species, the role in tissue damage, and their detoxication and toxication. Reactive species shown are hydrogen peroxide, superoxide, and hydroxyl radical.

However, under certain circumstances, such as in the presence of transition metal ions, hydroxyl radicals may result from either the **Haber–Weiss reaction** or the **Fenton reaction**, which cause lipid peroxidation and cell injury (Fig. 6.10).

Hydroxyl radicals are cytotoxic and can be involved in the production of further active oxygen species such as singlet oxygen.

If a large amount of the toxic substance is present, then the detoxication processes present are overwhelmed. Excess superoxide is produced, reduced GSH and NADPH are depleted, and hydroxyl radicals and singlet oxygen are formed. This is the condition known as oxidative stress (see also below this chapter). As well as causing lipid peroxidation, ROS will also cause DNA damage and damage to proteins.

Covalent Binding To Macromolecules

As well as free radicals, other reactive intermediates, usually electrophilic species, may be produced by metabolism. These reactive intermediates can interact with proteins and other macromolecules and bind covalently to them. Studies have shown that there is a correlation between both the amount and site of binding and tissue damage. Therefore it was suggested that covalent binding to critical macromolecules was a possible cause of the cellular injury. While this may be true with regard to mutations and tumor induction where the target macromolecule, DNA, is known, with other types of toxic response, the target molecule and the mechanism is less clear. Rather, it may simply be that covalent binding to macromolecules is an indication of the production of a reactive intermediate, which is also a necessary step in the mechanism of cytotoxicity. Studies with paracetamol have shown that treatments, which protect against cytotoxicity, do not alter covalent binding in hepatocytes and that more covalent binding may occur with nontoxic analogues of paracetamol. However, the measurements of binding are relatively nonspecific as the target molecule and intracellular site are mostly unknown. Studies with paracetamol *in vitro* have shown that the binding to protein is mainly directed toward sulfydryl groups. Investigation of the covalent binding of the hepatotoxic compound acetylhydrazine to protein *in vivo* revealed that binding to cellular organelles shows different profiles and that these changed after pretreatments, which increased the toxicity. Therefore, although covalent binding to cellular protein is a useful marker of activation, unless the specific target(s) are known it cannot be assumed that this is a direct cause or indicator of cytotoxicity. Clearly, binding to critical sites on proteins could conceivably alter the function of those proteins such as inhibiting an enzyme or damaging a membrane, but equally binding may be to noncritical sites and be of no toxicological consequence.

Thiol Status

As already discussed in chapter 4, reactive intermediates can react with reduced GSH either by a direct chemical reaction or by a GSH-transferase–mediated reaction. If excessive, these reactions can deplete the cellular GSH. Also, reactive metabolites can oxidize GSH and other thiol groups such as those in proteins and thereby cause a change in thiol status. When the rate of oxidation of GSH exceeds the capacity of GSH reductase, then oxidized glutathione (GSSG) is actively transported out of the cell and thereby lost. Thus, reduced GSH may be removed reversibly by oxidation or formation of mixed disulfides with proteins and irreversibly by conjugation or loss of the oxidized form from the cell. Thus, after exposure of cells to quinones such as menadione, which cause oxidative stress, GSH conjugates, mixed disulfides, and GSSG are formed, all of which will reduce the cellular GSH level.

The role of GSH in cellular protection (see below) means that if depleted of GSH, the cell is more vulnerable to toxic compounds. However, GSH is compartmentalized, and this compartmentalization exerts an influence on the relationship between GSH depletion or oxidation and injury. The loss of reduced GSH from the cell leaves other thiol groups, such as those in critical proteins, vulnerable to attack with subsequent oxidation, cross-linking, and formation of mixed disulfides or covalent adducts. The sulfydryl groups of proteins seem to be the most susceptible nucleophilic targets for attack, as shown by studies with paracetamol (see chap. 7), and are often crucial to the function of enzymes. Consequently, modification of thiol groups of enzyme proteins, such as by mercury and other heavy metals, often leads to inhibition of the enzyme function. Such enzymes may have critical endogenous roles such as the regulation of ion concentrations, active transport, or mitochondrial metabolism. There is evidence that alteration of protein thiols may also be a crucial part of cell injury. Thus, changes to isolated hepatocytes caused by paracetamol can be prevented, and even reversed, by the use of dithiothreitol, which reduces oxidized thiols. Similarly, data on the use of *N*-acetylcysteine as an antidote for paracetamol poisoning *in vivo* indicate the importance of thiol status in cell injury (see chap. 7 for a more detailed discussion).

Enzyme Inhibition

Although inhibition of enzymes may be a common consequence of exposure to toxic compounds, there are examples of where the inhibition of a single endogenous enzyme is the critical primary event. Thus, the inhibition of cytochrome aa_3 by cyanide leads to cell death by blocking cellular respiration, and the blockade of Krebs' cycle by the inhibition of aconitase by fluorocitrate similarly inhibits respiration (see chap. 7 for more details). Obviously, there will be various sequelae to this enzyme inhibition, such as depletion of ATP and other vital endogenous molecules. The effect of the inhibition will depend on the importance of the enzyme to the particular cell type. Thus, inhibition of cellular respiration by cyanide is especially critical in cells of the CNS. There are a number of compounds, which inhibit protein synthesis, either specifically at particular points such as **tetracycline** or more indiscriminately by damage to the endoplasmic reticulum such as carbon tetrachloride. This may cause cellular injury, especially steatosis (see below), but cell death is not an inevitable consequence. Several enzymes may be inhibited by one toxic compound such as **hydrazine**, for example, exposure to which inhibits phosphoenol pyruvate carboxy kinase, and many pyridoxal phosphate-requiring enzymes such as the transaminases, which will cause a variety of effects on intermediary metabolism. Another example is bromobenzene, which inhibits Na^+/K^+ ATPase, GSSG reductase, and glucose-6-phosphate dehydrogenase. Clearly, inhibition of each of these enzymes will have a different effect; inhibition of the Na^+/K^+ ATPase will affect the active transport of ions into and out of the cell, leading to an imbalance; inhibition of GSSG reductase will exacerbate depletion of reduced GSH and inhibition of glucose6-phosphate dehydrogenase—the first enzyme in the pentose phosphate pathway—will reduce the production of NADPH and hence influence the reduction of GSSG and metabolic pathways such as those catalyzed by cytochromes P-450.

Ischemia

The cessation or reduction of the supply of blood containing oxygen and nutrients such as glucose to a tissue leads to damage and cell death if prolonged. The lack of oxygen, hypoxia (or **anoxia**, if total), may be a specific effect, as in the case of methemoglobinemia caused by **nitrite**, for example, or carbon monoxide poisoning discussed in detail in chap. 7, or may simply be due to a reduction of blood flow. The latter is believed to underlie the testicular necrosis, which follows acute doses of cadmium. The blood supply to the testis arrives via the vessels of the pampiniform plexus, and if these are restricted by vasoconstriction, as caused by cadmium or by surgical ligation, then the testis suffers ischemic necrosis. Ischemia may also be a secondary event and occur because of the swelling of cells in tissue with subsequent reduction in blood flow. For example, damage to the sinusoidal cells of the liver caused by **pyrrolizidine alkaloids** leads to swelling and reduction in liver blood flow, and hence ischemic necrosis. The centrilobular region is especially susceptible as it receives less oxygenated blood.

Effects on Gene Expression

Chemicals can initiate responses by affecting gene expression at several levels, transcription, and intra- or extracellular signaling.

The binding of transcription factors to nucleotide sequences, which facilitates gene transcription, can be influenced by chemicals. For example, cadmium binds to a metal-binding protein factor, MFF-1, in place of zinc and so induces metallothionein synthesis. This, as it happens is a detoxication, as metallothionein binds cadmium.

TCDD binds to the AhR and initiates a number of responses as well as induction of CYP1A1, including induction of UDP-glucuronosyl transferases, GSH transferases, and stimulation of apoptosis in thymocytes, resulting in thymic atrophy.

The responses are all mediated via TCDD-AhR complex binding to ARNT (see chap. 5).

Ligand-activated transcription factors are also important in embryogenesis. TCDD also causes malformations via this mechanism, as do retinoids and glucocorticoids. The importance is indicated by the fact that mice without the AhR are nonresponsive to any of these effects of TCDD.

Another example is afforded by fibrate drugs and phthalate esters, which bind to the PPAR, in mimicry of polyunsaturated fatty acids.

In those species, which are responsive (i.e., have a functioning receptor) such as the rat, treatment with drugs, which interact with the PPARα receptor such as clofibrate, will cause a number of effects, such as induction of a number of enzymes, increased cell growth and turnover, and liver tumors in almost all the animals as a direct result of interaction with the receptor and changes in gene transcription. This will be discussed in more detail in chapter 7.

Transcription factors also can be modulated by indirect means such as the effect of chemicals on extracellular cell surface receptors and intracellular signaling networks.

Thus, in this way, chemicals can cause changes in signal transduction, which can also lead to changes in gene transcription.

For example, chemicals such as lead (Pb^{2+}) or stressors such as reactive oxygen species (ROS) and UV light can modulate this control. Thus, from Figure 6.11 it can be seen by stopping inhibitory control [eg. ROS which blocks action of protein phosphatases (PTP)] a chemical can be mitogenic and increase cell division. Alternatively stimulation [e.g., low concentrations of Pb^{2+} stimulate protein kinase C (PKC)], is also mitogenic.

Lead has similarities to calcium, one of the normal second messengers. Other substances can also stimulate PKC, such as phorbol esters and the fungal toxin **fumonisin**, which mimic diacylglycerol, another second messenger.

The c-fos and c-jun proteins when phosphorylated bind to the cyclin D gene, which influences the cell division cycle.

Cell cycle control involves phosphorylation (activation) by kinases and dephosphorylation (inhibition) by phosphatases, so interference with either can alter signaling pathways

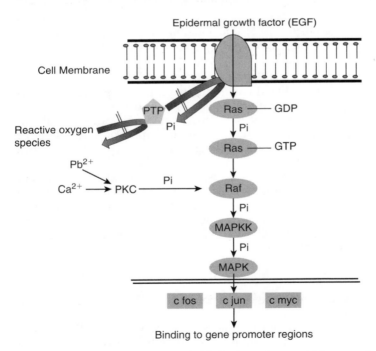

Figure 6.11 An example of a signal transduction pathway, which can control cell division. EGF is a ligand for the TK receptor, which is a cell membrane receptor. Binding to this causes phosphorylation allowing inactive Ras G protein (with GDP bound) to be converted to active Ras (with GTP bound). This activates protein kinase Raf and the MAPK phosphorylation cascade. The phosphorylated MAPK moves into the nucleus and phosphorylates the transcription factors (c-fos, c-jun, and c-myc). This allows them to bind to promoter regions of genes leading to increased cell division. PTP (a protein kinase) has an inhibitory action by dephosphorylating the tyrosine kinase inhibiting cell division. PKC, promotes phosphorylation of Raf and so stimulates the cascade. It is stimulated by calcium, a second messenger. *Abbreviations*: TK, tyrosine kinase; PKC, protein kinase C; MAPK, MAP kinase; EGF, epidermal growth factor; GDP, Guanine diphosphate.

and gene expression. For example, some phosphatases have sulphydryl residues (SH) residues at the active site, sensitive to thiol-oxidizing agents such as arsenite and hydrogen peroxide.

The potent toxin microcystin [blue-green algae inhibit one of the phosphatases (PP2A)] increases mitogenic activity and so is a tumor promoter at low-level exposures. The liver toxicity at high levels of acute exposure is probably also due to disturbances in protein phosphorylation.

Clearly, modulation of the signals outside the cell such as hormones, by drugs and chemicals can also change gene expression. If thyroid hormone (thyroxine) production is inhibited, such as by the pesticides amitrole and ethylenethiourea, or enzyme inducers such as phenobarbital increase thyroid hormone metabolism and excretion, levels of the hormone fall. Then thyroid-stimulating hormone (TSH) secretion is stimulated, stimulating the thyroid to produce more thyroxine, causing increased cell division, goiter, and cancer. Factors, which increase thyroid hormone levels would have the reverse effect and lead to thyroid atrophy.

Effects on Regulation of Excitable Cells
Specialized cells such as neurons and muscle cells are electrically excitable and controlled by transmitter and modulator substances. Chemicals can affect the regulation of the activities of such cells. This can occur by (*i*) alterations in a neurotransmitter, (*ii*) receptor function, (*iii*) intracellular signal transduction, or (*iv*) signal-terminating processes.

Chemicals, which affect neurons, such as the Puffer fish toxin **tetrodotoxin**, which blocks voltage-gated Na^+ channels in motor neurons, will also affect other cells such as skeletal muscle cells, causing paralysis.

Other chemicals are agonists such as those that directly affect the γ-butyric acid receptor (GABA$_A$) receptor in neurons. Drugs such as barbiturates and benzodiazepines are agonists (activators) for this receptor and cause neuronal inhibition and sedation at therapeutic doses but coma and depression at higher doses. Conversely, antagonists (inhibitors) for this receptor such as the insecticides lindane and cyclodienes and picrotoxin inhibit neurons and may cause tremor and convulsions at toxic doses.

Other toxicants, including some drugs, will indirectly affect receptors. For example, isoniazid and other hydrazides and hydrazines indirectly inhibit the GABA$_A$ receptor by inhibiting the synthesis of GABA, causing convulsions at high doses (see chap. 7).

Another important example is the nicotinic acetylcholine receptor, which is activated by the agonist nicotine causing muscular fibrillation and paralysis. Indirect effects can also occur. For example, organophosphates and other acetylcholinesterase inhibitors increase the amount of acetylcholine and thereby overstimulate the receptor, leading to effects in a number of sites (see chap. 7). Alternatively, **botulinum toxin** inhibits the release of acetylcholine and causes muscle paralysis because muscular contraction does not take place (see chap. 7).

Chemicals can also activate intracellular signal transduction pathways. For example, the voltage-gated Na^+ channels, which amplify signals generated by ligand-gated cationic channels, are activated by some toxicants such as DDT, leading to overexcitation in insects.

Chemicals can also interfere with the removal of the signal such as the level of a cation inside the cell. The signal is terminated by export of the cation through channels or transporters. Inhibition of these processes will prolong the signal and therefore the effect. Digitalis glycosides, which inhibit the Na^+K^+ ATPase so that Na^+ levels rise in cardiac cells, is an example (see chap. 7).

As well as excitable cells, other cell types such as liver cells may also have receptors, which can be affected by chemicals, although usually with less dramatic results. Thus, liver cells have α_1 adrenergic receptors, activation of which causes metabolic effects such as increased glycogenolysis and increased intracellular Ca^{2+}. The muscarinic receptor in exocrine gland cells is influenced by acetylcholine, hence excess secretions occur with acetylcholinesterase inhibitors such as organophosphates (see chap. 7).

6.3.2 Secondary Events
Changes in Membrane Structure and Permeability
Both the plasma membrane and internal membranes associated with organelles may be damaged by toxic compounds. Chemicals such as detergents, strong acids and alkalies, and snake venoms, which contain hydrolytic enzymes, can also *directly* damage the plasma membrane.

As already discussed, this may be due to lipid peroxidation, which alters and destroys membrane lipids. As many enzymes and transport processes are membrane bound, this will affect the function of the organelle as well as the structure. Sulfydryl groups in membranes may be targets for **mercuric ions** in kidney tubular cells and for **methyl mercury** in the CNS, for example. The results are changes in membrane permeability and transport and subsequent cell death. Structural damage can be detected by electron microscopy as disruption of the endoplasmic reticulum, for example, or swelling of the mitochondria. The result of mitochondrial damage could be alterations in intracellular Ca^{2+} concentrations (see below). The result of damage to the plasma membrane might be an influx or efflux of ions and other endogenous substances, leading to swelling of the cell due to an influx of water (e.g., see sect. "Hydropic Degeneration," below). However, before permeability changes occur, reversible structural changes such as blebbing of the cell surface can be detected. This structural change is visible by light microscopy in isolated cells *in vitro* and is also detectable *in vivo*. It can be considered a tertiary change and is discussed more fully below. Changes in the plasma membrane can be detected *in vitro* as leakage of ions such as K^+, leakage of enzymes such as lactate dehydrogenase, and uptake of dyes such as Trypan blue. Changes in the lysosomal membrane may have important sequelae, as this may lead to leakage of degradative enzymes (see below).

Changes in the Cytoskeleton

The cytoskeleton may be specifically damaged by toxic compounds such as phalloidin and the **cytochalasins**. Thus, **phalloidin**, which is a toxic component of certain poisonous mushrooms, binds to actin filaments and stabilizes them. They are thus unable to depolymerize, and this in some way leads to release of Ca^{2+} from an intracellular compartment. As the cytoskeleton is associated with the plasma membrane, damage to it may underlie the appearance of cell surface blebs, which rapidly occur after exposure of hepatocytes to phalloidin. The intracellular level of calcium also seems to be an important factor in the functioning of the cytoskeleton (see below). An increase in the level of Ca^{2+} causes a dissociation of actin microfilaments from α-actinin, a protein involved with binding the actin cytoskeletal network to the plasma membrane. Actin-binding proteins are also involved in the binding of the cytoskeleton to the plasma membrane, and these may be cleaved by proteases activated by calcium. The consequence of these changes will be to weaken the attachment of the plasma membrane to the cytoskeleton, and hence, blebs may form. However, not all cytoskeletal changes are due to changes in Ca^{2+} homeostasis.

Some toxicants damage the microtubules such as the colchicines and the metabolite of hexane, 2-5-hexanedione.

The natural toxin microcytin damages microfilaments by acting as a protein phosphatase inhibitor, thus increasing the phosphorylation of the proteins.

Mitochondrial Damage and Inhibition of Mitochondrial Function

The mitochondrion is often a target for toxic compounds, and effects on the organelle can be seen with the electron microscope. Thus, **hydrazine** causes swelling of mitochondria after acute doses, and prolonged treatment leads to the formation of mega-mitochondria. As structure and function are closely linked, compounds, which affect function, may cause changes in structure such as swelling or contraction. This may in some cases be due to fluid changes in the organelle. If intake of fluid is excessive, and it moves into the inner compartment from the outer, then this inner compartment expands and the cristae unfold. The membranes may then rupture. Contraction of the mitochondria is associated with increases in the ADP/ATP ratio, such as occurs after anoxia or with uncouplers of oxidative phosphorylation. Prolonged contraction leads to deterioration of the inner membrane and high amplitude swelling results. Eventually contraction becomes impossible, and rupture and deterioration of the membranes results. The mitochondria are crucial to the cell, and inhibition of the electron transport chain, such as caused by cyanide, leads to rapid cell death. Damage to the mitochondria may also allow release of Ca^{2+} and so increase the levels in the cytosol, although the role of the mitochondrion in Ca^{2+} homeostasis may only be a minor one. Swelling of mitochondria and disruption of structure often occur before cell death and necrosis. Some compounds, such as **2,4-dinitrophenol**, act by uncoupling oxidative phosphorylation, that is, the production of ATP is uncoupled from electron transport. Other compounds affect mitochondrial function by inhibiting the electron transport chain at one or more specific sites, such as the toxic metabolite of MPTP, which inhibits complex I (see chap. 7). The toxic metabolite of **hexachlorobutadiene** is believed to be nephrotoxic because of inhibition of mitochondria function in the proximal tubular cells.

Recently, it has become apparent that a specific change in the mitochondria, the mitochondrial permeability transition, could be responsible for loss of ability to control calcium levels and may precede cell death, both apoptotic and necrotic. Thus, opening of the mitochondrial channel and therefore dissipation of the electrochemical gradient leading to a depletion of ATP levels and increase in calcium could precede both apoptosis and necrosis. Mitochondrial swelling, resulting from pore opening, may lead to the release of factors, which trigger part of the apoptotic program. Increased mitochondrial calcium levels seem also to result from damage to pyridine nucleotides caused by oxidative stress.

The proportion of mitochondria undergoing permeability transition and pore opening may determine whether apoptosis or necrosis occur.

Damage to the Endoplasmic Reticulum

As the smooth endoplasmic reticulum is the site for the oxidative metabolism of many foreign compounds, it is vulnerable to damage from reactive metabolites such as epoxides and free radicals. Short-lived, reactive intermediates with only a narrow radius of action will obviously damage the immediate vicinity. Thus, with **carbon tetrachloride**, damage to both smooth and rough endoplasmic reticulum occurs, leading to disruption of functions, such as metabolism and protein synthesis, of the whole organelle. One particular function of the endoplasmic reticulum, which is important in terms of cellular homeostasis, is calcium sequestration. Compounds that damage the endoplasmic reticulum such as carbon tetrachloride are known to inhibit this function of sequestrating calcium.

Depletion of ATP and Other Cofactors

ATP is a crucial intermediate for cells to maintain normal activities; without a minimum level, the cell will die as systems will fail. ATP is the energy currency of the cell and is required for the synthesis of many substances such as macromolecules for structural and functional purposes, which the cell needs, but also for processes such as cell division, maintenance of the correct ionic balance, muscular and electrical activity, ciliary movement, membrane transporters, and specific ion channels.

ATP may be produced by either substrate level phosphorylation or oxidative phosphorylation. However, the majority of the ATP is produced by oxidative phosphorylation in the mitochondria.

Depletion of ATP is caused by many toxic compounds, and this will result in a variety of biochemical changes. Although there are many ways for toxic compounds to cause a depletion of ATP in the cell, interference with mitochondrial oxidative phosphorylation is perhaps the most common. Thus, compounds such as **2,4-dinitrophenol**, which uncouple the production of ATP from the electron transport chain, will cause such an effect, but will also cause inhibition of electron transport or depletion of NADH. Excessive use of ATP or sequestration are other mechanisms, the latter being more fully described in relation to ethionine toxicity in chapter 7. Also, DNA damage, which causes the activation of poly(ADP-ribose) polymerase (PARP), may lead to ATP depletion (see below). A lack of ATP in the cell means that active transport into, out of, and within the cell is compromised or halted, with the result that the concentration of ions such as Na^+, K^+, and Ca^{2+} in particular compartments will change. Also, various synthetic biochemical processes such as protein synthesis, gluconeogenesis, and lipid synthesis will tend to be decreased. At the tissue level, this may mean that hepatocytes do not produce bile efficiently and proximal tubules do not actively reabsorb essential amino acids and glucose.

Depletion of other cofactors such as UTP, NADH, and NADPH may also be involved in cell injury either directly or indirectly. Thus, the role of NADPH in maintaining reduced GSH levels means that excessive GSH oxidation such as caused by certain quinines, which undergo redox cycling, may in turn cause NADPH depletion (see below). Alternatively, NADPH may be oxidized if it donates electrons to the foreign compound directly. However, NADPH may be regenerated by interconversion of NAD^+ to $NADP^+$. Some quinones such as **menadione**, **1,2-dibromo-3-chloropropane** (DBCP), and **hydrogen peroxide** also cause depletion of NAD, but probably by different mechanisms. Thus, with menadione, the depletion may be the result of synthesis of NADPH at the expense of NAD; with hydrogen peroxide and DBCP, the depletion is thought to be a consequence of PARP activation.

There are a number of consequences of ATP depletion:

1. Loss of ATP means that cells cannot properly maintain the ionic balance in the cell and Na^+ levels rise.
2. Because of the loss of ionic balance, the water content of the cell increases (swelling or hydropic degeneration and membrane blebs occur).
3. Increased use of substrate level phosphorylation increases pyruvate and lactate levels and decreases NADH levels. Glucose levels fall.
4. ADP accumulates, along with phosphate, H^+, and Mg^{2+}. Hence, the cell pH drops as a result of this and of the increased pyruvate and lactate.
5. Excitable cells (muscle, nerve) fail due to loss of sufficient energy.

However, two events occur as a result, which diminish *initially* the pathological consequences. These are that the phosphate combines with Ca^{2+} to form insoluble salt, so limiting any rise in cytosolic free Ca^{2+}, and the drop in pH limits the activity of hydrolytic enzymes such as phospholipase and the mitochondrial permeability transition. These events will be discussed later in this chapter.

Changes in Ca^{2+} Concentration

Perhaps the most important cellular changes described recently in terms of mechanisms of cell injury and death are those involving changes in the concentration of calcium. It has long been known that calcium is in some way involved in cell death and that it accumulates in damaged tissue. More recently, changes in the intracellular distribution of this ion have been implicated in the mechanisms underlying the cytotoxicity of many different, although not all, toxic compounds in different tissues. Thus, the toxicity of **carbon tetrachloride** and **paracetamol** to liver cells, the toxicity of **lead** and **methyl mercury** and the toxicity of **TCDD** to cardiac and thymus cells are all believed to involve Ca^{2+}.

Calcium is an important element in mammalian cells, and levels of the ion are often critical in toxicological processes. It is worthwhile reviewing its roles in the cell.

Calcium ion, Ca^{2+}, has two main roles:

1. It acts as a second messenger.
2. It activates various enzymes and other proteins.

Therefore, the cell is very vulnerable to changes in intracellular Ca^{2+} concentration, which are maintained at a very low level, 100 nM (0.1 μM) relative to the external concentration, 1 mM. To maintain this 10,000-fold difference requires energy. There are specific pumps, which pump Ca^{2+} out of the cell, or into the lumen of the endoplasmic reticulum where concentrations may reach 100 μM, or into the mitochondria, which is a lower-affinity system and is normally less important.

It should be noted, however, that changes in the Ca^{2+} level can be either a cause or a consequence of damage. Therefore, any toxicant or other insult that affects these pumps directly or indirectly (such as by damaging the plasma membrane) can change the Ca^{2+} level and upset the balance.

For example, the toxic organotin compound, tri-*n*-butyl tin, damages the endoplasmic reticulum and interacts with thiols in the Ca^{2+} pump such that the intracellular Ca^{2+} concentration rises to 500 to 600 μM. This is shown in Figure 6.12.

Interference with any of these processes may be caused by toxic compounds and can alter calcium homeostasis. This can allow an influx of Ca^{2+}, inhibition of export of Ca^{2+} out of the cell, or a release of Ca^{2+} from compartments within the cell. The result of each of these will be a rise of intracellular Ca^{2+}, which can cause a variety of damaging events.

Interference with Ca^{2+} homeostasis may occur in a variety of ways:

1. Inhibition of Ca^{2+} ATPases. Compounds, which react with sulfydryl groups, will inhibit the ATPases, which are involved in calcium transport. Thus, intracellular

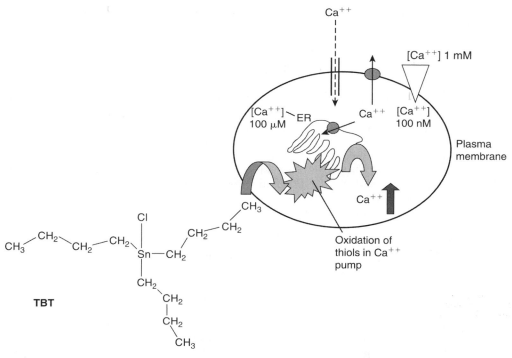

Figure 6.12 The cellular toxicity of TBT caused by damage to the thiols of the Ca^{2+} pump. This leads to dramatic mobilization of calcium from the ER. The filled circles represent ATP-dependent Ca^{2+} transporters. *Abbreviations*: TBT, tri-n-butyltin chloride; ER, endoplasmic reticulum.

movement and efflux of calcium will be altered. Consequently, if the plasma membrane Ca^{2+} transporting ATPase is inhibited, Ca^{2+} will tend to enter the cell because of the high concentration outside, but will not be effectively pumped out. Similarly, if the ATPase in the endoplasmic reticulum is inhibited, this organelle will fail to sequester Ca^{2+}, and so the free cytosolic concentration will again rise. The reactive aldehyde product of lipid peroxidation, 4-hydroxynonenal, will inhibit sequestration of Ca^{2+} by the endoplasmic reticulum, giving another mechanism, whereby lipid peroxidation may lead to cell injury and death. Carbon tetrachloride and bromobenzene cause release of Ca^{2+} from the endoplasmic reticulum. Quinones and peroxides inhibit the efflux of Ca^{2+} from the cell.

2. Damage to the plasma membrane allows increased influx of Ca^{2+}. The result of this will be the same as inhibition of the efflux, a rise in cytosolic free calcium. Paracetamol and carbon tetrachloride cause this increased influx.

3. Depletion of ATP will also lead to a rise in intracellular Ca^{2+}, presumably as a result of the reduction in activity of the ATPases and a reduction in other metabolic activity. The cytosolic free calcium may also rise because of release from mitochondrial stores as caused by cadmium, MPTP, and uncouplers. It seems that the crucial event is a sustained rise in cytosolic free calcium, and there are various consequences that arise from this event:

    1. Alterations in the cytoskeleton. The cytoskeleton depends on the intracellular Ca^{2+} concentration, which affects actin bundles, the interactions between actin and myosin and α-tubulin polymerization. The effect of increases in Ca^{2+} on the cytoskeletal attachments to the plasma membrane and the role of the cytoskeleton in cellular integrity have already been mentioned (see above). If the cytoskeleton is damaged or disrupted or its function altered by an increase in Ca^{2+}, then blebs or protrusions appear on the plasma membrane (see below). As well as an increase in Ca^{2+}, oxidation of, or reaction with sulfydryl groups, such as alkylation or arylation, for example, may disrupt the cytoskeleton, as thiols

are important for its integrity. Similarly, direct interaction with the cytoskeleton as occurs with **phalloidin** will also lead to cellular damage.

2. Ca^{2+}-activated phospholipases. These enzymes are found in biological membranes where they catalyze the hydrolysis of membrane phospholipids, and Ca^{2+} is required as a cofactor. Phospholipase A_2 enzymes are involved in detoxifying phospholipid hydroperoxides, which may occur during lipid peroxidation. However, prolonged action of these enzymes leads to breakdown of the membrane and release of compounds such as lysophospholipids, which are cytotoxic, and arachidonic acid, which may be further metabolized to substances, which are inflammatory mediators. Prolonged increases in cytosolic Ca^{2+}, therefore, will activate these enzymes and lead to breakdown of membranes such as the plasma membrane and mitochondrial membrane. This inevitably results in cytotoxicity. The breakdown of the mitochondrial membrane, releasing lysophospholipids and fatty acids can damage the mitochondria.

3. Ca^{2+}-activated proteases. These enzymes, also known as calpains, are extralysosomal. They are involved in normal cell functions such as enzyme activation and membrane remodeling. When activated by increases in cytosolic Ca^{2+}, these proteases especially target cytoskeletal and membrane proteins, including receptors, which are vulnerable proteins. This results in cytotoxicity.

4. Ca^{2+}-activated endonucleases. These enzymes are involved with the normal process of apoptosis or "programmed cell death" (see below) and cleave DNA into fragments. This destruction of DNA results in cell death. Other enzymes involved in DNA fragmentation may also be activated by Ca^{2+}.

   Activation of these enzymes may also lead to DNA single- and double-strand breaks initiating apoptosis and necrosis. This occurs after exposure to the reactive metabolite of paracetamol, for example.

5. Ca^{2+} cycling into and out of the mitochondria leads to NAD depletion and a fall in ATP. The entry of Ca^{2+} into the mitochondria dissipates the potential difference across the mitochondrial membranes and so inhibits the function of ATP synthase, which relies on the charge difference across the membrane (Fig. 6.13 and 7.60). Export of Ca^{2+} from the mitochondrial matrix may occur and be stimulated by some chemicals. However, this will lead to repeated cycling, which damages the membrane and further compromises ATP synthesis. The export of Ca^{2+} also uses up ATP as a result of the Ca^{2+} ATPases involved. Hence ATP levels fall.

6. Ca^{2+} activation of gene expression. The MAP kinases (c-fos, c-myc, and c-jun) are activated as is also the Fas death receptor pathway.

7. An effect secondary to the activation of enzymes by increased calcium levels can be increased production of reactive oxygen and nitrogen species. Thus, activation of mitochondrial dehydrogenases increases NADH production and electron transport, yet increased calcium uncouples ATP synthesis, and the excess electron generates superoxide. Calcium also activates nitric oxide synthetase.

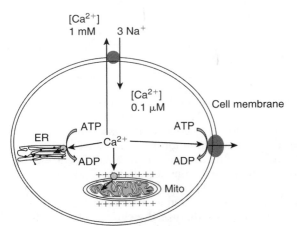

Figure 6.13 The mechanisms for the elimination of Ca^{2+} from the cytoplasm: sequestration into the ER via the Ca^{2+}-ATPase system; sequestration into the mitochondria (Mito) via the Ca^{2+} uniporter; transport outside the cell via the Ca^{2+}-ATPase system; ion gradient–driven transport outside the cell via the Ca^{2+}/Na^+ exchanger. *Abbreviations*: ER, endoplasmic reticulum.

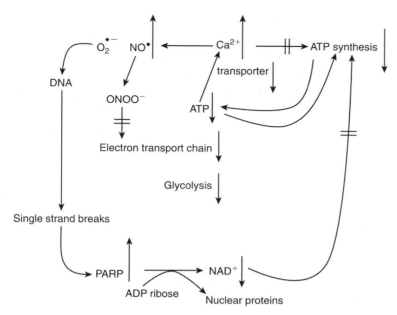

Figure 6.14 Interrelationships between primary events underlying toxicity showing how there is amplification. The double lines indicate inhibition of a pathway. Vertical arrows indicate an increase or decrease in the substance or process.

These are interrelated in such a way as to amplify the original insult as shown in Figure 6.14.

Thus, an initial drop in ATP is followed by increases in Ca^{2+}, which inhibits ATP synthase and increases ROS and reactive nitrogen species (RNS) formation via xanthine oxidase. These inhibit thiol-dependent Ca^{2+} transport. The reactive molecules can also inhibit the electron transport chain (by reacting with Fe at the active sites) and enzymes in glycolysis, notably glyceraldehyde 3-phosphate dehydrogenase, leading to further losses of ATP. The depleted ATP exacerbates the intracellular Ca^{2+} increase as a result of reduced transport out and sequestration into the endoplasmic reticulum.

Finally, the peroxynitrite formed from the RNS and ROS can cause single-strand breaks in DNA, and this activates PARP, which transfers ADP-ribose from NAD^+ to nuclear proteins. This consumes NAD^+, which means that NADH levels are compromised, and hence ATP is again affected. The resynthesis of NAD^+ also consumes more ATP.

Therefore, ATP levels will reach a point when the cell cannot survive, Ca^{2+} levels are too high, and the plasma membrane blebs and ruptures.

The eventual result may depend on the cell type but will certainly lead to what can be termed "tertiary events" (see below).

It now seems that cytosolic calcium may not play a central role in the initiation of oxidative injury as changes in calcium homeostasis occur well *after* the appearance of other indications of cellular injury. However, mitochondrial lesions do occur early on in the time course of oxidative cellular damage, and calcium may indeed play a role in these. Changes in concentrations of calcium will, however, result in the activation of signaling mechanisms and alterations in cellular structure and in gene expression. Such alterations may in some instances play a critical role in cellular toxicity. For example, increases in cytosolic Ca^{2+} inhibit mitochondrial function.

DNA Damage and Poly ADP-ribosylation
When the target molecule is DNA, then a mutation may result, which could lead to dysfunction in the cell affected or in a daughter cell or a tumor. This will be discussed in more detail in sections 6.7 and 6.8 in this chapter.

Compounds such as hydrogen peroxide and alkylating agents such as dimethyl sulfate, which cause single-strand breaks in DNA, cause the activation of the enzyme PARP. This enzyme catalyzes ADP-ribosylation, which is posttranslational protein modification, involving the addition of ADP-ribose to amino acids. It is also involved in polymerization reactions and, through modulation of DNA ligase activity, with DNA repair. The result of activation of this enzyme is the cleavage of the glycosidic link in NAD to yield ADP-ribose and to release nicotinamide. Therefore, there is a loss of NAD, and with severe DNA damage, this depletion of NAD is sufficient to lead to cell death. This may indeed be a protective mechanism, a form of cellular euthanasia, which ensures that cells with extensively damaged DNA do not replicate and so cannot pass on any errors, which might be encoded in the altered DNA.

Lysosomal Destabilization

At one time it was believed that release of lysosomal enzymes was the mechanism underlying cell injury and death. However, although this occurs during cell damage, it is generally a late event, occurring after the point of no return has been reached by the cell, rather than a cause of damage. There are situations, however, when lysosomal damage may be an initiating event. For example, lysosomal damage is involved in the toxicity of silica particles, which are engulfed by macrophages in the lungs and are then taken up into lysosomes, and with gentamycin, which is taken up into lysosomes in the proximal tubular cells. Similarly, the hydrocarbons, which bind to α_2-μ-globulin are taken up into lysosomes. In these cases, the presence of the foreign substance leads to rupture of the lysosomes, and the hydrolytic enzymes released damage and destroy the cell.

6.3.3 Tertiary Events

Steatosis/Fatty Change

The accumulation of fat is a common cellular response to toxic compounds, which is normally reversible. Usually triglycerides accumulate, although sometimes phospholipids accumulate, as occurs after exposure to the drug chlorphentermine (see chap. 2). Steatosis is particularly common in the liver as this organ has a major role in lipid metabolism (Fig. 6.15). The lipid may appear in the cell as many small droplets or as one large droplet. Interference with lipid metabolism can occur at several points:

1. Inhibition of excretion of lipid from the cell. This is the most common cause of steatosis, although there are several possible reasons for the inhibition. Thus,

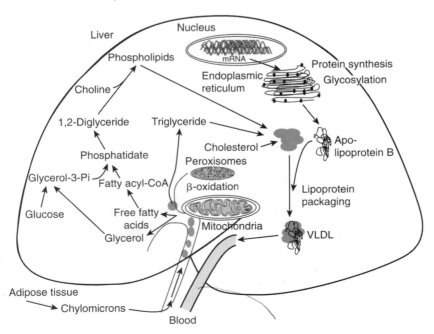

Figure 6.15 Pathways involved with lipid synthesis and catabolism in, and transport out of, the liver.

inhibition of protein synthesis will block the synthesis of the lipid acceptor protein (apoprotein) required for the transport of lipid out of the cell as very low-density lipoprotein (VLDL) (Fig. 6.15). There are numerous points at which protein synthesis can be inhibited: at the point of DNA transcription, during RNA translation, and during assembly of the apoprotein. Thus, **puromycin** and **tetracycline** both cause steatosis by inhibiting protein synthesis. Carbon tetrachloride causes steatosis by damaging the endoplasmic reticulum and Golgi apparatus where protein synthesis and assembly take place. **Ethionine** (see chap. 7), which depletes ATP, may cause steatosis by inhibiting protein synthesis as ATP is required for this. Also, ATP will be required for the transport of the lipoprotein complex out of the hepatocytes, and the lack of ATP will compromise this process.

2. Increased synthesis of lipid or uptake. Increased synthesis of lipid may be the cause of fatty liver after **hydrazine** administration as this compound increases the activity of the enzyme involved in the synthesis of diglycerides. Hydrazine also depletes ATP and, however, inhibits protein synthesis. Large doses of ethanol will cause fatty liver in humans, and it is believed that this is partly due to an increase in fatty acid synthesis. This is a result of an increase in the $NADH/NAD^+$ ratio and therefore of the synthesis of triglycerides. Changes in the mobilization of lipids in tissues followed by uptake into the liver can also be another cause of steatosis.

3. Decreased metabolism of lipids. Decreased mitochondrial oxidation of fatty acids is another possible cause of ethanol-induced steatosis. Other possible causes are vitamin deficiencies and the inhibition of the mitochondrial electron transport chain.

Phospholipidosis

Many drugs are now known to cause this adverse effect, which often becomes apparent during the preclinical safety evaluation.

It seems that for drugs to cause accumulation of phospholipids, the necessary physicochemical characteristic is the presence of both hydrophilic and lipophilic parts to the molecule, as exemplified by chlorphentermine (see chap. 3) (chap. 5, Fig. 1). They contain a hydrophobic ring structure and a hydrophilic side chain with a positively charged (cationic) amine group. Such molecules are known as cationic amphipathic drugs or CADs. Other drugs, all in use, known to cause phospholipidosis are amiodarone, chloroquine (chap. 5, Fig. 1), tafenoquine, and gentamycin.

The hydrophobic section can associate with the lipid fraction and the cationic section with the ionic part of the phospholipids.

In the case of chlorphentermine, it is believed to associate with the ionic portion of the phospholipids. Phospholipids are diglycerides with the remaining hydroxyl group of the glycerol esterified with a phosphate to which is attached a charged moiety such as choline. The long fatty acid chain is hydrophobic, the phosphate-choline is hydrophilic. They are particularly important constituents of membranes.

The clinical significance of phospholipidosis is related to secondary damage to tissue structure or impaired function possibly at the cellular level, for example, reduced immunological response. In the case of amiodarone, the phospholipidosis in the lung causes cough and breathing difficulties.

Hydropic Degeneration

The term "hydropic degeneration" describes the swelling of a cell due to the intake of water. It is a reversible change but may often precede irreversible changes and cell death. The osmotic balance of the cell is maintained by active processes, which control the influx and efflux of ions. The intracellular Na^+ and Ca^{2+} concentrations are maintained at a lower level than that in the extracellular fluid by an active process requiring ATP, whereas K^+ is maintained at a higher concentration inside the cell. However, the system depends on the integrity of the membrane and transport systems and on the supply of ATP. Inhibition of the ATPase enzyme involved in the active transport or reduction of the supply of ATP by metabolic inhibitors will upset the balance, allowing the concentration of ions in the cell to increase. The osmotic pressure thus created will cause water to enter the cell with swelling. This will lead to damage to the cell and organelles if it is excessive. Thus, the mitochondria and endoplasmic reticulum will swell and may become damaged.

Blebbing of the Plasma Membrane
This process is an early morphological change in cells often seen in isolated cells *in vitro* but also known to occur *in vivo*. The blebs, which appear before membrane permeability alters, are initially reversible. However, if the toxic insult is sufficiently severe and the cellular changes become irreversible, the blebs may rupture. If this occurs, vital cellular components may be lost and cell death follows. The occurrence of blebs may be due to damage to the cytoskeleton, which is attached to the plasma membrane as described above. The cause may be an increase in cytosolic Ca^{2+}, interaction with cytoskeletal proteins, or modification of thiol groups (see below).

Cellular Destruction and Cell Death
Earlier in this chapter we discussed the critical secondary events occurring following the primary mechanisms of toxicity. Depending on the severity, these events can be reversible or can lead to cell death.

However, there are two distinct patterns of morphological change, which occur before cell death, apoptosis, and necrosis. Apoptosis is an orderly process of cell shrinkage and condensation, blebbing, and phagocytosis. Necrosis, by contrast, is a disorderly process involving swelling and then rupture of membranes, dissolution of organized structures, and total loss of homeostasis and ATP. It has been proposed in some publications that these are two separate processes with little evidence for commonality but that the dose of the toxic chemical determines the process that occurs. Others seem to feel that these two processes are part of a continuum only separated by dose with mitochondrial permeability transition and calcium ion activation of hydrolytic enzymes underlying the progression to both apoptosis and necrosis.

Thus, in the section above, we discussed the potentially catastrophic events that can take place when there is interaction and amplification between a number of key events.

The results of these events can be an increase in the permeability of the inner mitochondrial membrane and then the opening of a mega-channel or mega-pore spanning both inner and outer mitochondrial membranes. This will allow most solutes (of molecular weight less than 1500) into or out of the mitochondria. Thus, protons enter, changing the import charge differential on the membranes, and ATP leaks out. Water also enters causing swelling. The ATP synthase enzyme changes and becomes an ATPase, so breaking down any remaining ATP. Any Ca^{2+} in the mitochondria can leak out into the cytoplasm. ATP depletion becomes critical as there is not enough for glycolysis or for any energy-requiring processes. This is the so called mitochondrial permeability transition or MPT.

This scenario can have three outcomes, recovery, apoptosis, and necrosis, the latter two of which are tertiary events. The crucial questions are why is this not always lethal to cell and why does it sometimes lead to apoptosis and at other times to necrosis? Many chemicals, including drugs, can cause both apoptosis and necrosis.

It seems that toxic chemicals will induce apoptosis after low doses or early after high doses but necrosis later after high doses. Both forms of cell death involve similar cellular and metabolic changes. The MPT seems to be crucial, as agents that block this process will actually prevent both apoptosis and necrosis (cyclosporin A, Bcl-2 overexpression).

It would seem that the availability of ATP in the cell is the crucial determinant of the outcome. Thus, studies have shown that when ATP is critically depleted, necrosis ensues, whereas if substrates are present, which allow ATP generation, apoptosis occurs. This is the theory proposed by Lemasters (1,2).

It seems that the *number of mitochondria* undergoing MPT is important.

At low levels of mitochondrial damage, removal of damaged mitochondria by lysosomes rescues the cell. If this is overwhelmed then apoptosis occurs; triggers such as cytochrome c (see below) start the apoptotic process. If most or all mitochondria are damaged, then the depletion of ATP is so extensive that the apoptotic process is not possible (ATP is required for the process), and the cell undergoes necrosis.

We perhaps often do not appreciate that when a cell suffers damage, it is not all damaged equally (in just the same way in an organ and tissue, some cells will be spared).

This is illustrated in the figure below (Fig. 6.16).

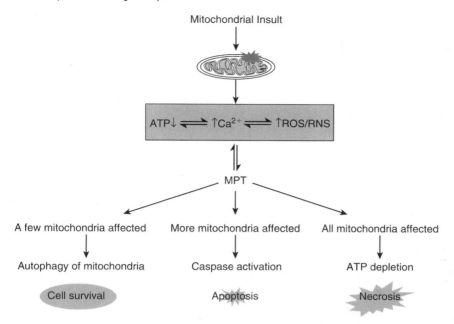

Figure 6.16   "Decision tree": the possible futures of a cell, which has undergone a toxic insult. *Abbreviations*: ROS, reactive oxygen species; RNS, reactive nitrogen species.

Apoptosis

Apoptosis has been termed "programmed cell death." It occurs as part of the normal maintenance and renewal of tissues. However, it is also stimulated by toxic chemicals but differs in significant respect from necrosis, although some of the triggers are probably the same. The crucial differences in the end result are that in apoptosis the plasma membrane remains intact; the DNA is fragmented in sections, not randomly; and phagocytes are encouraged to engulf the apoptotic cell. Consequently, no inflammation occurs. However, apoptosis is an active process, requiring energy in the form of ATP.

If there is too much damage such that the level of ATP is insufficient and/or the machinery of apoptosis is damaged, then necrosis will ensue. Also the exact pathway may depend on the cell type and the particular trigger signal.

Although not all cell death involves mitochondrial dysfunction, it often does. The various cellular changes and sequelae and effects on and changes in the mitochondria have been described above and the critical "decision points" between recovery, necrosis, and apoptosis have been outlined. What then occurs in the orderly cell death process of apoptosis?

Following the MPT described above, another crucial event in apoptosis is the release of cytochrome c. This will cause the cessation of the electron transport chain as it is a crucial link between complex III and IV (Fig. 6.7). Consequently, ATP synthesis will stop, and furthermore, the electrons in the chain, which are not reducing oxygen to water, will instead lead to the production of superoxide $O_2^{\bullet-}$. This of course can cause damage, and the loss of the ability to produce ATP will start to compromise the cell, potentially leading to necrosis. However, the cytochrome c, and maybe other proteins released from the damaged mitochondria, is a signal, which initiated the apoptotic process (Fig. 6.17). The released cytochrome c binds to an adaptor protein Apaf-1 along with ATP. This process causes proteolytic cleavage of procaspase-9 releasing active caspase-9. This cysteine protease is one of a group of such proteases, which have procaspase, inactive forms, but when active, attack proteins at specific places (asparagine). Some active caspases (2, 8, 9) activate procaspases. Certain of the active caspases (3, 6, 7) attack and clip intracellular proteins and activate or inactivate them. The caspase-catalyzed hydrolysis is an essential component of the apoptotic process and is directly or indirectly responsible for the changes that take place.

These events include the inactivation by proteolysis, of PARP, so DNA repair does not take place, but significantly, ATP is not wasted. Detachment of chromatin from the nuclear

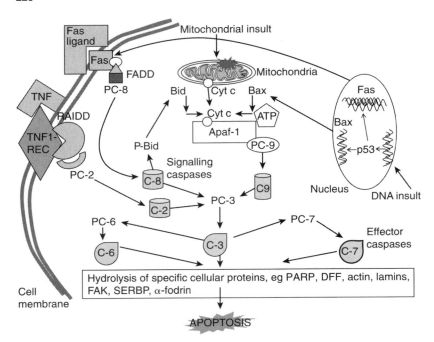

Figure 6.17   The intracellular signaling leading to apoptosis initiated by mitochondrial damage, DNA damage, or stimulation of Fas or TNF-1 receptors. For full explanation see text.

scaffold allows chromatin condensation. Hydrolytic activation of a DNA fragmentation factor leads to fragmentation of nuclear DNA. Endonuclease G cleaves chromatin DNA into nucleosomal fragments. Caspase-activated nucleases cleave DNA into 200 base pair fragments.

Proteins involved in cellular architecture such as actin, laminins, and α-fodrin are clipped, allowing the cytoskeleton to be disassembled. This affects the plasma membrane and results in the formation of apoptotic bodies. Macrophages can remove these by phagocytosis.

Inactivation of the enzyme involved in cell adhesion (focal adhesion kinase) allows the cell to separate from the extracellular matrix. Changes to the cell surface, resulting in phosphatidyl serine externalization on the plasma membrane and external accumulation of sterols, allow the apoptotic cell to be recognized and eventually removed by phagocytosis.

The initial process involving the mitochondria formation of the pore (MPT) and loss of cytochrome c are controlled by a family of proteins Bcl-2. Some of the constituent proteins such as Bax, Bad, and Bid facilitate the processes, whereas others such as Bcl-2 and Bcl-XL inhibit the processes.

The Bax, Bad, and Bid proteins or death promoters may act directly on the mitochondria. However, the death suppressors combine with the death promoters and form inactive dimers, thus neutralizing the effects. Therefore, the control depends on the relative amount of these proteins.

As can be seen from Figure 6.17 as well as mitochondrial injury, other events can trigger apoptosis and involve the caspases and Bcl-2 proteins. Thus, DNA damage can trigger the process as can interaction of ligands with the Fas receptor, or TNF-1 receptor also, leading to mitochondrial involvement. Oxidative stress will also trigger apoptosis, and xenobiotics can also directly damage ion channels and pumps, allowing influx of ions such as calcium.

Damage to DNA by drugs such as doxorubicin as well as ionizing radiation can stimulate apoptosis. This is because cells with damaged DNA are a risk to the organism, because if they have undergone significant mutations and are allowed to undergo clonal expansion, an initiated cell could form tumors. So they should be repaired or removed.

Following DNA damage, gene activation, involving the c-myc gene, triggers the cell cycle to move from G0 to G1, but further stages of the cycle are blocked by p53 (which involves the detection of DNA strand breaks) until repair has taken place. If repair fails, then the Bax gene is activated and apoptosis proceeds. However, apoptosis may be blocked by the Bcl-2 gene (death suppressor gene) by inhibiting the Bax gene.

The DNA damage causes stablization and activation of p53 protein, which is a transcription factor that can increase expression of Bax, one of the death promoters, which can start the process. This is a useful target for anticancer drugs, as they can stimulate apoptosis in cancer cells by acting on p53. However, they may also induce apoptosis in normal, rapidly dividing cells.

Stimulation of death receptors on the plasma membrane, such as TNF and Fas as well as DR3, DR4, and DR5, may be a trigger. The TNF receptor has an extracellular domain, a membrane-spanning domain, and an intracellular domain. The latter is also known as the death domain.

Stimulation of Fas and TNF-1 receptors directly activates caspases 2 and 8 , but Fas ligands can also activate Bid and so cause the mitochondria to enter the apoptotic process. T lymphocytes (cytotoxic lymphocytes) express the Fas ligand and so can cause activation of the Fas-caspase system in target cells.

Some chemical-mediated cell death involves this system also. For example, the spermatocyte death caused by **mono-(2-ethylhexyl)phthalate** or **2,5-hexanedione** (toxic metabolite of hexane) is due to apoptosis. This is induced in the spermatocytes by Fas ligand expressed on the surface of the Sertoli cells in which the microtubules have been damaged by the chemicals. As Sertoli cells are essential for support of the spermatocytes, damaged Sertoli cells are a "liability."

Necrosis (4)

In contrast to apoptosis, necrosis has been called accidental cell death. When irreversible damage occurs, a cell may reach the point of no return, and cell death ensues. This is followed by a series of degenerative changes, including hydrolysis of cellular components and denaturation of proteins. Unlike apoptosis, the plasma membrane undergoes a permeability change, causing changes in intracellular ions, DNA undergoes random digestion, the cell lyses, and then inflammation occurs.

This cellular necrosis, which is the death of cells forming part of living tissue, is characterized by changes such as swelling of the mitochondria and the endoplasmic reticulum, the appearance of vacuoles, and the accumulation of fluid. These changes may occur prior to necrosis and be reversible, or they may presage cell death and necrosis. Abrupt and extensive expansion of the mitochondria with disruption of the structure often occurs immediately prior to necrotic change. The endoplasmic reticulum may also undergo dilation and disruption. Both would allow calcium to leak out as well as other important molecules. Which organelle shows damage first may depend on the mechanism. Thus in hypoxia, the mitochondria show damage before the endoplasmic reticulum, whereas with carbon tetrachloride-induced cellular injury, the endoplasmic reticulum is damaged first. Cells undergoing necrosis also show changes in the nucleus in which the chromatin becomes condensed (pyknosis) and may then become fragmented (karyorrhexis). Alternatively, in some cases the nucleus simply fades and dissolves (karyolysis).

Biochemically, following the opening of the mitochondrial mega-pore and the MPT, if sufficient mitochondria are damaged to this level, then the cell will die by necrotic processes. Once the pore has opened, ATP can flow out and solutes such as Ca^{2+} can flow in.

The Ca^{2+} activates proteases, phospholipases, and endonucleases, leading to loss of membrane connections and more ion changes. If sufficient mitochondria are damaged beyond the point of no return such that cellular ATP has fallen to the point that ion channels are unable to function nor glycolysis or other ATP-requiring processes, then the cell will be unable to maintain its homeostasis. Reactive oxygen and nitrogen species will be produced from the electron transport chain, causing more damage to mitochondria and DNA. Consequently, NAD levels will also fall because mitochondrial damage will diminish supply by the citric acid cycle and because of PARP activity in relation to DNA repair. The remaining mitochondria will fail because of the calcium increasing in the surrounding cytoplasm. Cell pH will fall because of increased anaerobic metabolism and so protons will add to the problem by also changing the charge on the mitochondrial membranes so dissipating the energy required for ATP production. Lack of ATP means more solutes can enter the cell, and apoptosis cannot take place. The cell will have insufficient ATP to function at all, cell membrane blebs will rupture and the contents leak out.

DNA fragmentation will occur, but this is random digestion. Lysosomes will also rupture spilling hydrolytic and other degradative enzymes contributing to the destruction and also inflammation.

6.3.4 Protective Mechanisms

Biological systems possess a number of mechanisms for protection against toxic foreign compounds, some of which have already been mentioned. Thus, metabolic transformation to more polar metabolites, which are readily excreted, is one method of detoxication. For example, conjugation of paracetamol with glucuronic acid and sulfate facilitates elimination of the drug from the body and diverts the compound away from potentially toxic pathways (see chap. 7). Alternatively, a reactive metabolite may be converted into a stable metabolite. For example, reactive epoxides can be metabolized by epoxide hydrolase to stable dihydrodiols.

Although reactive intermediates can wreak havoc in a cell if left unchecked, the cell has various ways of removing or detoxifying many of them.

For electrophiles, the general mechanism of detoxication is conjugation with GSH. Quinones and hydroquinones are often reactive intermediates, and the enzyme DT diaphorase will carry out a two-electron reduction of these.

Reactive aldehydes and ketones can be detoxified by reduction to alcohols by aldehyde dehydrogenase.

Metals, such as cadmium and other heavy metals, can be bound by the small protein metallothionein and detoxified. They can also react with and be detoxified by GSH.

Nucleophiles are detoxified by conjugation at the nucleophilic group. Acetylation is the means of detoxifying amino ($-NH_2$) or hydrazine ($-NHNH_2$) or sulfonamido ($-SO_2NH_2$) groups.

These conjugations avoid the production of reactive intermediates.

Reactive intermediates can also effectively be detoxified by covalent reaction with noncritical proteins (such as plasma albumin).

Hypochlorous acid generated by neutrophils in the oxidative burst is detoxified by the amino acid taurine, which combines with it to produce taurochloramine.

Protein Toxins

Toxicants, which are proteins, may not be detoxified by the means so far described. They may however be degraded by proteases, removed by antibodies if antigenic, and recognized as foreign. Also those toxicants that rely on disulfide bonds for structure and activity can be inactivated by the enzyme thioredoxin, which reduces these bonds to sulfhydryl.

Glutathione

Probably the most important protective mechanisms involve the tripeptide GSH (chap. 4, Fig. 59). This compound is found in most cells, and in liver cells, it occurs at a relatively high concentration, about 5 mM or more. There are three pools of GSH: cytosolic, mitochondrial, and nuclear. GSH structure is unusual for a peptide in the glutamyl, and cysteine residues are not coupled via a peptide bond, hence the molecule is resistance to peptidase attack. It has a nucleophilic thiol group, and it can detoxify substances in one of three ways:

1. Conjugation catalyzed by a GSH transferase
2. Chemical reaction with a reactive metabolite to form a conjugate
3. Donation of a proton or hydrogen atom to reactive metabolites or free radicals, respectively.

The first two of these are discussed in chapter 4, and there are specific examples in chapter 7. The products are either excreted directly into the bile or further metabolized and excreted into the urine as cysteine or *N*-acetylcysteine conjugates. There are, however, examples of GSH conjugates being involved in toxicity as indicated in chapters 4 and 7.

The third protective role of GSH involves the reduction of reactive metabolites, the GSH becoming oxidized in the process (Fig. 6.18). If the reactive metabolite is a peroxide such as

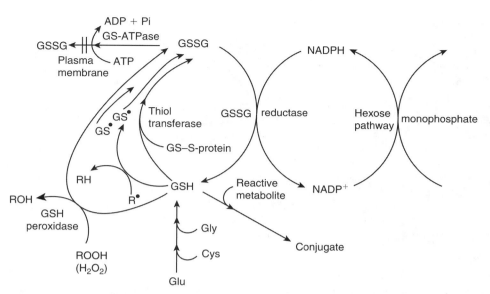

Figure 6.18 The various protective roles of reduced GSH and the relationship with GSSG. The NADPH may be regenerated by the pentose phosphate shunt. *Abbreviations*: GSH, glutathione; GSSG, oxidized glutathione.

hydrogen peroxide or an organic peroxide, the GSH becomes oxidized and the peroxide is reduced to an alcohol. This reaction is catalyzed by GSH peroxidase, which is a selenium-containing enzyme, located in the mitochondria and cytosol. Other reactive metabolites may chemically oxidize GSH to GSSG and in turn are reduced. Alternatively, GSH may donate a hydrogen atom to a free-radical intermediate and be converted into a GSH radical (thiyl radical), which may then react with another GSH radical to form GSSG or abstract a hydrogen atom from other substances to form new radicals and GSH (Fig. 6.18). Thus, the result of these types of reactions may be the oxidation of GSH, which may then suffer one of two fates. Under normal conditions, GSSG is reduced back to GSH via GSSG reductase, an enzyme that requires NADPH (Fig. 6.18). NADPH supply must also be maintained, and this involves the pentose phosphate shunt or hexose monophosphate pathway. The ratio of GSH to GSSG is usually about 100:1 in the cell.

However, in conditions of excess oxidation of GSH, such as oxidative stress induced by a toxic chemical, there may be insufficient NADPH and the enzyme-mediated reduction will be unable to cope with the production of GSSG. Then GSSG is exported out of the cell in an active process, which uses ATP and a translocase enzyme located in the plasma membrane (Fig. 6.18). GSH conjugates can also be transported out of the cell by this mechanism. For GSH, which is depleted by conjugation or lost as GSSG transported out of the cell, there needs to be resynthesis. This is controlled by the availability of cysteine and the other two amino acids. The supply of cysteine, for example, can be a limitation. GSH synthesis involves the enzyme γ-glutamylcysteine synthetase, which is the rate-limiting step. This enzyme is upregulated on increased demand. GSH is also used to reduce protein sulfydryls and maintain them in the reduced state. This is catalyzed by GSH thiol transferase (Fig. 6.19). *N*-acetylcysteine will also help to maintain SH groups in the reduced state and can also scavenge radicals and reactive metabolites.

Heat-Shock/Stress Proteins
The response of cells to stress (including ROS and reactive metabolites) includes the upregulation of synthesis of stress proteins, which play a role in cell survival and thus serve a protective function. These are at a low level in the normal cell but help to maintain cell homeostasis.

One group is the metallothioneins, which bind metals, another is the heat-shock proteins (hsps). These latter ones are called "molecular chaperones" and are involved in, for example,

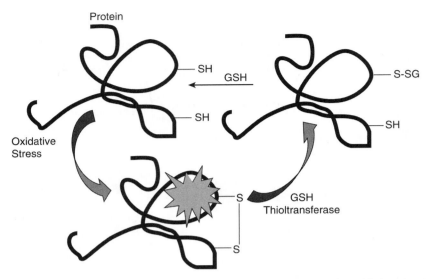

Figure 6.19   The role of GSH and GSH thiotransferase in repairing the oxidation of protein sulfydryl groups after oxidative stress.

the folding of proteins. There are a number of them designated by their molecular weight. For example, hsp100 has a molecular weight of 100 kDa. The other groups are hsp90, hsp70, hsp60, and hsp27. Reactive chemicals and oxidative stress may damage vulnerable and critical proteins, and another means of protection is to "mend" the structure using hsps. This is a kind of "damage limitation."

Of course, the stress proteins themselves may be targets for reactive metabolites or ROS, in which case the protective ability may be lost. For example, a reactive metabolite(s) of halothane, the hepatotoxic anesthetic drug (see chap. 7), binds to hsp60 and hsp70, which are mainly mitochondrial proteins.

Metallothioneins

These proteins are important for binding potentially toxic metals such as cadmium, mercury, and lead, which all bind to sulfydryl groups. Consequently, the binding and removal of these metals are protective functions. Metallothioneins are markedly induced by cadmium exposure and the small protein, rich in SH groups, can then sequester the metal. They also may have a protective role in oxidative stress and protect redox-sensitive processes. The protein also has a role in cadmium nephrotoxicity (see chap. 7).

Superoxide Dismutase

Free radical intermediates or other reactive intermediates may donate electrons to oxygen-forming active oxygen species such as superoxide anion radical, $O_2^{\bullet-}$, which can cause cellular damage (see above).

Mammalian systems have enzymes, superoxide dismutases (SODs), which dismutate two molecules of superoxide into one of hydrogen peroxide and one of oxygen. The enzymes are located in the cytosol and mitochondria. The location in the mitochondria is important because ROS are often produced there, and one form of SOD is inducible, as a result of oxidative stress.

The hydrogen peroxide produced by SOD is then removed by either catalase action or GSH peroxidase. The products are water and oxygen or water and GSSG (Figs. 6.10 and 6.18).

As indicated above, however, hydrogen peroxide may yield hydroxyl radicals in the presence of metal ions such as Fe^{2+}.

Vitamin E

Lipid radicals and other radicals may be removed by a number of endogenous compounds as well as GSH. One is vitamin E, (which includes α-tocopherol) a lipophilic substance. It can react with and neutralize lipid radicals and hydroperoxides, in the process becoming a free radical itself. This α-tocopheryl radical is relatively stable and can then be converted back into the α-tocopherol by vitamin C (ascorbate) or react with more radicals and become α-tocopherol quinone:

$$LO_2^{\bullet} + \alpha\text{-TH} \rightarrow LOOH + \alpha\text{-T}^{\bullet}$$

$$\alpha\text{-T}^{\bullet} + AscH^- \rightarrow \alpha\text{-TH} + Asc^{\bullet-}$$

or

$$\alpha\text{-T}^{\bullet} + LO_2^{\bullet} \rightarrow LOOH + \alpha\text{-TQ}$$

As α-tocopherol is lipophilic, it can easily enter a lipid environment, such as the plasma membrane, where unsaturated lipids readily form radicals (Fig. 6.20).

Other compounds that may have a protective role are vitamin K, cysteine, and ascorbate. These compounds act as alternative hydrogen donors in preference to the allylic hydrogen atoms of unsaturated lipids.

Lipid hydroperoxides can be removed by reaction with GSH catalyzed by GSH peroxidase. The enzyme phospholipase A_2 has been proposed to have a role in the detoxication of phospholipid hydroperoxides by releasing fatty acids from peroxidized membranes.

Macromolecular Repair

Damage to proteins, lipids, and DNA can all be repaired.

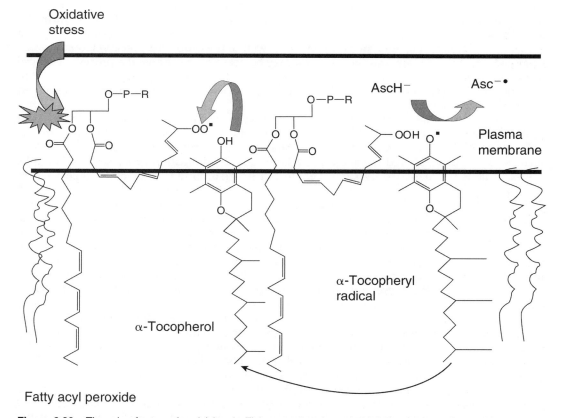

Figure 6.20 The role of α-tocopherol (vitamin E) in removing damaging lipid peroxides from membranes. The α-tocopherol is regenerated by ascorbate (vitamin C).

Thus, oxidation of SH groups in proteins to produce P-SS-P or P-SO, for example, can be repaired by reduction using the enzyme thioredoxin, which in turn is reduced by NADPH. As indicated above, GSH and thioltransferase can also be used. P-SS-G (GSH protein adducts) can be repaired by reduction, using the enzyme glutaredoxin. GSH and NADPH are required.

As already indicated when protein structure is damaged, hsps may repair it. As well as reduction by vitamin E as described, lipids damaged by oxidation to peroxides can be broken down by phospholipase A_2 and the fatty acid peroxide replaced.

Although nuclear DNA is well protected by histones, there are a number of repair systems for DNA. These will be discussed in the section of genotoxicity. Mitochondrial DNA, however, does not have histones for protection or efficient repair mechanisms.

Tissue Repair

An important factor in toxic injury, which has not been widely considered, is tissue repair. It seems that repair, initiated during tissue damage caused by chemicals, for example, is also a dose-related phenomenon. Thus, there will be a threshold for the initiation of repair and a maximum repair response such that doses above this will be more toxic.

Furthermore, initiation of this repair process by prior treatment of an animal with a small sub-toxic dose will afford protection against a second, larger dose. This may be part of the mechanism underlying the tolerance to the hepatotoxicity of carbon tetrachloride as well as the destruction of cytochrome P-450 (see chap. 7). Effects on tissue repair may also be important in interactions between compounds. Thus, exposure to another agent such as phenobarbital may stimulate tissue repair and decrease the eventual toxicity of carbon tetrachloride.

6.4 PHARMACOLOGICAL, PHYSIOLOGICAL, AND BIOCHEMICAL EFFECTS

These three types of toxic effects will be considered together as they are often all closely interrelated. Thus, toxic foreign compounds can affect the homeostasis of an organism by altering basic biochemical processes. These effects may often be reversible if inhibition of an enzyme or binding to a receptor is involved. A well-defined dose-response relation is often observed with such effects, and the endpoint may be death of the animal if a process of central importance is affected. These types of response include the **exaggerated** or **unwanted pharmacological responses** to clinically used drugs, which are the most commonly observed types of drug toxicity in humans.

Two distinct bases for these types of effects may be distinguished: **pharmacokinetic** and **pharmacodynamic**. Pharmacokinetic-based toxic effects are due to an increase in the concentration of the compound or active metabolite at the target site. This may be due to an increase in the dose, altered metabolism, or saturation of elimination processes, for example. An example is the increased hypotensive effect of **debrisoquine** in poor metabolizers where there is a genetic basis for a reduction in metabolic clearance of the drug (see chap. 5). Pharmacodynamic-based toxic effects are those where there is altered responsiveness of the target site, perhaps due to variations in the receptor. For example, individual variation in the response to **digitoxin** means that some patients suffer toxic effects after a therapeutic dose (see below and chap. 7). The inhibition of enzymes, blockade of receptors, or changes in membrane permeability, which underlie these types of effects often rely on reversible interactions. These are dependent on the concentration of the toxic compound at the site of action and possibly the concentration of an endogenous substrate if competitive inhibition is involved. Therefore, with the loss of the toxic compound from the body, by the processes of metabolism and excretion, the concentration at the site of action falls, and the normal function of the receptor or enzyme returns. This is in direct contrast to the type of toxic effect in which a cellular structure or macromolecule is permanently damaged, altered, or destroyed by a toxic compound. In some cases, however, irreversible inhibition of an enzyme may occur, which, if not fatal for the organism, will require the synthesis of new enzyme, as is the case with organophosphorus compounds, which inhibit cholinesterases.

It should also be noted that a profound physiological disturbance such as prolonged anoxia may result in irreversible pathological damage to sensitive tissue. There are many different types of response within this category with a variety of mechanisms, and it is beyond the scope of this book to examine all of these in detail. However, a brief illustration of some of the various types of response will be given, and some examples will be considered in more detail in chap. 7.

6.4.1 Anoxia
Lack of oxygen in the tissues may be due to respiratory or circulatory failure or absence of oxygen. Thus, the first situation may arise if a toxic compound affects breathing rate via central control, such as the drug **dextropropoxyphene** when taken in overdoses, or by effecting respiratory muscles such as **botulinum toxin** (see chap. 7). The second situation arises when a toxic compound inhibits oxygen transport. The classic example of this is carbon monoxide, which binds to hemoglobin in place of oxygen (see chap. 7 for more details). Another example is the oxidation of hemoglobin by nitrite; the methemoglobin produced does not carry oxygen.

6.4.2 Inhibition of Cellular Respiration
This type of effect can occur in all tissues and is caused by a metabolic inhibitor such as azide or cyanide, which inhibits the electron transport chain. Inhibition of one or more of the enzymes of the tricarboxylic acid cycle such as that caused by fluoroacetate (Fig. 6.7) also results in inhibition of cellular respiration (for more details of cyanide and fluoroacetate see chap. 7).

6.4.3 Respiratory Failure
Excessive muscular blockade may be caused by compounds such as the cholinesterase inhibitors. Such inhibitors, exemplified by the organophosphate insecticides such as malathion (chap. 5, Fig. 12) (see also chap. 7) and **nerve gases** (e.g., isopropylmethylphosphonofluoridate), cause death by blockade of respiratory muscles as a result of excess acetylcholine accumulation. This is due to inhibition of the enzymes normally responsible for the inactivation of the acetylcholine (see chap. 7). Respiratory failure may also result from the inhibition of cellular respiration by cyanide, for example, or central effects caused by drugs such as dextropropoxyphene.

6.4.4 Disturbances of the CNS
Toxic substances can interfere with normal neurotransmission in a variety of ways, either directly or indirectly, and cause various central effects. For example, cholinesterase inhibitors such as the **organophosphate insecticides** cause accumulation of excess acetylcholine. The accumulation of this neurotransmitter in the CNS in humans after exposure to toxic insecticides leads to anxiety, restlessness, insomnia, convulsions, slurred speech, and central depression of the respiratory and circulatory centers.

6.4.5 Hyper-/Hypotension
Drastic changes in blood pressure may occur as a toxic response to a foreign compound, such as the hypotension caused by hydrazoic acid and **sodium azide**. There may be various mechanisms involved such as vasodilation, badrenoceptor blockade or altered water balance. Antihypertensive drugs will clearly cause dangerous hypotension if given in large doses, but such an exaggerated response may also be due to reduced metabolism in some individuals, as in the case of debrisoquine (see chap. 5).

6.4.6 Hyper-/Hypoglycemia
Changes in blood sugar concentration can be caused by foreign compounds, and this may involve a variety of mechanisms. Drugs such as **tolbutamide**, a sulfonylurea, are used therapeutically to lower blood sugar levels. **Streptozotocin**, which destroys the pancreatic β-cells, which produce insulin, causes hyperglycemia indirectly by reducing insulin levels. **Hydrazine** the industrial chemical causes first hyperglycemia, as a result of glycogen mobilization, because of the hepatic effects and then hypoglycemia as glycogen stores are depleted and gluconeogenesis is inhibited. Similarly, salicylate poisoning causes first hyper- and then hypoglycemia as a result of mobilization of glycogen due to an increase in adrenaline followed by a depletion of available stores (see chap. 7 for more details).

6.4.7 Anesthesia
The induction of unconsciousness may be the result of exposure to excessive concentrations of toxic solvents such as carbon tetrachloride or vinyl chloride, as occasionally occurs in industrial situations (solvent narcosis). Also, volatile and nonvolatile anesthetic drugs such as halothane and thiopental, respectively, cause the same physiological effect. The mechanism(s) underlying anesthesia is not fully understood, although various theories have been proposed. Many of these have centered on the correlation between certain physicochemical properties and anesthetic potency. Thus, the oil/water partition coefficient, the ability to reduce surface tension, and the ability to induce the formation of clathrate compounds with water are all correlated with anesthetic potency. It seems that each of these characteristics are all connected to hydrophobicity, and so the site of action may be a hydrophobic region in a membrane or protein. Thus, again, physicochemical properties determine biological activity.

6.4.8 Changes in Water and Electrolyte Balance
Certain foreign compounds may cause the retention or excretion of water. Some compounds, such as the drug furosemide, are used therapeutically as diuretics. Other compounds causing diuresis are ethanol, caffeine, and certain mercury compounds such as mersalyl. Diuresis can be the result of a direct effect on the kidney, as with **mercury** compounds, which inhibit the reabsorption of chloride, whereas other diuretics such as ethanol influence the production of antidiuretic hormone by the pituitary. Changes in electrolyte balance may occur as a result of excessive excretion of an anion or cation. For example, salicylate-induced alkalosis leads to excretion of Na^+, and ethylene glycol causes the depletion of calcium, excreted as calcium oxalate.

6.4.9 Ion Transport
Alteration of the movement of ions, such as potassium, in heart tissue may be a toxic response to some foreign compounds. For example, the cardiac glycosides, such as the drug digitoxin, block the uptake of potassium into cardiac muscle cells via the Na^+ K^+ ATPase pump. Inhibition of the Na^+ K^+ pump indirectly leads to increased Ca^+ and Na^+ in the heart cell. As Ca^+ in cardiomyocytes is the signal involved in contraction of cardiomyocytes, contractions are increased. Although useful in the treatment of heart failure, large doses cause cardiotoxicity such as arrythmias and bradycardia. This is discussed in more detail in chapter 7. Diuretics may also cause low plasma potassium with potentially dangerous effects on heart function.

6.4.10 Failure of Energy Supply
Toxic compounds, which interfere with major pathways in intermediary metabolism, can lead to depletion of energy-rich intermediates. For example, fluoroacetate blocks the tricarboxylic acid cycle, giving rise to cardiac and CNS effects, which may be fatal (see chap. 7). Another example is cyanide (see chap. 7).

6.4.11 Changes in Muscle Contraction/Relaxation
This type of response may be caused by several mechanisms. For instance, the muscle relaxation induced by succinylcholine, discussed in more detail in chapter 7, is due to blockade of neuromuscular transmission. Alternatively, acetylcholine antagonists such as **tubocurarine** may compete for the receptor site at the skeletal muscle end plate, leading to paralysis of the skeletal muscle. **Botulinum toxin** binds to nerve terminals and prevents the release of acetylcholine; the muscle behaves as if denervated, and there is paralysis. This will be discussed in more detail in chapter 7.

6.4.12 Hypo-/Hyperthermia
Certain foreign compounds can cause changes in body temperature, which may become a toxic response if they are extreme. Substances such as **2,4-dinitrophenol** and salicylic acid will raise body temperature, as they uncouple mitochondrial oxidative phosphorylation. Thus, the energy normally directed into ATP during oxidative phosphorylation is released as heat.
Substances that cause vasodilation may cause a decrease in body temperature.

6.4.13 Heightened Sensitivity

Certain foreign compounds may cause an increase in sensitivity of a particular tissue to endogenous substances such as catecholamines. For instance, exposure to some **halogenated hydrocarbons**, such as the fluorinated and chlorinated compounds used in aerosol spray cans and fire extinguishers, may cause heightened sensitivity to catecholamines. The heart tissue is sensitized by exposure to the halogenated solvents, and then if exposure is followed by a fright or other situation—when catecholamines such as adrenaline are released into the body—the heart may fail. A number of cases of sudden death from heart failure in teenagers and young, and otherwise healthy adults, have occurred because of this type of toxicity. This type of toxic effect became known particularly as it occurred during glue sniffing or **solvent abuse**. However, only certain halogenated solvents cause this type of toxicity.

6.5 DEVELOPMENTAL TOXICOLOGY—TERATOGENESIS

6.5.1 Introduction

Alterations to the development of the embryo, fetus, or neonatal animal, either morphological or functional, and caused by chemicals is termed "developmental toxicology." The result may involve interference with normal growth, homeostasis, development, differentiation, and/or behavior.

Teratology/teratogenesis is that aspect concerned with the production of malformations.

Teratogenesis involves interference with the normal development of either the embryo or fetus in utero, giving rise to abnormalities in the neonate. This interference may take many forms, and there is therefore no general mechanism underlying this type of response. Many of the toxic effects described elsewhere in this book may be teratogenic in the appropriate circumstances.

Teratogenic agents may be drugs taken during pregnancy; radiation, both ionizing and nonionizing; environmental pollutants; chemical hazards in the workplace; dietary deficiencies; and natural contaminants.

Although mutations occurring in germ cells may give rise to abnormalities in the neonate, such as Down syndrome, teratogenicity is normally confined to the effect of foreign agents on somatic cells within either the developing embryo or fetus and the consequent effects on that individual, rather than inherited defects. However, the effects of foreign compounds on germs cells will also be considered within the context of teratogenesis in this section.

Embryogenesis is a very complex process involving cell proliferation, differentiation, migration, and organogenesis. This sequence of events (Fig. 6.21) is controlled by information transcribed and translated from DNA and RNA, respectively, (Fig. 6.38) in a time-dependent manner.

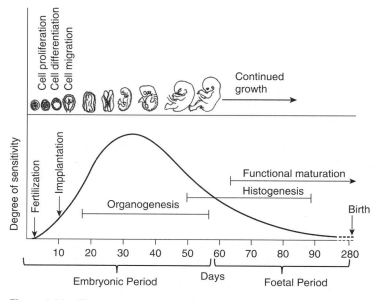

Figure 6.21   The stages of embryogenesis.

Although embryogenesis is not fully understood and is a highly complex process, with the knowledge available, certain predictions may be made regarding teratogenesis:

1. The sequence of events in embryogenesis could be easily disturbed, its interrelation-ships readily disrupted, and such interference could be very specific.
2. The timing of the interference with the process of embryogenesis would be very important to the final expression of the teratogenic effect.

Both of these predictions are found to be correct, and the teratogenic effect may be out of all proportion to the initiating event, if this event occurs at a critical time. This teratogenic effect might simply be a slowing down in the development of one group of cells, which are then out of synchrony with the overall process of embryonic development. For example, a deficiency of folic acid in pregnant rats may cause reno-uteral defects in the neonate because migration of certain primordial cell groups is disturbed by this deficiency. The cell groups are then not able to attain their correct position at a later stage.

A further prediction might be that any effect or foreign compound, which interfered with biochemical pathways, particularly transcription and translation, damaged macromolecules, caused a deficiency of essential cofactors, caused a reduction in energy supply, or produced direct tissue damage, would probably be teratogenic. Thus, ionizing radiation, which causes mutations and other effects, will readily cause birth defects if the pregnant animal is irradiated at the appropriate time. Preexisting mutations derived from the maternal or paternal germ cells may also result in malformations, of course.

The mechanisms underlying teratogenesis are therefore many and varied, and true understanding will only come with greater understanding of the remarkable but complex process of embryogenesis. However, the characteristics of teratogenesis can be examined, and general mechanisms described.

6.5.2 Characteristics of Teratogenesis

Selectivity
Teratogens interfere with either embryonic or fetal development, but often do not affect the placenta or maternal organism. They are therefore selectively, either embryotoxic or fetotoxic, giving rise to manifestations of such toxicity up to and including death, with subsequent abortion. The most potent teratogens may be regarded as those with no observable toxicity to the maternal system, but causing malformations in the fetus rather than death, such as the now infamous drug thalidomide. Clearly, most teratogens will cause fetal death at high-enough doses and probably maternal toxicity also. Certain compounds, such as **colchicine**, which cause fetal death, do not cause malformations, however.

Genetic Influences
Observations that species and strain differences exist in the susceptibility to certain teratogens suggest that genetic factors may be involved in teratogenesis. Similarly, it seems clear that in some cases at least a teratogen may increase the frequency of a naturally occurring abnormality. These genetic factors may simply be differences in the maternal metabolism or distribution of the compound, which lead to variation in the exposure to the ultimate teratogenic agent.

It is also clear from the previous discussion, however, that the instructions for the complex series of events, which constitute organogenesis, are probably coded in DNA, and conse-quently, mutations or damage to DNA will be expected to cause certain abnormalities if that information is transcribed and translated. Also, genetic susceptibility to mutation or chromo-some breakage may be factors. For example, the vitamin antagonist **6-aminonicotinamide** causes cleft palate and chromosomal abnormalities in mice. The chromosomal abnormalities are found in the somatic cells of a number of tissues, leading to faulty mitosis. This is supported by the fact that in some cases only one or two embryos are malformed out of several in multiparous animals, and this indicates that the embryonic genotype may sometimes be an important factor. However, although many mutagens and carcinogens are teratogenic, not all are, suggesting that genetic factors are not always involved in the underlying mechanisms of teratogenesis.

Susceptibility and Development Stage

The susceptibility of both an embryo and a fetus to a teratogen is variable, depending on the stage of development when exposure occurs. For gross anatomical abnormalities, the critical periods of organogenesis are the most susceptible to exposure, whereas other types of abnormality may have other critical periods for exposure.

After fertilization, the cells divide, giving rise to a blastocyst. During this time, there is little morphological differentiation of cells, except that some are located on the surfaces and others internally. The development of the blastocyst gives an internal cavity (the blastocoele) and hence further surfaces and positional differences. However, there are few specific teratogenic effects, which occur at this time, the major effects being death or overall developmental retardation.

The appearance of the embryonic germ layers, the ectoderm, endoderm, and mesoderm, is the next stage, with the gross segregation of cells into groups. Damage at this stage may be associated with specific effects.

The chemical determination, which precedes structural differentiation, is still not understood, but it is probable that this period is sensitive to interference.

Organogenesis, which is the segregation of cells, cell groups, and tissues into primordia destined to be organs, is particularly sensitive to teratogens although not exclusively so. Histological differentiation occurs concurrently with, and continues after, organogenesis, as does acquisition of function. Both of these stages may be susceptible to teratogens, and exposure to them may lead to defects, although generally these defects are not gross structural ones.

The sensitive period for induction of malformations is the 5- to 14-day period in the rat and mouse and the third week to the third month in humans. This is illustrated in Figure 6.22 for the teratogen **actinomycin D**. The later period of fetal development, like the initial proliferative stage, is less susceptible to specific effects and an "all-or-none" type of response is usually seen, such as either death or no gross effect.

Specificity

Different types of teratogens may give similar abnormalities if given during the same critical periods, and conversely, the same teratogen given at different times may produce different effects. Therefore, the particular abnormality may represent interference with a specific developmental process (Fig. 6.23). Although there is a wide variety of possible biochemical effects or structural changes at the molecular level, these may be manifested as relatively few types of abnormal embryogenesis, and hence similar defects may result from different teratogens. Thus, excessive cell death, incorrect cellular interaction, reduced biosynthesis of

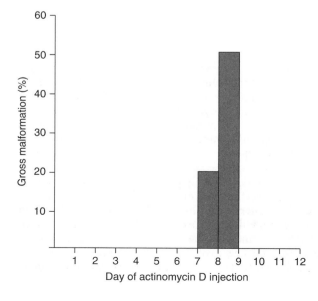

Figure 6.22 Critical timing of the teratogenic effects of actinomycin D in the rat. The histogram shows the percentage of gross malformations among surviving fetuses after a dose of 75 µg/kg, i.p. *Source*: From Ref. 6.

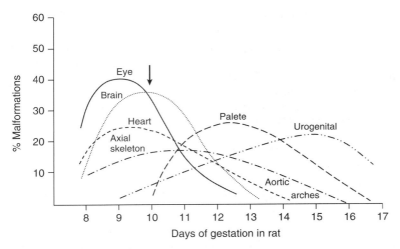

Figure 6.23   Periods of peak sensitivity to teratogens in the rat. A teratogen administered at the time shown by the arrow would cause a mixture of malformations. It would particularly damage the eyes and brain but would have little or no effect on the palate. *Source*: Adapted from Ref. 7.

RNA, DNA, or protein might be the primary manifestations of the teratogenic effect, but all give rise to the impaired migration of cells or cell groups. The eventual effect might conceivably be too few cells or cell products at a particular site for normal morphogenesis or maturation to proceed.

Manifestations of Abnormal Development
The final consequences of abnormal development are death, malformation, growth retardation and functional disorder. Each of these consequences follows exposure at different stages. Thus fetal death may result from exposure at an early or late stage of development to high concentrations of the toxin, without evidence of malformation. This may be due to substantial damage to the undifferentiated cells in the early stages or interference with some physiological mechanism in the late stage of development.

Malformations tend to occur following exposure during organogenesis, whereas functional disorders would be expected at later stages. The period of fetal growth is susceptible, and agents acting at this time may cause growth retardation. Growth retardation may reflect a variety of functional deficiencies, and both of these may occur without any sign of malformations.

Access to the Embryo and Fetus
For foreign compounds, in contrast to ionizing and other radiation or mechanical force, the route of access is via the maternal body. This is either through the fluids surrounding the embryo or via the blood, following the formation of the placenta. The blood from the maternal circulation enters a sinus into which the vessels of the embryo protrude like fingers (Fig. 6.24). Consequently, there is rapid and efficient exchange between maternal and embryonic blood for most foreign substances, provided they are not bound to large proteins and have a molecular weight less than about 1000. Some proteins do, however, seem to cross the placenta, and phagocytosis may play a role here.

Lipophilic compounds will cross the placenta most readily by passive diffusion, whereas those ionized at plasma pH will generally not, unless they are substrates for a carrier-mediated transport system. Most foreign compounds will enter the embryonic bloodstream by passive diffusion, and the exposure of the embryo of fetus will therefore depend on the concentration of the compound in the maternal bloodstream and the blood flow, which increases as pregnancy progresses. The ability of the maternal organism to metabolize and excrete the compound, reflected in the plasma level and half-life, will therefore be a major determinant of

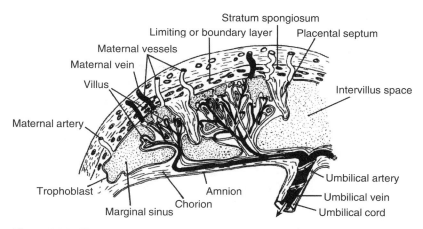

Figure 6.24   The structure and blood supply of the mammalian placenta. *Source*: From Ref. 8.

the exposure of the embryo or fetus to the compound. Similarly, metabolites of the compound may also enter the embryonic bloodstream provided they have the necessary physicochemical characteristics.

However, the embryo or fetus may have the ability to metabolize the compound itself or further metabolize metabolites. This is especially the case in humans where the fetal liver at mid-gestation has 20% to 40% of adult activity for phase 1 reactions. If the metabolites produced by the embryo or fetus are polar or large, they may become trapped inside the embryonic system as they will be unable to diffuse across the placenta into the maternal bloodstream. Similarly, if the compound or metabolites entering the embryo bind to or react with proteins or other macromolecules, this process may effectively entrap the compound inside the embryonic system. Either of these processes may cause a concentration gradient to exist across the placenta and will prolong embryonic or fetal exposure to the compound.

Dose Response

It is generally accepted that there is a true no-effect level for teratogenesis, certainly as far as death and malformations are concerned, because of the presence and influence of the maternal organism. Thus, it is clear that a critical number of cells may have to be damaged before the lethality or malformation becomes apparent. For potent teratogens, the dose-response curve will be steep and displaced from the dose-response curve for the maternal organism. Three different types of dose-response relationship can be identified, depending on the endpoint or manifestation, which is measured. Thus, in the first type, malformation, growth retardation, and lethality may each show dose-response curves, which are displaced (Fig. 6.25), each being a separate manifestation, dependent on a different mechanism. For example, thalidomide is only embryolethal at doses several times higher than those required to produce malformations. Alternatively, as in the second type, one or more of these manifestations may show a similar dose-response curve (Fig. 6.25), as is the case with **actinomycin D** where lethality and malformations occur at the same doses (see Fig. 7.72). Thus, the responses may be different degrees of manifestation of the initial insult. In the third type, malformations may not occur at all, simply growth retardation and lethality, the latter showing a steeper dose-response curve than the former (Fig. 6.25). As with the first type, these two responses are separate manifestations of the insult. Thus, the drugs **chloramphenicol** and **thiamphenicol**, which inhibit mitochondrial function, either cause growth retardation or embryolethality, which shows a very steep dose-response curve. There is no basis for selective effect, and therefore malformations in particular organs and tissues do not occur. However, it is clear that the relationships between lethality, malformation, and growth retardation are complex and will vary with the type of teratogen, dose, and time of dosing. Most foreign compounds will be embryotoxic if given in sufficient doses at an appropriate time.

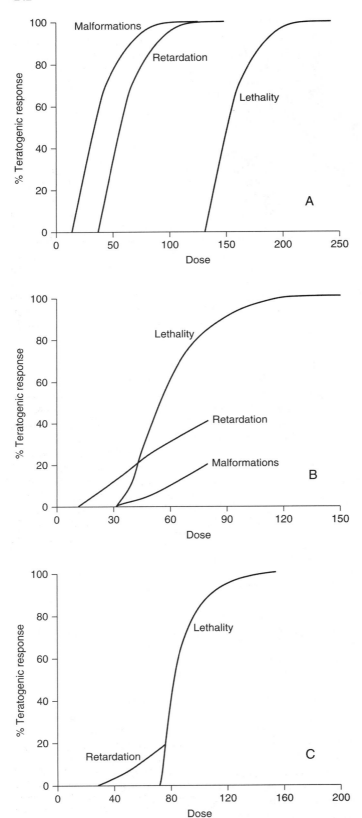

Figure 6.25 Dose-response relationship patterns for different types of teratogen. For explanation see text. *Source*: Adapted from Ref. 9.

6.5.3 Mechanisms of Teratogenesis

Because the process of development of the embryo and then the fetus is a complex series of events, it is clear that many different mechanisms may cause the disruption that underlies the teratogenic effects of foreign compounds. In many cases, the initiating events probably occur at the subcellular or molecular level in the embryo, but may be only detectable as malformations or cell death.

The various known and postulated underlying mechanisms are discussed below.

Cytotoxicity

Direct damage to cells leading to cell death and necrosis is one suggested cause of teratogenesis. Thus, cytotoxic compounds, such as the alkylating agent **N-methyl-N'-nitro-N-nitrosoguanidine** (MNNG), are teratogenic. This compound causes a spectrum of malformations, embryolethality, and growth retardation (Fig. 6.26). Thus, cell death and reduced proliferation specifically in the target tissue, which may be a limb bud or tissue destined to be an organ, will give rise to a malformation in that tissue if there is insufficient time for replacement. There are many underlying causes of cytotoxicity as already discussed. Generalized cell death is more likely to lead to embryolethality, if extensive, or growth retardation if the embryo is at an early stage where cells can be replaced by compensatory hyperplasia. After a single exposure to MNNG, fetuses within a single litter may show all three outcomes, and this therefore is an example of the second type of dose-response relationship (see above). In the case of MNNG, the necrosis of limb bud cells, or interference with or damage to DNA, may be involved in the teratogenicity.

Receptor-Mediated Teratogenicity

These are highly specific mechanisms, which will not involve extensive embryolethality or growth retardation, but a well-defined structural malformation. An example, where the receptor and mechanism are partly understood, is of the **glucocorticoids**, which are involved in normal growth and differentiation of embryonic tissue. After doses of glucocorticoids leading to exposure of the embryo to high levels relative to the normal physiological concentration, malformations such as cleft palate occur. There is a cytosolic receptor for glucocorticoids, and the level of these receptors correlates with the susceptibility of the animal to teratogenicity.

A foreign compound, which may act in this way is the herbicide **nitrofen** (2,4-dichloro-4'-nitro diphenyl ether), which causes a variety of malformations lethal to the neonate. There is no growth retardation or embryolethality at doses, which are teratogenic, however (Fig. 6.27), and therefore this exhibits the first type of dose-response relationship (see above). The

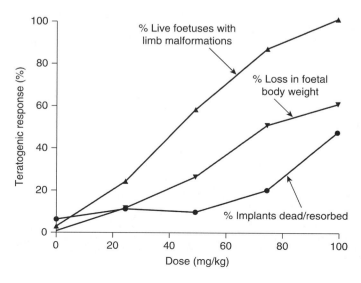

Figure 6.26 The dose-response patterns for teratogenicity caused by the cytotoxic agent MNNG. Pregnant mice were treated on day 11 with MNNG. *Abbreviation*: MNNG, N-methyl-N'-nitro-N-nitrosoguanidine. *Source*: From Ref. 9.

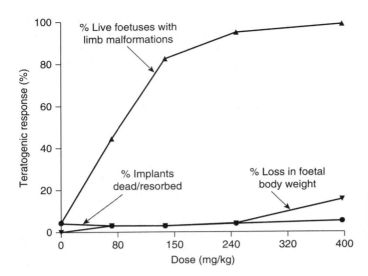

Figure 6.27 The dose-response patterns for teratogenicity caused by the herbicide nitrofen (2,4-dichloro-4′-nitro diphenyl ether). Pregnant rats were treated on day 11 with nitrofen. *Source*: From Ref. 9.

mechanism may involve the production of a metabolite, which interferes with the thyroid hormone status. The metabolite, which results from reduction and hydroxylation of the parent compound, was detected in embryonic tissue and cross-reacted with antibodies to tri-iodothyronine (T3). Thus, the metabolite may have T3 activity and exert this in the embryo.

Mutation

It is clear that mutations are the basis of many errors of development; maybe 20% to 30% of those occurring in humans are due to heritable mutations in the germ line. Mutations may occur in the genetic material of either the germ cells or the somatic cells. Germ cell mutations are heritable, whereas effects in somatic cells are only apparent in the fetus exposed. Somatic cell mutations will probably not lead to gross manifestations of damage such as malformations, unless a sufficiently large number of progeny cells are affected. Consequently, somatic cell mutations are probably an infrequent cause of fetal abnormality. Some somatic mutations, however, may be incompatible with the viability of the affected cell or group of cells. This might be less damaging than a viable mutant cell, which was the ancestor of an important organ or structure, however.

Not all mutagens are teratogens, but the cytotoxic alkylating agent MNNG mentioned above is a potent teratogen and causes replication-dependent mutations. As cell proliferation is important in embryogenesis, a mutation, which is induced in a sufficient number of critical cells and results in cell death or altered growth characteristics, for example, is potentially able to cause a malformation. Clearly, the nature of the mutation and the number and position of the cells affected will all be important factors.

Chromosomal Aberrations

Aberrations in chromosomes or chromatids, which are sometimes microscopically visible, may arise during mitotic division when newly divided chromosomes fail to separate or do so incorrectly. The absence of a chromosome is usually lethal, and an excess is often poorly tolerated, giving rise to serious defects. Aberrations of the sex chromosomes are more readily tolerated, however. Chromosome aberrations may be caused by foreign compounds as indicated in the section on mutagenesis (see chap. 6). However, those cells with aberrations seem to be rapidly eliminated and so may contribute to cell death rather than a heritable mutation.

Mitotic Interference

As embryogenesis involves extensive cellular proliferation, any interference with this process is potentially teratogenic. Interference with spindle formation, inhibition or arrest of DNA synthesis, or the incorrect separation of chromatids all fall into this category. Inhibition of DNA synthesis, such as that caused by **cytosine arabinoside**, slows or arrests mitosis, which cannot progress beyond the S-phase. Thus, those areas of the embryo, which show extensive cellular proliferation, are the most susceptible to both necrosis and subsequent malformation from cytotoxic compounds such as cytosine arabinoside.

Inhibition of spindle formation such as that caused by **vincristine** or **colchicine** stops separation of chromosomes at anaphase (see chap. 6). Proper separation of chromatids may not occur because of "stickiness" or bridging between the chromatids. Clearly, interference with mitosis, and hence cell proliferation, is an important cause of teratogenic effects.

Interference with Nucleic Acids
A number of antineoplastic and antibiotic drugs, which interfere with nucleic acid function, are teratogenic. Such effects as changes in replication, transcription, the incorporation of bases, and translation occurring in somatic cells are nonheritable and may be embryotoxic. However, inhibition of DNA synthesis or DNA damage does not seem to be as important as cell death, and significant inhibition of DNA synthesis alone usually seems to be insufficient to cause malformations. Interference with protein synthesis is generally lethal to the embryo rather than malformation causing.

Examples of teratogenic compounds, which interfere with nucleic acid metabolism, are cytosine arabinoside, which inhibits DNA polymerase, **mitomycin C**, which causes cross-linking, and **6-mercaptopurine**, which blocks the incorporation of the precursors, adenylate and guanylate. **Actinomycin D**, a well known teratogen, intercalates with DNA and binds deoxyguanosine, interfering with RNA transcription and causing erroneous base incorporation. **8-Azaguanine** is a teratogenic analogue of guanine.

Compounds such as **puromycin** and **cycloheximide**, which block rRNA transfer, and **streptomycin** and **lincomycin**, which cause misreading of messenger RNA (mRNA), block protein synthesis and are therefore often embryolethal.

Substrate Deficiency
If the requirements for the growth and development of the fetus are withheld, a disruption to these processes may occur and damage may ensure. Deficiencies in essential substrates, such as folic acid, may be caused by dietary lack or by substrate analogues.

Failure in the placental transport of essential substrates may be teratogenic and can be caused by certain compounds such as azo dyes. This has, however, only been demonstrated in rodents because of the inverted yolk sac type of placenta such animals have.

Deficiency of vitamins, such as folic acid, is highly teratogenic, as essential synthetic metabolic pathways are blocked or reduced. This may be caused by the administration of specific vitamin analogues or antagonists as well as by a failure in supply.

Deficiency of Energy Supply
The rapidly proliferating and differentiating tissue of the embryo would be expected to require high levels of energy, and therefore interference with its supply, not surprisingly, may be a teratogenic action. A deficiency of glucose due to dietary factors or due to hypoglycemia, which may be induced by foreign compounds, is teratogenic. Interference in glycolysis, such as that caused by **6-aminonicotinamide**, inhibition of the tricarboxylic acid cycle as caused by fluoroacetate (see chap. 7), and impairment or blockade of the terminal electron transport system as caused by hypoxia or cyanide, all cause abnormal fetal development. For example, chloramphenicol and thiamphenicol both interfere with the mitochondria and cause ATP depletion, inhibition of mitochondrial respiration, and cytochrome oxidase activity. They cause embryolethality and growth retardation rather than malformations. This may be because they affect all cells nonspecifically, and hence slow growth either overall or at higher doses is lethal to a sufficient number of cells to lead to death of the embryo. The dose-response curve for embryolethality is very steep (0–100% mortality between 100 and 125 mg kg^{-1} for **thiamphenicol**), depending on a threshold of critical energy depletion. This is the third type of dose-response relationship, as discussed where malformations do not normally occur.

Inhibition of Enzymes

Teratogens in this category are also included in some of the previous categories. Enzymes of central importance in intermediary metabolism are particularly vulnerable, such as dihydrofolate reductase, which may be inhibited by folate antagonists, giving rise to a deficiency in folic acid. **6-Aminonicotinamide**, which inhibits glucose-6-phosphate dehydrogenase, an important enzyme in the pentose phosphate pathway, is a potent teratogen.

5-Fluorouracil (chap. 3, Fig. 6), an inhibitor of thymidylate synthetase, and cytosine arabinoside, which inhibits DNA polymerase, are both teratogens, which have already been referred to.

Changes in Osmolarity

Various conditions and agents may change the osmolarity within the developing embryo and thereby disrupt embryogenesis. Thus, induction of hypoxia may cause hypo-osmolarity, which leads to changes in fluid concentrations. This causes changes in pressure, and the consequent disruption of tissues may result in abnormal embryogenesis. Other agents causing osmolar changes are hypertonic solutions, certain hormones, and compounds such as the azo dye, **Trypan blue**, and **benzhydryl piperazine**. Trypan blue has been widely studied and causes fluid changes in rodent embryos, leading to malformations of the brain, eyes, vertebral column, and cardiovascular system. It may also have other effects, however, such as inhibition of lysosomal enzymes, which may interfere with release of nutrients from the yolk sac, impairing fetal nutrition. Benzhydryl piperazine causes orofacial malformations in rat embryos, possibly the result of the edema induced in the embryo. It has also been suggested that changes in embryonic pH may underlie the teratogenicity of certain compounds.

Membrane Permeability Changes

Changes in membrane permeability might be expected to lead to osmolar imbalance and fetal abnormality. This is hypothetical, as there are no real examples of such a mechanism, although high doses of **vitamin A** are teratogenic and may cause ultrastructural membrane damage.

Changes in Maternal and Placental Homeostasis

As well as direct effects on the embryo or fetus, foreign compounds can also be teratogenic by influencing the maternal organism or the placenta. Thus, maternal malnutrition or protein deficiency, or a deficiency in one or more vitamins or minerals may lead to effects on the embryo and fetus. As might be expected, maternal malnutrition will tend to cause growth retardation. However, vitamin A and folic acid deficiencies cause malformations and embryolethality as well as growth retardation. Trypan blue is believed to be teratogenic to animals with a yolk sac placenta by interfering with the nutrition of the embryo (see above).

A permanent, but partial, decrease in placental blood flow will lead to growth retardation in the embryo. A total cessation of blood flow for a short time (3 hours) during organogenesis will cause malformations. Thus, foreign substances, which cause vasoconstriction in the placenta, may give rise to effects on the embryo indirectly. Exposure of pregnant rabbits to **hydroxyurea** rapidly results in a large reduction in placental blood flow (77%), which may be the cause of craniofacial and cardiac hemorrhages.

6.5.4 Role of Metabolic Activation

There is clear evidence from many different sources that the metabolism of compounds may be involved in their teratogenic effects, as will be seen in the final chapter in the discussion of thalidomide and diphenylhydantoin teratogenicity. The embryo and fetus of some species clearly have metabolic activity toward foreign compounds, which may be inducible by other foreign compounds.

Thus, fetal liver from primates has a more well-developed metabolic system for xenobiotics than does that from rodents and rabbits, for example. This may be due to the late development of the smooth endoplasmic reticulum and therefore of cytochrome(s) P-450 in the latter species. The use of metabolic inducers and inhibitors *in vivo* and the use of metabolizing systems with embryo or limb bud culture *in vitro* have all indicated that for some teratogens,

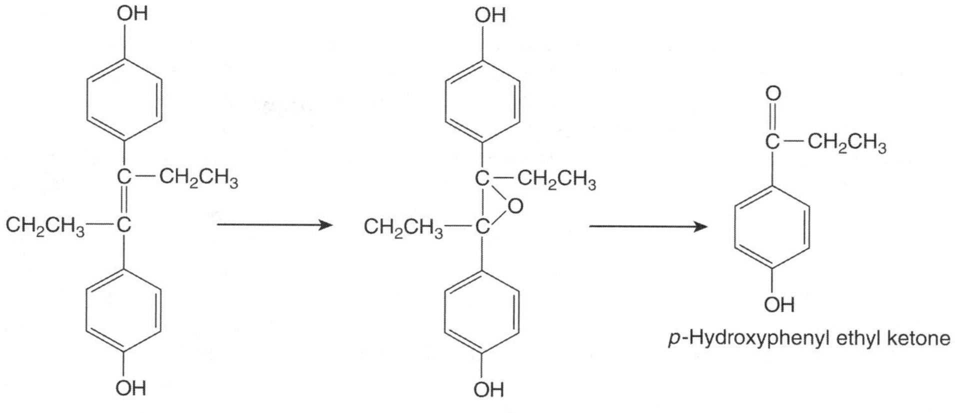

Figure 6.28 Metabolism of diethylstilboestrol via an epoxide intermediate. This potentially reactive intermediate may show an affinity for the estradiol receptor and thereby accumulate in oestrogen target organs. This may facilitate reaction with DNA in these organs. *Source*: From Ref. 10.

metabolism is involved. Metabolic activation in the maternal organism or in the embryo itself may both occur and have different roles in the eventual toxicity.

A good example of a compound that is a teratogen and requires metabolic activation is the anticancer drug **cyclophosphamide**, which has been studied extensively both *in vivo* and *in vitro*.

However, just as with other toxic effects, either the parent drug or a metabolite may be responsible for embryotoxicity, but it is often difficult to predict which, without substantial metabolic and biochemical data being available.

6.5.5 Transplacental Carcinogenesis
Brief mention should be made of the phenomenon of transplacental carcinogenesis. This is the induction of cancer in the offspring by exposure of the pregnant female. It may not occur in the fetus or even in the neonate, but may be evident only in adulthood. The best known example of this is the appearance of vaginal cancer in human females born of mothers given the drug **diethylstilboestrol** (Fig. 6.28) during pregnancy. The vaginal cancer did not appear in the female offspring until puberty.

6.5.6 Male-Mediated Teratogenesis
This is a controversial area with regard to humans where there is currently little hard data. Theoretically it is possible for a foreign compound to cause mutations in male germ cells, which result in malformations or the development of abnormal offspring. This is similar to the situation in which inherited mutations or chromosomal aberrations lead to the birth of abnormal offspring, such as occur in Down syndrome, for example, where an extra chromosome occurs (Trisomy 21).

Although this has been shown to occur in experimental animals after exposure of males to foreign compounds such as **cyclophosphamide**, there is only inconclusive evidence that this occurs in humans. Thus, studies of exposure of human males to **vinyl chloride, dibromo-chloropropane,** and anesthetic gases, for example, have revealed only equivocal evidence of developmental toxicity in the offspring. There now seems to be some evidence that the leukemia occurring in children, which appears to be clustered around nuclear fuel-reprocessing plants such as Sellafield in the United Kingdom, may be due to paternal exposure to radiation.

6.6 IMMUNOTOXICITY

It is beyond the scope of this book to consider the subject of immunology in any detail, but a brief introduction will be useful. Further details can be found in the books, chapters, and references in the References section.

The immune system functions to protect the organism against infection, foreign proteins, and neoplastic cells. It is organized into primary lymphoid tissue such as bone marrow and thymus and secondary lymphoid tissues such as spleen and lymph nodes. The cells of the immune system are leukocytes and a specific type produces immunoglobulins.

Leukocytes are derived from stem cells in the bone marrow. These cells become functionally mature cells: granulocytes, lymphocytes, and macrophages. Lymphocytes divided into T cells (thymus derived) and B cells (bursa equivalent), depending on the site of maturation.

The T cells may be subdivided into T-helper and T-cytotoxic cells. The B lymphocytes give rise to plasma cells, which are responsible for the production of antibodies (immunoglobulins). The immune system uses two mechanisms, a nonspecific resistance and specific, acquired immunity. The nonspecific mechanism does not require prior contact and involves macrophages, neutrophils, eosinophils, natural killer (NK) cells, and dendritic cells. Some of these cells are phagocytic cells and can eliminate antigens by engulfing them or by causing lysis or stimulating apoptosis.

The specific mechanism involves the lymphocytes and also macrophages and requires prior contact with the foreign substance. Macrophages and dendritic cells present the antigen to T cells. Following this initial contact and the establishment of memory of the foreign substance or organism, the second contact evokes an immune response. However, there are links between the specific and nonspecific mechanisms such as the presentation of antigens by lymphocytes to macrophages and the activation of macrophages by lymphokines produced by sensitized T lymphocytes.

Thus, the immune system is spread among different organs and cell types. The system is capable of expression of an almost limitless variety of antigen receptors and will cause effects all over the body and at different sites.

There are two main types of immune response: cell mediated and humoral. Cell-mediated immunity involves specifically sensitized thymus-dependent lymphocytes. Humoral immunity involves the production of antibodies (immunoglobulins) from lymphocytes or plasma cells. The mechanisms will be discussed in more detail below.

Immunotoxicity is a relatively new area of mechanistic toxicology, and some aspects are still rather poorly understood.

Chemical-Induced Immunotoxicity
The immune system can be affected by drugs and chemicals in four ways:

Immunosuppression
Immunostimulation
Hypersensitivity
Autoimmunity

6.6.1 Immunosuppression
Both the nonspecific and specific components of the immune system can be suppressed by chemicals, including drugs. It involves the suppression of maturation and development of immune cells. Both T and B cells develop in the bone marrow and thymus. This involves a complex series of changes in relation to antigen receptors and recognition. Chemicals can affect these processes, leading to a decrease in the number of mature T and B cells. This will result in inhibition of both the humoral and cellular responses.

If the damage is to the bone marrow where all the blood cells develop, this causes pancytopenia. In this case, both T- and B-precursor cells are damaged so that the system is compromised as well as other blood cells, including neutrophils.

Benzene is such a chemical that damages the bone marrow, and aplastic anemia results. This has several effects, one of which is a reduction in the lymphocyte as well as red cell population and so pancytopenia and therefore immunosuppression results. This leads to an increased susceptibility to infection and inhibited immunoglobulin production. These effects have been detected in both experimental animals and humans occupationally exposed to benzene. Thus, shoe workers in Turkey and China have been found to suffer aplastic anemia.

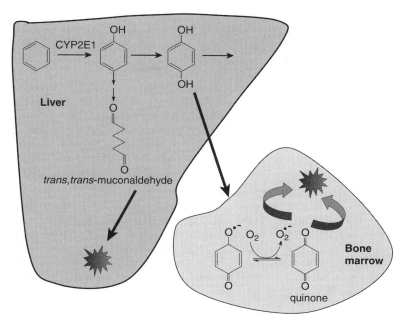

Figure 6.29   Metabolic activation of benzene to reactive metabolites, which may damage bone marrow and other tissues.

With chronic, higher level exposure, workers may suffer leukemia. **Benzene** is metabolized in the liver to a hydroquinone, which in the bone marrow is converted by the action of myeloperoxidase to the quinone and semiquinone products, which are reactive and therefore toxic. An unsaturated aldehyde, muconaldehyde is also a reactive metabolite, which could be involved in the toxicity (Fig. 6.29).

Drugs that cause damage to proliferating cells such as those in bone marrow, will cause a similar effect to benzene. Thus cyclophosphamide, the anticancer drug, inhibits the clonal expansion of T- and B-cell precursors in the bone marrow and causes anemia.

Some chemicals can affect other parts of the immune system. For example, polychlorinated hydrocarbons such as dioxins, dibenzofurans, and polychlorinated biphenyls (PCBs) damage the thymus, which is a lymphoid organ, producing mature T lymphocytes from the precursor cells, which are produced in the bone marrow. The result is depletion of the T cells in the thymus.

Thus, human exposure to PCBs in Japan (Yusho accident) and China has been associated with increased respiratory infections and decreased levels of immunoglobulins in serum. In animals exposed to these compounds, there is atrophy of both primary and secondary lymphoid organs, lower circulating immunoglobulins, and decreased antibody responses after exposure to antigens. Similarly, the exposure of both humans and farm animals to polybrominated biphenyls, which occurred in Michigan in 1973, resulted in depressed immune responses.

A particularly potent immunosuppressive chemical is **TCDD, dioxin**. This inhibits the differentiation of T cells by damaging the epithelial cells in the cortex of the thymus. These cells are involved in the maturation of T cells. A receptor is involved with this toxic effect, the AHR receptor, to which TCDD binds very strongly. The receptor is expressed in the thymus. Mice, which to do not express this receptor, do not show this particular toxic effect of TCDD even at 10 times higher doses.

The mechanism seems to involve activation of Ca^{2+} endonucleases, which causes apoptosis in the thymocytes.

Both CD4 and CD8 cells are affected but only $CD4^+/CD8^-$ and $CD4^-/CD8^+$ and not $CD4^+/CD8^+$.

There are many other compounds that have been shown to cause immunosuppression such as organophosphorus compounds, ozone, metals, and organotin compounds. Dialkyltin compounds seem selectively toxic to the thymus and hence affect T-cell function without affecting B-cell function or causing myelotoxicity or toxicity to non-lymphoid tissue.

Some drugs are designed to specifically interfere with the immune system, to protect against transplant rejection. These are immunosuppressive drugs such as cyclosporin and rapamycin. Cyclosporin interferes with the maturation of T and B cells.

Also, it interferes with a chain of events, which initiates the immune response. This involves interaction of T-helper cells with the antigen-presenting cells (APCs), which causes activation, proliferation, and differentiation of T cells. A chain of events, which is necessary for this, includes receptor interactions and interaction with cytokines and intracellular transduction molecules.

6.6.2 Immunoenhancement

This is the reverse of immunosuppression and is the situation where the immune response is enhanced. Some drugs can cause this such as levamisole and interferon.

Some of the new biological drugs are proteins specifically design to have stimulatory effects on the immune system. For example, **TGN 1412**, a novel superagonist anti-CD28 monoclonal antibody directly stimulates immune response and expands T cells. It activates and expands type 2 T-helper cells and $CD4^+$ and $CD25^+$ regulatory T cells. However, in the clinical trials in 2006 of this new drug in human volunteers, the subjects suffered major adverse effects. These included multi-organ failure and were caused by cytokine release syndrome. Hence, increased capillary permeability and cell leakage along with metabolic changes (lactic acidosis) and various physiological changes all occurred, followed by inflammation and pathological damage to organs. The circulatory changes resulted in gangrene in at least one of the victims.

This is believed to have occurred because the dosage used was too high, and this resulted in an overstimulated immune system. It can be seen from this example that the immune system can do immense damage to the organism if it overreacts.

6.6.3 Hypersensitivity

One of the functions of the immune system is to protect the body from foreign invasion by dangerous pathogens (e.g., bacteria) recognized as "not self." It does this by using a mixture of the innate immune and specific acquired immune systems.

Immunotoxicity of this type can result from faulty recognition of not self or an overreaction. Alternatively, an unwanted or exaggerated immune response toward a harmless antigen may occur and could be severe (e.g., fatal anaphylactic shock in response to peanuts).

Hypersensitivity reactions are usually idiosyncratic and difficult, if not impossible, to predict in safety evaluation studies in experimental animals or humans.

For the immune system to operate, there are three basic requirements:

1. Recognition of an "agent" as foreign or not self
2. Mount an appropriate response
3. Retain a memory of the foreign agent

To rapidly recognize and attack and preferably destroy or inactivate a foreign invader, the system has to have a template in its memory derived from an earlier exposure. Consequently, immune reactions require *at least two exposures* and sometimes more. The first exposure is the sensitization stage, followed by an elicitation stage when a subsequent exposure(s) occurs. Frequently, repeated or long-term exposure is necessary.

For the agent or antigen to be recognized as nonself, usually a molecular weight of greater than 100 and probably about 3000 Da is necessary. With the exception of the newest type of drugs, which are protein derived, such as monoclonal antibodies (e.g., TGN1412, see above), most traditional drugs are small molecules. There are, of course, many larger molecules to which we are exposed such as plant and animal proteins in our diet, but these are generally not presented to the immune system and do not in most individuals cause problems. Some individuals are sensitive, however, such as to "foreign" proteins in milk and those on the outside of peanuts.

In order to explain the stimulation of an immune response by small drug molecules, Landsteiner in the 1930's showed that small molecules *could* stimulate an immune reaction *if* combined with a protein. This became the hapten hypothesis, that small molecules could be

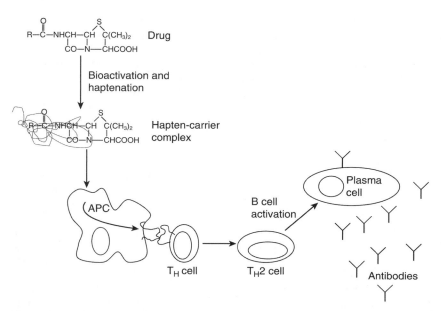

Figure 6.30 General mechanism of interaction of drug metabolite or breakdown product (hapten) with protein and production of antibodies, which recognize the subsequent antigen produced in the APC. T-cells and B-cells are types of lymphocytes. *Abbreviation*: APC, antigen presenting cell.

recognized by the immune system if attached to a larger molecule such as an endogenous protein (Fig. 6.30). The system was therefore recognizing an altered protein as not self. The small foreign molecule is known as the hapten, which forms the protein-hapten complex with cellular or serum proteins. More than one molecule may be attached to the endogenous protein and the "epitope" density can be an important determinant of the immunological response. The molecule that becomes the hapten could be the drug itself or a breakdown product or a reactive metabolite, which can bind to or "haptenize" the protein.

However, although these are the basics of the mechanism, there are additional factors, which must be considered.

One such factor is the formation and presentation of the hapten-protein complex to the immune system shown in general terms in Figure 6.30. The initial exposure to the drug and formation of the hapten-protein complex, the sensitization stage, involves the presentation to T-helper cells by APCs, which process the hapten-protein complex. The interaction of the T-helper cell with the APC leads to the activation and proliferation and differentiation of the T cell. Other factors such as cytokines and adhesion molecules may also be involved in this process. The T-helper cell must recognize that the hapten-protein complex is foreign. The activated T-helper cell then either differentiates into a T-helper 1 (T_H1) cell, which promotes T-cell-mediated responses (type IV) or into a T_H2 cell, which stimulates antigen-specific B cells to produce antibodies. There may be no reaction because by the time the antibodies are produced some days later, the antigen may well have disappeared. However, after a subsequent exposure, the elicitation stage, the memory cells and antigen-specific antibodies will be able to start an immediate response.

However, although this is the underlying mechanism, there are different outcomes. Thus, hypersensitivity reactions can result in one of four types of immune response:

Types I to III are humoral, being mediated by antibodies.
Type IV is cell mediated, mediated by T cells.
The details of the mechanisms will be discussed below.

Hypersensitivity reactions types I to III involve antibodies for recognition and as part of the process of removal of the antigen.

Antibodies are immunoglobulins, proteins with a molecular weight of 150,000 (IgD, IgG) to 890,000 (IgM), which are synthesized in response to an antigen. There are five classes of immunoglobulins, some of which are detectable in plasma: IgA, IgD, IgE, IgG, and IgM. All the immunoglobulins have certain common structural features, the light and heavy polypeptide chains, which have regions of variable and constant structure. The variable region retains the same number of amino acids, but the particular amino acids vary. The antibody molecule is structurally symmetrical and has binding sites for antigens on the variable sections of both light and heavy chains. More than one antibody may be produced in response to a single antigen. The plasma immunoglobulin fraction is often raised after exposure to an antigen, and is one indicator that an immune response has occurred.

Type I Hypersensitivity Reactions

In response to the attachment of the antigen to the IgE antibodies, the macrophages release mediators such as histamine, which causes release of further mediators such as cytokines and leukotrienes. These inflammatory mediators cause the dilation and increased permeability of blood vessels and constriction of smooth muscles.

The antibody, a homocytotropic antibody, circulates in the bloodstream, but has a high affinity for the surface of mast cells and binds to receptors on the surface (Fig. 6.31). This type of reaction, which occurs quickly after reexposure, underlies reactions in the respiratory system (asthma, rhinitis), skin (urticaria), gastrointestinal tract (food allergies), and vascular system (anaphylactic shock). Type I reactions can be severe, causing difficulty in breathing, loss of blood pressure, anoxia, edema in the respiratory tract, and bronchospasm, which may prove fatal.

Thus, the result can be a variety of symptoms of different severity, ranging from mild skin rashes to occasionally fatal anaphylactic shock.

Type II Hypersensitivity Reactions

IgG or IgM antibodies direct the immune response toward the antigen located on a cell (e.g., a red blood cell or thrombocyte). Macrophages, NK cells, and neutrophils are recruited by the antibodies to the site of the antigen on the cell surface and destroy the cell by phagocytosis or lysis. Additionally, complement activation will damage the cell (Fig. 6.32). The result, for example, where red cells are the targets is hemolytic anemia.

Leukocytopenia and thrombocytopenia would occur if other blood cells were targets. The agglutination of cells may also stimulate the complement cascade. This is a cascade of physiologically active proteins, which results in the release of kinins, histamine, and lysosomal enzymes, which may cause lysis of cells. Consequently, the result can be severe if significant loss of blood cells occurs either through lysis or removal of agglutinated cells by the spleen. For example, administration of the drug aminopyrine can cause severe granulocytopenia as a result of agglutination and removal of leukocytes.

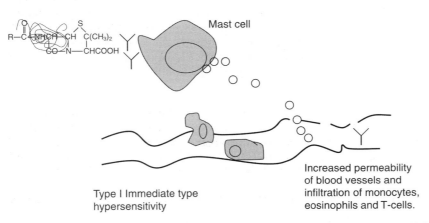

Figure 6.31   The basis of type I hypersensitivity reactions. The antigens cross-link with IgE antibodies, which are attached to mast cells.

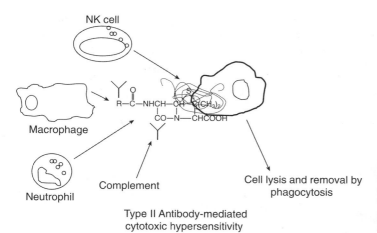

Figure 6.32 The basis of type II hypersensitivity reactions. The antigen is part of a cell, which becomes the target for macrophages, NK cells, neutrophils, and complement. *Abbreviation*: NK, natural killer.

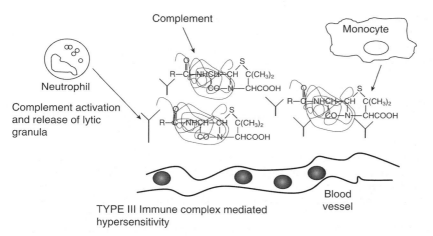

Figure 6.33 The basis of type III hypersensitivity reactions. These are mediated by immune complexes formed between antigen and IgG antibodies, which accumulate in capillaries and in tissues.

Type III Hypersensitivity Reactions (Arthus)

Freely circulating IgG antibodies form complexes with soluble antigens, which accumulate because they are not removed from the circulation and tissues by phagocytosis. These immune complexes are then deposited in places such as the kidney glomeruli, joints, and capillaries where they cause damage by blockage and thrombosis, by cytotoxic cells such as neutrophils, and activation of the complement system, which is stimulated by the complexes (Fig. 6.33). This type of immune response causes symptoms such as systemic lupus erythematosus in which there can be inflammation of the capillaries, joint and muscle inflammation, and kidney damage (glomerulonephritis). Serum sickness is another manifestation.

Type IV Hypersensitivity Reactions

This type of immune response is unlike the other three in that antibodies are not involved, the response being initiated by sensitized T_H1 cells. These are activated by contact with APCs. This stimulates them to secrete cytokines, which recruit macrophages, neutrophils, and eosinophils causing inflammation at the site of exposure. The cytokines also recruit and activate cytotoxic T cells, which can destroy the cell such as when an altered cell membrane is the antigen (Fig. 6.34).

Contact hypersensitivity caused by drugs and other chemicals is of this type, resulting in eczema and erythema.

As already indicated, the complex can be formed simply by the interaction of an unstable drug or chemically reactive breakdown product with a protein. This therefore could occur in

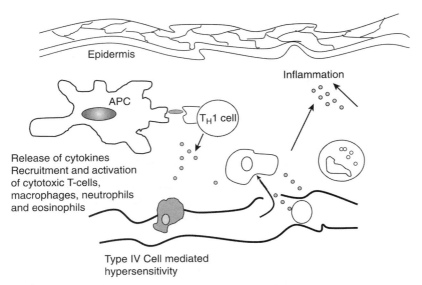

Figure 6.34 The basis of type IV hypersensitivity reactions. This is cell mediated rather than antibody mediated.

any tissue, such as blood or skin with proteins that exist in those environments. These therefore would most likely be extracellular hapten-protein complexes. Alternatively, the drug might be metabolized within the cells of the liver or other tissue to a reactive metabolite, which binds to protein nearby. This would therefore be an intracellular complex. For the immune system to recognize the complex as an antigen, however, it needs to be made aware of its existence, that is, presented to the immune system. This occurs differently for intracellular compared with extracellular hapten complexes.

Extracellular antigens are detected by APCs, such as lymphocytes, macrophages, and dendritic cells in interstitial fluid and blood. These detect the hapten and engulf the whole antigenic complex. Then, when inside the APC, the complex is partly dismantled and peptides attached to proteins similar to immunoglobulins, known as MHC II. The modified peptide-hapten complex is moved to the surface of the APC and presented as a complex with MHC to T-helper cells (CD4$^+$), which activates and instructs the APC to make antibodies to the hapten and also B cells (memory cells with "memory" of the hapten) to proliferate. These events lead to types I to III responses.

Intracellular antigens could be modified internal proteins, which are continuously removed by the cell, the structure altered and attached to MHC I. This takes place in the rough endoplasmic reticulum. The protein/peptide-hapten fragments are then presented to the external surface of the cell membrane as a complex with the MHC I. Then cytotoxic T cells (CD8$^+$) accept the protein/peptide and destroy the cell. This mechanism gives rise to a type IV response.

The choice of an immune response mediated by T_H1 or T_H2 cell depends on a number of factors and in particular on cytokines (Fig. 6.35). Thus, if mast cells, APC and T-regulatory cells are involved in secreting cytokines such as IL4, then T_H2 differentiation will be stimulated. Conversely, if activated dendritic cells and NK cells are involved in secreting IL2 and IFNγ, then T_H1 differentiation will be stimulated.

However, simply the production of a hapten and then an antigen and presentation to the immune system does not necessarily mean that an immune response will occur. Other factors are important, such as the epitope density already mentioned.

There is some tolerance in the immune system, otherwise it could react unnecessarily to any foreign, large molecule such as those in food or the air (there are people who do indeed show hypersensitivity reactions to such things, hence hay fever, gluten intolerance, and peanut allergy). If only the antigen is present, the T cell may well ignore it.

It seems therefore that usually the immune system, specifically the naive T cells, needs to get two signals.

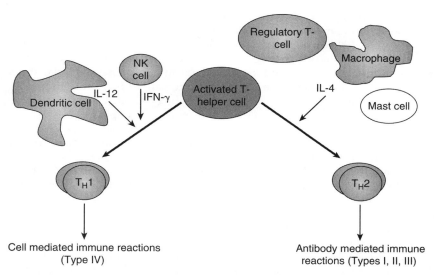

Figure 6.35 The factors regulating T_H1 versus T_H2 responses. The choice between a T_H1 versus a T_H2 fate is influenced by the cytokines excreted locally by other immune cells.

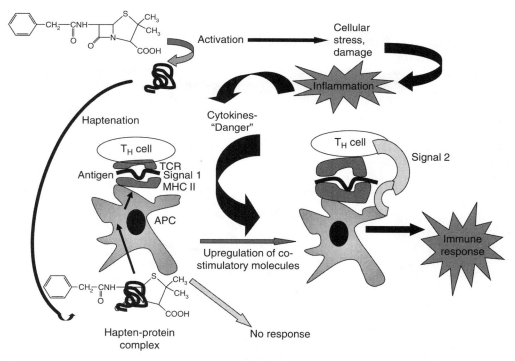

Figure 6.36 Induction of danger signals. Inflammation due to cellular stress, physical damage, or the presence of pathogens can lead to induction of co-stimulatory molecules on the APC. This increases the risk of activation of an immune response directed toward the drug-conjugate. *Abbreviation*: APC, antigen-presenting cells.

One signal is as described above, the APC-T-cell receptor interaction, which is antigen specific. The second is signal indicating that the foreign substance is indeed potentially dangerous. This is the so-called danger hypothesis (Fig. 6.36).

Thus, if the drug or a metabolite causes damage to tissue or cellular stress, for example, inflammatory cytokines will be released. These could form part of the second, "danger" signal. The second signal also acts on the T cell via APCs and costimulatory receptors (B7, e.g., which interacts with CD28 on the T cell).

However, even with both signals, there are still individual factors, which determine susceptibility and so most immune-mediated drug toxicities are rare and hence idiosyncratic, occurring in 1 in 10,000 to 1 in 100,000 patients. There could be a range of factors, including genetic differences in the metabolism of the drug, which might affect the amount of hapten formed. The tolerance in the immune system and the production and response to cellular stress caused by the drug or metabolite are other factors. The cytokine pattern is another important factor, which is influenced by route of exposure, dose, and type of antigen. Note that there could be more than one antigen produced from a drug and more than one hapten.

Also, it is known that there are genetic factors that influence the choice between T_H1 and T_H2 responses because of increased IgE responsiveness as a result of genes influencing cytokines.

Furthermore, there are many unanswered questions regarding idiosyncratic drug toxicity mediated by immune mechanisms, and the above explanations are not always sufficient. Some cases do not fit into the classical hapten hypothesis model. For example, antidrug antibodies are not always detectable. Immune reactions can take a period of months or even years to develop (e.g., hydralazine see below).

One mechanism that has been identified is the p-i concept (Fig. 6.37).

Thus, immune responses to drugs usually occurs as a result of the production of a carrier-hapten complex, followed by uptake, processing, and presentation by APC to T cells.

However, in the p-i concept there is a pharmacological interaction (low affinity) between the drug and the peptide-MHC molecule on the surface of the APC, with weak binding but sufficient to activate T cells, which have receptors for the MHC.

Drugs such as **lidocaine** and **sulfamethoxazole** seem to be able to cause hypersensitivity reaction by this mechanism.

6.6.4 Characteristics of Immunological Reactions

Dose response relationship

There may be no clear dose-response relationship for immune responses, as the magnitude of the response is dependent on the type of reaction of the endogenous immune system, not on the concentration of the foreign compound. However, there may be a relationship between likelihood of a response occurring and size of response with size of exposure and frequency of

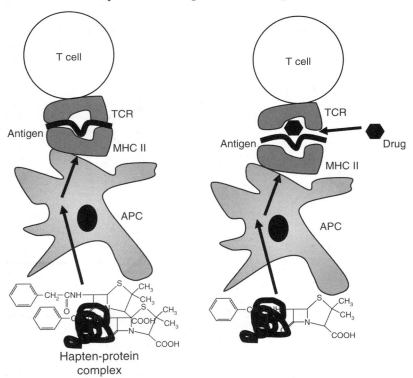

Figure 6.37   The pharmacological interaction p-i concept.

exposure. A relationship between the *frequency* of occurrence of an immunotoxic response and the exposure certainly occurs with some drugs such as hydralazine (see below).

Exposure

Repeated or chronic exposure is required for immunotoxic reactions as the immune system must first be sensitized as described above. The exposure to the compound need not be continuous and in fact repeated, discontinuous exposure is often more potent.

There is no generalized, characteristic exposure time or sequence for immunotoxic reactions. In halothane hepatitis (see below), the number of exposures seems to be important, with about four being optimal. With **hydralazine**-induced **Lupus syndrome,** exposure for a considerable length of time (average of 18 months), during which there is repeated exposure, is generally required for the development of the immune response. With some types of immune response, exposure to minute amounts of the antigen, such as might occur with an environmental pollutant, may be sufficient to elicit a severe response in a sensitized individual.

Specificity

The nature and extent of the immunological reaction do not generally relate directly to the chemical structure of the foreign compound, which is acting as a hapten. Thus, many different compounds may elicit the same type of reaction. For example, **nickel, *p*-phenylenediamine**, ***p*-aminobenzoic acid**, and **neomycin** all cause dermatitis. The antigen produced may have some bearing on the type of response, but there are many other factors. Thus, it can be stated that chemically similar foreign compounds may produce very different immunotoxic reactions and conversely different types of substance may elicit the same type of reaction. However, once sensitized to a particular foreign substance, other chemically similar substances may show cross-reactivity and precipitate the same immune response.

However, the exact chemical structure is often a crucial factor in the antigenicity, and small changes in structure may therefore remove antigenicity.

Site of Action

Various target organs may be involved in the immune response, but this usually depends on the type of reaction rather than on the distribution of the foreign substance. The many substances, which cause immune responses, may cause anaphylactic reactions, giving rise to asthma and various other symptoms as described above. The site of exposure to the foreign compound may not necessarily be the lungs, however. Similarly, a common immune response is urticaria, or the formation of wheals on the skin, which can occur when exposure has been via the oral route. Thus, the target organ is generally due to the particular response rather than the circumstances of exposure or distribution of the compound. However, there are exceptions to this such as the type IV cell-mediated immune reactions where the cell is altered by the foreign compound and is then a target. In the case of halothane hepatitis, the liver is the target for metabolic reasons (see below for more detail). Another exception is the situation where the foreign compound is chemically reactive and reacts with proteins in the skin or respiratory system. These altered proteins may then become targets for an immune response at the site of exposure, although there may well be other sites involved as well. Thus, the final response is determined by the type of immune mechanism involved.

Thus, immune-mediated responses can be immediate or delayed and localized or widespread. The response can be restricted to the area of exposure or can be systemic. Similar compounds may cross-react or produce very different responses. Many different foreign compounds can cause an immunotoxic response: drugs such as penicillin, halothane, and hydralazine, industrial chemicals such as trimellitic anhydride and toluene di-isocyanate, natural chemicals such as pentadecylcatechol found in poison ivy, food additives such as tartazine, and food constituents such as egg white (albumen).

However, these may not be immunotoxic in all exposed individuals, and sometimes chemically similar compounds are not immunotoxic. Also, some highly reactive compounds and reactive metabolites of compounds such as paracetamol (see chap. 7) do not seem to be immunotoxic despite reacting with protein.

These are currently areas of obscurity in immunotoxicology, but the importance of this aspect of toxicology is increasing in view of the growing realization that possibly many adverse effects in humans are mediated by the immune system.

6.7 GENETIC TOXICITY

6.7.1 Introduction

Damage to or changes in the genetic material and machinery involved in cell division can be caused by chemicals. Such changes may be irreversible or reversible, that is, repaired, and may have no significant consequences or can lead to serious diseases such as cancer or congenital malformations. For there to be an adverse effect on the phenotype depends on the extent of change and where it occurs. Thus, it may occur in a region of DNA where there are no genes or critical codes, or the change may be of no significance or even beneficial.

There are three types of effect: **mutagenesis**, **aneugenesis**, and **clastogenesis**, and all three are implicated in cancer development. However, genetic toxicity is not essential for cancer development as a result of chemical exposure.

Mutagenesis is the production of a mutation by loss, addition, or alteration of individual bases or base pairs or a small number of base pairs in the DNA molecule.

Aneuploidy is the acquisition or loss of complete chromosomes, and clastogenicity is the induction of chromosomal aberrations as a result of loss, addition, or rearrangement of parts of chromosomes.

To understand the significance of these changes, it is necessary to understand the basic structure and function of DNA and its role in heredity.

DNA is a double-helical molecule, which contains information stored in a linear chemical code comprising pairs of bases arranged in triplets. The units of the DNA molecule are nucleotides. These are constructed from one of four bases plus a sugar phosphate. The four bases are either purines (adenine and guanine) or pyrimidines (thymine and cytosine). The sugars of adjacent nucleotides are linked together via the phosphate group and thus the sugar-phosphates form a polymer, which is the backbone of each strand of the double helix. The two single strands are linked together by bonding between bases, but this is not random, adenine (A) pairs to thymine (T) and guanine (G) to cytosine (C). Therefore, each strand is complementary to the other (Fig. 6.38).

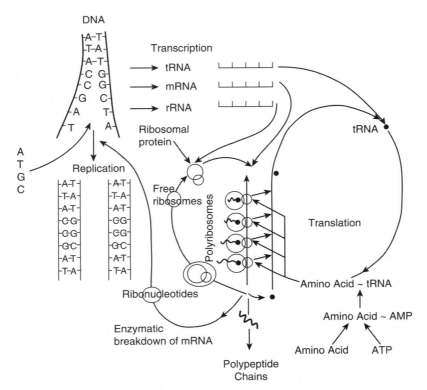

Figure 6.38 Basic processes involved with replication of DNA and the transcription and translation of its encoded information into RNA and thence protein. *Source*: Modified from Ref. 11.

The DNA molecule forms the basis of the chromosomes, which reside in the nucleus of cells, and a small amount of DNA is located in the mitochondria. The chromosomes also have other molecules associated with them such as histone proteins.

The relatively weak links between the complementary bases in the DNA strands allow the helix to split at cell division and be replicated to yield an identical pair of DNA molecules as each half acts as a template. Thus, DNA is able to undergo *replication*.

However, the function of the DNA is to encode information, which is used for the running of the cell via the production of proteins such as enzymes.

This occurs by the processes of *transcription* in which complementary RNA is produced from specific sections of DNA and then *translation* of this message into protein (Fig. 6.38).

RNA is a polymer, very similar to DNA but not identical. The differences are that one base is different, thymine is replaced with uracil, the sugar is ribose instead of deoxyribose, and the polymer is single, rather than double stranded. However, weak bonds can form between complementary bases or groups of bases. Most of the RNA transcripts are mRNA, which guides the synthesis of proteins.

Each triplet code, the *codon*, in the messenger RNA specifies a different amino acid, but there are more possible codons (64) than amino acids (20), and several codons can specify the same amino acid. To each codon, one molecule of transfer RNA becomes attached via a complementary *anticodon.* These transfer RNA molecules carry a single, specific amino acid. In this way a protein, comprising a chain of amino acids, can be built up (Fig. 6.38).

The process takes place in the cell on the rough endoplasmic reticulum and is carried out by ribosomes, which consist of 2 ribosomal RNA molecules and about 50 proteins. The ribosome attaches itself to the end of the mRNA, moves along it, and effectively stitches together the amino acids into a chain by adding them one at a time.

As DNA is a large molecule, its information is broken down into smaller units with each mRNA molecule specifying a protein and corresponding to a gene, which is a specific region of DNA.

However, in a molecule of DNA, there are many genes specifying thousands of proteins so the expression of individual genes is regulated. Thus, the cell makes proteins it needs. This is controlled by stretches of regulatory DNA, which are interspersed along the section, which codes for a protein. These are noncoding regions, which bind to specific protein molecules, which control the rate of transcription. Other sections of DNA are also noncoding, specifying instructions such as start and stop transcription. The coding regions, *exons*, are interspersed with noncoding regions, *introns*, which are still transcribed into RNA. However, this section of RNA is lost and degraded after RNA splicing, which serves to connect two exons together.

DNA is a very long molecule and, in the cells of eukaryotes such as mammals, is divided into pairs of *homologous* chromosomes (one inherited from the maternal and one from the paternal organism) plus the sex chromosomes (X and Y), which are nonhomologous. In humans, there are 22 pairs of homologous chromosomes plus the two sex chromosomes (46 in all).

Each chromosome contains DNA, some of which is genes, but it is tightly packaged into chromatin by proteins specialized for folding. There are also proteins (enzymes) necessary for DNA replication and repair and proteins that bind to DNA, such as histones. These are responsible for the organization of chromosomes into nucleosomes.

Although the DNA is tightly packaged, the genes can still be transcribed, which takes place during interphase in the cell cycle (Fig. 6.39). DNA replication also takes place during this phase (in S phase). Cell division takes place during M phase (mitosis).

Some mutagens may not be able to cross the nuclear membrane and are only active at mitosis when the nuclear material is in the cytoplasm.

It is beyond the scope of this book to describe the processes is detail, and the student is referred to the references for extra, detailed explanation.

Replication of DNA shows great fidelity, because of the repair of errors that occur during the formation of complementary DNA strands, and new DNA molecules are recognized and corrected. The erroneous bases, which are the result of substitutions or chemical modification, are excised by repair enzymes before the mistake is encoded in the daughter molecule. If the mutation is not recognized and repaired, however, the false information is transcribed into RNA, and hence the error or mutation is expressed as an altered protein, resulting from perhaps only a single amino acid change. Such a slight change may drastically alter a protein if

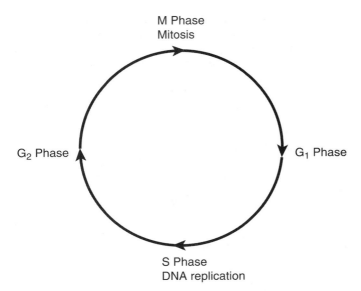

Figure 6.39 The phases of the cell cycle. G_1, S, and G_2 are all parts of interphase. In M phase, the nucleus and then cytoplasm divide.

it occurs at a crucial site in the molecule. For example, in some of the inherited metabolic anemias, this is the case, with a single amino acid change in the hemoglobin molecule being sufficient to impair its function drastically. However, it is clear that only mutations recognized as errors will be repaired and therefore not transcribed. A mutation in the DNA molecule, if not recognized, may therefore be transcribed into mRNA, the wrong amino acid specified and translated into a mutant protein. This will occur provided the original mutation is compatible with the transfer RNA codes. Alternatively, the protein may be terminated prematurely and be too short and therefore unable to function.

If a mutation or major chromosomal change occurs and is not repaired, it may not be compatible with the continued existence of the cell and as already discussed above, cell death may occur by apoptosis. However, if the mutation is not lethal and not repaired, when the cell divides and the DNA replicates, the mutation will be passed on to the daughter cells and "fixed" in the genome. If the mutation is simply an altered base, at replication it will be corrected in one half of the molecule but compounded in the other.

For a simple mutation involving small changes in the code, the consequence will be of variable severity because of the following:

1. The code for amino acids is degenerate. So there may be several codes for any one amino acid. Consequently, a mutation may be hidden, and no change in the amino acid incorporated will be seen.
2. A change in the code may change an amino acid in an inconsequential position such as in a structural or terminal portion of the protein.
3. There are "nonsense" codes, which do not specify amino acids but specify instructions such as "start" or "stop" transcription or translation. A mutation in such a code might therefore have a profound effect. Alternatively, mutations may give rise to a nonsense code, causing erroneous instructions to be carried out such as the premature termination of a protein. Obviously, mutations of this kind may have a much more profound effect than a single change in an amino acid.

6.7.2 Mutagenesis

The majority of the information in DNA is not associated with specific genes and does not specify phenotype. This is the so-called junk or noncoding DNA with only about 1.5% of mammalian DNA having sequence information for proteins and about 2% encoding information about gene expression and other functions.

Therefore, random mutations will mostly affect the junk DNA and be probably of no consequence, that is, neutral. Mutations in the coding sequences, which will include genes for specific proteins, will potentially cause adverse changes in the gene and therefore the phenotype. However, mutations could also be advantageous and possibly neutral. For example, if only one amino acid in the protein is changed, this may have no influence on the structure or function of the protein or even improve it. Furthermore, as indicated above, the triplet code is not absolute with sometimes several triplets coding for the same amino acid.

Assuming access to the nucleus, chemicals can interact with DNA in a variety of ways. This can include incorporation and direct chemical reaction with parts of the DNA molecule. The main types are given below:

1. Direct interaction with DNA:
   a. Alkylation of bases
   b. The addition of bulky adducts
   c. Formation of cross-links

2. Primary DNA damage:
   a. Removal of a purine or pyrimidine base(s) forming apurinic or apyrimidinic sites
   b. DNA-DNA cross-links
   c. DNA-protein cross-links
   d. Formation of pyrimidine dimers
   e. Interacalation
   f. DNA strand breaks (either single or double)
   g. Oxidative DNA damage caused by ROS

3. Base substitutions (transitions or transversions)
4. Base additions or deletions

6.7.3 Direct Interaction with DNA

Alkylating agents or other chemically reactive electrophilic intermediates can react covalently with DNA and alkylate or form adducts with individual DNA bases on vulnerable sites, on guanine especially (Fig. 6.40). Electron-rich atoms such as N and O are the most common targets. For example, N^7 and O^6 of guanine and the N^3 and N^7 of adenine are the most reactive. Dimethylnitrosamine methylates the O^6 and N^7 of guanine. Vinyl chloride, however, alkylates the N^1 of adenine and N^3 of cytidine as well as the N^7 of guanine. The formation of adducts will in some cases affect the ability of the bases to form hydrogen bonds across the DNA double helix and so cause erroneous pairing at replication and base substitution (see below).

Larger alkylating agents can cross-link between strands or within one strand of the DNA.

Larger reactive metabolites can form bulky adducts on bases, such as benzo[a]pyrene (with N^2 of guanine) and aflatoxin (with N^7).

The particular site and type of base alteration will be dependent on the particular reactive metabolite.

Modifications to the base(s) of DNA with large adducts can cause misreading of the code during translation or replication and therefore mutations or affect the action of polymerases or alternatively can stimulate DNA repair, which could be error prone. Alkylation of bases may cause mispairing at DNA replication and so mutations in the daughter cells.

Thus, if these kinds of alterations to the DNA molecule are not repaired correctly, then changes to the genetic code can result and become fixed in the replicated DNA after mitosis

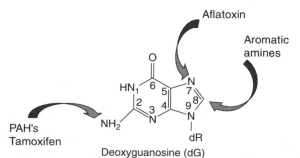

Figure 6.40 Sites of electrophilic attack on guanine. PAHs are polycyclic aromatic hydrocarbons such as benzo(a)pyrene.

(replication in somatic cells) or meiosis (replication in germ cells). However, some adducts are repaired quickly, others not, and some are unstable. Some adducts cause mutations, others do not.

Therefore, the formation of alkyl DNA or bulky DNA adducts can be regarded as structural modifications of DNA and so are potentially *premutagenic* lesions.

6.7.4 Primary DNA Damage

There are other kinds of DNA damage/changes, which can lead to mutations and are also premutagenic lesions therefore.

The production of apurinic and apyrimidinic sites can both be caused by alkylating agents. When, for example, the N^7 of guanine is alkylated, this causes the bond from base to sugar to become labile and so the base can be lost completely. Thus, apurinic sites are formed. Insertion of an incorrect base at this site can then cause a mutation. Loss of purines and pyrimidines can also occur spontaneously.

Cross-links and pyrimidine dimers can cause misreading or incorrect functioning of DNA. These dimers form between any two adjacent pyrimidines (C or T) on one strand.

Chemicals, which have planar molecules such as acridine, and the anticancer drugs vincristine and vinblastine can intercalate in the DNA molecule. This distorts the structure and so alters the function of the molecule, possibly leading to mutations via base insertions (see sect. 6.7.7 below). This may also interfere with the excision repair processes during meiosis.

Strand breaks , both single and double, can be induced by chemicals with obvious potential mutagenic consequences.

Similarly, oxidative damage can result from exposure to chemicals and cause small changes in DNA bases (e.g., formation of 8-oxodeoxyguanosine, 8-oxo-dG), which are potentially promutagenic (Fig. 6.41). 8-Oxo-guanine can be detected in urine. Exposure to some metals such as Cr (VII) and Ni (II) can cause the oxidation of bases and therefore potential mutations.

ROS can also deaminate 5-methylcytosine and cytosine to give thymine and uracil (Fig. 6.42).

It should be noted that endogenously produced ROS causes DNA damage and indeed accounts for far more such damage that exogenous mutagens.

Thus, mitochondria and peroxisomes will both produce ROS as part of normal metabolism, which if not detoxified can interact with DNA (both mitochondrial and nuclear). Mitochondrial DNA is especially vulnerable because of the lack of histones, surrounding the molecule, inefficient repair, and the close proximity to the site production.

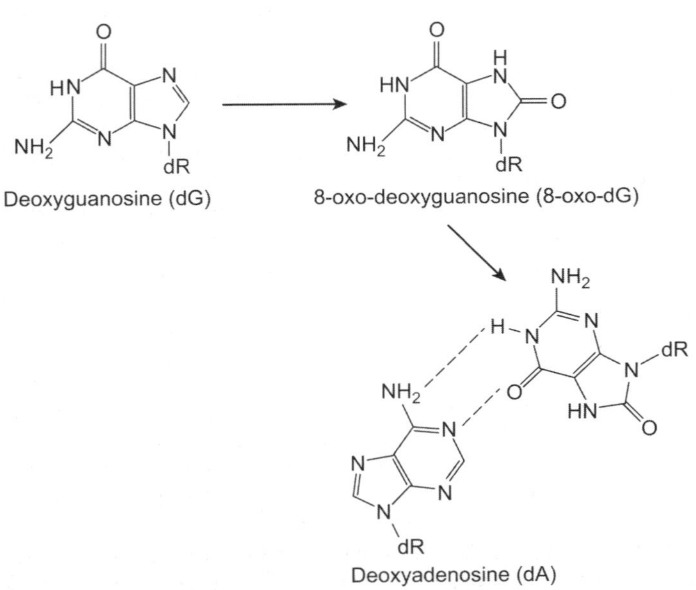

Deoxyguanosine (dG) → 8-oxo-deoxyguanosine (8-oxo-dG)

Deoxyadenosine (dA)

Figure 6.41 Oxidation of deoxyguanosine to 8-oxo-dG, such as might occur as a result of exposure to reactive oxygen species. 8-oxo-dG pairs erroneously with deoxyadenosine instead of deoxycytosine as shown. *Abbreviation*: 8-oxo-dG, 8-oxo-deoxyguanosine.

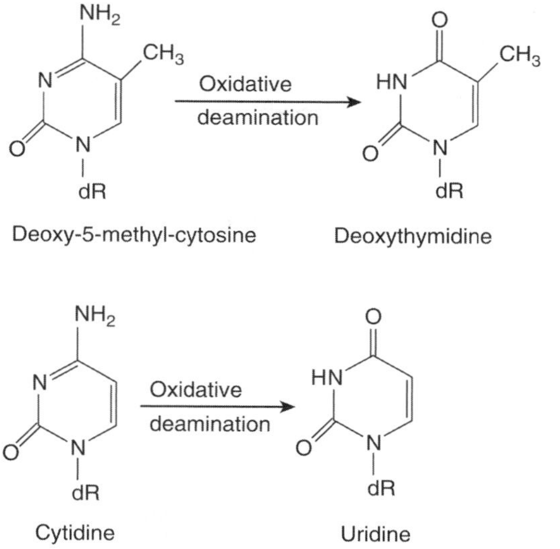

Figure 6.42 Oxidation of deoxy-5-methylcytosine to deoxythymidine and cytidine to uridine. This may be caused by reactive oxygen species or oxidizing agents such as nitrous acid. d5mC is present in methylated CpG sequences.

Inflammation can also be responsible for oxidative damage to DNA via inflammatory cells such as neutrophils and other phagocytic cells. These produce oxidizing agents such as hypochlorite (OCl^-), superoxide ($O_2^{\bullet -}$), nitric oxide (NO), and hydrogen peroxide (H_2O_2). All of these can potentially oxidize or interact with DNA bases and cause promutagenic effects such as deamination, oxidation, single and double-strand breaks, and apurininc and apyrimidinic sites.

Chemicals, which cause repeated inflammation (e.g., asbestos), could lead to increased numbers of mutations. Indeed, DNA in inflamed tissue has been shown to have more 8-oxo-dG (Fig. 6.41) than normal tissue.

6.7.5 Gene Mutations

These are small changes in the DNA sequence, which can occur in either somatic cells or germ cells. They are either additions or deletions of bases or substitutions of bases.

The base substitutions can be either a change from one purine or pyrimidine to another, which is a *transition* or a change of a purine for a pyrimidine and vice versa, which is a *transversion*. A chemical change in the base or formation of an adduct, which changes the nature of the base(s), could cause a substitution at replication. The position of the adduct on the particular DNA base would determine the type of mutation.

The addition or deletion of one or a few (but not 3) base pairs leads to a frameshift mutation. Large adducts, which are inside the double helix, can distort the DNA structure and are more likely to cause frameshift mutations.

Some chemicals can cause mutations at any stage of the cell cycle, but most are more active at S phase.

With germ cells, although the mechanism of production of gene mutations is similar to somatic cells, the relationship between exposure and timing of DNA replication is important. In males, the spermatogonial stem cells are especially vulnerable because they are present throughout the reproductive life of the animal. However, in mammals only a small fraction of the spermatogonial stem cells are in S phase, so the probability of DNA repair taking place prior to DNA replication is high. Each time a spermatogonial stem cell divides, it produces a differentiating spermatogonium and another stem cell. Therefore, the stem cells can accumulate damage. Late spermatids and sperm, produced after meiosis has taken place, lack DNA repair and so are particularly vulnerable to the induction of mutations.

In the case of oogenesis, the primary oocyte arrests after the first meiosis, before birth, and the next S phase is not until after fertilization. Therefore, the oocyte is more resistant to chemical mutagens.

These changes in bases and base pairing can lead to errors at replication or a complete stop to transcription. They may only be point mutations, affecting one base or could be block mutations, which can affect several genes. The result, if the error is not repaired, will be an altered genotype, but the phenotype of the cell may not alter depending on the site of the change. Thus, gene mutations may be recessive or dominant, they may be in important genes (e.g., a tumor suppressor gene) or be silent (until perhaps relocated by chromosomal translocation) or be neutral or lethal to the cell, in which case there will probably be no serious consequences in normal somatic cells.

6.7.6 Base Substitutions

Base substitutions, either transitions or transversions, could occur as a result of replication of an altered template. This may be the result of the production of an altered base such as the formation of an adduct or chemically changed base, perhaps as a result of oxidative damage. For example, formation of 8-oxo-dG is the most common DNA lesion caused by oxidation and can be a promutagenic lesion as it can erroneously pair with adenosine during replication (Fig. 6.41). This means that the original base pair, G:C, will become first 8-oxo-G:A, then T:A. Therefore, a G to T transversion has occurred. Another example is deamination of 5-methylcytosine to thymine (Fig. 6.42) at CpG sites, which results in G:C to A:T transitions.

The deamination of 5-methylcytosine occurs more frequently than that of other bases but the product, thymine, is of course a normal base and therefore will not be recognized by the repair system (Fig. 6.42). This represents a major source of point mutations in DNA.

Such changes can occur spontaneously or from exposure to chemicals. Cytidine can be deaminated to uridine (Fig. 6.42) and adenine to hypoxanthine (Fig. 6.43) by exposure of DNA to the oxidizing agent nitrous acid.

For instance, if such a substitution is induced in the tobacco mosaic virus genome, changes in the viral protein can be detected, which correlate with the mutation. This has been made possible because the viral protein has been characterized. Thus, it is found that the amino acid threonine is replaced by isoleucine. It is also clear that several triplet codes in the nucleic acid specify the amino acid threonine, namely ACA, AGG, ACC, and ACU, emphasizing the degeneracy of the code.

Changing the cytosine to uracil, such as by deamination, in these codes gives AUC, AUA, and AUU, the codes that specify isoleucine. This mutation is therefore a C to U transition. The effect of a base-pair substitution on replicating DNA can be seen in Figure 6.43, using the example of the deamination of adenine to hypoxanthine, which results in a mutagenic transition of the type A:T to G:C. It can be seen that only one of the product DNA molecules has the altered base pair GC, and therefore only one of the daughter cells is affected. However, this may not necessarily be the case, as some experimental evidence suggests that before replication, the partner of the abnormal base may be excised and replaced. Therefore, both strands of the molecule will contain the mutagenic transition.

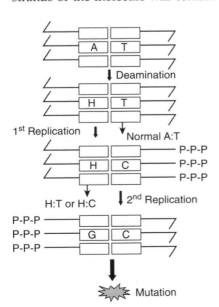

Figure 6.43 Mechanism of a mutagenic transformation by deamination of adenine to hypoxanthine. Hypoxanthine (Fig. 4.29) pairs like guanine, and this results in a transition A:T→G:C. *Source:* Adapted from Ref. 12.

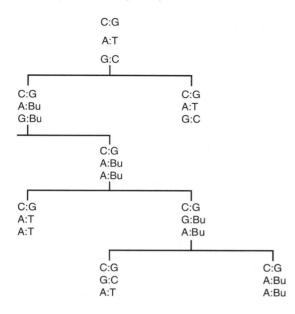

Figure 6.44 Replication of a nucleic acid in the presence of the base analogue bromouracil. After the first replication, Bu enters opposite G, giving a G:C→A:T transition. If G enters opposite Bu already incorporated, an A:T→G:C transition results. *Abbreviations*: Bu, bromouracil; G, guanine. *Source*: Adapted from Ref. 12.

The expression of the mutant genotype depends in turn on certain factors such as the turnover rate of the product proteins, the rapidity of cell division, and, in diploid organisms, the dominance or recessivity of the mutation. Similar types of mutation are caused by mutagens, which alkylate bases, such as the alkyl nitrosamines, which alkylate the N^7 and O^6 of guanine (Fig. 6.40). This may lead to a G:C to A:T transition. Another example is afforded by the mutagen and carcinogen vinyl chloride, which is discussed in more detail in the next unit. This compound causes base-pair substitutions, as a result of alkylation of N1 of adenine and the N^3 of cytidine. The alkylation, addition of the etheno moiety, gives rise to an imidazole ring. This causes adenine to pair, like cytidine and cytidine, to become similar to adenine. The result of these alkylations are A-T to C-G and C-G to A-T transversions.

The second type of base-pair substitution, incorporation of an abnormal base analogue, only occurs at replication of the DNA molecule. An example of such a mutagen is the thymine analogue, 5-bromouracil (Bu). This compound is incorporated extensively into the DNA molecule during replication, being inserted opposite adenine in place of thymine (Fig. 6.44). 5-Bromouracil first forms the deoxyribose triphosphate derivative, a prerequisite for incorporation. If 5-bromouracil is incorporated opposite A, it seems that the A:Bu pair will function as A:T, and the organism continues to grow and divide. Therefore, this change does not seriously affect the function of the DNA.

Bu may pair with G, however, as if it were C (Fig. 6.44), with the consequence that G:C to A:T and A:T to G:C transitions result. The first type of transition occurs because of erroneous incorporation and the second results from erroneous replication against the incorporated Bu. Other base analogues, some of which are used as anticancer drugs, may produce lethal mutations in mammalian cells, particularly in cancerous cells, which are involved in rapid DNA synthesis.

6.7.7 Frameshift Mutations

This type of mutation arises from the addition or deletion of a base to the nucleic acid molecule. This puts the triplet code of the genome out of sequence. Consequently, transcription of the information is erroneous when it is carried out distal to the mutant addition or deletion. In illustration of this, the following code makes sense when read in groups of three:

HOW CAN RAT EAT

After the deletion or insertion of a base, however, the code makes no sense:

HOW CAN THR ATE AT...

The transcription of information is therefore generally seriously affected by a frameshift mutation, as large parts of the code may be out of register. Thus in the sequence

ACA AAG AGU CCA UCA

threonine lysine serine provaline serine

If the **A** is removed, the code becomes

ACA AGA GUC CAU CA...

threonine lysine valine histidine

It is clear that in a long-chain protein produced from such a template, substantial errors may occur.

However, if the addition of a base is followed by a deletion in close proximity or vice versa, production of proteins with partial function may occur.

Frameshift mutations may be produced by various mechanisms. For instance, intercalation of planar molecules, such as acridine within the DNA molecule, allows base insertions at replication. Acridine-type compounds may also induce mutations of the frameshift type by interfering with the excision and repair processes, which correct errors occurring during unequal crossover between homologous chromosomes at meiosis. Such unequal crossover gives rise to errors of the frameshift type.

Crossover, which involves the exchange of homologous segments between chromatids at meiosis and between homologous chromatids at mitosis, occurs normally in diploid organisms, by the breaking and reunion of chromatids or by switching of the replication process, a mechanism known as copy choice. The broken ends of the chromatids, which are formed during this process tend to reunite with other broken ends. (Fig. 6.45). Any disruption to this system may therefore potentially cause a frameshift mutation. Also, gross alterations of chromosome structure and the complete loss of sections of DNA may occur if the broken section is not attached to the centromere and the chromatids do not move properly at meiosis.

The effects of small changes to the base code such as base transitions and frameshift mutations are illustrated in Table 6.1. It can be seen that the consequences in translation vary with the particular change.

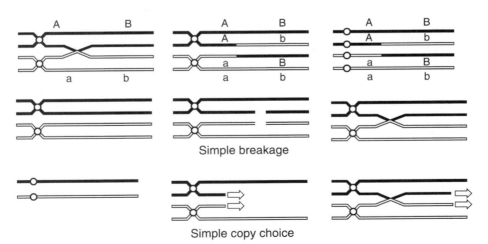

Figure 6.45 Crossover at meiosis. Two homologous chromosomes are shown aligned in the top line. Each consists of two chromatids, joined together at the centromere. A and B represent genes on one chromosome, and a and b the corresponding alleles on the other chromosome. After recombination two new gene combinations are apparent. The middle and bottom lines represent other possible methods of recombination. *Source*: From Refs. 12 and 13.

Table 6.1 Different Effects of Small Changes to the DNA Base Code

DNA triplet code	RNA triplet code	Amino acid specified
AAA (wild type)	UUU	Phenylalanine
AAG (base transition)	UUC	Phenylalanine (neutral mutation)
AGA (base transition)	UCU	Serine-wrong amino acid (missense mutation)
ATT (frameshift)	UAA	Stop codon-no amino acid formed (nonsense mutation)

6.7.8 Clastogenesis

This is the damage to chromosomes resulting in large scale aberrations, including strand breaks and displacements, terminal and interstitial deletions of segments, rearrangements and exchange between chromosomes, production of fragments such as micronuclei, acentric fragments, dicentric chromosomes, and ring chromosomes.

Damaged DNA can give rise to errors of DNA replication such as deletions and exchanges of chromatids. Where exposure to certain compounds occurs in G_1 phase (Fig. 6.39) with damage replicated in S phase, only one chromatid would be affected. Where exposure occurs in S or G_2 phase, both may be affected depending on the compound. Mis-ligation of strand breaks or interactions of regions being repaired by nucleotide excision repair can also lead to chromosome damage. Incorrect joining can give rise to chromosomal exchanges within and between chromosomes. Failure to rejoin double-strand breaks gives rise to deletions from which micronuclei and other fragments can arise. Such effects could be the result of erroneous repair.

Certain compounds, including many of those able to induce base-pair transformations, cause the breakage of chromosomes. Alkylating agents, both monofunctional and difunctional, cause chromosome breakage, the latter type probably by causing cross-linking across the DNA molecule. Inhibition of DNA repair may be another cause of this type of mutation. Thus, mechlorethamine exposure causes broken chromosomes and detached terminal fragments. It causes chromosomes to undergo a breakage-fusion-bridge cycle.

Other mutagens will also cause chromosome breaks such as bromouracil (see above) and the antibiotic streptonigrin, which is a very potent chromosome breaker.

Inhibition of repair can be a major part of the process of chromosome breakage, for example, the chemical hydroxylamine will inhibit the repair of radiation-induced strand breaks.

Chromosome breakage may therefore be linked to other mechanisms of mutagenesis in some fundamental manner. The correction of errors involves breakage of the DNA molecule for both the "cut and patch" type of repair and for sister-strand exchange (Fig. 6.46).

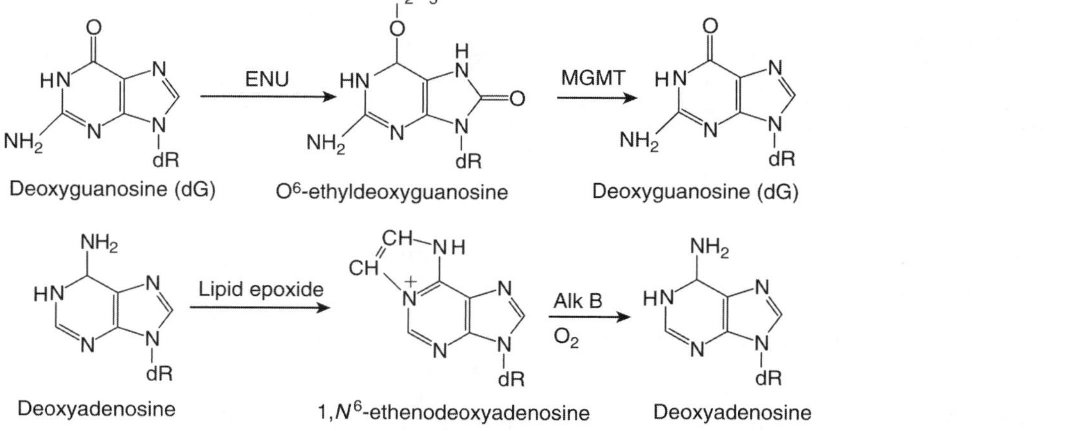

Figure 6.46 Representation of two mechanisms of repair of DNA damage such as ultraviolet light–induced dimerization. The upper line represents cut and patch repair, the lower sister-strand exchange. Thick lines represent newly synthesized DNA. *Source*: From Ref. 12.

Interference with these processes could clearly result in breakage in one or both chains of the DNA molecule, which it is now accepted forms the backbone of the chromatid. Large deletions may also occur after breakage of the chromosome if the repair processes are compromised.

Sister chromatid exchange (Fig. 6.46) occurs in S phase and is presumed to be a result of errors in the replication process.

However, even major chromosome abnormalities can become established in the germ line, and so agents capable of inducing breaks are able to cause heritable effects.

6.7.9 Aneugenesis

This involves interference with the processes of mitosis or meiosis and can lead to major changes in chromosome number called aneuploidy. During the normal M phase of the cell cycle, when mitosis takes place prior to separation of the daughter cells, sister chromosomes, or in meiosis, homologous chromosomes, line up in a plane of the metaphase plate. The chromosomes then become associated with proteins, α- and β-tubulin, heterodimers of which polymerize to form the microtubules. These microtubules create fibers, which form the metaphase spindle, a bipolar arrangement, each half of which migrates to the opposite pole of the dividing cell. The fibers extend from the centrosomes to the chromosomes. The spindle fibers should pull the sister chromatid pairs apart, and therefore each chromatid moves toward the appropriate centrosomes. Thus, after mitosis the partition should ensure that each daughter cell has the right chromosomes in the correct number.

Clearly, this complex process must be controlled by the cell such that there are checkpoint controls, which make sure that there are two centrosomes and two half spindles, that the correct chromatids associate with the correct half-spindle and that separation does not occur until all of the chromatids are in the correct place. Failure of any part of this system could lead to the wrong number of chromosomes in one daughter cell. This uneven segregation is known as chromosomal nondisjunction. A number of chemicals, including some drugs, will interfere with this process. Thus, spindle formation can be disrupted, so the chromosomes may segregate in a random fashion.

Colchicine is a specific spindle poison, which binds to tubulin, and inhibits its polymerization. Consequently, colchicine blocks mitosis, causing aneuploidy, the unequal partition of chromosomes, and metaphase arrest.

The result of nondisjunction, loss of one chromosome, for example, may be incompatible with survival of the cell, so aneugens are often cytotoxic. This obviously only occurs in cells, which are actively dividing, therefore such compounds find use as anticancer drugs. Two examples are vincristine and vinblastine, two vinca alkaloids, derived from a plant.

These bind to the α- and β-tubulin heterodimers, which block the GTP binding site and prevent the addition of further heterodimers so the whole microtubular spindle cannot form.

Other chemicals, which affect tubulin polymerization or spindle microtubule stability, are podophyllotoxin and the drugs, paclitaxel, benomyl, griseofulvin, nocodazole, and colecimid.

The unequal partition of chromosomes or nondisjunction is a serious effect if the affected daughter cells survive. Down syndrome or Trisomy 21 is the result of chromosome nondisjunction in humans, those affected having 47 chromosomes instead of 46. This unequal partitioning of chromosomes can occur at mitosis in germ cells or during meiosis in the production of sperm or ova.

6.7.10 DNA Repair

There are several levels of protection of cellular DNA from reactive intermediates such as ROS and reactive chemical metabolites. The first line of defense as already discussed previously is detoxication of the reactive agent by reaction with GSH and other antioxidants. However, if electrophilic or oxidizing exogenous or endogenous metabolites escape detoxication and damage DNA, then repair of the damage is possible. So repair is a backup for detoxication. If repair is either not possible because damage is too extensive or does not occur, the cell does have the option of apoptosis—programmed cell death to save the organism from possibly cancerous cells developing, serious mutations, or other adverse effects. If the damage is repaired before replication and cell division, no heritable changes will occur.

Mutations caused by chemicals are mostly due to errors of DNA replication on a damaged template. Repair therefore is critical as if this occurs efficiently and before replication, the mutation rate will be low. This could be affected by cell cycle checkpoint genes, which stop the cell from entering S phase (Fig. 6.39).

As already indicated, the repair processes may inadvertently contribute to errors leading to mutation and more serious endpoints. Some chemicals can affect the repair process, so increasing mutation rates. Lack of a repair system also leads to increased mutations and cancer as occurs in patients suffering from the disease xerodermum pigentosum. These individuals are unable to remove thymidine dimers formed by exposure to ultraviolet light in the DNA of skin cells. They consequently suffer a very high incidence of skin cancer.

During the process of DNA replication, the normal rate of error is remarkably low, but errors still occur. Consequently, the cell has a system for repairing errors (Fig. 6.46) occurring during cell division when chromosomal DNA sequences are replicated by one of a series of polymerases. This is carried out by mismatch repair enzymes, which monitor and detect miscopied new DNA sequences. However, these repair processes are generally for normal bases put in the wrong place or the wrong number of bases in a repeat sequence.

Cellular DNA is however, attacked by both internal, endogenous agents such as ROS and exogenous agents such as radiation and chemicals. Indeed, it is now clear that as a result of the use of sensitive analytical techniques the majority of DNA damage and mutations occurring in cells are due to endogenous agents arising from normal metabolism rather than exogenous agents.

The DNA molecule is to an extent protected by its structure, with the bases internalized, so that the nucleophilic sites are less-readily accessible. But damage still occurs.

As we have seen, damage causing changes to the normal bases, if not repaired, leads to changes to the base code. The cell needs to repair these damage errors as well as mismatch errors.

Basically, DNA repair is either error free, returning the DNA to its original undamaged state, or error prone, giving improved but still altered DNA.

The processes generally involve the following (Fig. 6.46):

1. Recognition of damage
2. Removal of damage
3. Repair-DNA synthesis
4. Ligation sticking together

The probability of DNA damage giving a permanent genetic alteration, a mutation, depends on the particular repair process, the rate of repair, whether repair occurs at all and misrepair.

Base Excision Repair
This occurs, for example, after oxidative damage (caused by either exogenous or endogenous agents).

After recognition, there are several stages.

1. A glycosylase enzyme recognizes and removes the damaged base to give an apurinic or apyrimidinic site.
2. Then a lyase and endonuclease cut the 5' and 3' sides of the strand to remove the deoxyribose.
3. DNA polymerase β repairs the section.
4. DNA ligase joins the ends (phosphate bonds).

Nucleotide Excision Repair
This is for the removal of bulky lesions such as polycyclic hydrocarbons attached to bases (e.g., benzo[a]pyrene) and requires distortion of the double helix. The system can recognize bulky adducts.

The steps involved are as follows:

1. Damage recognition by the nucleotide excision repair multiprotein complex.
2. Cleavage up and downstream of the damage.

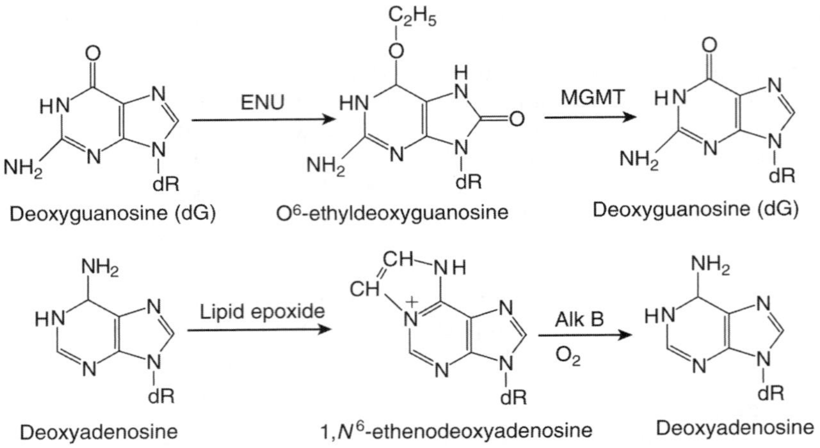

Figure 6.47 The removal of small alkyl groups from bases by O^6-MGMT or Alk B. *Abbreviations*: MGMT, methylguanosine methyl transferase; ENU, ethylnitroso urea.

3. Removal of a 25- to 30-nucleotide fragment.
4. DNA polymerase–mediated replacement of the removed section.
5. DNA ligase joins the ends.

If there is damage in the actively transcribed strand, this is a priority, the nucleotide excision repair complex is coupled to the RNA polymerase and may blockade the latter.

Removal of Alkyl Groups O^6- Methyl Guanine DNA Methyltransferase Repair (MGMT)
This protects against alkylating agents such as ethylnitrosourea, which will ethylate DNA bases (Fig. 6.47). The methyl or ethyl group is transferred to a cysteine in the enzyme, regenerating the normal base.

There is also an enzyme for removal of methyl and ethyl groups from bases by oxidizing them to their respective aldehydes-the Alk B repair enzyme. Lipid epoxides, which may be produced in inflammatory tissue and can yield alkylated bases, will also be repaired by this enzyme (Fig. 6.47). Metabolic activation of vinyl chloride will also yield the same adduct (see chap. 7).

The carcinogenic industrial chemical vinyl chloride produces the same alkylated base, with a C_2 unit across the exocyclic and ring nitrogen atoms of adenine (see chap. 7).

DNA Double-Strand Breaks
This type of damage can be repaired but if not will have serious consequences for the cell. If not repaired, the cell may enter apoptosis and die. Alternatively, the damage will lead to a check on the progression of the cell cycle. Thus, progression from G_1 phase to S phase is blocked by such damage, and in S phase (Fig. 6.39), DNA replication is halted. DNA strand breaks can be repaired by either homologous recombination or nonhomologous end joining.

The homologous recombination can take place more easily if chromosomes have replicated but not divided and cannot take place in G_1 phase when there are no sister chromatids.

Homologous Recombination Repair
Homologous recombination is a normal process in both somatic cells and germ cells, but repair of strand breaks can occur by the same process.

It involves several stages and the use of the sister chromatid:

1. Production of a 3′ end to the strand(s) by exonuclease.
2. Invasion of the sister chromatid, which has unwound to give complementary strands.

3. Elongation of broken strands by DNA polymerase.
4. Separation of chromatids and pairing of strands of damaged chromatid.
5. Elongation of strands to fill in remaining gaps and restoration of the helix with a DNA ligase.

Nonhomologous End Joining

Unlike the foregoing, this process has no template, and so it tends to occur in G_1. The cut/damaged ends of the strands are simply fused. Hence, there may be loss of several base pairs compared with the original. It is therefore error prone. The initial part of the process is similar to homologous recombination with production of a 3' end by a exonuclease, but then the strands are brought together to allow base pairing, if possible. Then the gaps are filled in and joined by a DNA ligase, and the double helix is reconstructed.

Strangely, this process does have a normal role in gene rearrangement for the production of antibodies and T-cell receptors.

6.7.11 Mutagenesis in Mammals

The effects of mutagens on mammalian cells *in vivo* are less-clearly defined than their effects on bacterial cells. When mutagens act on mammalian germ cells, inherited defects may arise, and when somatic cells are exposed to mutagens, cellular organization may be disrupted, giving rise to tumors. It now seems clear that many mutagens are carcinogens, and most, but not all, carcinogens are mutagenic (see below).

Major mutagenic changes in somatic cells will clearly lead to total disruption of DNA-directed cellular organization, possibly cell death but probably an inability to grow and divide.

Mutagenic attack in rapidly dividing cells, such as in embryonic tissue, will obviously be particularly serious, giving rise to defects in growth and differentiation, leading to birth defects. Consequently, mutagens are often also teratogens, although some are directly cytotoxic, and it may be this rather than their mutagenicity, which causes the teratogenic effect.

However, both carcinogenesis and teratogenesis are discussed separately, and therefore attention will be confined to the susceptibility of the germ cell line.

In the humans, the primordial germ cells are differentiated from about the sixth week of gestation and are consequently susceptible from then onward. In the female, production of the primary oocytes, which involves the first meiotic division, occurs in fetal life. These primary oocytes do not mature into ova until puberty, with the second meiotic division yielding one ovum from each primary oocyte (Fig. 6.48).

In the male, the stem cells or primary spermatogonia undergo mitotic division regularly, one of the products becoming the primary spermatocyte, which after two meiotic divisions gives rise to four spermatozoa. This continues throughout childhood, but before puberty the sperm degenerate. In the female organism, chromosomes replicate about 70 times in the production of an ovum, while in the male 950 replications take place in the production of a sperm. In both cases, however, only two continuous meiotic divisions have occurred, involving one possible crossover of chromatids.

It is therefore clear that susceptibility to the action of mutagens on the mammalian germ line is different in the female and the male. In the female, mutagens, which act on replicating DNA or on the process of mitosis, have effects on the fetal germ tissue and possibly on meiosis occurring at maturation of the ovum. Mutagens, which act on non-replicating DNA, are clearly active throughout the lifetime of the female organism. As the greatest number of mitoses take place in the fetus, exposure to mutagens should be particularly avoided during this time, the first trimester of pregnancy. This is also the period of greatest teratogenic susceptibility. In testing for mutagenic effects in the female mammalian organism, therefore, continuous exposure of the animal to the mutagen may be important, particularly with mutagens, which act at the meiosis, which occurs during the short period of the maturation of the ovum.

In the male, mutagens, which act on replicating DNA or on the process of mitosis, do so in the fetus and during the period of reproductive life. Mutagens acting on non-replicating DNA are also active throughout fetal and reproductive life. Agents, which are mutagenic at meiosis will be so during the period of maturation of the spermocytes to produce sperm.

As would be predicted, there are large differences in mutagenicity depending upon the time of treatment. This reflects the sensitivity of the various stages of spermatogenesis or

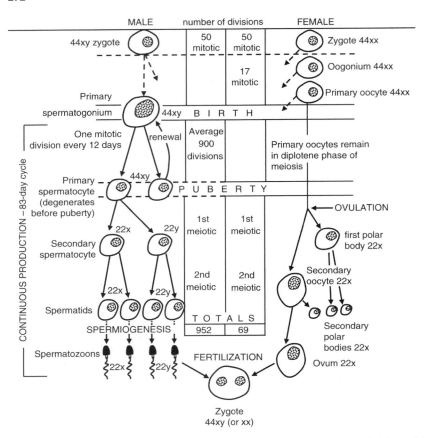

Figure 6.48 Gametogenesis in humans showing the number of nuclear divisions giving rise to the gametes. Each average fertilizing spermatozoon is the result of 900 divisions of the primary spermatogonium.

oogenesis. Therefore, the time after maturation when the organism is exposed to the mutagen and affected by it may give some indication of which stage is sensitive.

6.7.12 Determination of Mutagenicity and its Relation to Carcinogenicity
The potential mutagenicity of a foreign compound may now be rapidly and easily determined using a variety of short-term tests, one of the best known being the Ames test. This test uses histidine-requiring mutant strains of *Salmonella* bacteria. If a foreign compound causes a mutation, the bacteria may then grow on histidine-free medium and can be readily detected. The test has been adapted to use liver homogenate, usually from rats or other experimental animals, so as to identify mutagens, which require metabolic activation. Major changes in chromosomes as in aneuploidization and clastogenesis can be detected by microscopy (see "Bibliography").

There are now many different *in vitro* mutagenicity tests employing different strains of bacteria and also *in vivo* tests designed to detect mutations in higher organisms and mammalian systems. It is beyond the scope of this book to discuss these, but details can be found in other texts.

The relationship of mutagenicity to carcinogenicity is discussed at the end of the next section.

6.8 CHEMICAL CARCINOGENESIS

6.8.1 Introduction
Cancer is a major disease in humans, affecting 25%, and is the cause of death in 20%. Other mammalian species also suffer cancer, including experimental animals such as rats and mice,

and there is a spontaneous rate of tumor appearance in different sites, which varies between the species.

It has been known for a long time that external agents (carcinogens) are capable of causing cancer, namely viruses, radiation, and chemicals, both natural and man-made. This information has been derived from both epidemiology and experiments in animals.

Probably the earliest such observation was by Sir Percival Pott, an English physician, in 1775. He noted that chimney sweeps, who tended to suffer from scrotal cancer, were also exposed to soot and tar. He correctly connected these two events. More recent research confirmed that coal tar and the **aromatic hydrocarbons** it contains will cause cancer of the skin in experimental animals.

Because cancer is usually a progressive disease, and although some types are treatable many are fatal, the study of cancer is of major importance.

Thus, the biochemical changes associated with exposure to carcinogens have been studied in detail, and consequently, the target site and the interactions of carcinogens with it is more clearly understood than in other areas of toxicology.

The number of chemical substances known to be carcinogenic in animals, including humans, is now large. It is striking that how diverse the substances are in terms of structure, ranging from metals to complex organic chemicals, and there is large variation in potency. For example, the artificial sweetener **saccharin** may cause tumors in experimental animals after large doses ($TD_{50} = 10$ g), whereas aflatoxin B_1 is an exquisitely potent carcinogen ($TD_{50} = 1$ mg). The variation in structure and potency suggests that more than one mechanism is involved in carcinogenesis.

It is also clear that apart from exposure to carcinogens, other factors such as the genetic predisposition of the organism exposed may also be important. Thus, patients with the genetic disease xeroderma pigmentosum are more susceptible to skin cancer. It has already been mentioned that the incidence of bladder cancer is significantly higher in those individuals who have the slow acetylator phenotype.

6.8.2 Mechanisms Underlying Carcinogenesis

Cancer is the unrestrained, malignant proliferation of a somatic cell. The result is a tumor or neoplasm, which is a progressively growing mass of abnormal tissue. The term "cancer" usually applies to malignant tumors rather than benign ones as the latter do not normally invade healthy tissue. A characteristic of cancer cells is the loss of growth control, which allows uncontrolled proliferation. It is clear that a permanent heritable transformation has occurred in the cell because, on division, the daughter cells possess the same characteristics. Thus, cancer cells have undergone a malignant transformation in which they lose their specific function and acquire the capacity for unrestrained growth. The process of development of a tumor after exposure is typically quite long, and there are modifying factors such as the frequency of exposure, sex of the animal, genetic constitution, age, and the host immune system. Chemical carcinogens may cause the following:

1. The appearance of unusual tumors
2. An increase in the incidence of normal tumors
3. The appearance of tumors earlier
4. Increased multiplicity

A neoplasm is defined as a heritably altered relatively autonomous growth. So its characteristics are (*i*) heritable nature and (*ii*) relative autonomy.

There are several different types of carcinogen:

1. DNA-damaging carcinogens. These are genotoxic/mutagenic agents, which react chemically with DNA to cause mutations.
   They may be (*i*) direct carcinogens not requiring metabolic activation such as alkylating agents; (*ii*) indirect procarcinogens, requiring metabolic activation such as **polycyclic hydrocarbons**; (*iii*) inorganic agents. The mechanisms involved are not always clear. For example, **cadmium** and **plutonium**; or (*iv*) radiation and oxidative DNA damage. Ionizing radiation or cellular processes such as lipid peroxidation.
2. Epigenetic carcinogens. These are (*i*) tumor promoters; mitogenic agents, causing proliferation of initiated cells. Examples are **phorbol esters** and **peroxisome proliferators**;

(*ii*) hormones. For example, estrogens such as **diethylstilboestrol**; (*iii*) solid state carcinogens. Plastic, **asbestos**; (*iv*) immunosuppressive agents. For example, **cyclosporin**.
3.   Progressor agents. For example, **arsenic salts**, **benzene**.

It is now generally recognized that the cancer is a multistage process with at least three distinct stages. These are

1.   initiation,
2.   promotion, and
3.   progression.

Furthermore, it is accepted that the *somatic cell-mutation theory* underlies chemical carcinogenesis. This theory presumes that there is an interaction between the carcinogen and DNA to cause a mutation (Fig. 6.49). This mutation is then fixed when the DNA divides. Further replication then provides a clone of cells with the mutation, which may become a tumor.

This theory fits much of the data available on carcinogenesis. Thus, such a mutation would account for the heritable nature of tumors, their monoclonal character, the mutagenic properties of carcinogens and the evolution of malignant cells *in vivo*. However, although many carcinogens are also mutagens, by no means all of them are. Thus, ethionine and thioacetamide are carcinogens, but do not seem to be mutagens. Conversely, however, some mutagens such as sodium azide and **styrene oxide** do not appear to be carcinogenic, indicating that a single mutation alone is not sufficient to cause a tumor to develop.

However, interaction of carcinogens with other macromolecules such as RNA and the proteins, which regulate gene function, may also be involved in the carcinogenic process. Interactions with RNA have been suggested as possibly leading to a permanently altered state of differentiation through changes in gene expression.

Work with the carcinogen **acetylaminofluorene** found that residues of the compound in ribosomal RNA may correlate more closely with liver tumor formation than residues in DNA. Direct interactions with the mechanisms of protein synthesis, or with DNA and RNA polymerase enzymes, can also be seen as possible mechanisms. For instance, a modification of the polymerase enzymes by a carcinogen, either directly or indirectly, could lead to the erroneous replication of DNA or RNA and hence the permanent incorporation of a mutation.

Also, it is known that there are protective mechanisms operating in cells in that the chemical or its reactive metabolites can be detoxified. Furthermore, the mutation can be removed by repair. Alternatively, the initiated cell can be removed by apoptosis. Thus, not all cells with a mutation will maintain that genetic change and survive long enough for DNA replication to take place.

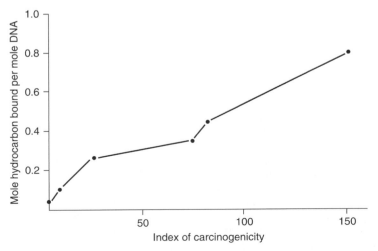

Figure 6.49 Relationship between the binding of carcinogens to DNA and carcinogenic potency. *Source*: From Ref. 14.

Consequently with initiation, metabolic activation, DNA repair, and replication are three important aspects of the process.

Once fixed by DNA replication however, then DNA damage is irreversible.

As some initiated cells can arise spontaneously, then clearly the processes of promotion and progression are crucial in the development of cancer.

Stages of the multistage process are as follows:

Initiation: This is an *irreversible* change, which results from a simple mutation in one or more cellular genes controlling key regulatory pathways.
Promotion: This is a *reversible* change, involving selective functional enhancement of signal transduction pathways in initiated cells.
Progression: This is an *irreversible* change and involves the continuing evolution of a basically unstable karyotype. Karyotype instability increases as the neoplasm grows.

The overall process is shown in Figure 6.50.

Classic Experiments with Dimethylbenzanthracene
Studies with the carcinogen dimethylbenzanthracene (**DMBA**) (a polycyclic hydrocarbon) have clarified some aspects of the mechanisms underlying carcinogenesis. Early studies found that after a single painting of mouse skin with DMBA, the *initiation* stage, multiple treatments with **tetradecanoyl-phorbol acetate** (TPA) (a *promoter*) would lead to the appearance of papillomas (Fig. 6.51). If treatment with TPA stopped, the papillomas would regress. If treatment continued, the papillomas developed into advanced papillomas, which once formed, remained without any further treatment. This was the *promotion* stage. Further TPA treatments could occasionally lead to the *progression* to carcinomas.

Treatment of the advanced papillomas with a second dose of DMBA without further TPA treatments also leads to progression to carcinomas (Fig. 6.51).

More recent studies have found that the first treatment with DMBA causes a mutation in the H-*ras* proto-oncogene, causing it to become an active oncogene.

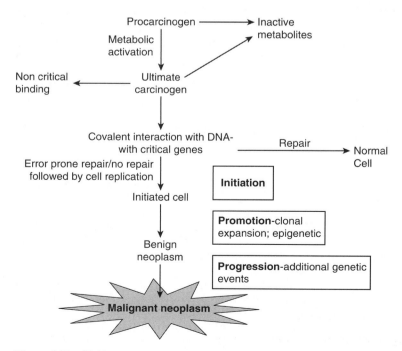

Figure 6.50 Multistage chemical carcinogenesis.

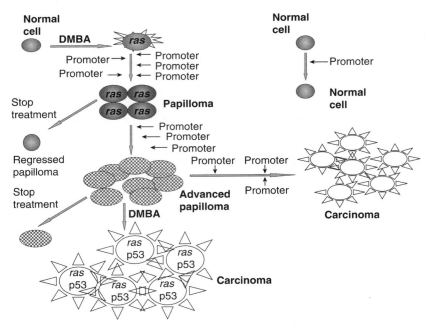

Figure 6.51 Scheme showing the initiation and promotion of skin papillomas and carcinomas after treatment with DMBA and promotion with TPA (Fig. 53). The initiation converts (mutation) proto-oncogene *ras* to oncogene and promotion stimulates growth and division, forming a papilloma. Further exposure to initiator DMBA, alters (mutation) tumor suppressor gene *p53* which collaborates with *ras* and causes carcinoma formation. See text for full explanation. *Abbreviation*: DMBA, dimethylbenzanthracene. *Source*: From Ref. 15.

The second treatment with DMBA causes a second mutation, this time in the tumor suppressor gene p53 (Fig. 6.51). These are the important mutations, there being many others.

Thus DMBA is an initiator, a potent mutagen. TPA is a promoter, and it appears to act by mimicking the signaling molecule diacylglycerol, which activates PKC- α. PKC stimulates transcription of several signaling pathways involving NF-κ, MAPK, and AP-1. The downstream effectors of PKC collaborate with the H-*ras* oncogene in some way to stimulate proliferation of the initiated cells. It is not clear why only the initiated cells are stimulated to yield a clone of initiated cells.

Within the general concept of the somatic cell mutation theory, it is now clear that there are a number of different mechanisms underlying carcinogenesis. Thus, the carcinogen can interact with a number of processes, not only causing a mutation, but influencing DNA repair and cell proliferation, for example. Furthermore, it seems cancers arise from an accumulation of several mutations.

6.8.3 Initiation

Evidence has accumulated that carcinogens such as **DMBA, 3-methylcholanthrene (3-MC) (another polycyclic hydrocarbon), methyl-nitro-nitroso-guanidine (MNNG)**, and **methyl nitroso urea (MNU)** directly damage DNA at the oncogene (H-*ras*) although causing different mutations. For example, MNU and MNNG cause G to A transitions on codon 12 of the H-*ras* gene. 3MC in contrast causes G to T transversions on codon 13 of the H-*ras* gene and A to T transversions on codon 61 of the same gene. DMBA also causes A to T transversions at codon 61.

However, it is clear from this that activation of an oncogene may only require a point mutation in the proto-oncogene sequence (e.g., H-*ras* , G to T in reading frame). This would mean that glycine was replaced with valine. All Ras oncoproteins have amino acid substitution in residues 12, 61, or 13.

Thus, the initiation process requires changes in gene expression with mutations in critical genes controlling cell division, and it seems it may require several such mutations. Hence, several years are needed for initiated cells to form tumors.

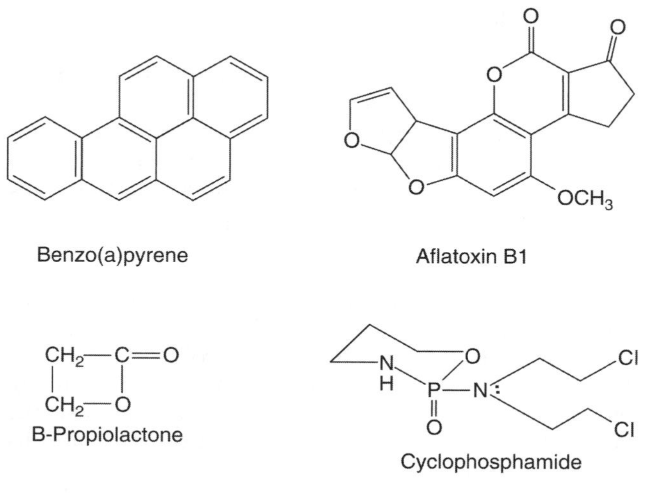

Figure 6.52 Structures of various carcinogens, which are also genotoxic initiators.

Examples of initiators include alkylating agents, **dimethylsulfate** and **β-propiolactone**, procarcinogens (requiring metabolic activation) such as **benzo(a)pyrene**, **aflatoxin**, and **cyclophosphamide** (Fig. 6.52) as follows:

Proto-oncogenes can be activated by mutations, chromosome translocations, gene amplifications, or promoter insertion.

As well as mutations caused directly by DNA damaging agents, mutations and other alterations in genetic material causing changes in gene expression can occur as a result of chemicals inhibiting DNA repair or causing cell proliferation and hence repeated DNA replication. As well as promoters, chemicals causing inflammation or cytotoxicity may also cause cell proliferation.

Oncogenes code for oncoproteins, which are mostly involved in signal transduction controlling/regulating cell growth, differentiation, or apoptosis. These pathways often involve modification of proteins by phosphorylation (e.g., of serine, tyrosine, threonine) by kinases.

For example, the growth factor receptor pathway involves an extracellular signal interacting with a transmembrane receptor. This then activates a tyrosine kinase causing phosphorylation of the receptor and allowing interaction with specific cellular proteins. This then sends a signal, which causes the expression of a particular gene. Oncoproteins alter this process.

Oncoproteins, for example, the Bcl-2 oncogene-protein can also cause inhibition of apoptosis, which removes one of the mechanisms by which the body can eliminate cells with significant DNA damage.

Cell cycle control is also important and is altered in cancer. For example, the protein pRb controls the cell cycle in cells and the level of phosphorylation of this protein is important. Some oncoproteins bind to and inhibit the activity of pRb, especially the hyperphosphorylated form found in the G_1 phase. So pRb is therefore inactivated and does not control the cell cycle.

Direct mutation in the Rb gene may also inactivate the protein product. Alternatively, the Rb promoter can be methylated and so inactivated.

Tumor suppressor genes, like oncogenes, also have a role in cell proliferation, but function as gatekeepers and govern the dynamics of cell proliferation.

However, activation of tumor suppressor genes gives a loss of function (the reverse of proto-oncogenes). Tumor suppressor genes usually control cell death and are negative regulators of cell growth. Such genes function to prevent appearance of abnormal cells such as those capable of forming tumors.

Thus, the tumor suppressor gene p53 regulates apoptosis. This is the gene most frequently mutated in cancer cells, which inactivates the gene. A point mutation in the reading frame of the p53 tumor suppressor gene gives growth-promoting capability.

As well as apoptosis, the p53 tumor suppressor gene causes DNA repair, blocks angiogenesis, and causes cell cycle arrest. As with oncogenes, tumor suppressor genes also have effects on cell cycle control.

Toxic compounds causing growth arrest (cell cycle arrest) or apoptosis also increase p53 protein.

Loss of tumor suppressor gene function can occur through mutation but also promoter methylation, which silences genes.

Although it is necessary to have both alleles altered to give complete loss of function, it is possible to have an intermediate position. If the second allele is altered, then loss of *heterozygosity* has occurred. This could be through chromosome non-disjunction or mitotic recombination.

6.8.4 Promotion

This stage involves alteration in gene expression and regulation via cell surface or cytosolic receptors. Most promoters affect gene expression via perturbation of the signal transduction pathways: tyrosine kinase, steroid, or G protein linked. The result is cell proliferation.

Indeed, in the experiment described above with the carcinogen DMBA, the production of papillomas in mice shows a sigmoid dose-response curve when plotted against the concentration of TPA. The dose-response curve for the interaction between the TPA and its receptor mirrors this dose-response curve almost exactly.

Promoters may well proliferate initiated cells that have formed spontaneously by erroneous cell division as well as those due to exposure to chemicals. Some promoters are tissue specific.

Examples of promoters are **TPA, TCDD, estradiol, nafenopin, cholic acid**, and **phenobarbital** (Fig. 6.53).

Preneoplastic cells, that is, those which have been initiated and which are selectively proliferated, may have altered mechanisms of cell cycle control. However, this selective enhancement of the cell cycle in initiated cells is not understood.

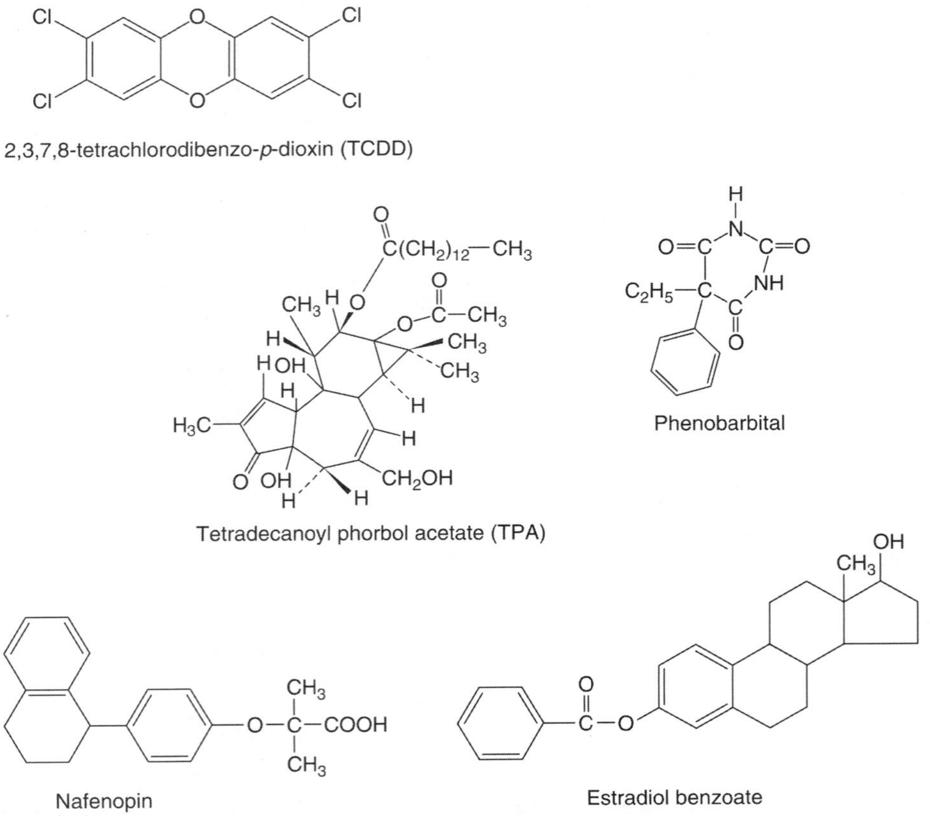

2,3,7,8-tetrachlorodibenzo-*p*-dioxin (TCDD)

Tetradecanoyl phorbol acetate (TPA)

Phenobarbital

Nafenopin

Estradiol benzoate

Figure 6.53 The structure of some known promoters.

Thus, promoters and initiators act together. Some promoters are irritant such as TPA and alcohol. Alcohol as consumed in spirits such as whisky can damage the epithelial cells of the throat causing the underlying stem cells to proliferate. However, these cells may already have mutations due to exposure to the polycyclic hydrocarbon in cigarette smoke such as 3-MC and benzo[a]pyrene, which are carcinogenic initiators. Hence the initiated cells can be encouraged to proliferate by alcohol exposure. Further damage to these proliferated cells by initiators could then cause a second mutation and continued use of alcohol would cause more promotion (i.e., proliferation), allowing progression to a tumor.

Thus, cytotoxic chemicals can act as promoters. Indeed chronic inflammation itself can lead to the proliferation of cells. TPA stimulates inflammation via PKC and TNF-α, which cause the accumulation of inflammatory cells.

Regression of cells after exposure to the promoter has been stopped because of apoptosis removing the altered cells.

6.8.5 Progression

It is necessary to have clonal expansion of initiated cells to gain a second mutation and so proceed to progression of the tumor. This may occur as a result of misrepair during cell division. Mitotic recombination and faulty segregation may contribute to loss of heterozygosity. This will allow mutant tumor suppressor genes to contribute to tumor progression. As repeated cell division shortens telomeric DNA, this ultimately leads to breakage-fusion-bridge cycles capable of causing mutagenesis. In the absence of normal p53 activity, this will lead to cancers.

Furthermore, exposure to cytotoxic or irritant chemicals causes inflammation, which generates ROS as a result of the inflammatory cells, which invade the area. These ROS can therefore potentially damage the DNA of the surrounding cells.

Thus, chemicals can contribute to and indirectly cause cancer as a result of stimulating either cell proliferation or inflammation or both. This includes hormones.

Some chemicals are progressor agents such as benzene, asbestos, and arsenic salts, for instance, and these could also be complete carcinogens. These may induce chromosomal breakage (clastogenesis) or progression could be spontaneous.

From studies in colon cancer, it seems that cells progressing from the normal to the tumor state accumulate mutations. That is, a neoplastic phenotype is acquired.

The changes include activation of the proto-oncogene (K-ras) and inactivation of three tumor suppressor genes (Fig. 6.54).

So tumor progression is also multistage, as each step involves disruption of key signaling processes and involves changes in both morphology and genes. This is illustrated in Figure 6.54.

But it is not so simple or clear, and some mutations seem irrelevant to the process and outcome. For example, it seems there is collaboration of mutant cellular genes. There are also some epigenetic steps such as hypomethylation and methylation of gene promoters.

Thus methylation of promoters yields repression, and derepression occurs through demethylation.

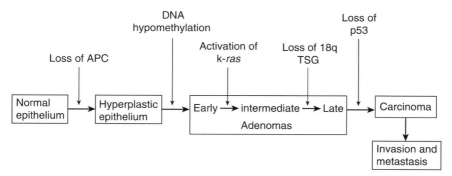

Figure 6.54 Tumor suppressor genes and colon carcinoma progression. APC (chromosome 5p) and *p53* (chromosome 17p) are TSGs. There is also loss of TSG or TSGs from chromosome 18q. *Abbreviations*: TSG, tumor suppressor gene; APC, antigen presenting cells. *Source*: From Ref. 15.

The hallmark and molecular characteristic of progression is karyotype instability. This is due to disruption of mitotic apparatus, alteration of telomere function, DNA hypomethylation, chromosome translocations and recombination, gene amplification, and gene transposition. There is also alteration of mismatch repair genes in some cancers.

Many initiating agents such as benzo[a]pyrene, discussed earlier, are complete carcinogens. Indeed most maybe, it might be only a matter of dose.

6.8.6 DNA Repair

The cell has the ability to recognize and repair damage to DNA as already discussed in the section on mutagenesis (see above). This must occur before cell division to be effective, otherwise there may be mispairing of bases, rearrangements and translocations of sections of DNA. The size of the DNA adducts will influence the amount of DNA repaired. Chemicals, which affect the repair process, may therefore increase the possibility of a neoplastic transformation. There are, however, differences in the rate of repair, depending on the type of repair and also species and tissue differences in this rate. For example, the damage to liver DNA caused by the carcinogen dimethylnitrosamine is more slowly repaired in the hamster than in the rat, and the hamster is more susceptible to tumors from this compound than the rat. There are also tissue differences in repair within the same species, which account for differences in susceptibility between tissues and organs. For example, the methylation of guanine at the O^6 position is more crucial than that at N^7, as excision repair of O^6 methylguanine is poor in certain tissues, such as the brain, in comparison with the liver and kidney. The brain is consequently more susceptible to tumor development.

6.8.7 Non-Genotoxic Mechanisms

Epigenetic carcinogens cause the appearance of tumors without a genotoxic effect. This includes promotion, immunosuppression, cocarcinogenicity, and cytotoxicity. This group probably involves diverse mechanisms. They may act by

1. exposure of a preexisting genetic abnormality;
2. impairment of DNA polymerase;
3. gene amplification;
4. alteration of chromosome composition; and
5. induction of a stable, altered state of gene expression.

Other types of carcinogen include the peroxisome proliferators. These are a diverse group of chemical substances, which may be potent rodent carcinogens, but are not mutagens. These substances include the hypolipidemic drugs such as **clofibrate** and various **phthalate esters** used as plasticizers. When given to rodents, these were found to cause an increase in the number of peroxisomes. Peroxisome proliferators not only cause an increase in the number of peroxisomes, but an increase in liver size due to hyperplasia. As well as these changes, it was later found they also caused liver tumors, and with some of the hypolipidemic drugs, the incidence of liver tumors may be as high as 100% in experimental animals. Certain peroxisomal enzymes involved in fatty acid oxidation (e.g., acyl CoA oxidase) produce hydrogen peroxide, and after proliferation of peroxisomes, induction of the enzyme leads to excess hydrogen peroxide, which may give rise to oxidative stress. An early suggestion to explain the carcinogenicity of these compounds is that the manifold increases in some of the enzymes involved in β-oxidation result in a large increase in the production of hydrogen peroxide. However, removal of this hydrogen peroxide depends on catalase, but this enzyme, although present in the peroxisome, is overwhelmed by the amount of hydrogen peroxide because catalase is only increased about twofold by peroxisome proliferators. Thus, there is an excess of hydrogen peroxide, which can diffuse out of the peroxisome into the cytoplasm and may reach the nucleus. The hydrogen peroxide and further products may therefore damage the DNA and other macromolecules. However, this does not seem to be the only or all of the mechanism, and it seems that hyperplasia is also important. Thus, increased cell proliferation may also be required as well as oxidative damage or some other type of damage to DNA. Thus, damaged cells may be promoted and progress to tumors by increased cell replication.

It is now known that the effect is mediated by a receptor, the PPARα receptor. Transgenic mice lacking this receptor do not show either the effect of increased peroxisome number and liver size or the liver tumors.

Peroxisomal proliferators may also inhibit apoptosis and are known to affect growth factor expression; therefore, these could be alternative or additional mechanisms.

However, peroxisomal-proliferating chemicals do not appear to cause the phenomenon in all species, and humans are believed to be a nonresponsive species. Thus, guinea pigs are refractory, and with humans given **ciprofibrate**, only limited peroxisomal proliferation was seen, and there was no increase in acyl CoA oxidase.

More recently, however, cynomologous monkeys have been found to show limited responses, including increased liver weight but no evidence of DNA damage.

Furthermore, peroxisomal proliferators show a clear dose threshold/no-effect level for both proliferation and the appearance of tumors. Therefore, in relation to assessment of the hazard and therefore risk to humans of peroxisomal proliferators, it seems that, although humans have a receptor, these chemicals do not pose a serious risk to humans.

The effects of the hypolipidemic drugs and the mechanisms underlying them will be discussed in more detail in chapter 7.

6.8.8 Cell Proliferation

Chemically induced cell proliferation can also be an important part of carcinogenesis and may be a mechanism of epigenetic carcinogenesis. The increase in cellular replication will lead to an increase in the frequency of spontaneous mutations and a decrease in the time available for DNA repair. During cell replication, the DNA molecule is unwound, allowing the transcription of certain genes, and hence the molecule is more vulnerable to attack by carcinogens. Tumor promotion and progression are associated with cell proliferation. The stimulus for cell proliferation might be direct mitogenic action due to compounds such as the peroxisome proliferators. Alternatively, cytotoxic compounds, especially when administered chronically, may cause regenerative cell growth. For example, the induction of α2-μ-globulin nephropathy by a variety of chemicals is associated with the appearance of renal tumors. The protein α2-μ-globulin is filtered by the kidney glomerulus and reabsorbed into the proximal tubules where it is degraded enzymatically by lysosomal enzymes. Such compounds as **1,2-dichlorobenzene**, **unleaded petrol**, and **dimethyl phosphonate**, which all cause both renal tumors and nephropathy, make α2-μ-globulin more resistant to enzymatic breakdown. This results in protein droplets and lysosomal overload, and the end result is necrosis of the proximal tubular cells. Thus, chronic exposure to such compounds results in increased cell proliferation. This is believed to cause promotion of neoplastic lesions, which may arise by spontaneous mutation.

In conclusion, it is clear that the mechanisms underlying chemical carcinogenesis are complex, involving numerous steps and more than a single gene mutation. It is also clear that the cell may protect itself from chemical carcinogens by repair of altered macromolecules, thereby resisting potent mutagens. Perhaps failure of these repair mechanisms leading to incorporation of mutations, coupled with epigenetic effects or the derepression of growth control, are some of the major factors involved in the causation of cancer by chemicals.

6.8.9 Testing for Carcinogenicity and Relation of Mutagenicity

As already discussed, there is a relationship between mutagenesis and carcinogenesis and the Ames technique is used to predict *potential* carcinogenicity. However, although it is clear that many mutagens, as detected in the Ames test, are also carcinogens, there are discrepancies. There are some mutagens, which are not carcinogens, and some substances shown to be carcinogenic in animals, which are not mutagenic.

The test is set up to be extremely sensitive to mutagenic activity and the incorporation of the liver homogenate allows for metabolically activated mutagens. However, the major route of metabolism of a mutagen *in vivo* may be detoxication rather than activation, which may account for some of the false positives. Furthermore, although the Ames and similar tests will indicate if a chemical is mutagenic (possibly after metabolic activation) and therefore is potentially an initiator, the chemical will not cause a malignant cancer unless it is in fact a complete carcinogen or is followed by exposure to a promoter or another stimulus such as inflammation.

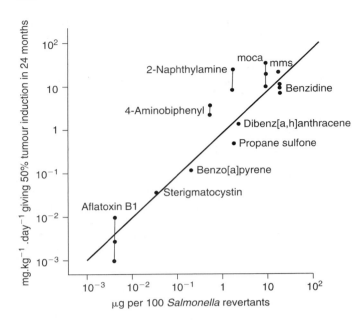

Figure 6.55 Log-log plot of relative carcinogenic potencies for a group of chemicals against mutagenic potency as determined by the Ames test. The potency is the amount of chemical giving tumors in 50% of the animals or 100 mutant colonies (revertants) of Salmonella bacteria. Aflatoxin B_1 is the most potent and benzidine the least potent. *Abbreviations*: Moca, 4,4'-methylene-*bis*-(2-chloroaniline); mms, methylmethane sulfonate. *Source*: From Ref. 16.

Consequently, the results of short-term mutagenicity tests must be interpreted with caution as regard potential carcinogenicity.

However, there is a reasonably good relationship between mutagenicity and carcinogenicity as shown by the graph in Figure 6.55.

The necessity for both initiation and promotion raises two important questions when considering testing for carcinogenicity. The current regulatory dogma has it that for a genotoxic carcinogen there will be no threshold, the dose response has to be presumed to pass through zero. However, the promotion stage of carcinogenesis involves proliferation in which the promoter binds to protein targets such as in signaling pathways. These types of responses show the classical sigmoid dose-response curve, which has a threshold and does not pass through zero. Therefore, a threshold for the overall process may well exist.

Furthermore, the common practice of testing at the maximum tolerated dose (MTD) is also suspect in view of the foregoing discussion. This is done because the chance of seeing a statistically significant increase in a tumor in an animal study is very slight because of the small numbers used. Therefore, the MTD is used to maximize the effect, but this often means that the doses used of the agent being tested are causing toxic effects in one or more organ or tissue. However, this as discussed earlier may cause either inflammation or cell death, which will stimulate proliferation, thus increasing the chances of cancers forming, either independently or in collaboration with the chemical agent administered.

Again, the cytotoxic effects of the chemical agent will also show a classic sigmoid dose response, and so at lower doses such cell death or inflammation will not occur. If as likely the expected human dose or exposure level is at a non-cytotoxic level, then tumors are much less likely.

For both these reasons, carcinogenicity testing at the MTD is questioned by many scientists (including Bruce Ames) and means that simply extrapolating from a high (i.e., MTD) dose to a low and more relevant dose is an overexaggeration of the risk. A good example of this is the situation, which occurred with the food additive (sweetener) saccharin, where only very high doses caused a slight increase in bladder tumors, probably because of irritation and inflammation.

SUMMARY

Toxic responses can be divided into six types: direct tissue lesions; pharmacological, physiological, and biochemical effects; teratogenesis; immunotoxicity; genetic toxicity; and carcinogenesis.

Toxicity can be detected as all or none responses: death; presence/absence of pathological damage; or graded responses: biochemical/ physiological changes; changes in normal status. **Direct tissue lesions** result in damage to a target organ. The target organ is determined by a number of factors: (*i*) function and position, (*ii*) blood supply, (*iii*) presence of uptake systems, (*iv*) pathways of intermediary metabolism, (*v*) absence of repair mechanisms, (*vi*) vulnerability to damage/disruption, (*vii*) biotransformation capabilities, and (*viii*) binding to particular macromolecules.

The liver is a major target organ because of factors (*i*), (*ii*), (*iv*), and (*vii*). The main effects are fatty liver, cytotoxicity (necrosis), cholestasis, cirrhosis, and tumors.

The kidney is a target because of function (excretion). The lung is a target because of position (absorption/site of exposure).

Most toxicity is due to the interaction between the *ultimate toxicant* and a *target molecule* via covalent bonding; non-covalent bonding, hydrogen abstraction, electron transfer, or an enzyme reaction.

The outcomes will depend both on the *attributes* and *function* of the target molecule.

The cellular injury, which underlies target organ toxicity, may result from various underlying events: primary, secondary, and tertiary events. Primary events result from initial damage, for example, lipid peroxidation, enzyme inhibition, covalent binding to crucial macromolecules, ischemia, and changes in thiol status.

Chemicals can initiate responses by affecting gene expression at several levels, transcription, intra- or extracellular signaling.

Chemicals can affect the regulation specialized cells (neurons, muscle cells) via transmitter and modulator substances.

Calcium is an important element in mammalian cells and levels of the ion are often critical in toxicological processes.

Secondary events result from primary events, for example, changes in membrane structure/permeability, mitochondrial damage, and lysosomal destabilization. Tertiary events are final observable manifestations, for example, fatty change and phospholipidosis, apoptosis, blebbing, and necrosis.

An increase in the permeability of the inner mitochondrial membrane, opening of a mega-channel, exacerbates ATP depletion, which becomes critical. This is the so called mitochondrial energy transition or MPT. When ATP is critically depleted, necrosis ensues whereas if substrates are present, which allow ATP generation, apoptosis occurs.

The response of cells to stress (including reactive oxygen species and reactive metabolites) includes the upregulation of synthesis of stress proteins: metallothioneins; hsp100, hsp 90, hsp70, hsp60, and hsp27.

Physiological, pharmacological, *and* **biochemical** responses do not usually lead to a tissue lesion, although there may be organ failure as a result. They may result from interactions of chemicals with receptors or specific enzymes leading to anoxia, inhibition of cellular respiration, respiratory failure, changes in pH, temperature, blood pressure or electrolyte balance, for example.

Teratogenesis is the specific interference with the development of the embryo or fetus.

There are four outcomes of this interference: death, malformations, functional deficiencies, and growth retardation. Because the development of the embryo and fetus involves a sequence of events, many chemicals and other agents can interfere with this, leading to one or more of the above four manifestations. There will be many mechanisms underlying the response, therefore such as interference with normal biochemical pathways as well as cytotoxicity.

It also follows that the timing of the interference will be crucial.

Another feature of teratogens is that they are usually selective for the embryo or fetus and are not toxic to the maternal organism. Because the development of the embryo is a sequence of events, the timing of the interference will often determine the nature of the outcome, that is, the type of malformation and whether it occurs.

Immunotoxicity can be caused by drugs and chemicals in one of four ways:

Immunosuppression
Immunostimulation
Hypersensitivity
Autoimmunity

The immune system functions to protect the organism against infection, foreign proteins, and neoplastic cells.

There are two main types of immune response: cell mediated and humoral. Cell-mediated immunity involves specifically sensitized thymus-dependent lymphocytes. Humoral immunity involves the production of antibodies (immunoglobulins) from lymphocytes or plasma cells.

Both the nonspecific and specific components of the immune system can be suppressed by chemicals including drugs. It involves the suppression of maturation and development of immune cells.

This is the reverse of immunosuppression, and is the situation where the immune response is enhanced.

Some of the new biological drugs are proteins, specifically designed to have stimulatory effects on the immune system.

One of the functions of the immune system is to protect the body from foreign invasion by dangerous pathogens (e.g., bacteria) recognized as not self. It does this by using a mixture of the innate immune and specific acquired immune systems.

Landsteiner in 1930s found that small molecules could stimulate an immune reaction if combined with a protein. This finding that small molecules could be recognized by the immune system if attached to a larger molecule such as an endogenous protein became the hapten hypothesis.

For the immune system to operate, there are three basic requirements:

1. Recognition of an agent as foreign or not self
2. Mount an appropriate response
3. Retain a memory of the foreign agent

Hypersensitivity reactions types I to III involve antibodies for recognition and as part of the process of removal of the antigen.

Antibodies are proteins known as immunoglobulins.

Type IV hypersensitivity reactions. This type of immune response is unlike the other three in that antibodies are not involved, the response being initiated by sensitized T_H1 cells.

Characteristics of immunological reactions are

dose-response relationship—not always clear;

exposure—repeated or chronic exposure is required for immunotoxic reactions as the immune system must first be sensitized;

specificity—the nature and extent of the immunological reaction does not generally relate directly to the chemical structure of the foreign compound, which is acting as a hapten; and

site of action—various target organs may be involved in the immune response, but this usually depends on the type of reaction rather than on the distribution of the foreign substance.

Genetic Toxicity

Damage to or changes in the genetic material and machinery involved in cell division can be caused by chemicals.

There are three types of effect: mutagenesis, aneuploidy, and clastogenicity

Mutagenesis is the production of a mutation by loss, addition, or alteration of individual bases or base pairs or a small number of base pairs in the DNA molecule.

Aneuploidy is the acquisition or loss of complete chromosomes, and clastogenicity is the induction of chromosomal aberrations as a result of loss, addition, or rearrangement of parts of chromosomes.

Mutations in the coding sequences, which will include genes for specific proteins, will potentially cause adverse changes in the gene and therefore the phenotype. Chemicals can interact with DNA in a variety of ways including adduct formation, base substitutions (transitions or transversions), and base additions or deletions.

Clastogenesis

This is the damage to chromosomes, resulting in large scale aberrations, including strand breaks and displacements, terminal and interstitial deletions of segments, rearrangements, and exchange between chromosomes.

Aneugenesis

This involves interference with the processes of mitosis or meiosis and can lead to major changes in chromosome number.

DNA Repair

There are several levels of protection of cellular DNA from reactive intermediates such as ROS and reactive chemical metabolites.

Chemical Carcinogenesis

Cancer is the unrestrained, malignant proliferation of a somatic cell producing a neoplasm or tumor defined as a heritably altered relatively autonomous growth. So its characteristics are (*i*) heritable nature and (*ii*) relative autonomy.

The *somatic cell mutation theory* underlies chemical carcinogenesis so there is an interaction between the carcinogen and DNA to cause a mutation.

There are several different types of carcinogen:

DNA-damaging carcinogens
Epigenetic carcinogens
Progressor agents

Cancer is a multistage process with three distinct stages as follows:

1. Initiation, an *irreversible* change due to a simple mutation in one or more cellular genes controlling key regulatory pathways
2. Promotion, a *reversible* change involving selective functional enhancement of signal transduction pathways in initiated cell
3. Progression, an *irreversible* change with the continuing evolution of a basically unstable karyotype

Epigenetic carcinogens cause tumors without a genotoxic effect and act via promotion, immunosuppression, cocarcinogenicity, and cytotoxicity.

Chemically induced cell proliferation also may be a mechanism of epigenetic carcinogenesis.

REVIEW QUESTIONS

1. List the different types of toxic response, which chemicals may cause.
2. List the reasons why an organ might be a target for toxicity.
3. List five ways in which toxicity may be detected.
4. Give the different types of toxic response shown by the liver.
5. Indicate which of the following are true: The area around the portal tract in the liver is
   a. the centrilobular region
   b. zone 1
   c. the periportal region
   d. the acinus
   e. a lobule

6. Which of the following are true? Compared with zone 1, zone 3 of the liver lobule has
   a. more oxygen
   b. more cytochrome P-450
   c. less triglyceride synthesis
   d. less glutathione

7. By which mechanisms may fatty liver may be caused?
   a. inhibition of protein synthesis
   b. increased synthesis of triglycerides
   c. decreased oxidation of fatty acids
   d. hydropic degeneration

8. Give an example of a chemical, which damages (a) the liver, (b) the kidney, (c) the lung, (d) the nervous system, (e) the testes, (f) the heart, and (g) the blood.

9. Which are the important primary events underlying cellular injury?

10. Explain the mitochondrial permeability transition and why it is important in cell toxicity.

11. List the ways of detecting damage to (a) the kidney and (b) the liver.

12. Explain how damaged macromolecules are repaired.

13. What are the toxic products of lipid peroxidation?

14. List the main characteristic features of teratogenesis.

15. Give two examples of teratogens with suspected mechanism.

16. Indicate two ways that chemicals can interact with the immune system.

17. Which of the following are involved with type III immune responses caused by chemicals?
    1. antigens
    2. immunoglobulins
    3. killer lymphocytes
    4. complement

18. What are the four characteristics of chemically induced immune reactions?

19. Explain the "danger hypothesis."

20. Give the three types of effect, which may result from interaction of chemicals with genetic material.

21. List the four types of mutagenic change.

22. Which simple bacterial test can be used to detect mutagens?

23. The carcinogenic process is not a single-stage process. How many stages has it and what are they?

24. There are three types of carcinogens. List these, and give an example of each.

25. Explain the somatic cell mutation theory.

REFERENCES

1. Lemasters JJ, Nieminen A-L, Qian T, et al. The mitochondrial permeability transition in cell death: a common mechanism in necrosis, apoptosis and autophagy. Biochim Biophys Acta 1998; 1366:177–196.
2. Qian T, Herman B, Lemasters JJ. The mitochondrial permeability transition mediates both necrotic and apoptotic death of hepatocytes exposed to Br-A23187. Toxicol Appl Pharmacol 1999; 154:117–125.
3. Timbrell JA. Biotransformation of xenobiotics. In: Ballantyne B, Marrs TC, Syversen T, eds. General and Applied Toxicology. 2nd ed. Oxford:Macmillan Reference, 1999.
4. Timbrell JA. The liver as a target organ for toxicity. In: Cohen GM, ed. Target Organ Toxicity. Boca Raton, FL: CRC Press, 1989.
5. Rappaport AM. Anatomical considerations. In: Schiff L, ed. Diseases of the Liver. Philadelphia: J.B. Lippincott, 1969.
6. Tuchmann-Duplessis H, Mercier-Parot L. Ciba Foundation Symposium on Congenital Malformations. In: Wolstenholme GEW, O'Connor CM, eds. Boston: Little Brown, 1960.
7. Wilson JG. Environment and Birth Defects. New York: Academic Press, 1973.
8. Gray H. Anatomy of the Human Body. In: Clemente CD, ed. 30th ed. Philadelphia: Lea & Febiger, 1985.
9. Manson JM. Teratogens. In: Klaassen CD, Amdur MO, Doull J, eds. Casarett and Doull's Toxicology. 3rd ed. New York: Macmillan, 1986.

10. Metzler M, McLachlan JA. Diethylstilbestrol metabolic transformation in relation to organ specific tumor manifestation. Arch Toxicol 1979; (suppl 2):275–280.
11. Watson JD. The Molecular Biology of Gene. New York: Benjamin WA, 1965.
12. Goldstein A, Aronow L, Kalman SM. Principles of Drug Action: The Basis of Pharmacology. New York: John Wiley, 1974.
13. Srb AM, Owen RD, Edgar RS. General Genetics. San Francisco: W.H. Freeman, 1965.
14. Brookes P. Quantitative aspects of the reaction of some carcinogens with nucleic acids and the possible significance of such reactions in the process of carcinogenesis. Cancer Res 1966; 26:1994–2003.
15. Weinberg RA. The Biology of Cancer. New York: Garland Science, 2007.
16. Meselson M, Russel K, et al. Comparisons of carcinogenic and mutagenic potency. In: Hiatt HH, Watson JD, Winsten JA, eds. Origins of Human Cancer, Book C: Human Risk Assessment. Cold Spring Harbor, NY: Cold Spring Harbor Laboratory Press, 1977.

BIBLIOGRAPHY

General

Albert A. Selective toxicity. A Classic Book With Many Interesting Examples of Toxicity and A Novel Approach. London: Chapman & Hall, 1979.
Aldridge WN. Mechanisms and Concepts in Toxicology. London: Taylor and Francis, 1996.
Boerlsterli UA. Mechanistic Toxicology. 2nd ed. Boca Raton: CRC Press, 2007.
Hayes AW, ed. Principles and Methods of Toxicology. 5th ed. Florida: CR Press, 2007. Various chapters.
Hodgson E, Smart RC, eds. Introduction to Biochemical Toxicology. 3rd ed. New York: Wiley Interscience, 2001. Various chapters.
Pratt WB, Taylor P, eds. Principles of Drug Action. New York: Churchill Livingstone, 1990.
Weinberg RA. The Biology of Cancer. New York: Garland Science, 2007.

Basic Mechanisms

Armstrong JS. Mitochondrial Membrane Permeabilization: the *sine qua non* for cell death. Bio Essays 2006; 28:252–260.
Caldwell J, Mills JJ. The Biochemical basis of toxicity. In: Ballantyne B, Marrs TC, Syversen T, eds. General and Applied Toxicology. Vol. 1. London: Macmillan, 2000.
Caro AA, Cederbaum AI. Oxidative stress, toxicology and pharmacology of CYP2E1. Annu Rev Pharmacol Toxicol 2004; 44:27–42.
Corcoran GB, Fix L, Jones DP, et al. Apoptosis: molecular control point in toxicity. Toxicol Appl Pharmacol 1994; 128:169–181.
Cotgreave IA, Morgenstern R, Jernstrom B, et al. Current molecular concepts in toxicology. In: Ballantyne B, Marrs TC, Syversen T, eds. General and Applied Toxicology. Vol. 1. 2nd ed. London: Macmillan.
Davies KJA. Oxidative stress, antioxidant defenses, and damage removal, repair and replacement systems. IUBMB Life 2000; 50:279–289.
Denecker G, Vercammen D, Declercq W, et al. Apoptotic and necrotic cell death induced by death domain receptors. Cell Mol Life Sci 2001; 58:356–370.
Dennehy MK, Richards KAM, Wernke GR, et al. Cytosolic and nuclear protein targets of thiol-reactive electrophiles. Chem Res Toxicol 2006; 19:20–29.
Gregus Z, Klaassen CD. Mechanisms of toxicity. In: Klaassen CD, ed. Cassarett and Doull's Toxicology, The Basic Science of Toxicology. 6th ed. New York: McGraw Hill, 2001.
Hansen JM, Go Y-M, Jones DP. Nuclear and mitochondrial compartmentation of oxidative stress and redox signaling. Annu Rev Pharmacol Toxicol 2006; 46:215–234.
Kaplowitz N, Fernandezcheca JC, Kannan R, et al. GSH transporters: molecular characterization and role in GSH homeostasis. Biol Chem Hoppe Seyler 1996; 377:267–273.
Kass GEN, Orrenius S. Calcium signaling and cytotoxicity. Environ Health Perspect 1999; 107:25–35.
Kehrer JP, Jones DP, Lemasters JJ, et al. Mechanisms of hypoxic cell injury. Toxicol Appl Pharmacol 1990; 106:165–178.
Kensler TW, Wakabayashi N, Biswal S. Cell survival responses to environmental stresses via the Keap-1-Nrf2-ARE pathway. Annu Rev Pharmacol Toxicol 2007; 47:89–116.
Nagelkerke JF, van der Water B. Molecular and cellular mechanisms of toxicity. In: Mulder GJ, Dencker L, eds. Pharmaceutical Toxicology. London: Pharmaceutical Press, 2006.
Orrenius S, Gogvadze V, Zhivotovsky B. Mitochondrial oxidative stress: implications for cell death. Annu Rev Pharmacol Toxicol 2007; 47:143–183.
Park BK, Kitteringham NR, Maggs JL, et al. The role of metabolic activation in drug induced hepatotoxicity. Annu Rev Pharmacol Toxicol 2005; 45:177–202.
Parkinson A. Biotransformation of Xenobiotics. In: Klaassen CD, ed. Cassarett and Doull's Toxicology, The Basic Science of Toxicology. 6th ed. New York: McGraw Hill, 2001.

This chapter has many examples of toxicity due to metabolism and metabolic activation.

Pumford NR, Halmes NC, Hinson JA. Covalent binding of xenobiotics to specific proteins in the liver. Drug Metab Rev 1997a; 29:39–57.

Pumford NR, Halmes NC. Protein targets of xenobiotic reactive intermediates. Annu Rev Pharmacol Toxicol 1997b; 37:91–117.

Raffray M, Cohen GM. Apoptosis and necrosis in toxicology: a continuum or distinct modes of cell death? Pharmacol Ther 1997; 75:153–177.

Reed DJ. Mechanisms of chemically induced cell injury and cellular protection mechanisms. In: Hodgson E, Smart RC, eds. Introduction to Biochemical Toxicology. 3rd ed. Connecticut: Appleton-Lange, 2001.

Trump BF, Berezesky IK. Calcium mediated cell injury and cell death. FASEB J 1995; 9:219–228.

Trump BF, Berezesky IK, Smith MW, et al. Relation between ion deregulation and toxicity. Toxicol Appl Pharmacol 1989; 97:6–22.

Wallace KB, Starkov AA. Mitochondrial targets of drug toxicity. Annu Rev Pharmacol Toxicol 2000; 40:353–388.

Weinberg RA. The Biology of Cancer. New York: Garland Science, 2007. Chapter 9, p53 and apoptosis: Master guardian and executioner.

Direct Toxic Action: Tissue Lesions

Acosta D, ed. Cardiovascular Toxicology. 2nd ed. New York: Raven Press, 1992.

Anthony DC, Montine TJ, Valentine WM, et al. Toxic responses of the nervous system. In: Klaassen CD, ed. Cassarett and Doull's Toxicology, The Basic Science of Toxicology. 6th ed. New York: McGraw Hill, 2001.

Ballantyne B. Peripheral sensory irritation: basics and applications. In: Ballantyne B, Marrs TC, Syversen T, eds. General and Applied Toxicology. Vol. 2. 2nd ed. London: Macmillan, 2000a.

Ballantyne B. Toxicology related to the eye. In: Ballantyne B, Marrs TC, Syversen T, eds. General and Applied Toxicology. Vol. 1. 2nd ed. London: Macmillan, 2000b.

Baskin SI, Behonick GS. Cardiac toxicology. In: Ballantyne B, Marrs TC, Syversen T, eds. General and Applied Toxicology. Vol. 1. 2nd ed. London: Macmillan, 2000.

Bloom JC, Brandt JT. Toxic responses of the blood. In: Klaassen CD, ed. Cassarett and Doull's Toxicology, The Basic Science of Toxicology. 6th ed. New York: McGraw Hill, 2001.

Braganza JM, Foster JR. Toxicology of the pancreas. In: Ballantyne B, Marrs TC, Syversen T, eds. General and Applied Toxicology. Vol. 1. 2nd ed. London: Macmillan, 2000.

Cohen DE, Rice RH. Toxic responses of the skin. In: Klaassen CD, ed. Cassarett and Doull's Toxicology, The Basic Science of Toxicology. 6th ed. New York: McGraw Hill, 2001.

Dekant W, Neumann HG, eds. Tissue Specific Toxicity—Biochemical Mechanisms. New York: Academic Press, 1992.

Fonum F. Neurotoxicology. In: Ballantyne B, Marrs TC, Syversen T, eds. General and Applied Toxicology. Vol. 1. 2nd ed. London: Macmillan, 2000.

Forge A, Harpur ES. Ototoxicity. In: Ballantyne B, Marrs TC, Syversen T, eds. General and Applied Toxicology. Vol. 1. 2nd ed. London: Macmillan, 2000.

Fox DA, Boyes WK. Toxic responses of the ocular and visual system. In: Klaassen CD, ed. Cassarett and Doull's Toxicology, The Basic Science of Toxicology. 6th ed. New York: McGraw Hill, 2001.

Glaister JR. Principles of Toxicological Pathology. London: Taylor & Francis, 1986.

Harman AW, Maxwell MJ. An evaluation of the role of calcium in cell injury. Annu Rev Pharmacol Toxicol 1995; 35:129–144.

Harvey P, ed. The Adrenal in Toxicology. London: Taylor & Francis, 1996.

Hermansky SJ. Cutaneous toxicology. In: Ballantyne B, Marrs TC, Syversen T, eds. General and Applied Toxicology. Vol. 1. 2nd ed. London: Macmillan, 2000.

Hinton RH, Grasso P. Hepatotoxicity. In: Ballantyne B, Marrs TC, Syversen T, eds. General and Applied Toxicology. Vol. 1. 2nd ed. London: Macmillan, 2000.

Kehrer JP. Systemic pulmonary toxicity. In: Ballantyne B, Marrs TC, Syversen T, eds. General and Applied Toxicology. Vol. 1. 2nd ed. London: Macmillan, 2000.

Kotsonis FN, Burdock GA, Flamm WG. Food toxicology. In: Klaassen CD, ed. Cassarett and Doull's Toxicology, The Basic Science of Toxicology. 6th ed. New York: McGraw Hill, 2001.

Lake BG. Mechanisms of hepatocarcinogenicity of peroxisome proliferating drugs and chemicals. Annu Rev Pharmacol Toxicol 1995; 35:483–507.

Lamb JC, Foster PMB. eds. Physiology and Toxicology of the Male Reproductive System. London: Academic Press, 1988.

Lock EA. Responses of the kidney to toxic compounds. In: Ballantyne B, Marrs TC, Syversen T, eds. General and Applied Toxicology. Vol. 1. 2nd ed. London: Macmillan, 2000.

Lansdown ABG. The toxicology of cartilage and bone. In: Ballantyne B, Marrs TC, Syversen T, eds. General and Applied Toxicology. Vol. 1. 2nd ed. London: Macmillan, 2000.

Mehendale HM, Thakore KN. Hepatic defences against toxicity: regeneration. In: McCuskey RS, Earnest DL, eds. Comprehensive Toxicology. Vol. 9, Hepatic and gastrointestinal toxicology. New York: Pergamon, 1997.

Mehendale HM, Roth RA, Gandolfi AJ, et al. Novel mechanisms in chemically induced hepatotoxicity. FASEB J 1994; 8:1285–1295.

Misulis KE, Clinton ME, Dettbarn W-D. Toxicology of skeletal muscle. In: Ballantyne B, Marrs TC, Syversen T, eds. General and Applied Toxicology. Vol. 1. 2nd ed. London: Macmillan, 2000.

Nagelkerke JF. Liver toxicity. In: Mulder GJ, ed. Dencker Pharmaceutical Toxicology. London: Pharmaceutical Press, 2006.

Nelson SD. Mechanisms of the formation and disposition of reactive intermediates that can cause acute liver injury. Drug Metab Rev 1995; 27:147–177.

Ramos KS, Melchert RB, Chacon E, et al. Toxic responses of the heart and vascular systems. In: Klaassen CD, ed. Cassarett and Doull's Toxicology, The Basic Science of Toxicology. 6th ed. New York: McGraw Hill, 2001.

Rappaport AM. The microcirculatory acinar concept of normal and pathological hepatic structure. Beitr Pathol 1976; 157:215–243.

Schnellman RG. Toxic responses of the kidney. In: Klaassen CD, ed. Cassarett and Doull's Toxicology, The Basic Science of Toxicology. 6th ed. New York: McGraw Hill, 2001.

Schiller CM, ed. Intestinal Toxicology, Target Organ Toxicology Series. In: Dixon RL, ed. New York: Raven Press, 1982.

Smith CV, Jones DP, Guenther TM, et al. Compartmentation of glutathione: implications for the study of toxicity and disease. Toxicol Appl Pharmacol 1996; 140:1–12.

Soni MG, Mehendale HM. Role of tissue repair in toxicological interactions among hepatotoxic organics. Environ Health Perspect 1998; 106(suppl 6):1307–1317.

Thomas JA, Thomas MJ. Endocrine toxicology. In: Ballantyne B, Marrs TC, Syversen T, eds. General and Applied Toxicology. Vol. 1. 2nd ed. London: Macmillan, 2000.

Treinen-Moslen M. Toxic responses of the liver. In: Klaassen CD, ed. Cassarett and Doull's Toxicology, The Basic Science of Toxicology. 6th ed. New York: McGraw Hill, 2001.

Turton J, Hooson J, eds. Target Organ Pathology. London: Taylor and Francis, 1998.

Van der Water B. Kidney toxicity. In: Mulder GJ, Dencker L, eds. Pharmaceutical Toxicology. London: Pharmaceutical Press, 2006.

Witschi HR, Last JA. Toxic responses of the respiratory system. In: Klaassen CD, ed. Cassarett and Doull's Toxicology, The Basic Science of Toxicology. 6th ed. New York: McGraw Hill, 2001.

Zimmerman HJ. Hepatotoxicity: The Adverse Effects of Drugs and Other Chemicals on The Liver. 2nd ed. Philadelphia: Lippincott-Williams & Wilkins, 1999.

Pharmacological, Physiological and Biochemical Effects

Albert A. Selective Toxicity. A classic book with many interesting examples and a novel approach. London: Chapman & Hall, 1979.

Albert A. Xenobiosis. London: Chapman & Hall, 1987.

Brunton L, Lazo J, Parker K, eds. Goodman & Gilman's The Pharmacological Basis of Therapeutics. New York: McGraw Hill, 2005.

Capen CC. Toxic responses of the endocrine system. In: Klaassen CD, ed. Cassarett and Doull's Toxicology, The Basic Science of Toxicology. 6th ed. New York: McGraw Hill, 2001.

Coulson CJ. Molecular Mechanisms of Drug Action. London: Taylor & Francis, 1994.

Dekant W, Vamvakas S. Glutathione-dependent bioactivation of xenobiotics. Xenobiotica 1993; 23:873–887.

Moreland DE. Effects of toxicants on electron transport and oxidative phosphorylation. In: Hodgson E, Levi PE, eds. Introduction to Biochemical Toxicology. 2nd ed. New York: Wiley Interscience, 2001.

Pratt WB, Taylor P, eds. Principles of Drug Action. New York: Churchill Livingstone, 1990.

Teratogenesis and Reproductive Toxicology

Finnell RH, Gelineau-van Waes J, Eudy JD, et al. Molecular basis of environmentally induced birth defects. Annu Rev Pharmacol Toxicol 2002; 42:181–208.

Gupta RC, Sastry BVR. Toxicology of the placenta. In: Ballantyne B, Marrs TC, Syversen T, eds. General and Applied Toxicology. Vol. 2. 2nd ed. London: Macmillan, 2000.

Juchau MR. Bioactivation in chemical teratogenesis. Annu Rev Pharmacol Toxicol 1989; 29:165–187.

Kacew S. Neonatal toxicology. In: Ballantyne B, Marrs TC, Syversen T, eds. General and Applied Toxicology. Vol. 2. 2nd ed. London: Macmillan, 2000.

McElhatton PR, Ratcliffe JM, Sullivan FM. Reproductive toxicology. In: Ballantyne B, Marrs TC, Syversen T, eds. General and Applied Toxicology. Vol. 2. 2nd ed. London: Macmillan, 2000.

Rogers JM, Kavlock RJ. Developmental toxicology. In: Klaassen CD, ed. Cassarett and Doull's Toxicology, The Basic Science of Toxicology. 6th ed. New York: McGraw Hill, 2001.

Ruddon RW. Chemical teratogenesis. In: Pratt WB, Taylor P, eds. Principles of Drug Action, The Basis of Pharmacology. New York: Churchill Livingstone, 1990.

Thomas MJ, Thomas JA. Toxic responses of the reproductive system. In: Klaassen CD, ed. Cassarett and Doull's Toxicology, The Basic Science of Toxicology. 6th ed. New York: McGraw Hill, 2001.

Tyl RW. Developmental toxicology. In: Ballantyne B, Marrs TC, Syversen T, eds. General and Applied Toxicology. Vol. 2. 2nd ed. London: Macmillan, 2000.

Wilson JG. Environment and Birth Defects. New York: Academic Press, 1973. A classic text.

Wilson JG, Fraser FC, eds. Handbook of Teratology. Vols 1–4. New York: Plenum Press, 1977. A reference text.

Witorsch RJ, ed. Reproductive Toxicology. New York: Raven Press, 1995.

Immunotoxicity

Boelsterli UA. Xenobiotic acyl glucuronides and acyl CoA thioesters as protein reactive metabolites with the potential to cause idiosyncratic drug reactions. Curr Drug Metab 2002; 3:439–450.

Burns-Naas LA, Meade BJ, Munson AE. Toxic responses of the immune system. In: Klaassen CD, ed. Cassarett and Doull's Toxicology, The Basic Science of Toxicology. 6th ed. New York: McGraw Hill, 2001.

Descotes J. An Introduction to Immunotoxicology. London: Taylor and Francis, 1999.

Kimber I, Dearman RJ. Evaluation of respiratory sensitization potential of chemicals. In: Ballantyne B, Marrs TC, Syversen T, eds. General and Applied Toxicology. Vol. 2. London: Macmillan, 2000.

Kretz-Rommel A, Rubin RL. Disruption of positive selection of thymocytes causes autoimmunity. Nat Med 2000; 6:298–305.

Laskin DL, Laskin JD. Role of macrophages and inflammatory mediators in chemically induced toxicity. Toxicology 2001; 160:111–118.

Miller K, Meredith C. Immunotoxicology. In: Ballantyne B, Marrs TC, Syversen T, eds. General and Applied Toxicology. Vol. 2. London: Macmillan, 2000.

Park BK, Kitteringham NR. Drug-protein conjugation and its immunological consequences. Drug Metab Rev 1990; 22:87–144.

Park BK, Kitteringham NR, Powell H, et al. Advances in molecular toxicology : towards understanding idiosyncratic drug toxicity. Toxicology 2000; 153:39–60.

Pratt WB. Drug allergy. In: Pratt WB, Taylor P, eds. Principles of Drug Action, The Basis of Pharmacology. New York: Churchill Livingstone, 1990.

Roitt IM, Brostoff J, Male D. Immunology. 3rd ed. London: Mosby-Year Book Europe, 1993.

Svensson C. Immunotoxicity. In: Mulder GJ, Dencker L, eds. Pharmaceutical Toxicology., London: Pharmaceutical Press, 2006.

Uetrecht J. Idiosyncratic drug reactions: current understanding. Annu Rev Pharmacol Toxicol 2007; 47:513–539.

Mutagenesis

Ames BN. Identifying environmental chemicals causing mutations and cancer. Science 1979; 204:587–593. A classic article.

Anderson D. Cytogenetics. In: Ballantyne B, Marrs TC, Syversen T, eds. General and Applied Toxicology. Vol. 2. 2nd ed. London: Macmillan, 2000.

Brusick D. Principles of Genetic Toxicology. 2nd ed. New York: Plenum Press, 1987.

Hellman B. Genotoxicity. In: Mulder GJ, Dencker L, eds. Pharmaceutical Toxicology. London: Pharmaceutical Press, 2006.

Preston RJ, Hoffmann GR. Genetic toxicology. In: Klaassen CD, ed. Cassarett and Doull's Toxicology, The Basic Science of Toxicology. 6th ed. New York: McGraw Hill, 2001.

Ruddon RW. Chemical mutagenesis. In: Pratt WB, Taylor P, eds. Principles of Drug Action, The Basis of Pharmacology. New York: Churchill Livingstone, 1990.

Tweats DJ, Gatehouse DG. Mutagenicity. In: Ballantyne B, Marrs TC, Syversen T, eds. General and Applied Toxicology. Vol. 2. 2nd ed. London: Macmillan, 2000.

Carcinogenesis

Cattley RE, Deluca J, Elcombe C, et al. Do peroxisome proliferating compounds pose a hepatocarcinogenic hazard to humans? Regul Toxicol Pharmacol 1998; 27:47–60.

Cooper CS, Grover PL, eds. Chemical Carcinogenesis and Mutagenesis II. Handbook of Experimental Pharmacology. Vol. 94. Berlin: Springer, 1990.

Esteller M. Aberrant DNA methylation as a cancer inducing mechanism. Annu Rev Pharmacol Toxicol 2005; 45:605–656.

Goodman JI, Watson RE. Altered DNA methylation : a secondary mechanism involved in carcinogenesis. Annu Rev Pharmacol Toxicol 2002; 42:501–525.

Klaunig JE, Kamendulis LM. The role of oxidative stress in carcinogenesis. Annu Rev Pharmacol Toxicol 2004; 44:239–267.

McGregor D. Carcinogenicity and genotoxic carcinogens. In: Ballantyne B, Marrs TC, Syversen T, eds. General and Applied Toxicology. Vol. 2. 2nd ed. London: Macmillan, 2000.

Miller EC, Miller JA. The metabolism of chemical carcinogens to reactive electrophiles and their possible mechanism of action in carcinogenesis. In: Searle CE, ed. Chemical Carcinogens. Washington D.C.: American Chemical Society, 1976. A seminal article by the original authors of this important concept.

Moggs JG, Orphanides G. The role of chromatin in molecular mechanisms of toxicity. Toxicol Sci 2004; 80:218–224.

Moggs JG, Goodman JJ, Trosko JE, et al. Epigenetics and cancer: implications for drug discovery and safety assessment. Toxicol Appl Pharmacol 2004; 196:422–430.

Pitot HC, Dragan YP. Chemical carcinogenesis. In: Klaassen CD, ed. Cassarett and Doull's Toxicology, The Basic Science of Toxicology. 6th ed. New York: McGraw Hill, 2001.

Ruddon RW. Chemical carcinogenesis. In: Pratt WB, Taylor P, eds. Principles of Drug Action, The Basis of Pharmacology. New York: Churchill Livingstone, 1990.

Smart RC, Akunda JK. Carcinogenesis. In: Hodgson E, Smart RC, eds. Introduction to Biochemical Toxicology. 3rd ed. New York: Wiley Interscience, 2001.

Waalkes MP, Ward JM, eds. Carcinogenesis. New York: Raven Press, 1994.

Williams GM. Mechanisms of chemical carcinogenesis and applications to human cancer risk assessment. Toxicology 2001; 166:3–10.

Techniques and Technologies

Ballantyne B, Marrs TC, Syversen T, eds. General and Applied Toxicology. Vol. 2. 2nd ed. London: Macmillan, 2000.

Hayes AW, ed. Principles and methods of toxicology. 5th ed. Florida: CR Press, 2007.

These Two Texts Both Have Sections on Techniques

Robertson D, Beuhausen S, Pennie W. Toxicopanomics: applications of genomics, proteomics and metabonomics in predictive and mechanistic toxicology. In: Hayes AW, ed. Principles and Methods of Toxicology. 5th ed. Florida: CR Press, 2007.

Stonard MD, Evans GO. Clinical Chemistry. In: Ballantyne B, Marrs TC, Syversen T, eds. General and Applied Toxicology. Vol. 1. London: Macmillan, 2000.

7 | Biochemical Mechanisms of Toxicity: Specific Examples

7.1 CHEMICAL CARCINOGENESIS

7.1.1 Acetylaminofluorene

This compound is a well-known carcinogen and one of the most widely studied, and therefore, this discussion must confine itself to the principles rather than details.

The study of acetylaminofluorene carcinogenicity has provided insight into the carcinogenicity of other aromatic amines and also illustrates a number of other important points. Acetylaminofluorene is a very potent mutagen and a carcinogen in a number of animal species, causing tumors primarily of the liver, bladder, and kidney. It became clear from the research that metabolism of the compound was involved in the carcinogenicity. The important metabolic reaction was found to be N-hydroxylation, catalyzed by the microsomal mixed-function oxidases, and this was demonstrated both *in vitro and in vivo*. Thus, N-hydroxyacetylaminofluorene (Fig. 7.1), the product, is a more potent carcinogen than the parent compound. The production of this metabolite *in vivo* was found to be increased ninefold by the repeated administration of the parent compound, a finding of particular importance considering the general use of single dose rather than multiple low-dose studies for evaluating the toxicity of compounds. This effect is presumably the result of induction of the microsomal enzymes involved in the production of N-hydroxyacetylaminofluorene. N-hydroxylation has since proved to be an important metabolic reaction in the toxicity of a number of other compounds. The N-hydroxy intermediate in particular is of importance for the carcinogenicity of a number of aromatic amino, nitro, and nitroso compounds. This intermediate may arise by reduction as well as oxidation (see chap. 4). However, N-hydroxyacetylaminofluorene is not the ultimate carcinogen, as it requires further metabolism to initiate tumor production. The N-hydroxylated intermediate is stable enough to be conjugated, and the N-O glucuronide of acetylaminofluorene is an important metabolite (chap. 4, Fig. 52).

Thus, it has been found that conjugates of N-hydroxyacetylaminofluorene are more potent carcinogens than the parent compound, and there is a wealth of evidence, at times confusing and conflicting, which indicates that both the sulfate and acetyl esters of N-hydroxyacetylaminofluorene are involved and that both may give rise to reactive **nitrenium** ions and **carbonium** ions. Thus, the metabolism of acetylaminofluorene may involve several different pathways as shown in Figure 7.1; some of the metabolites are mutagenic and some are also cytotoxic and carcinogenic. Thus, sulfate conjugation is clearly a requirement for both the cytotoxicity and tumorigenicity of N-hydroxyacetylaminofluorene and results in DNA adducts. The cytotoxicity is believed to be due to binding to DNA rather than protein. However, some data suggest that deacetylation also occurs to yield N-hydroxyaminofluorene, which is subsequently sulfated to yield the ultimate carcinogen. Deacetylation of acetylaminofluorene itself may also occur and could be an alternative route to N-hydroxyaminofluorene. Whether sulfation of N-hydroxyacetylaminofluorene is also involved is as yet unclear. The sulfate conjugate is not mutagenic when formed in situ in a *Salmonella in vitro* test system and has low carcinogenicity when applied to the skin of experimental animals. Similarly, the acetate ester of N-hydroxyacetylaminofluorene, which is carcinogenic, has been suggested as giving rise to a reactive nitrenium ion, which could be responsible for the DNA adduct isolated after *in vivo* exposure and the nucleoside adduct detected *in vitro*. Alternatively, the N-acetyltransferase enzyme (N,O-arylhydroxamic acid acyltransferase) is known to catalyze a transfer of the acetyl group from the nitrogen atom to the oxygen to yield the O-acetyl ester of N-hydroxyaminofluorene. This compound is mutagenic, is highly reactive, and spontaneously reacts with macromolecules. It can give rise

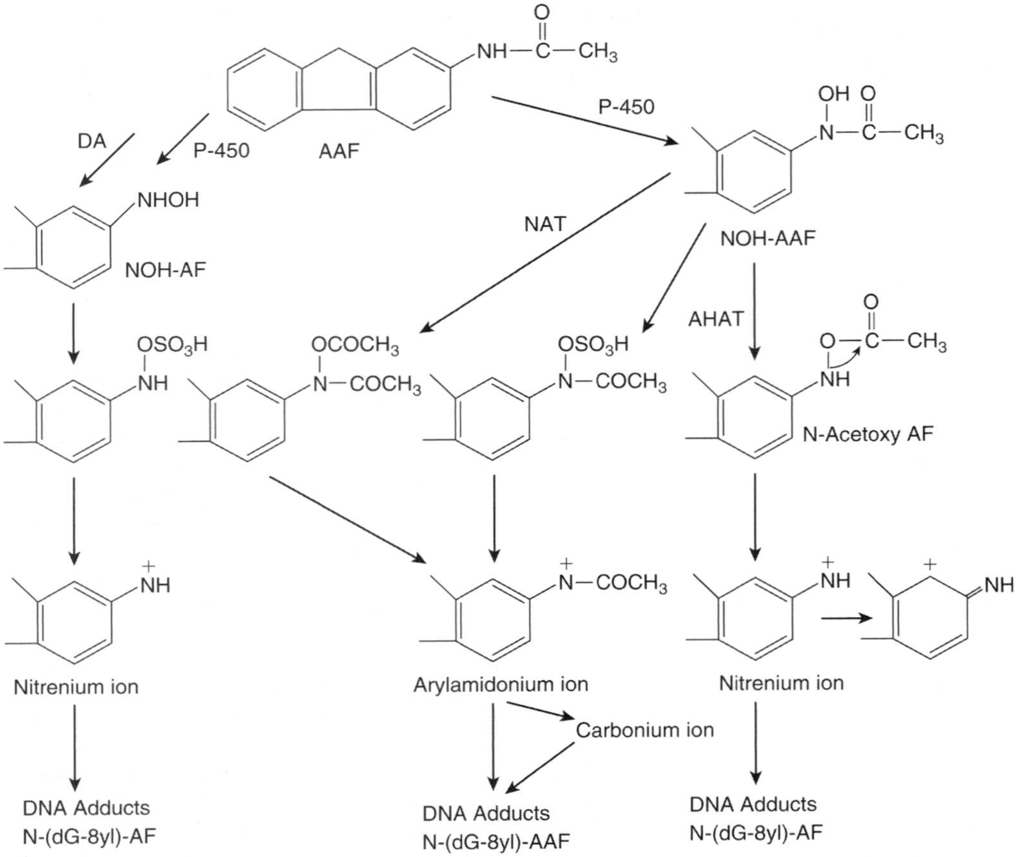

Figure 7.1 Some of the possible routes of metabolic activation of AAF. *Abbreviations*: AAF, acetylamino-fluorene; N-OH-AF, *N*-hydroxyaminofluorene; N-OH-AAF, *N*-hydroxyacetylaminofluorene; *N*-acetoxy AF, *N*-acetoxyaminofluorene; N-(dG-8yl)-AF, *N*-deoxyguanosinyl-aminofluorene; N-(dG-8yl)-AAF, *N*-deoxyguanosinyl-acety-laminofluorene; P-450, cytochrome(s) P-450; DA, deacetylase; NAT, *N*-acetyltransferase; AHAT, N, O-arylhydroxamic acid acyltransferase.

to a reactive nitrenium ion and will form the same DNA adduct as the sulfate conjugate, namely, *N*(deoxyguanosin-8-yl)-2-aminofluorene. It is noteworthy that this adduct has lost the acetyl group, and therefore deacetylation must occur at some point and may be a prerequisite. In fact guinea pig microsomal deacetylase will metabolize both acetylaminofluorene and *N*-hydroxyacetylaminofluorene to products that bind to DNA *in vitro*. Inhibition of the deacetylase decreases the mutagenicity of *N*-hydroxyacetylaminofluorene.

The reactive nitrenium or carbonium ions postulated to be produced will react with nucleophilic groups in nucleic acids, proteins, and sulfydryl compounds such as glutathione (GSH) and methionine. The arylation of DNA by acetylaminofluorene has been demonstrated *in vivo* and *in vitro*. The involvement of sulfate conjugation brings other factors into play. Depletion of body sulfate reduces, and supplementation with organic sulfate increases the carcinogenicity of acetylaminofluorene. The production of covalent adducts between acetylaminofluorene and cellular macromolecules *in vivo* can be shown to be correspondingly decreased and increased by manipulation of body sulfate levels.

Although the N-O glucuronide has a lower chemical reactivity and carcinogenicity than the sulfate conjugate, the glucuronide may be responsible for the production of bladder cancer by acetylaminofluorene. Furthermore, the N-O sulfate conjugate is not the only ultimate carcinogen, as acetylaminofluorene induces tumors in tissues without sulfotransferase activity. For example, the mammary gland has no deacetylase and does not carry out sulfate conjugation. However, a single application of *N*-hydroxyacetylaminofluorene yields tumors at the site of application. This may be due therefore to the presence of the N,O-acetyltransferase enzyme or another means of activation.

N-hydroxylation is not the only or major route of metabolism *in vivo*, nor is it the only reaction catalyzed by the microsomal enzymes. Ring hydroxylation is the major route of metabolism, the products of which are not carcinogenic. These alternative routes of metabolism are inducible *in vivo* by pretreatment with agents such as phenobarbital and 3-methylcholanthrene (see chap. 5).

Glucuronidation of the resulting hydroxyl derivatives is also induced by phenobarbital pretreatment. Consequently, pretreatment of animals with both these agents reduces the carcinogenicity of acetylaminofluorene. This illustrates the difficulty of predicting the effect of environmental influences on toxicity when multiple metabolic pathways are involved.

A well-documented species difference in susceptibility to acetylaminofluorene carcinogenicity is the resistance of the guinea pig. This affords an interesting illustration of the role of species differences in metabolism as a basis for species differences in susceptibility to toxicity. The guinea pig is not resistant to the carcinogenicity of the metabolite N-hydroxyacetylaminofluorene, however, indicating that this species has low activity for N-hydroxylation. In addition, the guinea pig has low sulfotransferase activity. A combination of these factors therefore confers resistance on the guinea pig. The low production of N-hydroxyacetylaminofluorene followed by low sulfate conjugation result in little of the ultimate carcinogen being produced.

The study of acetylaminofluorene carcinogenesis therefore provides many insights into the factors affecting chemical carcinogenesis. A wealth of other data not discussed here are available, which confirms the occurrence of covalent interactions between the ultimate carcinogenic metabolite of acetylaminofluorene and nucleic acids to yield covalent conjugates (Fig. 7.1). It remains to be seen how these conjugates initiate the process of carcinogenesis.

Clearly, the mechanism(s) underlying the carcinogenesis of acetylaminofluorene is very complex, and sometimes apparently conflicting and confusing data may reflect the fact that several different metabolic pathways are involved, which are more or less predominant in different tissues and in different animal species. It has been suggested, for instance, that the sulfate ester of N-hydroxyacetylaminofluorene is responsible for cytotoxicity and cell death, cell proliferation, and therefore promotion, whereas an N-hydroxyaminofluorene metabolite is responsible for initiation. Thus, both pathways would be necessary (Fig. 7.1).

7.1.2 Benzo[a]pyrene

The polycyclic aromatic hydrocarbons constitute a large group of compounds, which include a number of carcinogens found originally in coal tar but which have since been detected in cigarette smoke, the exhaust fumes from internal combustion engines, and smoke from other processes involving the burning of organic material. Benzo[a]pyrene is one of the most intensely studied compound as it is an extremely potent carcinogen. Although chemically stable, *in vivo* polycyclic aromatic hydrocarbons undergo a wide variety of metabolic transformations catalyzed by the microsomal mixed-function oxidases as illustrated for benzo[a]pyrene (Fig. 7.2). These are mainly hydroxylations occurring at the various available sites on the aromatic rings and conjugations of the hydroxyl groups with glucuronic acid or sulfate. The majority of these hydroxylation reactions probably proceed through an epoxide intermediate, as discussed in chapter 4. The prostaglandin synthetase system may also be involved in the production of quinols and phenols via free radicals.

Initially, particular attention was focused on the epoxides of the so-called **K region**. As in the case of benzo[a]pyrene and certain other polycyclic aromatic hydrocarbons, these were more carcinogenic than the parent compound. The K region had attracted particular interest, as it is electronically the most reactive portion of the polycyclic aromatic hydrocarbon molecule. However, with other carcinogenic polycyclic aromatic hydrocarbons, this was not found to be the case. It now seems that the ultimate carcinogen is an epoxide of a dihydrodiol metabolite, where the epoxide is adjacent to the so-called **bay region** (Fig. 7.2).

Although a number of the epoxides and **diol epoxides** are mutagenic, the 7,8-dihydrodiol 9,10-oxide, shown in Figure 7.2, is believed to be the ultimate carcinogen. It should be noted that there are several possible **diastereoisomers** of this metabolite, but as the action of epoxide hydrolase yields a *trans* dihydrodiol and the epoxide ring produced by the cytochrome P-450 may be *cis* or *trans*, there are two diastereoisomers produced metabolically. Thus, benzo(a)pyrene

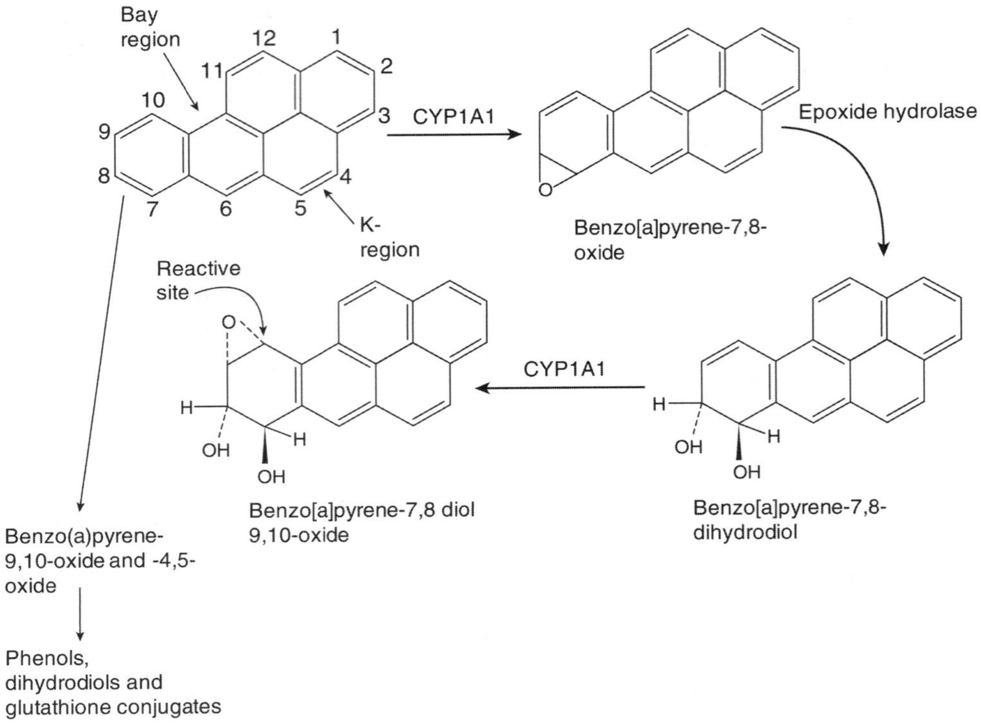

Figure 7.2 The metabolic activation of benzo[a]pyrene by cytochrome P-450 1A1 to a diol epoxide metabolite, a mutagen. This is believed to be the ultimate carcinogenic metabolite. Other routes of metabolism also catalyzed by cytochrome P-450 give rise to the 9,10, and 4,5 oxides and subsequent metabolites namely phenols, diols, and glutathione conjugates. The reactive site (carbon atom) on the metabolite is indicated.

is metabolized stereoselectively by CYP1A1 to the (+)-7R,8S-oxide (Fig. 7.2), which in turn is metabolized by epoxide hydrolase to the (−)-7R,8R-dihydrodiol. This metabolite is further metabolized to (+)benzo[a]pyrene, 7R,8S-dihydrodiol, 9S,10R-epoxide, in which the hydroxyl group and epoxide are *trans* and which is more mutagenic than other enantiomers. The (−)-7R,8R-dihydrodiol of benzo[a]pyrene is 10 times more tumorigenic than the (+)-7S,8S enantiomer. The benzylic carbon of the epoxide ring is the electrophilic site, but in this case, the configuration may be more important for tumorigenicity than the chemical reactivity. The 9,10-epoxide is stabilized by hydrogen bonding across the ring with the hydroxyl group of the diol, hence allowing it to move from the endoplasmic reticulum and enter the nucleus. The fact that the molecule is planar is also probably a factor in the interaction with DNA. The flat structure will allow intercalation of the hydrocarbon within the DNA molecule, thereby facilitating the interaction of the reactive intermediate with the DNA bases.

Benzo[a]pyrene is known to form adducts with DNA bases, such as on the exocyclic amino group of guanine (N^2) but also on the O^6 and the ring nitrogen atoms. It seems the epoxide next to the bay region is less readily detoxified because of steric hindrance, and so, when the epoxide opens, the electrophilic carbon atom at position 10 is reactive toward nucleophilic atoms such as the nitrogen and oxygen of the base guanine.

The reactive diolepoxide reacts with guanine residues in DNA (Fig. 7.3) on the exocyclic nitrogen or the ring nitrogens or the O^6, which is the more mutagenic product. Taking the adduct with the N^2 exocyclic nitrogen, this atom is normally involved in the base pairing with cytidine, but when the nitrogen is adducted to benzo[a]pyrene, this reaction is impossible, and so if not repaired, at replication of this section of DNA, pairing is only possible with adenosine (Fig. 7.4). At the next replication, the adenosine will pair with thymine, and a mutation will result. Therefore, the formation of adducts with guanine would be expected to lead to G:C to T:A transversions, where the G becomes a T after replication. This is what is found experimentally in mutations induced by benzo[a]pyrene. It is implicated in causing

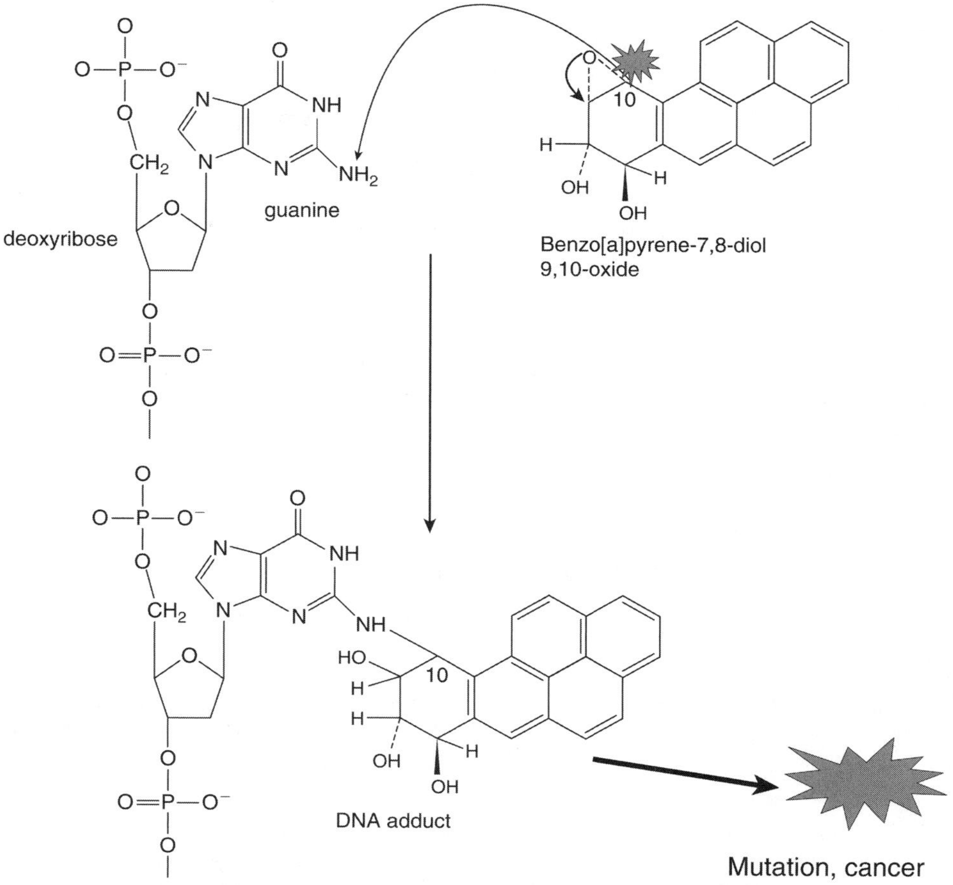

Figure 7.3 The interaction (arylation) of guanine on the N^2 position with the reactive metabolite of benzo[a] pyrene, leading to an adduct believed to be responsible for mutation and cancer.

mutations in the tumor suppressor gene p53. Indeed, in lung carcinomas in smokers, 33% of the mutant p53 alleles carried this transversion.

Of course, if the adducts are formed in sections of DNA that are "junk" DNA or spacing sections, there may be no consequence, even if not repaired. However, if the unrepaired DNA adduct(s) is in a section of DNA that is part of a tumor suppressor gene or a proto-oncogene, then more serious changes may result such as the initiation of a tumor. In fact, benzo[a]pyrene does cause mutations (G to T transversions) in the 12th codon of one of the ras proto-oncogenes. This could convert the proto-oncogene into an active oncogene.

It is postulated that this ultimate carcinogen reacts covalently with nucleic acids, producing nucleic acid adducts. It has been demonstrated that benzo[a]pyrene reacts covalently with nucleic acids *in vitro*, provided that the microsomal enzyme systems necessary for activation are present, and also in whole cell systems. The 7,8-dihydrodiol metabolite of benzo [a]pyrene binds more extensively to DNA after microsomal enzyme activation than does benzo [a]pyrene or other benzo[a]pyrene metabolites, and the nucleoside adducts formed from the 7,8-dihydrodiol of benzo[a]pyrene are similar to those obtained from cells in culture exposed to benzo[a]pyrene itself. Furthermore, the synthetic 7,8-diol-9,10-epoxides of benzo[a]pyrene are highly mutagenic in mammalian as well as in bacterial cells.

Other studies of DNA adducts formed *in vivo* with benzo[a]pyrene and using cells in culture have also indicated that the 7,8-diol-9,10-epoxides are responsible for most of the covalent binding to nucleic acids, even though the K region 4,5-epoxide is highly mutagenic. The 7,8-epoxide for benzo[a]pyrene and the 7,8-dihydrodiol are carcinogenic, but the 4,5- and 9,10-epoxides are not. These findings all point toward the conclusion that the further metabolism of the 7,8-epoxide of benzo[a]pyrene by epoxide hydrolase to the dihydrodiol

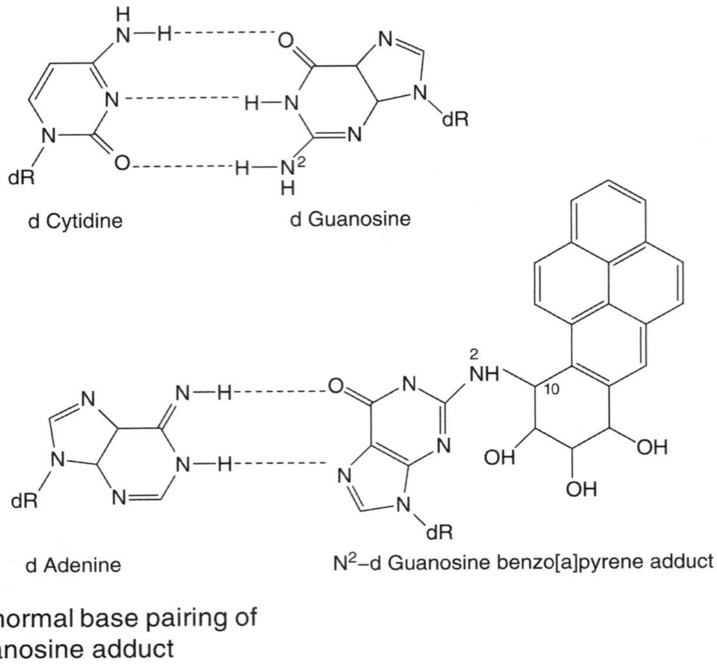

Abnormal base pairing of
guanosine adduct

Figure 7.4 The formation of a benzo[a]pyrene adduct alters the normal base pairing of guanosine with cytidine. The adduct pairs instead with the imine form of adenosine, leading to a base pair transversion after replication.

metabolite and then further oxidation of the dihydrodiol by the mixed-function oxidases to the diolepoxide are the necessary steps for production of the ultimate carcinogen. These metabolic transformations are illustrated in Figure 7.2. Other possible reactive intermediates, which have been postulated, are radicals and radical cations.

There are a number of other metabolic pathways available for benzo[a]pyrene, including the important detoxication of the ultimate reactive metabolite, the diol-epoxide by glutathione conjugation. Other metabolites of benzo[a]pyrene known to be produced are the 3,6-quinone and semiquinone. These metabolites are cytotoxic and cause DNA strand breaks and are also mutagenic. They also inhibit the cytochromes P-450-mediated oxidation of benzo[a]pyrene and of the dihydrodiol, thus inhibiting the production of the ultimate carcinogen. Reduction and subsequent glucuronidation removes the quinone and semiquinone and consequently decreases their cytotoxicity. By removing the quinone, however, glucuronidation increases the mutagenicity of benzo[a]pyrene and the covalent binding of metabolites to DNA. Thus, the products of one metabolic pathway influences another, and glucuronidation is effectively both route for detoxication and responsible for increasing toxicity.

It is clear that the effects of induction or inhibition of the metabolism will be complex because of the large number of possible metabolic pathways through which benzo[a]pyrene may be metabolized. For instance, the microsomal enzyme inducer 5,6-benzoflavone inhibits the carcinogenicity of benzo[a]pyrene in mouse lung and skin, whereas inhibitors such as SKF 525A may increase the tumor production from certain polycyclic hydrocarbons.

7.1.3 Dimethylnitrosamine

Many nitrosamines are potent carcinogens, and dimethylnitrosamine is one of the most intensely studied of this group of compounds. It is a hepatotoxin, a mutagen, and a carcinogen and also has an immunosuppressive effect. Single doses cause kidney tumors, whereas low-level chronic exposure results in liver tumors. It may also cause tumors in the stomach, esophagus, and central nervous system (CNS). Although dimethylnitrosamine is evenly distributed throughout the mammalian body, acute single doses cause centrilobular hepatic necrosis, indicating that metabolism is a factor in this toxicity. The major route of metabolism

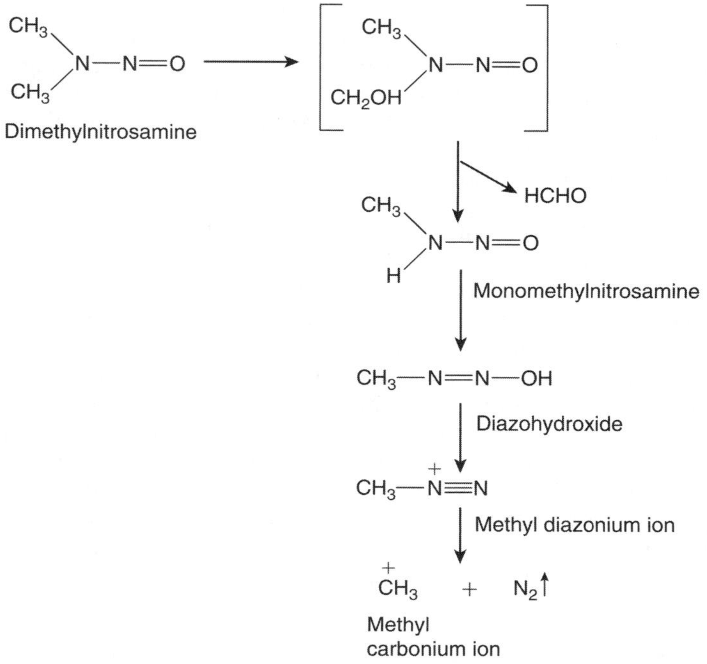

$$CH_3 \diagdown N-N{=}O \qquad \left[\begin{array}{c} CH_3 \diagdown \\ CH_2OH \diagup \end{array} N-N{=}O \right]$$

Dimethylnitrosamine

→ HCHO

$$CH_3 \diagdown N-N{=}O$$
H

Monomethylnitrosamine

$$CH_3-N{=}N-OH$$

Diazohydroxide

$$CH_3-\overset{+}{N}{\equiv}N$$

Methyl diazonium ion

$$\overset{+}{C}H_3 \quad + \quad N_2\uparrow$$

Methyl
carbonium ion

Figure 7.5 Metabolism of dimethylnitrosamine to the reactive carbonium ion intermediate responsible for methylation of nucleic acids and thought to be the ultimate carcinogen.

accounting for about 67% *in vivo* is demethylation to monomethylnitrosamine (Fig. 7.5). This reaction is catalyzed by cytochromes P-450 but may involve more than one step, and the second reaction may not be catalyzed by cytochromes P-450. There are two isoenzymes of cytochrome P-450 involved in the first oxidation step, a low-affinity form and a high-affinity form. The first intermediate, which has been postulated to be a free radical, may lead to the production of either nitrite, methylamine, and formaldehyde or *N*-hydroxymethylnitrosamine as shown in Figure 7.3. Like many *N*-hydroxymethyl compounds, this intermediate rearranges to yield the monomethyl nitrosamine, which then rearranges to give methyldiazohydroxide, methyldiazonium ion, and finally, **methyl carbonium ion**. The latter metabolite is a highly reactive **alkylating agent**, which will methylate nucleophilic sites on nucleic acids and proteins. Other routes of metabolism have also been postulated.

The acute toxicity is reduced in animals pretreated with the microsomal enzyme inducers 3-methylcholanthrene and phenobarbital, and these pretreatments also protect animals from the carcinogenic effects. Phenobarbital and 3-methylcholanthrene reduce demethylation. However, inhibition of metabolism by certain compounds also protects animals against the toxic and carcinogenic effects of dimethylnitrosamine. These data are consistent with there being other routes of metabolism, some of which may be detoxication pathways, which may explain the protective effects of some microsomal enzyme-inducing agents. Dimethylnitrosamine is a potent methylating agent, and the degree of **methylation of DNA** *in vivo* in various tissues correlates with the susceptibility of those tissues to tumor induction.

Thus, metabolism is also important for the carcinogenicity. The major site of DNA methylation is the N^7 position of guanine. However, there is also methylation on the O^6 position of guanine, and this correlates better than N^7 methylation with both the carcinogenicity and mutagenicity of dimethylnitrosamine. This O^6 methylation produces mispairing of guanine with thymidine, causing a GC to AT base transition. The ability of the cell to remove this error before cell division is a critical factor in the susceptibility of a particular tissue to the development of tumors. The methylation of DNA, however, is greater in the liver than in any other organ, including the kidney, after a single dose of dimethylnitrosamine; yet such single doses rarely cause hepatic tumors, but do induce tumors in the kidney. Therefore, it seems that the liver may initially have greater resistance to the carcinogenicity of dimethylnitrosamine, because of protective mechanisms, but that these may be compromised by repeated doses of the carcinogen, thereby allowing tumor induction in the liver after chronic exposure.

Treatment of neonatal animals, or those subjected to partial hepatectomy, with single doses of dimethylnitrosamine, leads to hepatic tumors, as in these cases the liver cells are actively dividing and are more susceptible. It has been found that the liver cells in culture are susceptible at particular stages in the cell cycle.

The metabolic activity of various tissues is also a determinant of susceptibility to tumor production. The gastrointestinal tract is resistant to dimethylnitrosamine carcinogenesis, even when the compound is given orally, as the metabolic activity of this organ is low as far as activation of dimethylnitrosamine is concerned.

The diet of the animal may also influence organ sensitivity to this carcinogen. For instance, protein-deficient diets reduce the toxicity of the compound but increase the incidence of kidney tumors. A concomittant increase in the methylation of nucleic acids in the kidney is observed under these conditions. It was concluded from this and other evidence that a protein-deficient diet decreased metabolism of dimethylnitrosamine in the liver, but did not decrease it in the kidney, and so the kidney and other organs were exposed to a higher concentration of unchanged carcinogen in the protein-deficient animals. Dimethylnitrosamine also interferes with the immune system, causing a depression of humoral immunity. There is evidence that this is due to a metabolite of dimethylnitrosamine, but does not involve the metabolic pathway(s) responsible for the carcinogenicity or hepatotoxicity.

7.1.4 Vinyl Chloride

Vinyl chloride, or vinyl chloride monomer (VCM), as it is commonly known, is the starting point in the manufacture of the ubiquitous plastic polyvinyl chloride (PVC). Vinyl chloride is a gas, normally stored as a liquid under pressure, which when heated polymerizes to yield PVC. This plastic was introduced a number of years ago, and there have been many workers exposed or potentially exposed to vinyl chloride during the course of their working lives. However, safety standards in factories and working practices have not always been as rigorous as they are today and were perhaps not always observed. In some cases, workers were required to enter reaction vessels periodically to clean them, despite the fact that they still contained substantial amounts of vinyl chloride. This was sufficient for some of the workers to be overcome by solvent narcosis. Thus, the acute poisoning caused elation followed by lethargy, loss of consciousness (at 70,000 ppm), hearing and vision loss (at 16,000 ppm), and vertigo (at 10,000 ppm).

Chronic exposure to vinyl chloride results in "vinyl chloride disease," which comprises of **Raynaud's phenomenon**, skin changes (akin to scleroderma), changes to the bones of the hands (acro-osteolysis), and liver damage in some cases. The bone changes, mainly to the distal phalanges, are due to aseptic necrosis following ischemic damage due to degeneration and occlusion of small blood vessels and capillaries. The liver may become enlarged, perhaps due to enzyme induction, and also fibrotic, although liver function tests are normal. It has been suggested that the vinyl chloride syndrome has an immunological basis, as circulating immune complexes are deposited in vascular epithelium. Also, raised plasma immunoglobulins and complement activation are features. This may result from reactive metabolites forming antigens by reacting with proteins.

Vinyl chloride is also suspected of having effects on the reproductive systems of both males and females. The most severe toxic effect, which probably resulted from repeated high-level exposure and was confined to the cleaners of the reaction vessels, was not immediately apparent. This was a type of liver tumor known as hemangiosarcoma, which was very rare and was observed only in epidemiological studies of workers in this industry. This type of liver tumor has now also been produced in experimental animals. Fortunately, the hygiene and safety standards applied to working with vinyl chloride are now stricter. However, these changes occurred with the benefit of hindsight, and with more foresight, the tragedy might have been avoided. Thus, vinyl chloride is now known to be both mutagenic and carcinogenic in experimental animals. **Hemangiosarcoma** is a tumor of the reticuloendothelial cells (sinusoidal cells), not hepatocytes, and gives rise to tumors of the hepatic vasculature.

Vinyl chloride is readily absorbed through skin and is rapidly eliminated, either unchanged or as metabolites. The metabolism is catalyzed by cytochromes P-450, and vinyl chloride induces its own metabolism but also destroys cytochrome P-450. The tumor incidence in rats correlates

Table 7.1 Correlation Between Exposure Concentration of VC, Metabolism, and Induction of Hepatic Angiosarcoma in Rats

Exposure concentration (ppm of VC)	mg L^{-1} of air	VC metabolized (μg)	Liver angiosarcoma (%)
50	128	739	2
250	640	2,435	7
500	1,280	3,413	12
2,500	6,400	5,030	22
6,000	15,360	5,003	22
10,000	25,600	5,521	15

Abbreviation: VC, vinyl chloride.
Source: From Ref. 1.

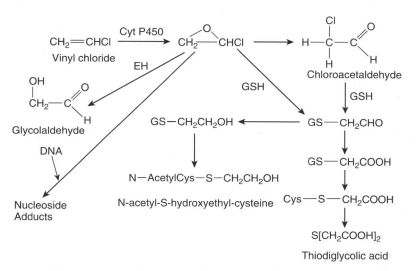

Figure 7.6 The metabolism of vinyl chloride. *Abbreviation*: EH, epoxide hydrolase.

with the amount metabolized rather than the exposure concentration (Table 7.1). Indeed, it seems that the metabolism is saturable, and hence the incidence of tumors reaches a maximum (Table 7.1). This indicates that the dose response for tumor incidence is not linear but reaches a maximum at around 20% incidence. However, it means that extrapolation from the maximum tumor incidence to determine a safe concentration is scientifically unsound without a knowledge of the dose-response curve. This example also shows the importance of metabolic data and illustrates the need for incorporation of such data into risk assessment models.

Monooxygenase-mediated metabolism gives rise to an epoxide intermediate and to chloroacetaldehyde (Fig. 7.6). The latter may in fact be derived from the epoxide. Other metabolites detected are *N*-acetyl-*S*(hydroxyethyl)cysteine, carboxyethyl cysteine, monochloroacetic acid, and thiodiglycolic acid. Both the epoxide and chloroacetaldehyde react with GSH and may cause depletion of the tripeptide. The epoxide also reacts with DNA and RNA, and both metabolites react with protein. The oxirane ring formed by oxidation is highly reactive and is almost certainly the reactive intermediate. Both the epoxide intermediate and chloroacetaldehyde are mutagenic, and the epoxide is known to bind to DNA at the N^7 position of deoxyguanosine and also to RNA to give 1,N^6-ethenoadenosine and 3,N^4-ethenocytidine.

The epoxide metabolite of vinyl chloride may be a substrate for epoxide hydrolase, which would yield glycolaldehyde. It has been shown that epoxide hydrolase blocks the reaction of vinyl chloride with adenosine, whereas alcohol dehydrogenase blocks binding to protein. In conclusion, it seems that the DNA binding is due to the epoxide, whereas the protein binding is probably due to chloroacetaldehyde. *In vitro* addition of GSH blocks the binding to protein, but not the binding to DNA. Using isolated rat hepatocytes, it was shown

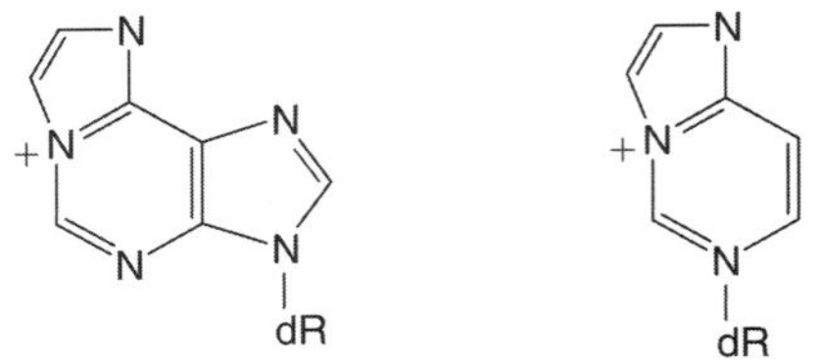

Etheno-deoxyadenosine   Etheno-deoxycytidine

Figure 7.7 The structures of the nucleoside adducts etheno-deoxyadenosine and ethenodeoxycytidine. *Abbreviation*: dR, deoxyribose.

that radiolabeled vinyl chloride was metabolized and became bound to protein and DNA, and the binding to DNA could be blocked with epoxide hydrolase. These experiments showed that the epoxide and other metabolites were stable enough to leave the hepatocyte. An epoxide hydrolase inhibitor did not affect binding to protein, and GSH transferase does not seem to be very important in the metabolism of the epoxide. The epoxide can rearrange to give chloroacetaldehyde, which can then bind to protein and GSH.

DNA adducts isolated from rat liver after chronic exposure to vinyl chloride and adducts with calf liver DNA added to *in vitro* preparations in which chloroethylene oxide or chloroacetaldehyde were formed were found to be the same. Thus, both ethenodeoxyadenosine and ethenodeoxycytidine were detected (Fig. 7.7). These adducts follow alkylation at N^1 of deoxyadenosine and N^3 of deoxycytidine. Similar adducts were detected with RNA. In both cases, an imidazole ring is formed by the addition to the nucleotide, which is coplanar and shields hydrogen-bonding sites. Ethenodeoxycytidine becomes similar to deoxyadenosine in configuration, and this would result in ethenodeoxycytidine simulating deoxyadenosine in base pairing. The imidazole ring resembles an alkyl group and blocks base-pairing sites. Overall, the changes might be expected to lead to dAdT to dC-dG transversions from ethenodeoxyadenosine and dC-dG to dA-dT transversions from ethenodeoxycytidine. These findings are all consistent with vinyl chloride, chloroacetaldehyde, and chloroethylene oxide causing base-pair substitutions.

The carcinogenicity of vinyl chloride was eventually predictable from the animal studies, but the occupational histories and case histories of the workers were vital in tracking down the cause of the hemangiosarcoma. The other toxic effects may be due, at least in part, to metabolism to toxic, reactive metabolites, which react with proteins. The lessons from this example are that safety standards need to be stringent in factories, and that animal studies are important in assessing potential toxicity and highlighting the type of toxic effect that might be expected. This should be known before human exposure occurs. As a consequence of this type of industrial problem, legislation is now in force in most major western countries, which deals specifically with industrial chemicals. For example, in the United Kingdom, all chemicals produced in quantities of greater than 1 tonne have to undergo toxicity testing, while strict occupational hygiene limits, known variously as occupational exposure limits (in United Kingdom) or threshold limit values (TLV) (in United States) for industrial chemicals are enforced. Vinyl chloride alkylates the N^1 of adenine and N^3 of cytidine as well as the N^7 of guanine producing an alkylated base, etheno deoxyadenosine for example. This has a C_2 unit across the exocyclic and ring nitrogen atoms of adenine.

From these mechanistic studies, specific biomarkers of effective exposure were proposed: the DNA adducts ethenodeoxyadenosine and ethenodeoxycytidine. These can be detected in liver biopsy and white blood cells and so can be used to monitor workers. Furthermore, more recently, a specific biomarker of response has also been detected. This is a mutant p21 ras protein, which results from the interaction with DNA and can be detected in the serum of workers exposed to vinyl chloride. The level of this protein detected in workers was found to have increased as exposure to vinyl chloride increased. Therefore, both these biomarkers can be used in risk assessment.

7.1.5 Tamoxifen
This drug is a good illustration of the problems and difficulties of risk benefit considerations with drugs. It also illustrates how understanding the mechanisms underlying toxicity can help in this regard.

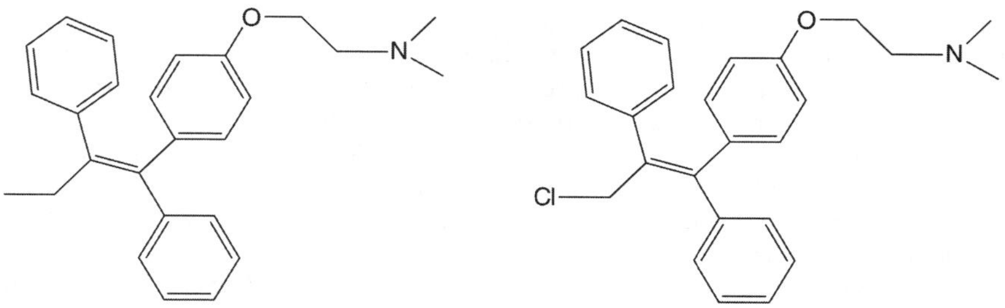

Figure 7.8 The structure of tamoxifen and the analogue toremifene.

Table 7.2 The Carcinogenic Effects of the Drug Tamoxifen in Rats

Site/effect	Male rats with tumors Dose (mg kg^{-1} day^{-1})				Female rats with tumors Dose (mg kg^{-1} day^{-1})			
	0	5	20	35	0	5	20	35
Hepatocellular								
Adenoma	1/102	8/51	11/51	8/50	1/104	2/52	6/52	9/52
Carcinoma	1/102	8/51	34/51	34/50	0/104	6/52	37/52	37/52
Metastatic to lung or lymph node	0/102	0/51	3/51	4/50	1/104	0/52	2/52	2/52
Mammary gland								
Fibroadenoma	1/102	0/51	0/51	0/50	16/104	0/52	1/52	0/52
Adenocarcinoma	0/102	0/51	0/51	0/51	9/104	0/52	0/52	0/52

Source: From Ref. 2.

Tamoxifen (Fig. 7.8) is an extremely effective anticancer drug used to treat breast cancer, which can also be used prophylactically to prevent the occurrence of breast cancer in women at high risk of developing the disease. It has antiestrogenic properties and is termed "**a selective estrogen receptor modulator**" (SERM). Because of this action, tamoxifen has other important effects, including stopping bone loss in osteoporosis and lowering low-density lipoproteins in blood as well as reducing breast cancer both actively and prophylactically.

It works against breast cancer because it is estrogen sensitive, and so, tamoxifen stops the growth of the cancer by blocking the estrogen receptor and depriving the cancer cells of the steroid stimulus. This causes the tumor to shrink. Unfortunately, however, although it blocks estrogen receptors in some tissues, it stimulates it in others, notably the uterus, and so can increase the risk of uterine cancer.

Furthermore, animal studies have revealed that tamoxifen is also capable of causing liver tumors in rats exposed to the drug. Research showed that tamoxifen was genotoxic *in vitro* and was also a genotoxic carcinogen *in vivo*. It was further found that the drug formed adducts with DNA (on the N^2 of guanine for example).

Other work in mice found that in this species, although adducts could be detected, these were at a lower level than in rats, and no hepatic tumors were observed. These observations therefore raised the questions: was tamoxifen hepatocarcinogenic in humans, and what was the mechanism? It was important to know the risk.

The treatment of rats, both male and female, for two years with tamoxifen revealed a dose-related increase in malignant liver tumors (Table 7.2). Thus, the table indicates that liver carcinomas caused by tamoxifen show a dose-related increase in both sexes, with a similar incidence, and metastases showed a slight but similar incidence. Research by various groups on the mechanisms underlying the carcinogenicity of tamoxifen has been extremely important in assessing the risk from this important drug and has allowed it to continue to be used.

The mechanism by which tamoxifen is carcinogenic is via metabolic activation to yield a reactive metabolite, which reacts with DNA. Study of the metabolism in different species has revealed the important metabolic pathways involved in both activation and detoxication.

Tamoxifen can undergo several routes of oxidative metabolism (Fig. 7.9). Thus, as expected aromatic hydroxylation to yield, the 4-hydroxy tamoxifen is catalyzed by cytochrome P-450. This metabolite is eliminated after conjugation. Alternatively, oxidation of the alkyl groups attached to the nitrogen atom, also catalyzed by cytochrome P-450, leads to dealkylation (Fig. 7.9). These are detoxication pathways.

However, as also shown in the Figure 7.9, tamoxifen can undergo oxidation on the α-carbon atom of the aliphatic chain. This is catalyzed by cytochrome P-450. The α-hydroxylated product can then undergo conjugation, and it is this pathway that is crucially involved in the carcinogenicity. In rats, the preferred route is not glucuronidation, as this species has little detectable glucuronosyl transferase activity toward this substrate. Instead, the rat conjugates the α-hydroxytamoxifen with sulfate, having high sulfotransferase activity toward the metabolite. The conjugate formed readily loses the sulfate group, forming a positively charged moiety, a **carbocation**, which is electrophilic. O-Sulfate conjugates can be good leaving groups, generating reactive electrophilic cations, if the cation produced is stabilized by resonance, as in this case.

The mouse has less sulfotransferase activity with respect to the α-hydroxytamoxifen, but more glucuronosyl transferase activity. The resulting glucuronic acid conjugation is a clear detoxication pathway, as a stable metabolite is formed, which can be readily eliminated. Hence, the mouse is much less susceptible, and although some tamoxifen is covalently bound to liver DNA, no hepatic tumors have been detected.

The question however is what happens in humans? Fortunately, it seems that humans have more glucuronosyl transferase activity than mice and much more than (100×) rats (in which it is negligible). Furthermore, humans have much lower levels (5×) of sulfotransferase activity with this substrate, and so humans are the opposite extreme to rats. In studies in women treated with tamoxifen, no DNA adducts were detected in tissue samples, and liver tumors have not been detected.

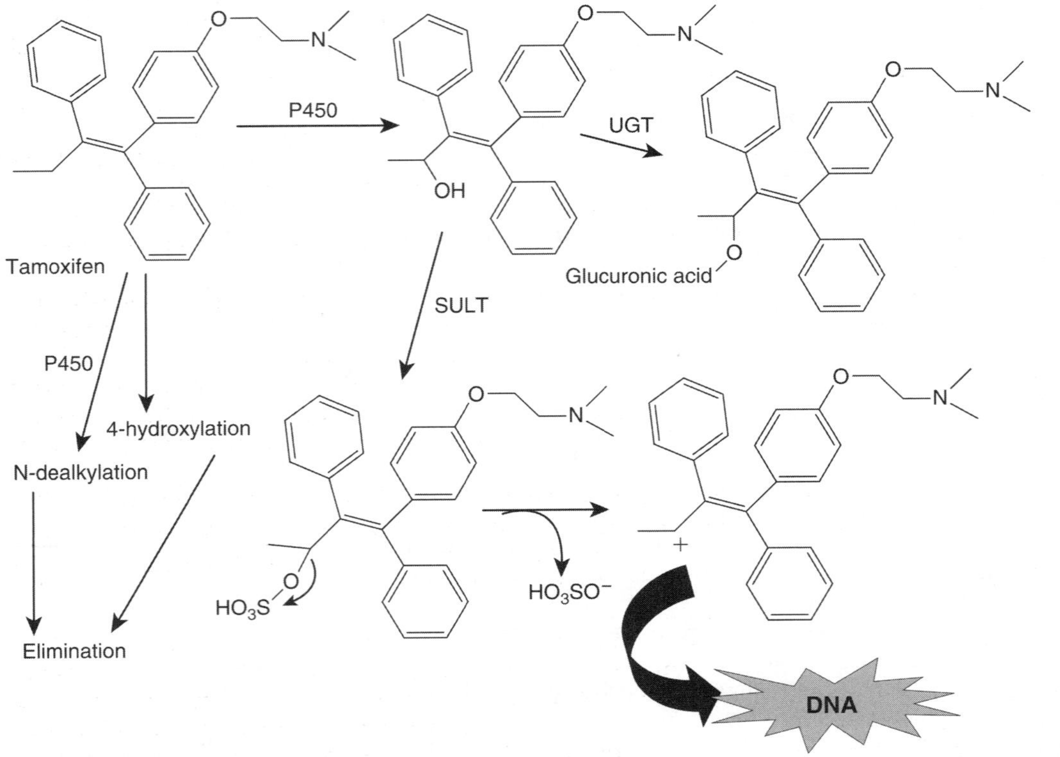

Figure 7.9 The metabolism and metabolic activation of tamoxifen. *Abbreviations*: SULT, sulfotransferase; UGT, UDP-glucuronosyl transferase.

This understanding allows a rational risk assessment of tamoxifen, which is that the risk of liver cancer is extremely low because of detoxication pathways and the lack of metabolic activation.

Furthermore, understanding this mechanism of toxicity has allowed a less-toxic analogue to be devised and used as a drug. This is toremifene (Fig. 7.8), which has similar pharmacological properties to tamoxifen. In this molecule, a chlorine atom is inserted on the aliphatic chain. This bulky halogen substituent reduces sulfation of the α-hydroxy metabolite, and its electronegativity reduces the cleavage of the sulfate from any conjugate that is formed. Hence, DNA adducts are not formed.

7.1.6 Clofibrate and Other Hypolipidemic Drugs

A number of chemicals, including some therapeutically important **antihyperlipdemic drugs** (e.g., clofibrate, ciprofibrate, bezafibrate, nafenopin, and Wy-14643) and plasticizers (e.g., diethylhexylphthalate), cause a phenomenon known as peroxisome proliferation. This phenomenon was discovered in 1960, and the importance of this became more apparent when in 1980 it was found that in rodents the same chemicals caused hepatic carcinomas.

However, these chemicals in general and the drugs in particular are not mutagenic in the Ames or other tests. They are therefore classified as nongenotoxic carcinogens.

Drugs such as clofibrate, fenofibrate, ciprofibrate, and nafenopin are examples of drugs in this category, which are used to treat high blood lipid and triglyceride levels. It is a particularly important finding with such drugs (and now others), and the study of the mechanisms involved have lead to an understanding of the mechanism of action of these drugs and new therapeutic targets.

In rats and mice, treatment with these drugs causes a number of biochemical and ultrastructural changes. At the gross pathology level, there is a very significant increase in liver size and weight (latter may increase 3–4 times normal in some cases). This is the result of both hypertrophy (increased cell size) and hyperplasia (increased cell number). The increase in cell size is primarily due to increased numbers and size of an organelle, the **peroxisome**. These organelles, although having some similarities with mitochondria, carry out the oxidation of very long-chain fatty acids (β-oxidation pathway), as well as being involved in cholesterol and glycerolipid metabolism. Biochemical pathways associated with the peroxisome are also increased. Thus, certain enzymes of the peroxisomal β-oxidation pathway specific to the organelle are markedly increased, such as acyl CoA oxidase, which is increased 15-fold. In some cases, after two years of treatment, the incidence of liver cancer in the animals may be as high as 100%. However, these effects are dose related, so there is a threshold.

The hepatic carcinogenicity does not occur in other species, and only some of the biochemical and ultrastructural effects are seen in semiresponsive species such as hamsters. Other species such as guinea pigs seem refractory.

More recently, however, in cynomologous monkeys, a nonhuman primate, some limited responses to these compounds have been seen with a twofold increase in liver weight, hepatocyte hypertrophy, an increase in peroxisomes (2.7×), and an increase in mitochondria. However, no DNA damage was detected. With humans taking ciprofibrate, only limited peroxisome proliferation was seen, and there was no increase in acyl CoA oxidase. Thus, the question is, "are humans at risk?" In order to understand this, the mechanism needs to be understood.

Research has now established that the peroxisome proliferators act on a receptor, called the **peroxisome proliferator-activated receptor (PPAR)**, discovered in 1990. There are now known to be three receptors: PPARα, PPARδ, and PPARγ. These are parts of the nuclear hormone superfamily. The PPARs are ligand-dependent transcription factors, which have different functions and tissue locations.

PPARα and PPARδ will bind drugs, causing lipid lowering, whereas PPARγ is involved in insulin sensitivity. PPARα is known to induce changes in the gene-encoding enzymes involved in lipid and lipoprotein metabolism. Thus, expression of the genes encoding the enzymes, such as acyl CoA oxidase, involved with peroxisomal β-oxidation are induced by peroxisomal proliferators such as clofibrate and ciprofibrate. Such drugs and other chemicals, which cause peroxisomal proliferation, are now known to be agonists for PPARα but may also activate PPARδ.

1. Is it known that the promoters of some of the genes involved in the observed responses to these drugs contain response elements (PPRE) for the PPARα? Removal by mutation of the PPRE leads to loss of transcriptional activation.
2. Transgenic mice lacking the PPARα and exposed to drugs known to cause peroxisome proliferation do not show any of the toxic effects and increased peroxisome numbers or liver size and crucially no evidence of liver tumors.

When PPARα binds a ligand such as clofibrate, it heterodimerises with another receptor RXRα, and after conformation changes and recruitment of transcriptional coregulators, the complex is able to modulate gene expression via PPREs. Natural ligands are fatty acids and derivatives, indicating that PPARs modulate lipid homeostasis. Lipid-lowering drugs may mimic these natural ligands. The fact that these drugs perform the same function in both humans and rodents suggests that the PPARα has the same function in all these species. This raises the question, "why are they not carcinogenic in humans?"

Mechanism of Carcinogenicity
There are a large number of target genes for PPARα, including both mitochondrial and peroxisomal β-oxidation enzymes, apolipoproteins, fatty acid transporters, lipoprotein lipase, CYP4A, and thioesterases. As PPARα is located in and particularly abundant in the liver, this is the target organ for peroxisomal proliferator chemicals.

The human, guinea pig, and hamster orthologues of the PPARα have been cloned and studied. They are activated by peroxisomal proliferators *in vitro* and contain the genes for PPRE.

Various chemicals, such as the fibrate (e.g., clofibrate, bezafibrate, and ciprofibrate) and other lipid-lowering drugs (e.g., Wy-14643) and phthalate plasticizers (e.g., diethylhexyl phthalate), bind to and activate PPARα and are also hepatocarcinogenic in rodents. However, different compounds bind with different affinities, from strong (e.g., Wy-14643) to weak (e.g., phthalates). Thus, it is believed that peroxisomal proliferators act via the receptor (PPARα) to cause some of the effects seen.

Research has shown clearly that PPARα is necessary for hepatocarcinogenesis in mice. Thus, the PPARα-agonist drug Wy-14643 causes a 100% incidence of liver tumors in wild-type mice, yet PPARα-null mice are resistant. This has been confirmed using the drug bezafibrate.

PPARα may also have a role in controlling apoptosis, cell proliferation, inflammation, and oxidative stress. Thus, peroxisomal proliferators also stimulate DNA synthesis, liver growth, and inhibit apoptosis.

However, growth regulatory pathways and signaling involved in cell proliferation may be independent of the receptor. There are several critical events thought to be involved in the hepatic carcinogenesis caused by PPARα agonists: (*i*) receptor binding and activation; (*ii*) induction of key target genes; (*iii*) increased cell proliferation and inhibition of apoptosis; (*iv*) oxidative stress causing DNA damage and/or increased signaling for cell proliferation; (*v*) and finally, clonal expansion of initiated liver cells. Evidence such as the studies with null mice (see above) supports the first critical event.

However, the role proliferation of the peroxisomes in the hepatocarcinogenesis may be coincidental rather than causal, as extent of the proliferation does not correlate well with the carcinogenicity.

Acyl CoA oxidase, which is induced up to at least 15× normal levels, catalyzes a specific step characteristic of β-oxidation in the peroxisome-producing hydrogen peroxide as a byproduct (Fig. 7.10). Other oxidases may also lead to increased hydrogen peroxide. Normally, this is removed and detoxified by the enzyme catalase, which breaks hydrogen peroxide down to water and oxygen. However, catalase is only increased about twofold after treatment with peroxisomal proliferators.

This could lead to oxidative stress as shown in the figure (Fig. 7.10), which can in turn cause cell proliferation and oxidative damage to DNA and protein.

Thus, as previously discussed, hydroxyl radical can initiate lipid peroxidation, generating free radicals as well as directly damaging DNA. Oxidative damage to proteins yields lipofuscin, which is known to accumulate in rodent liver treated with peroxisomal proliferators. Treatment with antioxidants will inhibit the liver cancer.

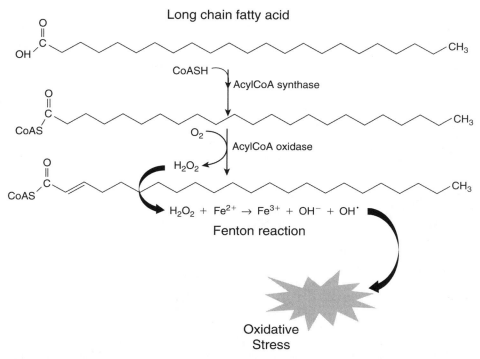

Figure 7.10 Part of the peroxisomal β-oxidation pathway showing the production of hydrogen peroxide by acyl CoA oxidase. Hydrogen peroxide can be detoxified by catalase to water and oxygen. However, in the absence of sufficient enzyme, it can be broken down to the highly reactive hydroxyl radical in the presence of metal ions such as Fe^{2+}.

It is suggested that PPARα activation may lead to DNA replication, cell growth, and inhibition of apoptosis via tumor necrosis factor α (TNF-α). Alternatively or additionally, the oxidative stress could activate the MAP kinase pathway, which will also influence cell growth (Fig. 7.11), but this might be independent of PPARα activation. However, the role of these events is still unclear.

The evidence for DNA damage in peroxisomal proliferator-treated animals is more equivocal with some studies, reporting no change in markers such as 8-hydroxydeoxyguanosine residues, whereas other studies indicate otherwise. Also, increased expression of DNA repair enzymes has been reported. The relationship between oxidative stress and signaling for cell proliferation however is not clear.

However, although there is uncertainty about the details of the connection between oxidative stress and liver cancer, there is a strong weight of evidence showing an association. The increased cell proliferation and inhibition of apoptosis that occur is also related to PPARα activation, as indicated from studies with PPARα null mice. However, the genes involved are not known. When such changes occur at the same time as DNA damage leading to mutation, the initiated cell(s) can then undergo proliferation. As previously discussed, this combination of events is necessary to lead to the production of a tumor. Thus, it is clear that the mechanism underlying the hepatocarcinogenesis may involve several processes.

One major question is whether humans are at risk from these chemicals. Although they clearly have the receptor for the lipid-lowering drugs to work, there is no evidence of increased risk of hepatic cancer from epidemiological studies. However, fibrates are relatively weak PPARα agonists.

Refractory species such as guinea pig seem to have fewer PPARα receptors in the liver, and observations suggest there are significant differences between rodents and humans in a number of aspects of the response. Recent studies in cynomologous monkeys treated with ciprofibrate have detected peroxisomal proliferation but only found slight changes in certain parameters [e.g., messenger RNA (mRNA) induction for acyl CoA oxidase] of minimal oxidative stress, and despite hypertrophy, cell proliferation was not detected.

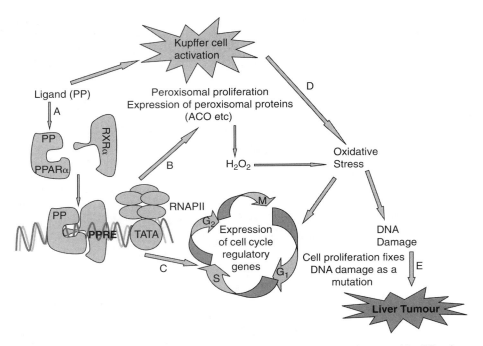

Figure 7.11 Possible mechanisms underlying the hepatocarcinogenesis caused by PPs. As a result of ligand activation (A), the PPARα modulates expression of target genes, including acyl CoA oxidase (ACO), via interaction with DNA and the PP response element (PPRE). This results in peroxisomal proliferation, which produces increased levels of hydrogen peroxide and oxidative stress. This can damage DNA and possibly stimulate the cell cycle to result in increased cell proliferation. Ligand activation of the PPARα may also cause inhibition of apoptosis and cell proliferation via direct changes in gene expression, but this is uncertain. PPARα ligands can also activate Kupffer cells leading to oxidative stress. The DNA damage (initiation) is fixed as mutations by cell replication, and further cell proliferation (promotion) eventually leads to liver tumors. TATA—sequence of DNA in promoter region, which is recognized and bound by a transcription factor. RNAP II-RNA polymerase II enzyme, which carries out transcription of section of DNA. *Abbreviations*: PP, peroxisome proliferator; RXR, retinoid X receptor; PPARα, peroxisome proliferator-activated receptor; PPRE, peroxisomal proliferator response element. *Source:* From Ref. 3.

Recent research using NMR has revealed novel noninvasive biomarkers for peroxisomal proliferation. These are two breakdown products of NAD detectable in plasma by HPLC. These reflect the increased demand and production of NAD for oxidation metabolism such as β-oxidation of fatty acids. This biomarker will be useful in studies in humans, especially clinical trials of more potent lipid-lowering drugs.

Very recent work using mice humanized for the PPARα has suggested that there are structural differences between the human and rodent receptor, which are responsible for the different susceptibility between the species.

7.2 TISSUE LESIONS: LIVER NECROSIS

7.2.1 Carbon Tetrachloride
This solvent, once used extensively in dry cleaning and even as an anesthetic is primarily hepatotoxic, causing two different types of pathological effect. The hepatotoxicity of carbon tetrachloride has probably been more extensively studied than that of any other hepatotoxin, and there is now a wealth of data available. Its toxicity has been studied both from the biochemical and pathological viewpoints, and therefore the data available provide particular insight into mechanisms of toxicity.

Carbon tetrachloride is a simple molecule which, when administered to a variety of species, causes centrilobular hepatic necrosis (zone 3) and fatty liver. It is a very lipid-soluble compound and is consequently well distributed throughout the body, but despite this, its major toxic effect is on the liver, irrespective of the route of administration. It should be noted that it does have other toxic effects, and there are species and sex differences in toxicity. Chronic administration or exposure causes liver cirrhosis, liver tumors, and also kidney damage. The reason for the liver being the major target is that the toxicity of carbon tetrachloride is dependent on metabolic activation by CYP2E1. Therefore, the liver becomes the target, as it contains the greatest concentration of cytochromes P-450, especially in the centrilobular region, which is where the damage is greatest. Low doses of carbon tetrachloride cause only fatty liver and destruction of cytochromes P-450. Interestingly, a low dose of carbon tetrachloride protects the animals against the hepatotoxicity of a subsequent larger dose because of this destruction of cytochromes P-450. The destruction of cytochromes P-450 occurs particularly in the centrilobular and mid-zonal areas of the liver. It is also selective for a particular isozyme, CYP2E1 in the rat, whereas other isozymes such as CYP1A1 are unaffected. The destruction of CYP2E1 seems to be influenced by the amount of oxygen available, being greater when more oxygen is available. This destruction therefore has been interpreted as due to the trichloroperoxy radical, which is more reactive than the trichloromethyl radical (see below). This is an example of suicide inactivation of an enzyme, i.e., carbon tetrachloride is a suicide substrate.

Although carbon tetrachloride was originally thought to be resistant to metabolic attack, it is now clear that it is metabolized by cytochromes P-450. As the microsomal enzymes are involved in the metabolic activation, pretreatment with enzyme inducers and inhibitors, respectively, increases and decreases the toxicity. However, in this instance, cytochrome P-450 is functioning in the reductive mode, catalyzing the addition of an electron, which then allows homolytic cleavage and the loss of a chloride ion and the formation of the trichloromethyl radical (Fig. 7.12). The resulting trichloromethyl radical may then undergo one of several reactions. It may abstract a hydrogen atom from a suitable donor such as the methylene bridges on polyunsaturated fatty acids or a thiol group. This will produce chloroform, which is a known metabolite of carbon tetrachloride (Fig. 7.12) both *in vitro* in the absence of oxygen and *in vivo*. The other product will be a lipid radical or thiol radical, depending on the source of the hydrogen atom (Figs. 9 and 18, chap. 6).

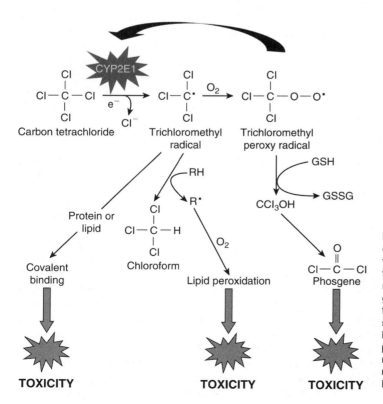

Figure 7.12 The microsomal enzyme-mediated metabolic activation of carbon tetrachloride to the trichloromethyl radical. This radical may either react with oxygen or abstract a hydrogen atom from a suitable donor (R) to yield a secondary radical, or react covalently with lipid or protein. If R is a polyunsaturated lipid, (R), a lipid radical (R) is formed, which can undergo peroxidation (see chap. 6, Fig. 9).

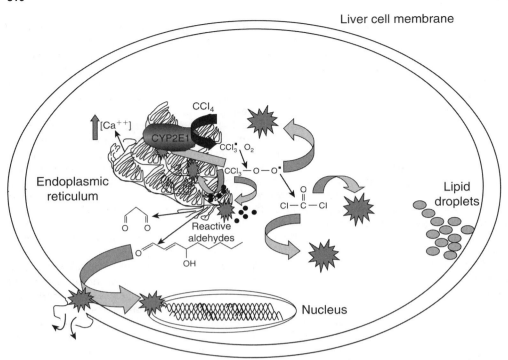

Figure 7.13 The sequence of events underlying carbon tetrachloride toxicity to the liver cell. Although the process starts in the smooth endoplasmic reticulum at CYP2E1, the destruction spreads throughout the cell.

The trichloromethyl radical is reactive, but it has been suggested that it is insufficiently reactive to account for the toxicity, and the **trichloromethyl peroxy radical** (Figs. 7.12 and 7.13) has been postulated as an alternative. Whichever is the reactive species, it will damage the immediate vicinity of the cytochromes P-450, including the enzyme itself and the endoplasmic reticulum (Fig. 7.13). Thus, the trichloromethyl-free radical covalently binds to microsomal lipid and protein and reacts directly with membrane phospholipids and cholesterol. The trichloromethyl peroxy radical may produce **phosgene** and electrophilic chlorine. However, various studies have suggested that although the initiating event may be the formation of the trichloromethyl radical, this is not the major cause of damage. The covalent binding to protein occurs in the absence of oxygen, but destruction of cytochromes P-450 and other enzymes of the endoplasmic reticulum requires the presence of oxygen. Rather, the production of a reactive radical metabolite is the start of a cascade of events, such as lipid peroxidation, thought to be the cause of the damage (chap. 6, Fig. 9) (Fig. 7.13). The lipid peroxidation process yields products that may spread the cellular damage beyond the endoplasmic reticulum. Thus, formation of lipid peroxides results in the breakdown of unsaturated lipids to give carbonyl compounds such as **4-hydroxynonenal** and other hydroxyalkenals. Such compounds are known to have many biochemical effects, such as inhibition of protein synthesis and inhibition of the enzyme glucose-6-phosphatase. 4-Hydroxynonenal will react with both microsomal protein and thiol groups and may inhibit enzymes by the latter process. There is, however, no overall depletion of GSH, although there is a reduction in certain protein thiol groups, which may be related to this. 4-Hydroxynonenal will also increase plasma membrane permeability *in vitro*. All of the effects of this reactive aldehyde are also caused by carbon tetrachloride itself. Perhaps the most significant effect of this product of lipid peroxidation is the reduction in calcium sequestration by the endoplasmic reticulum, which leads to a rise in cytosolic-free calcium (Fig. 7.13). The calcium pump seems to be very sensitive to lipid peroxidation and substances that interact with thiol groups.

The first events occurring after a toxic dose of carbon tetrachloride can be observed or detected biochemically around the endoplasmic reticulum. Within one minute of dosing, carbon tetrachloride is covalently bound to microsomal lipid and protein in the ratio of 11:3. Conjugated dienes, indicators of lipid peroxidation, can be detected in lipids within five minutes.

These changes in lipids may reflect a transient stage in the alteration of the polyunsaturated fatty acids in the endoplasmic reticulum, but they are not dose related. Within 30 minutes of dosing, protein synthesis is depressed, reflecting changes in the ribosomes and rough endoplasmic reticulum, and loss of ribosomes can be detected by electron microscopy. Also, the cytochromes P-450 content of the cell and its activity are decreased. Other indicators of damage to the endoplasmic reticulum are inhibition of the enzyme glucose-6-phosphatase and of calcium sequestration.

Between one and three hours after dosing with carbon tetrachloride, triglycerides accumulate in hepatocytes, detectable as fat droplets, and there is continued loss of enzyme activity in the endoplasmic reticulum. Cellular calcium accumulates, and the rough endoplasmic reticulum becomes vacuolated and sheds ribosomes (Fig. 7.13). The smooth endoplasmic reticulum shows signs of membrane damage and eventually contracts into clumps. At later time points, lysosomal damage may become apparent when the centrilobular cells are damaged, cells may begin to show intracellular structural modifications, and eventually the plasma membrane ruptures (Fig. 7.13). It is now generally accepted that although metabolic activation to a reactive radical may be the first event, the production of lipid radicals and subsequent chain reactions leading to reactive products of lipid peroxidation are the important mediators of cellular toxicity. There may be many targets for these reactive products, but interference with calcium homeostasis seems to be one of the most important targets. Other targets will be membrane lipids, which may fragment during peroxidation, and thiol-containing enzymes. Clearly, there are many biochemical and structural changes occurring during carbon tetrachloride hepatotoxicity, and a single critical event that leads to cellular necrosis may not be easily identified, although the crucial events probably include increased permeability and eventual disruption of the plasma membrane. Fatty liver may have a simpler basis in the inhibition of protein synthesis, which is known to result in the decreased production of the lipoprotein complex responsible for the transport of lipids out of the hepatocyte (see chap. 6).

The damage to the endoplasmic reticulum leads to loss of ability to synthesize proteins. This is carried out in the rough endoplasmic reticulum, whereas CYP2E1 activation takes place nearby in the smooth endoplasmic reticulum.

The overall consequences are shown in terms of stages in Figure 7.14. Repeated doses of carbon tetrachloride lead to fibrosis and eventually cirrhosis, which involves the deposition of collagen and proliferation of fibroblasts as part of the repair process and inflammatory response. This has been shown to involve cytokines, in particular TNF-α, a proinflammatory cytokine. Thus, knockout mice lacking the TNF-α receptors showed much less fibrosis, but the acute hepatotoxic response was not affected.

$$CCl_4 \rightarrow CCl_3^{\bullet} + Cl^-$$

$$CCl_3^{\bullet} + RSH \rightarrow RS^{\bullet}, R\text{—}S\text{—}CCl_3, CHCl_3$$

$$CCl_3^{\bullet} + \text{Protein, Unsaturated lipid} \rightarrow \text{Covalent binding}$$

$$CCl_3^{\bullet} + \text{Polyunsaturated lipid} \rightarrow \text{Lipid peroxidation}$$

Primary disturbances

Lipid peroxidation $\rightarrow$ Membrane damage, enzyme inactivation, aldehyde and peroxide products

Secondary disturbances

Aldehydes and lipid peroxidation $\rightarrow$ Increased capillary permeability
Increased platelet aggregation
Protein cross linking
Reaction with SH
Decreased DNA synthesis
Decreased enzyme activities

Tertiary disturbances

Figure 7.14 The sequence of cellular events following the metabolism of carbon tetrachloride to a reactive free radical.

7.2.2 Valproic Acid

This is a widely used antiepileptic drug, which occasionally causes liver dysfunction. There are two types of dysfunction, the first type is a mild transient elevation of the transaminases (ALT, AST), which resolves; the second type is of a more severe hepatotoxicity, manifested as **fatty liver** with jaundice and necrosis, which may lead to fatal liver failure.

The incidence of the severe form is between 1 in 20,000 and 1 in 40,000 in adults, although the incidence is much higher in children (1 in 5000). The fatty liver is a "visible" symptom of dysfunction, not necessarily a cause of liver failure, although it can be. Valproic acid is similar to a fatty acid and therefore can become incorporated into fatty acid metabolism. This involves formation of an acyl CoA derivative and also a carnitine derivative. However, this depletes both CoA from the intramitochondrial pool and carnitine and so compromises the mitochondria and reduces the ability of the cell to metabolize short-, medium-, and long-chain fatty acids via β-oxidation (Fig. 7.15).

Furthermore, valproic acid is metabolized by dehydrogenation, a reaction catalyzed by CYP2C9, to a fatty acid analogue (chap. 4, Fig. 29). This fatty acid analogue is then incorporated into the β-oxidation pathway in the mitochondria (Fig. 7.15). However, the metabolite analogue is reactive, as it has conjugated double bonds. This reactive intermediate disrupts β-oxidation by damaging the enzymes involved in the process. It also depletes mitochondrial GSH, allowing still more damage to occur (Fig. 7.15).

Therefore, the metabolite interferes with mitochondrial function and decreases the production of ATP and other important cofactors such as NADH and FADH. Therefore, after repeated use of the drug, mitochondrial integrity is reduced and cellular and overall liver fat oxidation is inhibited. Consequently, fat accumulates, seen as microvesicular steatosis. Electron microscopy shows swollen mitochondria and damaged mitochondrial structures. The accumulated lipid may encourage lipid peroxidation and oxidative stress to occur, causing further damage.

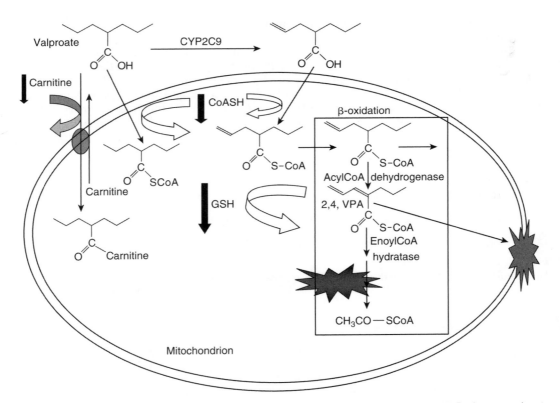

Figure 7.15 The interaction between valproate and the mitochondrial β-oxidation system. Dark arrows denote depletion. Filled in circle is the carnitine transporter. 2,4, VPA: 2,4, diene-VPA CoA ester. This reactive metabolite damages the enzyme enoyl CoA hydratase and the mitochondrial membranes and depletes GSH, as indicated.

7.2.3 Nucleoside Analogues

These drugs can also cause mitochondrial damage, dysfunction of fat metabolism and liver failure.

Nucleoside analogues are drugs used to treat HIV and hepatitis. One such drug **fialuridine** and other drugs of this type have caused severe hepatic dysfunction. This dysfunction was characterized by fatty liver and fatal liver failure. Fialuridine caused fatal damage in 5 of 12 patients in early clinical trials. Fialuridine inhibits DNA polymerases. However, there is also DNA in the mitochondria [mitochondrial DNA (mtDNA)].

mtDNA may be copied at any time during the cell cycle, depending on the energy demands of cell. mtDNA has 13 genes, some of which code for some of the enzymes for mitochondrial metabolism.

Fialuridine causes inhibition of mtDNA replication. This inhibition and also mutations of mtDNA, which can result from treatment with the drug, lead eventually to reduced *numbers* of mitochondria.

Consequently, repeated exposure to nucleoside analogues such as fialuridine leads to impaired function of the cell because of the reduced number of functioning mitochondria.

In tissues such as of the liver, mitochondrial activities such as fatty acid metabolism are compromised, and the liver eventually fails with potentially fatal consequences.

7.2.4 Paracetamol

Paracetamol (acetaminophen) is a widely used analgesic and antipyretic drug that is relatively safe when taken at prescribed therapeutic doses. However, it has become increasingly common for overdoses of paracetamol to be taken for suicidal intent. In the United Kingdom, for example, around 200 deaths a year result from overdoses of paracetamol. This has prompted some to call for changes in its availability, with newspaper headlines such as "Doctors demand curbs on 'killer' paracetamol" (Sunday Times, London, 14th November, 1993). When taken in overdoses, paracetamol causes primarily centrilobular hepatic necrosis, but this may also be accompanied by renal damage and failure.

By measurement of the blood level of paracetamol in overdose cases, it is possible to estimate the likely outcome of the poisoning, and hence determine the type of treatment. Measurement of the blood level of paracetamol and its various metabolites at various times after the overdose showed that the half-life was increased several folds (Table 7.3), and the patients who sustained liver damage had an impaired ability to metabolize paracetamol to conjugates (Fig. 7.16).

Paracetamol causes centrilobular hepatic necrosis in various species, although there are substantial species differences in susceptibility (Fig. 7.17). Experimental animal species susceptible to paracetamol have been used to study the mechanism underlying the hepatotoxicity. It was found that the degree of hepatic necrosis caused by paracetamol was markedly increased by pretreatment of animals with microsomal enzyme inducers and, conversely, reduced in animals given inhibitors of these enzymes (Table 7.4). Similarly, humans who are exposed to enzyme-inducing drugs are more susceptible to the hepatotoxic effects of paracetamol, and resistance to these effects has been described in a patient taking the drug cimetidine, which is an inhibitor of the microsomal enzymes.

Table 7.3 Mean Plasma Concentration and Half-Life of Unchanged Paracetamol in Patients with and Without Paracetamol-Induced Liver Damage

Patients	Plasma paracetamol half-life (hr)	Plasma paracetamol Concentration (μg mL^{-1})	
		4 hr after ingestion	12 hr after ingestion
No liver damage (18)	2.9 ± 0.3	163 ± 20	29.5 ± 6
Liver damage (23)	7.2 ± 0.7	296 ± 26	124 ± 22

Numbers of patients in parentheses.
Source: From Ref. 4.

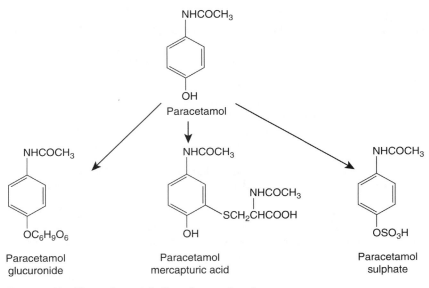

Figure 7.16 The major metabolites of paracetamol.

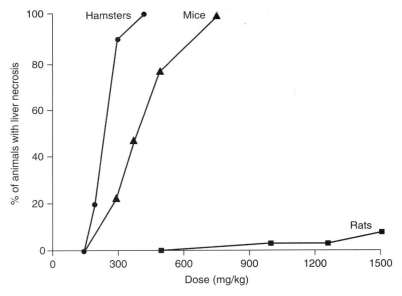

Figure 7.17 Species differences in the dose-response relationship for paracetamol-induced liver damage. *Source*: From Refs. 5 and 6.

In studies in experimental animals, there was a lack of correlation between tissue levels of unchanged paracetamol and the necrosis. Furthermore, when radiolabeled paracetamol was administered, several important findings were noted:

1. The radiolabel was bound to liver protein covalently and to a much greater extent than in other tissues such as the muscle (Table 7.4).
2. Autoradiography revealed that the binding was located primarily in the centrilobular areas of the liver, which also suffered the damage.
3. The covalent binding to liver protein was inversely related to the concentration of unchanged paracetamol in the liver.
4. The covalent binding was increased and decreased by inducers and inhibitors of the microsomal enzymes respectively (Table 7.4).

Table 7.4 The Effect of Various Pretreatments on the Severity of Hepatic Necrosis and Covalent Binding of Paracetamol to Mouse Tissue Protein

| | | | Covalent binding of ^{3}H-paracetamol (nmol mg^{-1} protein) | |
| | | | | |
Treatment	Dose of paracetamol (mg kg^{-1})	Severity of liver necrosis[a]	Liver	Muscle
None	375	1–2+	1.02 ± 0.17	0.02
Piperonyl butoxide[b]	375	0	0.33 ± 0.05	0.01
Cobaltous chloride[b]	375	0	0.39 ± 0.11	0.01
A-[b]Napthylisothiocyanate	375	0	0.11 ± 0.06	0.01
Phenobarbital[c]	375	24+	1.6 ± 0.1	0.02

Severity of necrosis: 1 + <6% necrotic hepatocytes; 2 + >6% <25% necrotic hepatocytes; 3 + >25% <50% necrotic hepatocytes; 4 + >50% necrotic hepatocytes.
[a]Data from Ref. 6.
[b]Microsomal enzyme inhibitor.
[c]Microsomal enzyme inducer.
Source: Data from Refs. 7 and 8.

5. The extent of binding was dose related and increased markedly above the threshold dose for hepatotoxicity (Fig. 7.18A).
6. The covalent binding to liver protein was time dependent, but that to muscle protein was not.

These findings suggested that a reactive metabolite of paracetamol was responsible for the hepatotoxicity and covalent binding rather than the parent drug.

More recent immunohistochemical studies using antiparacetamol antibodies have shown that covalent binding of a paracetamol metabolite occurs in the damaged centrilobular regions of human liver after overdoses.

The metabolic profile is straightforward and, as indicated by the urinary metabolites (Fig. 7.16), mainly involves conjugation with glucuronic acid and sulfate, with a small amount excreted as a mercapturic acid or N-acetylcysteine conjugate (abut 5% of the dose in humans). None of these metabolites is chemically reactive and therefore not likely to react with hepatic protein. However, the excretion of the mercapturic acid was found to be dose dependent and decreased in experimental animals after heptotoxic doses (Fig. 7.18B). There was also a corresponding depletion of hepatic GSH to around 20% or less of the normal level (Fig. 7.18A). Thus, there was an apparent relationship between the metabolism, covalent binding to liver protein, and hepatic GSH, which resulted in the scheme proposed in Figure 7.19. After overdoses of paracetamol, the normally adequate detoxication of the reactive metabolite of paracetamol by GSH was overwhelmed and hepatic GSH levels were depleted, allowing the reactive metabolite to damage the liver cell. Synthesis of new GSH is inadequate to cope with the rate of depletion. There is thus a marked dose threshold for toxicity, which occurs when the hepatic GSH is depleted by 80% or more of control levels. Studies *in vitro* have established that the reactive intermediate is produced by cytochromes P-450, requires NADPH and molecular oxygen, and is inhibited by carbon monoxide. Using ^{18}O, it was established that as no ^{18}O was incorporated, an epoxide is not the reactive intermediate.

The reactive intermediate is believed to be **N-acetyl-p-benzoquinone imine** (NAPQI); although the preceding step(s) is not yet clear, N-hydroxyparacetamol is not the precursor. Using human liver microsomes, it has been found that cytochromes P-450 are responsible for the production of a reactive metabolite, which binds to microsomal protein. The metabolic activation certainly involves CYP2E1 as mice, which are null for the cyp2e1 gene (cyp2e1 −/−) are resistant to what would normally be toxic doses of paracetamol. However, other isoforms of cytochrome P-450 are also important, including CYP1A2 and CYP3A4. Using knockout mice, it was shown that CYP2E1 as well as CYP1A2 were involved in the metabolic activation. The different isoforms may become more important at higher doses.

(A)

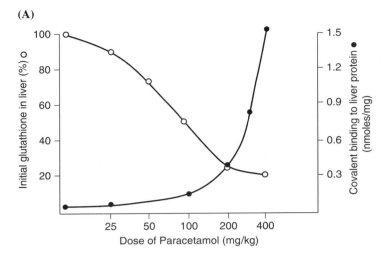

(B)

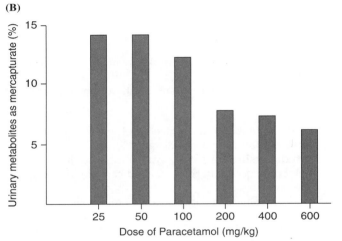

Figure 7.18 (**A/B**) Relationship between hepatic glutathione, covalent binding of radio-labeled paracetamol to hepatic protein, and urinary excretion of paracetamol mercapturic acid after different doses of paracetamol. *Source*: From Ref. 9.

It should also be mentioned that prostaglandin synthetase can activate paracetamol (Fig. 7.20) to reactive metabolites. Although probably not the primary route of activation in the liver, it has been suggested that this could be important in the kidney (which can also be damaged in paracetamol overdose).

Seventy percent of the covalent binding in the liver is to cysteine residues in hepatic proteins through the 3-position on the benzene ring. Furthermore, this binding does not seem to be random and is to specific proteins in both mouse and human liver *in vivo*, and in microsomes and isolated hepatocytes *in vitro*. Some of the binding seems to be similar in both mice and humans. Thus, both species show binding of a paracetamol metabolite to various proteins, but especially to a 58-kDa protein, which has been termed "APAP-binding protein" and is also a target for other reactive metabolites. Although its function is unknown, it has been suggested that it may serve as a transcription factor that senses electrophilic metabolites. It seems to be translocated to the nucleus after reaction with the reactive intermediate and therefore may bind to specific response elements in the promoter regions of stress genes. Such a chain of events might then trigger rescue pathways or cell death pathways.

Several other cellular target proteins for the reactive metabolite of paracetamol have also been detected and identified, namely, formyl tetrahydrofolate dehydrogenase, glyceraldehyde-3-phosphate dehydrogenase (GAPDH), and GSH transferase, all cytosolic proteins.

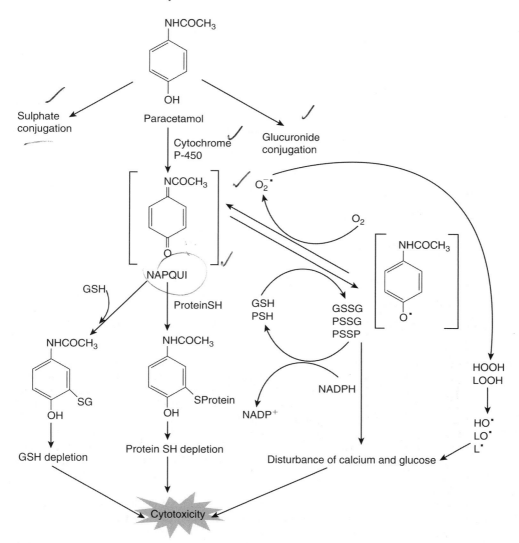

Figure 7.19 Proposed metabolic activation of paracetamol to a toxic, reactive intermediate *N*-acetyl-*p*-benzoquinone imine (NAPQI). This can react with glutathione (GSH) to form a conjugate or with tissue proteins. Alternatively, NAPQI can be reduced back to paracetamol by glutathione, forming oxidized glutathione (GSSG).

Mitochondrial proteins are also targets, namely, glutamate dehydrogenase and aldehyde dehydrogenase. Several protein targets are found in the endoplasmic reticulum, namely, calreticulin, thiol protein disulfide oxidoreductase, another GSH transferase, and glutamine synthetase. The activity of some of these enzymes was significantly decreased by the binding of the reactive metabolite, and this may contribute to the impaired mitochondrial function and other effects observed in paracetamol toxicity. However, as previously mentioned, the critical nature of the macromolecule will influence the outcome of binding.

The role of covalent binding is still unclear as studies *in vitro* in isolated hepatocytes and *in vivo* have shown that certain agents can protect against the cytotoxicity of paracetamol or reduce the cytotoxicity, despite that fact that substantial covalent binding has occurred. Indeed, covalent binding does not necessarily lead to hepatotoxicity, as shown by studies with the regioisomer of paracetamol, 3-hydroxyacetanilide, which binds to liver proteins in rodents but is not hepatotoxic. Although the role, if any, of covalent binding to protein in the development of the necrosis is not yet clear, there has been no demonstration of hepatotoxicity without binding.

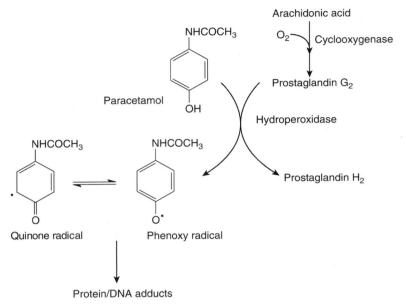

Figure 7.20 Activation of paracetamol by prostaglandin synthetase system to reactive radical intermediates.

However, the reactive metabolite will cause other changes as well as binding to protein. Thus, NAPQI will react both chemically and enzymatically with GSH to form a conjugate and will also oxidize it to GSSG and in turn be reduced back to paracetamol. This cyclical process may explain the occurrence of extensive depletion of GSH. NADPH will also reduce NAPQI and in turn be oxidized to NADP, although reduction via GSH is probably preferential. NADPH oxidation may also result from GSSG reduction via GSH peroxidase (Fig. 7.18).

Analogues of paracetamol, which are unable to undergo covalent binding to protein, are still hepatotoxic and can undergo a redox reaction with GSH. However, oxidative stress has not been demonstrated *in vivo*, and there are differences between *in vivo* data and that obtained in isolated hepatocytes.

The depletion of GSH and NADPH will allow the oxidation of protein sulfhydryl groups, which may be an important step in the toxicity. Thus, GSH and protein sulfhydryl groups, such as those on Ca^{2+}-transporting proteins, are important for the maintenance of intracellular calcium homeostasis. Thus, paracetamol and NAPQI cause an increase in cytosolic calcium, and paracetamol inhibits the Na^+/K^+ ATPase pump in isolated hepatocytes.

The mitochondria are also damaged by the reactive intermediate and the oxidative stress. Hence, cellular ATP levels will fall, which will lead to necrotic cell death in many cases.

The reactive metabolite starts a cascade of events, which lead to hepatotoxicity such as increases in cytosolic Ca^{2+}. This will activate a number of processes such as endonucleases, which attack double-stranded DNA and cause strand breaks (Fig. 7.21).

As well as the formation of NAPQI and covalent binding to target proteins, there are various other possible metabolic pathways, including deacetylation, reactive oxygen species (ROS), and semiquinone radical formation, which may or may not play a role in the hepatotoxicity. The importance of and interrelationships between covalent binding to particular hepatic proteins, cyclical oxidation and reduction of GSH, and oxidation of protein thiol groups and intracellular calcium level are currently unclear. These events are not mutually exclusive, and so it is possible that all are a series of necessary events occurring at particular stages in the development of paracetamol hepatotoxicity.

However, covalent binding to protein is still believed to be the important event in the toxicity. The current concepts are shown in Figure 7.21.

Studies using a combination of techniques have revealed a sequence of events starting within 15 minutes of a toxic dose, with GSH depletion and mitochondrial changes detected by electron microscopy. The depletion of GSH leaves the mitochondria and the cell, in general, vulnerable. Mitochondria without adequate GSH will generate ROS, which will cause more

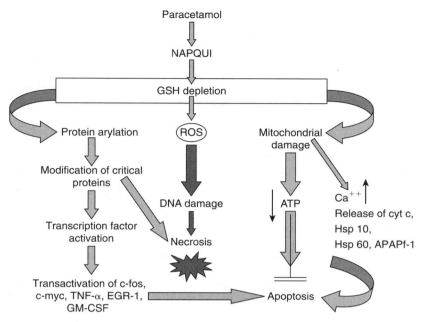

Figure 7.21 The cascade of events in the liver following a toxic dose of paracetamol. Hsp10, Hsp 60 are heat shock proteins, which have a protective function. cyt c is ctyochrome c. TNF-α, EGR-1, and GM-CSF are all genes involved with apoptosis. c-fos, APAF-1, and c-myc are also genes.

damage, such as DNA damage, and critical proteins within the mitochondria will be damaged also. Thus, ATP synthetase has been shown to be a target for NAPQUI and would also be damaged by ROS and was shown to be decreased in mitochondria from paracetamol-treated mice. This will lead to a dramatic fall in ATP, thereby allowing Ca^{2+} to rise, and a change of the inner mitochondrial membrane potential. These events will cause the release of proapoptotic factors such as cyt-c, Hsp 10, and Hsp 60. The ROS will switch on certain genes such as c-myc, c-fos, EGR-1, and TNF-α. The activation of c-fos, possibly a simple response to injury, may represent an attempt by the cell at rescue.

The increase in GM-CSF may be due to effects on the Kupffer cells, and this is known to trigger EGR-1, which in turn triggers TNF-α (Fig. 7.21). The latter is involved in necrosis, but all three are also known to trigger apoptosis, which may or may not take place as discussed earlier.

If the ATP depletion is too great, then apoptosis cannot proceed. Kupffer cell activation involving IFNγ (interferon) and TNF-α is part of an inflammatory response in which nitric oxide and various cytokines are produced.

Hsp 10 and Hsp 60 are chaperone proteins located in the mitochondria, which have a protective function but are released when the mitochondria are damaged, hence the mitochondrial level drops after paracetamol.

Other changes noted were the loss of certain enzymes and other proteins as well as ATP synthetase already mentioned, including a number involved with β-oxidation and glycolysis (e.g., thiolase and GAPDH). These changes would tend to reduce ATP generation from fatty acid oxidation and glycolysis.

GAPDH is inhibited to more than 80 % by a toxic dose of paracetamol. It seems likely that the initiating events are chemical interactions and protein damage and that gene changes may arise later, as a response and as part of the repair processes.

It now seems that an overdose of paracetamol causes a massive chemical stress, which causes an immediate adaptive defense response in the liver cell, which senses danger via redox-sensitive transcription factors. A number of mechanisms are involved, including the release, as a result of the stress, of a transcription factor Nrf-2 from its binding with Keap 1, a cytoplasmic inhibitor. Nrf-2 translocates to the nucleus and with other activators binds to an antioxidant-response element. This leads to transcription of a number of genes, so producing a number of antioxidant proteins and enzymes. Thus, GSH synthesis, GSH transferase, and glucuronyl transferases are all initially increased, along with heme oxygenase. However, components of this defense system can be dislocated, as amounts of the paracetamol metabolite overwhelm the system and damage critical proteins. It is the critical proteins that are damaged, as already indicated, which is the important event. Thus, damage to and inhibition of γ-glutamyl cysteine synthetase will occur and will compromise the cell's ability to replenish depleted GSH. Similarly, inhibition of ATPases occurs, which will compromise the ability of the cell to maintain ionic homeostasis.

Covalent binding to protein, initially observed by radiolabeling the paracetamol, has now been further studied by western blotting and proteomics. This has revealed that binding to some proteins is critical (critical protein hypothesis) and that this binding occurs in different parts of the cell at about the same time. Thus, some protein targets are in the endoplasmic reticulum, but others are in other compartments. In certain cases, the binding compromises the activity of the protein. For example, GAPDH is 80% inhibited by binding of NAPQUI to a cysteine residue at the active site. The inhibition of this cytosolic enzyme contributes to the rapid depletion of ATP (80% at 2 hours).

The inhibition of membrane Ca^{2+}/Mg^{2+} ATPase leads to a derangement of Ca^{2+} levels, which will damage the mitochondria and hence also indirectly contribute to ATP depletion. Inhibition of γ-glutamyl cysteine synthetase reduces the ability of the liver cell to synthesize new GSH, so reducing its ability to protect itself. Overall, some 17 enzymes have been shown to be inhibited ex vivo after treatment of animals and another 14 are known to have bound paracetamol and may or may not be inhibited.

However, as well as the changes going on in the hepatocyte, other events contribute to the hepatic necrosis such as the Kupffer cells. These release chemical mediators that contribute to the pathological response. Changes to the microcirculation of the liver also seem to play a part as indicated by the role of nitric oxide.

The dramatic species differences in the hepatotoxicity of paracetamol (Fig. 7.17) are due to the differences in metabolic activation, as indicated by *in vitro* studies using liver microsomal preparations and hepatocytes from various species. The activity of these microsomal preparations, using covalent binding to protein as an endpoint, approximately correlated with the *in vivo* toxicity. The sensitivity of the rat to paracetamol is increased by pretreatment with phenobarbital, as it increases the activity of the cytochromes P-450 system responsible for the metabolic activation to NAPQI, and therefore more is metabolized by this pathway. This is also the case in humans, whereas in the hamster, phenobarbital pretreatment decreases the hepatotoxicity as a result of an increase in glucuronic acid conjugation, and hence detoxication, because of induction of glucuronosyl transferase. Pretreatment of hamsters with a polycyclic hydrocarbon such as 3-methylcholanthrene, however, increases the toxicity. Another factor that will increase the toxicity is a reduction in detoxication pathways such as by depletion of GSH with compounds such as diethyl maleate (chap. 4, Fig. 63) and inhibition of sulfate and glucuronic acid conjugation by treatment with salicylamide. The latter treatment diverts more paracetamol through the toxic pathway.

Studies in human subjects have revealed that there is considerable variation between individuals in their rate of metabolism of paracetamol via the toxic, oxidative pathway. Thus, at one end of the frequency distribution, some individuals, approximately 5% of the particular population studied, metabolized paracetamol at a rate similar to hamsters and mice. Other individuals at the opposite end of the frequency distribution have rates of oxidation about one-quarter of the highest rates, and these individuals are probably more akin to rats, the resistant species.

Although the investigation of the mechanism underlying paracetamol hepatotoxicity has been of intrinsic toxicological interest, there has also been a particularly significant benefit that has arisen from this work. This is the development of an antidote that is now successfully used for the treatment of paracetamol overdose. The antidote now most commonly used is N-acetylcysteine, although **methionine** is also used in some cases, as it can be given orally. There are various mechanisms by which N-acetylcysteine may act:

1. It promotes the synthesis of GSH used in the conjugation of the reactive metabolite NAPQI.
2. It stimulates the synthesis of GSH used in the protection of protein thiols.

3. It may relieve the saturation of sulfate conjugation, which occurs during paracetamol overdose.
4. It may itself be involved in the reduction of NAPQI.
5. It may reduce oxidized protein thiols.

Although originally thought to be effective only when administered up to 10 to 12 hours after an overdose, significantly, *N*-acetylcysteine may be effective when given 15 hours or more after an overdose. Clearly, the majority of the metabolic activation and covalent binding to protein will have taken place by this time, and therefore, it is unlikely that it will be functioning by reacting with the NAPQI. Therefore, it has been proposed that *N*-acetylcysteine may also protect the liver cells against the subsequent changes that occur after the reaction of the reactive metabolite with cellular constituents. Methionine, however, is not effective when given at such later times, possibly because synthesis of GSH from methionine is inhibited or compromised by paracetamol toxicity. This is possibly due to inhibition of the thiol-containing enzymes that are involved in the synthetic pathway.

7.2.5 Bromobenzene

Bromobenzene is a toxic industrial solvent that causes centrilobular hepatic necrosis in experimental animals. It may also cause renal damage and bronchiolar necrosis. The study of the mechanism underlying the hepatotoxicity of bromobenzene has been of particular importance in leading to a greater understanding of the role of GSH and metabolic activation in toxicity.

The metabolism of bromobenzene is complex, but the pathway thought to be responsible for the hepatotoxicity is that leading to 4-bromophenol, postulated to proceed via the 3,4-oxide (Fig. 7.22), a pathway catalyzed by the microsomal monooxygenase system. An alternative metabolic pathway is the formation of 2-bromophenol via the 2,3-oxide. The use of strains of mice with different degrees of microsomal monooxygenase activity toward epoxidation at the 2,3- and 3,4-positions showed a correlation between the activity for 3,4-epoxidation and centrilobular necrosis. It was also observed that radiolabeled bromobenzene became covalently bound to liver protein in the necrotic tissue of the centrilobular area. One of the major urinary metabolites of bromobenzene is a mercapturic acid in which *N*-acetylcysteine is conjugated on the 4-position (Fig. 7.22). This results from an initial conjugation of the 3,4-oxide, a reactive metabolite, with GSH, followed by enzymatic removal of the glutamyl and glycinyl residues. It was observed that pretreatment of rats with the microsomal enzyme inducer phenobarbital resulted in greater hepatotoxicity, increased excretion of the GSH conjugate from the 3,4-oxide, and greater covalent binding to liver protein. Inhibitors of the microsomal enzymes decreased hepatotoxicity and covalent binding to liver protein.

Investigation of the level of GSH in the liver of animals revealed that bromobenzene caused a dose-dependent depletion of the tripeptide (Fig. 7.23A), which coincided with a decrease in the excretion of the mercapturic acid and an increase in covalent binding (Fig. 7.23B). These observations culminated in the hypothesis that GSH protects the hepatocyte against the reactive metabolite bromobenzene 3,4-oxide by conjugating with it, chemically, enzymatically, or both. After an hepatotoxic dose, however, there is sufficient of the reactive metabolite to deplete the available GSH from the liver. The reactive metabolite is therefore not detoxified by conjugation with GSH and is able to react with cellular macromolecules such as protein. Studies *in vitro* revealed that the addition of GSH to the incubation mixture reduced the amount of covalent binding to protein and the production of hydroxylated metabolites, yet the overall amount of metabolism remained the same. More recently, it has been shown *in vitro* that these reactive epoxides react at different rates with different proteins. Thus, the 3,4-oxide reacts with histidine groups on microsomal protein more avidly than the 2,3-oxide that preferentially reacts with cysteine residues on hemoglobin. However, although the 2,3-oxide appears to be chemically more stable than the 3,4-oxide, the latter has been detected in the blood *in vivo* and in the incubation medium of hepatocytes *in vitro* after exposure to bromobenzene. These data mean that the stability of the bromobenzene 3,4-oxide would allow it to leave the endoplasmic reticulum where it is formed and reacts with macromolecules in various parts of the cell. In confirmation of this are data that indicate that the conjugation with GSH is catalyzed by a cytosolic GSH transferase. This apparent stability perhaps seems at odds with the concept of chemically reactive metabolites, but there may be an optimal reactivity for

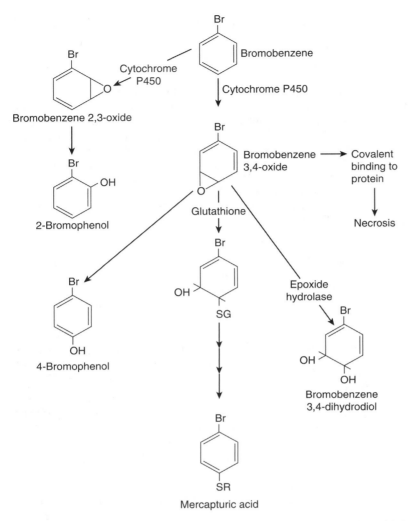

Figure 7.22 Metabolism of bromobenzene. The bromobenzene 2,3-oxide and 3,4-oxide may undergo chemical rearrangement to the 2- and 4-bromophenol, respectively. Bromobenzene 3,4-oxide may also be conjugated with glutathione, and in its absence react with tissue proteins. An alternative detoxication pathway is hydration to the 3,4-dihydrodiol via epoxide hydrolase.

toxic metabolites—not too great to result in indiscriminate reactions with macromolecules, yet sufficient to react with and modify certain proteins.

As well as detoxication via reaction with GSH, the reactive 3,4-epoxide can be removed by hydration to form the dihydrodiol, a reaction that is catalyzed by epoxide hydrolase (also known as epoxide hydratase). This enzyme is induced by pretreatment of animals with the polycyclic hydrocarbon 3-methylcholanthrene, as can be seen from the increased excretion of 4-bromophenyldihydrodiol (Table 7.5). This induction of a detoxication pathway offers a partial explanation for the decreased hepatotoxicity of bromobenzene observed in such animals. A further explanation, also apparent from the urinary metabolites, is the induction of the form of cytochrome P-450 that catalyzes the formation of the 2,3-epoxide. This potentially reactive metabolite readily rearranges to 2-bromophenol, and hence there is increased excretion of 2-bromophenol in these pretreated animals (Table 7.5).

The metabolites 2- and 4-bromophenol can also be metabolized by a further oxidation pathway to yield catechols and quinones, some of which are cytotoxic and potentially hepatotoxic. Thus, *in vitro* studies have indicated that bromoquinones and bromocatechols may be responsible for some of the covalent binding to protein and reaction with GSH.

(A)

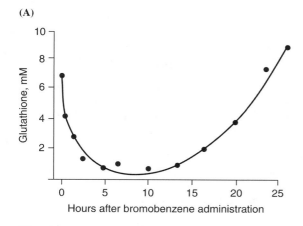

(B)

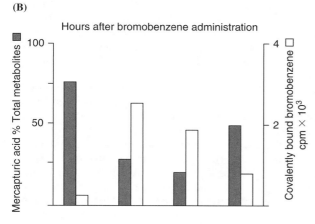

Figure 7.23 (**A/B**) Relationship between hepatic glutathione, covalent binding of radiolabeled bromobenzene to hepatic protein and urinary excretion of bromophenyl mercapturic acid in rats. Animals were given bromobenzene (10 mmol/kg, i.p.) and then radiolabeled bromobenzene at various times thereafter. *Source*: From Ref. 10.

Table 7.5 Effect of 3-Methylcholanthrene (3-MC) Pretreatment on the Urinary Metabolites of Bromobenzene in Rats

Total urinary metabolites (%)

Metabolites	Untreated	3-MC treated
4-Bromophenylmercapturic acid	72	31
4-Bromophenol	14	20
4-Bromocatechol	6	10
4-Bromophenyldihydrodiol	3	17
2-Bromophenol	4	21

Dose of bromobenzene 10 mmol kg^{-1} to rats.
Source: From Ref. 11.

However, administration of primary phenolic metabolites does not cause hepatotoxicity. At least seven GSH conjugates have been identified as metabolites of bromobenzene and its primary phenolic metabolites.

The mechanism of hepatotoxicity is therefore currently unclear. It has been suggested that lipid peroxidation is responsible rather than covalent binding to protein. Arylation of other low molecular weight nucleophiles such as coenzyme A and pyridine nucleotides also occurs and may be involved in the toxicity. Bromobenzene is known to cause the inhibition or inactivation of enzymes containing SH groups. It also causes increased breakdown of phospholipids and inhibits enzymes involved in phospholipid synthesis. Arylation of sites on plasma membranes that are crucial for their function may also occur. However, it seems likely that the toxicity is a mixture of such events.

Bromobenzene is also nephrotoxic but because of different mechanisms. This toxicity is due to the production of reactive GSH conjugates. This is discussed in more detail later in this chapter.

Thus, bromobenzene hepatotoxicity is probably the result of metabolic activation to a reactive metabolite, which covalently binds to protein and other macromolecules and other cellular molecules. It may also stimulate lipid peroxidation, and biochemical effects, such as the inhibition of SH-containing enzymes, may also play a part.

7.2.6 Isoniazid and Iproniazid

Isoniazid and iproniazid are chemically similar drugs having different pharmacological effects; they may both cause liver damage after therapeutic doses are given. Isoniazid is still widely used for the treatment of tuberculosis, but iproniazid is now rarely used as an antidepressant.

The major routes of metabolism for isoniazid are acetylation to give acetylisoniazid, followed by hydrolysis to yield isonicotinic acid and acetylhydrazine (Fig. 7.24). The acetylation of isoniazid in human populations is genetically determined and therefore shows a bimodal distribution (see chap. 5). Thus, there are two acetylator phenotypes, termed "rapid and slow acetylators," which may be distinguished by the amount of acetylisoniazid excreted or by the plasma half-life of isoniazid.

Mild hepatic dysfunction, detected as an elevation in serum transaminases, is now well recognized as an adverse effect of isoniazid and occurs in 10% to 20% of patients. Possibly, as many as 1 % of these cases progress to severe hepatic damage, and it has been suggested that this latter, more severe form, of hepatotoxicity may have a different underlying mechanism. However, the greater incidence of hepatotoxicity reported in rapid acetylators has since been questioned. It seems that the incidence of the mild form of isoniazid hepatotoxicity is not related to the acetylator phenotype, but the incidence of the rarer, more severe form is more common in slow acetylators.

A recent study of isoniazid-induced liver dysfunction was investigated in patients genotyped for NAT2. This study revealed a *clear* increase in both the incidence and severity of hepatic dysfunction in those genotyped as slow acetylators.

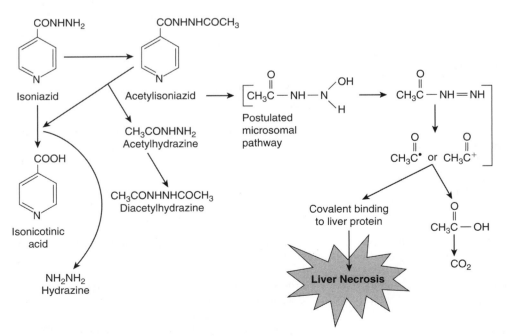

Figure 7.24 Metabolism of isoniazid. The acetylhydrazine released by the hydrolysis of acetylisoniazid is further metabolized to a reactive intermediate thought to be responsible for the hepatotoxicity.

Table 7.6 Effect of Various Pretreatments on the Hepatotoxicity, Covalent Binding, and Metabolism to CO_2 of Acetylisoniazid and Acetylhydrazine

	Acetylisoniazid (200 mg kg^{-1})			Acetylhydrazine (20 mg kg^{-1})		
Pretreatment	Necrosis[a]	Covalent binding (nmol mg protein^1)[b]	$^{14}CO_2$[b] (% dose)	Necrosis	Covalent binding (nmol mg protein^1)[b]	$^{14}CO_2$[b] (% dose)
None	0 – +	0.20	10	0 – +	0.15	29
Pb[c]	+ +	0.31	12	+ + +	0.19	35
Pb + CoCl$_2$[d]	0	0.15	4	+	0.09	22
Pb + BNPP[e]	0	0.11	4	+ + +	0.23	37

Severity of necrosis: 1 + <6% necrotic hepatocytes; 2 + >6% <25% necrotic hepatocytes; 3 + >25% <50% necrotic hepatocytes; 4 + >50% necrotic hepatocytes.
[a]Data from Ref. 12.
[b]Data from Ref. 13.
[c]Microsomal enzyme inducer (phenobarbital).
[d]Microsomal enzyme inhibitor (cobalt chloride).
[e]Acylamidase inhibitor (bis-*p*-nitrophenyl phosphate).

However, *initial* suggestions that patients with the rapid acetylator phenotype (see chap. 5) were more susceptible to such damage prompted a study of the relationship between metabolism and toxicity.

Animal studies revealed that both acetylisoniazid and acetylhydrazine were hepatotoxic, causing centrilobular necrosis in phenobarbital-pretreated animals. Conversely, pretreatment of animals with microsomal enzyme inhibitors reduced necrosis. Inhibition of the hydrolysis of acetylisoniazid (Fig. 7.24) with **bis-p-nitrophenylphosphate** (chap. 5, Fig. 38), an acyl amidase inhibitor, reduced its hepatotoxicity, but not that of acetylhydrazine. Further studies revealed that metabolism of acetylhydrazine by the microsomal monooxygenases was responsible for the hepatotoxicity, resulting in the covalent binding of the acetyl group to liver protein. These data indicated that acetylhydrazine was the toxic metabolite produced from acetylisoniazid, and therefore isoniazid. Acetylhydrazine required metabolic activation via the microsomal enzymes to a reactive acylating intermediate that would react with protein (Table 7.6).

The proposed activation of acetylhydrazine involves N-hydroxylation, followed by loss of water to yield **acetyldiazine**, an intermediate that would fragment to yield acetyl radical or acetyl carbonium ion (Fig. 7.24). GSH was not depleted by hepatotoxic doses of acetylhydrazine, indicating that unlike bromobenzene or paracetamol toxicity, it does not have a direct protective role.

The role of acetylhydrazine in isoniazid hepatotoxicity is complex, as the production of the toxic intermediate involves several steps. Further study has indicated that acetylhydrazine is detoxified by further acetylation to diacetlhydrazine (Fig. 7.24) and that isoniazid may interact with acetylhydrazine metabolism. It is therefore clear that the relative rates of production, detoxication, and activation of acetylhydrazine are very important determinants of the hepatotoxicity of isoniazid.

The acetylation of isoniazid and acetylhydrazine are both subject to genetic variability, so the production and detoxication pathways are both influenced by the acetylator phenotype. The hepatotoxicity of isoniazid is therefore dependent on a complex interrelationship between genetic factors and individual variation in the pharmacokinetics of isoniazid. Studies of the pharmacokinetics of acetylhydrazine in human subjects after a dose of isoniazid have revealed that although the peak plasma concentration of acetylhydrazine is higher in rapid acetylators, when the plasma concentration of acetylhydrazine is plotted against time, the area under the curve is greater in slow acetylators. Thus, the overall exposure to acetylhydrazine is greater in the slow acetylators. Another possible toxic metabolite is hydrazine itself, which causes fatty liver and may cause liver necrosis in some animal species. This metabolite arises by metabolic hydrolysis of isoniazid, but the amount of isoniazid that is metabolized by this route is difficult to estimate as the other product of the hydrolysis, isonicotinic acid, can also arise from acetylisoniazid (Fig. 7.24). Hydrazine has been detected in the plasma of human subjects treated with isoniazid and was found to accumulate in slow, but not rapid, acetylators. Indeed, the production of hydrazine has been reported to be increased in patients also taking rifampicin.

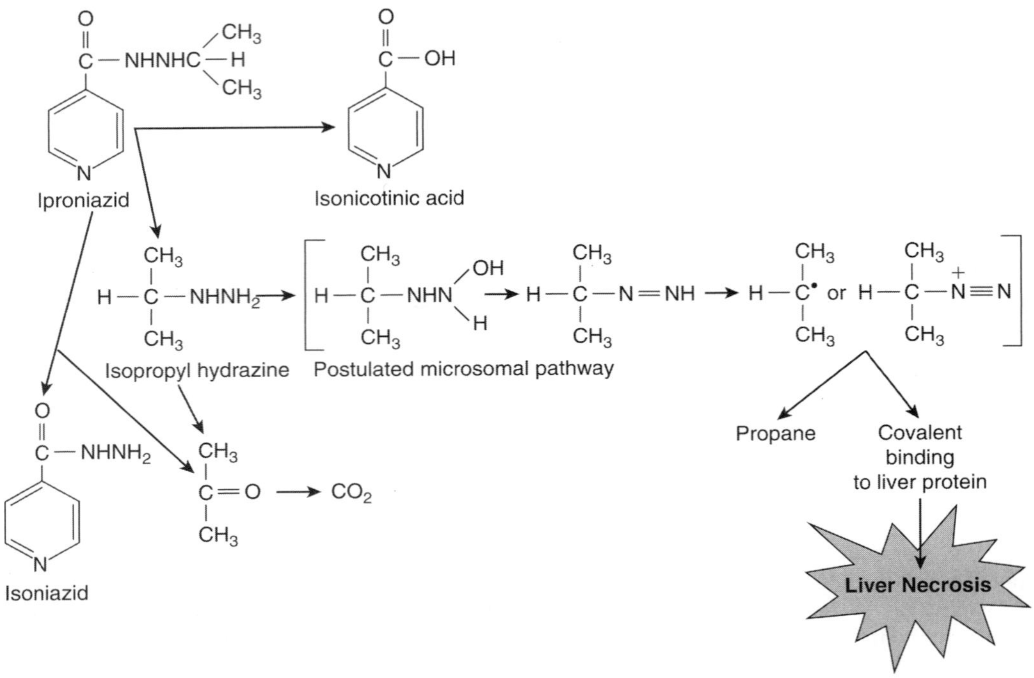

Figure 7.25 Metabolism of iproniazid. The isopropylhydrazine moiety released by hydrolysis is further metabolized to a reactive intermediate thought to be responsible for the hepatotoxicity.

The role of hydrazine in isoniazid hepatotoxicity, if any, remains to be clarified.

Iproniazid hepatotoxicity has a similar basis to that of isoniazid toxicity. Hydrolysis of iproniazid yields isopropylhydrazine (Fig. 7.25), which is extremely hepatotoxic in experimental animals. Isopropylhydrazine is metabolically activated by the microsomal enzymes to a reactive alkylating species, which covalently binds to liver protein. One of the products of this metabolic route has been shown to be propane gas, the presence of which indicates that the isopropyl radical may be the reactive intermediate. This was further suggested by double-labeling experiments, which showed that the whole isopropyl moiety is bound to protein without loss of hydrogen atoms (Fig. 7.25). Sulfydryl compounds reduce the covalent binding to protein *in vitro*, and *S*-isopropyl conjugates have been isolated from preparations *in vitro*, which confirms the role of the isopropyl group as an alkylating species. However, sulfydryl compounds are not depleted *in vivo* by hepatotoxic doses of isopropylhydrazine.

Thus, the hepatotoxicity of iproniazid and isoniazid may involve the alkylation or acylation of tissue proteins and other macromolecules in the liver. How these covalent interactions lead to the observed hepatocellular necrosis is at present not understood.

7.2.7 Microcystins

These natural toxins are heptapeptides produced by **cyanobacteria**, which are associated with algal blooms. These substances are a hazard to wild and farm animals and sometimes humans who come in contact with contaminated water. There are a number of these toxins, some of which such as microcystin LR are hepatotoxic, causing damage to both hepatocytes and endothelial cells. The toxins have some unusual structural features, incorporating three D-amino acids and two very unusual ones, namely, methyldehydro alanine (Mdha) and amino-methoxy-trimethyl-phenyl-decadi-enoic acid (Adda) (Fig. 7.26).

Furthermore, the structure of microcystin includes an electrophilic carbon atom, (Fig. 7.26), which is part of the Mdha amino acid. If microcystin is ingested from contaminated water, for example, it is taken up into the liver by an organic anion transporter (OAT) system and therefore is concentrated in the liver. The structure of the microcystins means they are able to associate with the enzymes protein phosphatases, such as PP-1, PP-2A, and PP-2B via hydrophobic and ionic interactions.

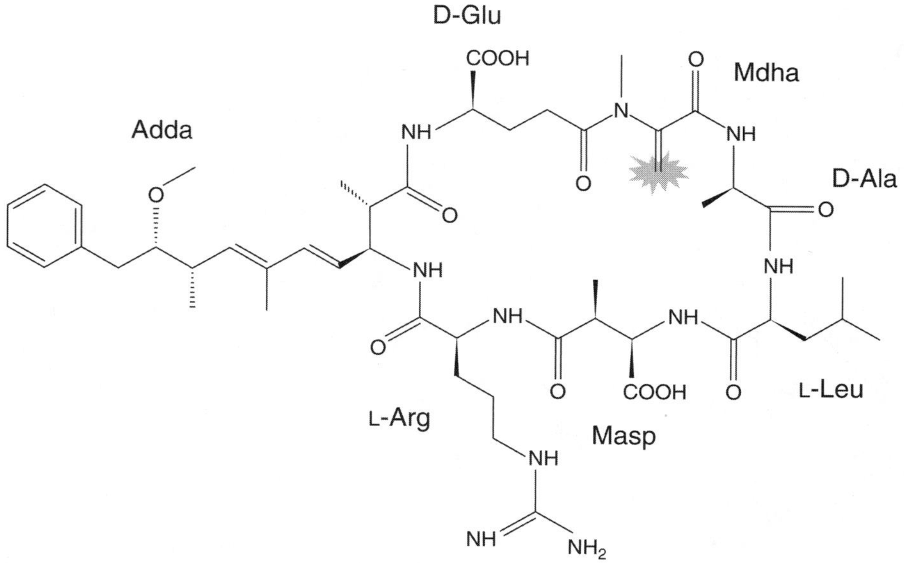

Figure 7.26 The structure of the hepatotoxic cyclic heptapeptide microcystin LR. L-Arginine and L-leucine are variable amino acids. The reactive unsaturated group is indicated by the star. *Abbreviations*: Adda, amino-methoxy-trimethyl-phenyl-decadienoic acid; Mdha, methyldehydro-alanine; Masp, methyl D-*iso*-aspartate; D-Glu, D-*iso*-glutamate; D-Ala, D-alanine.

However, this association allows the electrophilic carbon to form a covalent bond with the SH group of cysteine in the structure of the enzyme. This inactivates the enzyme. Protein phosphatases are enzymes involved in the regulation of proteins by the addition and removal of phosphate groups. This reversible **phosphorylation** is essential to the function and integrity of many proteins. When inactivated, certain proteins can be hyperphosphorylated, leading to inactivation or structural breakup. For example, cytoskeletal proteins become hyperphosphorylated, leading to collapse and as a consequence cell death. Change in phosphorylation status also affects MAP kinases and therefore the MAPK cascade, and this may be responsible for the apoptosis that is observed as another toxic effect. Lower concentrations affect the protein dynein, which, with ATP, drives vesicles along microtubules in the cell. Hyper-phosphorylation stops this process.

Microcystins are extremely potent inhibitors of protein phosphatases, active at the level of 0.2 nM (2×10^{-10} M). Coupled with the active uptake into the liver, this means that microcystins are extremely toxic. As well as being hepatotoxic, they are also tumor promoters.

7.3 TISSUE LESIONS: KIDNEY DAMAGE

7.3.1 Chloroform
Unlike the halogenated hydrocarbons discussed below, chloroform is nephrotoxic following metabolic activation via the microsomal enzyme system. Chloroform is also hepatotoxic, and again this involves cytochromes P-450-mediated activation, although in male mice the kidney is susceptible at doses that are not hepatotoxic. There are considerable species and strain differences in susceptibility to chloroform-induced nephrotoxicity, and males are generally more susceptible than females. Thus, in male mice, chloroform causes primarily renal damage, while in female mice it causes hepatic necrosis. The renal damage caused by chloroform is necrosis of the proximal tubular epithelium, which may be accompanied by fatty infiltration, swelling of tubular epithelium, and casts in the tubular lumen. The damage can be detected biochemically by the presence of protein and glucose, a decrease in the excretion of organic ions, and an increased blood urea (BUN).

The mechanism is believed to involve metabolic activation in the kidney itself. Thus, when radiolabeled chloroform was given to mice, in the kidney, the radiolabel was localized in

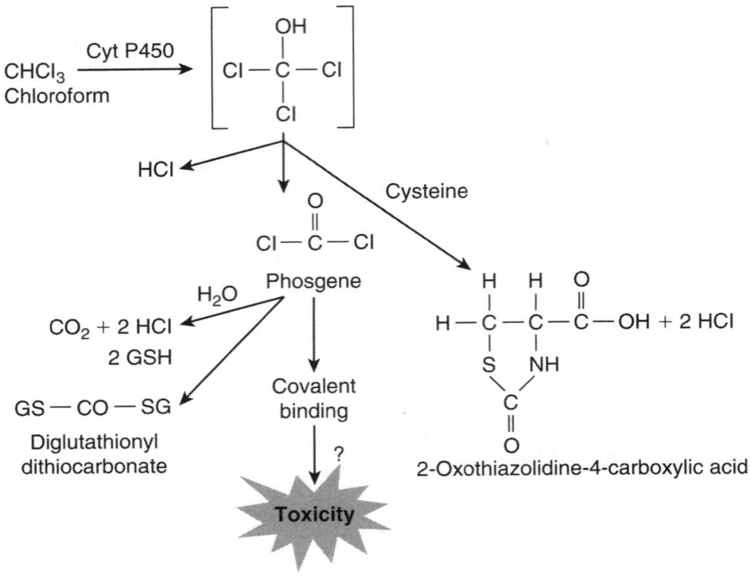

Figure 7.27   The proposed metabolic activation of chloroform.

the tubular cells that were necrotic. Certain microsomal enzyme inducers such as 3-methylcholanthrene decreased the nephrotoxicity but not hepatotoxicity of chloroform, and phenobarbital pretreatment had no effect on nephrotoxicity but increased hepatotoxicity. Pretreatment with polybrominated biphenyls, however, increased toxicity to both target organs and also increased mixed-function oxidase activity in both. *In vitro* studies have shown that microsomal enzyme-mediated metabolism of chloroform to CO_2 occurs in renal tissue and is associated with cytotoxicity and covalent binding of metabolites to microsomal protein. The sex difference is maintained *in vitro* with cortical slices from male mice showing a toxic response and metabolizing chloroform to a reactive metabolite that binds to protein, but tissue from female mice was neither susceptible nor did it metabolize chloroform. These findings are consistent with the lower level of cytochromes P-450 in the kidneys of female mice when compared with males. The finding that *deuterated chloroform* $CDCl_3$ was less nephrotoxic than the hydrogen analogue indicated that cleavage of the CH bond was involved in the metabolism, was a rate-limiting step, and was necessary for the nephrotoxicity. Studies using rabbits indicated that phosgene was produced via a cytochrome P-450-mediated reaction. Addition of cysteine reduced covalent binding and yielded a conjugate 2-oxothiazolidine-4-carboxylic acid, and in rabbit renal microsomes, this was increased by prior induction with phenobarbital. These data, in conjunction with studies on the hepatotoxicity of chloroform, suggested that oxidative metabolism to phosgene was responsible for the nephrotoxicity in mice and rabbits (Fig. 7.27).

7.3.2 Haloalkanes and Alkenes
A variety of halogenated alkanes and alkenes such as **hexachlorobutadiene, chlorotrifluoro-ethylene, tetrafluoroethylene**, and **trichloroethylene** (Fig. 7.28) are nephrotoxic. Studies have shown that metabolic activation is necessary for toxicity, but this does not involve cytochromes P-450. Thus, **hexachlorobutadiene** (HCBD) is a potent nephrotoxin in a variety of mammalian species, and the kidney is the major target.

The compound damages the pars recta portion of the proximal tubule with the loss of the brush border. The result is renal failure detected as glycosuria, proteinuria, loss of concentrating ability, and reduction in the clearance of inulin, *p*-aminohippuric acid, and tetraethylammonium ion.

The mechanism seems to involve first the formation of a GSH conjugate in the liver, catalyzed by the microsomal GSH *S*-transferase. The hepatic metabolite is then secreted into

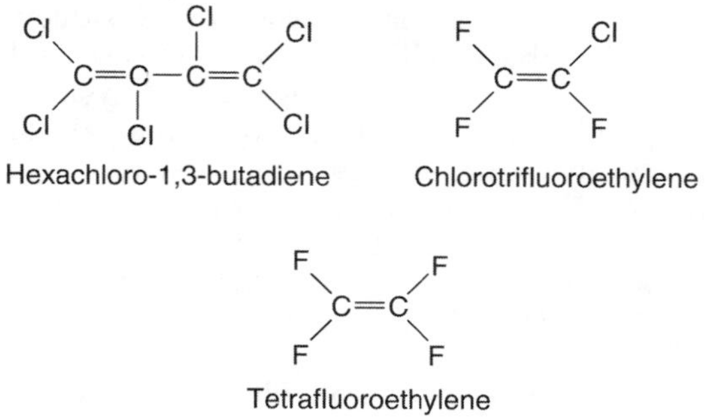

Figure 7.28 The structures of some nephrotoxic halogenated compounds.

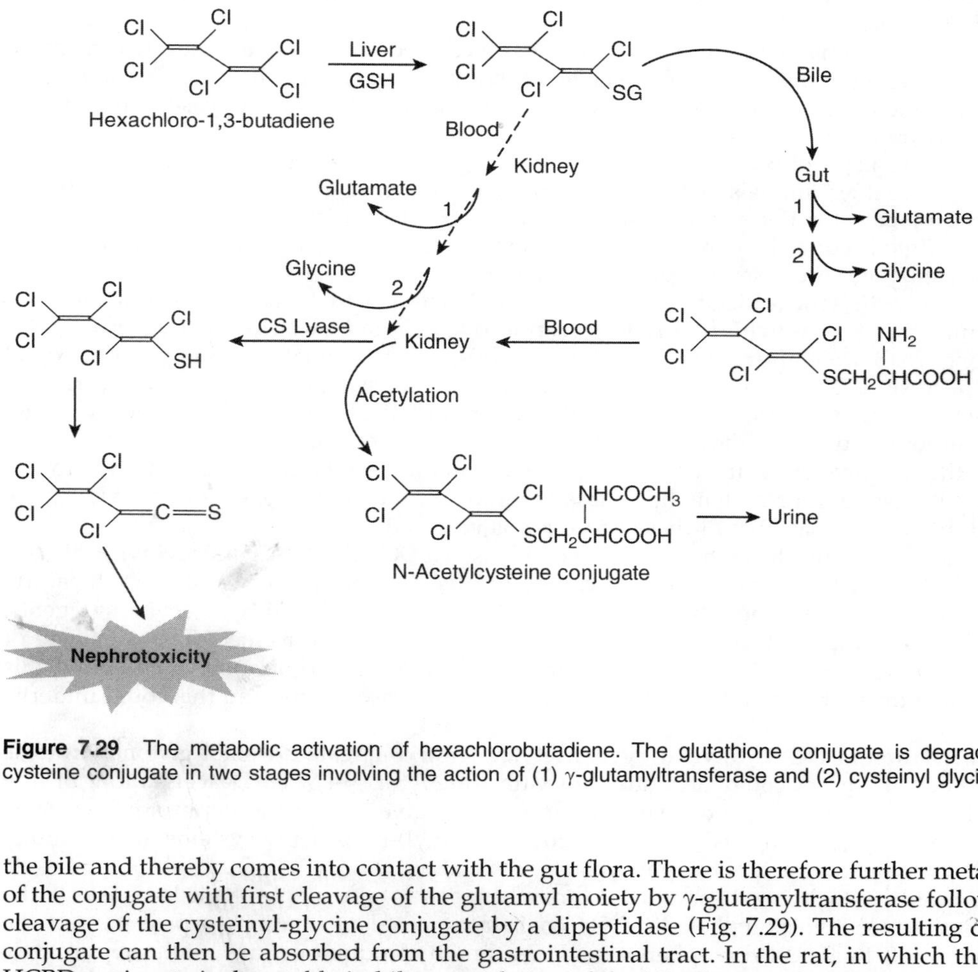

Figure 7.29 The metabolic activation of hexachlorobutadiene. The glutathione conjugate is degraded to a cysteine conjugate in two stages involving the action of (1) γ-glutamyltransferase and (2) cysteinyl glycinase.

the bile and thereby comes into contact with the gut flora. There is therefore further metabolism of the conjugate with first cleavage of the glutamyl moiety by γ-glutamyltransferase followed by cleavage of the cysteinyl-glycine conjugate by a dipeptidase (Fig. 7.29). The resulting cysteine conjugate can then be absorbed from the gastrointestinal tract. In the rat, in which the GSH-HCBD conjugate is detectable in bile, cannulation of the bile duct prevents the nephrotoxicity of HCBD. Thus, *enterohepatic recirculation* may play an important role in the toxicity. The cysteine conjugate is then absorbed from the gastrointestinal tract and transported via the blood to the kidney where it can either undergo acetylation and be excreted as a mercapturic acid or be further metabolized by the enzyme cysteine conjugate **β-lyase** (C-S lyase) to a reactive thiol. The other products of this reaction are a keto acid and ammonia. β-Lyase is present in the liver and kidney. In the kidney, it occurs in the cytosol, mitochondria, and brush border membranes, especially in the proximal tubules, including the pars recta (S3) section where damage particularly occurs.

An alternative fate for the GSH conjugate is transportation via the blood to the kidney, filtration out of the blood and in the brush border of the tubular cells glutamyltransferase, and cleavage of the conjugate by a dipeptidase to yield the cysteine conjugate. The cysteine conjugate is then taken up by the amino acid transporter system into the proximal tubular cell where toxicity occurs. The result of this is then basically the same as the other scenario.

The reactive thiol/thioketene produced by the β-lyase is an alkylating fragment, which binds to protein, DNA, and GSH. The fact that one of the locations of C-S lyase is in mitochondria may explain why this organelle seems to be damaged. Damage to the respiratory chain will lead to depletion and a shortage of ATP, which is vitally necessary for the activity of the kidney in terms of active uptake and secretion.

The sensitivity of the kidney is partly due to its ability to concentrate filtered metabolites in the tubular lumen and the active uptake processes (e.g., OAT 1), which can then reabsorb the metabolite into the tubular cells.

The conjugate of hexachlorobutadiene inhibits mitochondrial respiration *in vitro*, and this may be the ultimate target in the kidney. Damage to DNA also occurs, and this could underlie some of the toxicity.

Low-level human exposure to HCBD has occurred in the area around a chemical plant. However, humans seem likely to be less susceptible to damage than rats in whom most of the experimental studies were done; as conjugation with GSH is less, β-lyase activity is less and N-acetyltransferase is less.

Thus, the nephrotoxicity and covalent binding of metabolites to renal protein can be reduced by treating animals with the organic acid probenecid, which competes with the cysteine conjugate for the active uptake system. The cytotoxicity of the GSH conjugate of HCBD *in vitro* is reduced by inhibitors of β-glutamyltransferase and β-lyase, indicating that both these enzymes are essential for the toxicity.

Trichloroethylene is metabolized similarly and gives rise to dichlorovinyl cysteine. It has been found that **S-(1,2-dichlorovinyl)-L-cysteine** (DCVC) and **S-(2-chloroethyl)-DL-cysteine** (CEC) (Fig. 7.30) are both nephrotoxic when administered to animals causing renal proximal tubular necrosis. CEC does not require β-lyase activation to be nephrotoxic, but can rearrange, possibly to a reactive episulfonium ion, by nucleophilic displacement of the chlorine atom. These compounds decrease the activity of the renal tubular anion and cation transport system.

If the β-lyase enzyme, or the renal basolateral membrane transport system, or γ-glutamyltransferase, or cysteinylglycinase is inhibited, the nephrotoxicity of DCVC can be reduced, indicating that each of these processes is involved.

The toxic thiol produced by the action of β-lyase on DCVC is **1,2-dichlorovinylthiol**. This may be further metabolized or may rearrange to give reactive products such as thioacetyl chloride or chlorothioketene. These are reactive and unstable alkylating or acylating agents, and hydrogen sulfide, which is toxic, has also been shown to be formed. The conjugate of hexachlorobutadiene DCVC and CEC all inhibit mitochondrial respiration *in vitro*, and this may be the ultimate target in the kidney. Damage to DNA also occurs, and this could underlie the toxicity.

Acetylation of DCVC to yield the N-acetylcysteine conjugate does not give nephrotoxic metabolites, as these conjugates are not substrates for β-lyase. Deacetylation of the N-acetylcysteine conjugates, however, may occur and give rise to the nephrotoxic cysteine conjugate. Similarly, analogues such as S-(1,2dichlorovinyl)-DL-α-methylcysteine, which cannot undergo β-elimination, are not nephrotoxic.

In addition to being hepatotoxic, bromobenzene is also nephrotoxic because of the production of reactive polyphenolic GSH conjugates, covalent binding to protein, and the production of ROS. 2-Bromophenol and 2-bromohydroquinone are both nephrotoxic metabolites of bromobenzene. Quinones are both oxidants and electrophiles, undergoing both one and two electron reduction and reaction with sulfhydryl groups such as GSH and

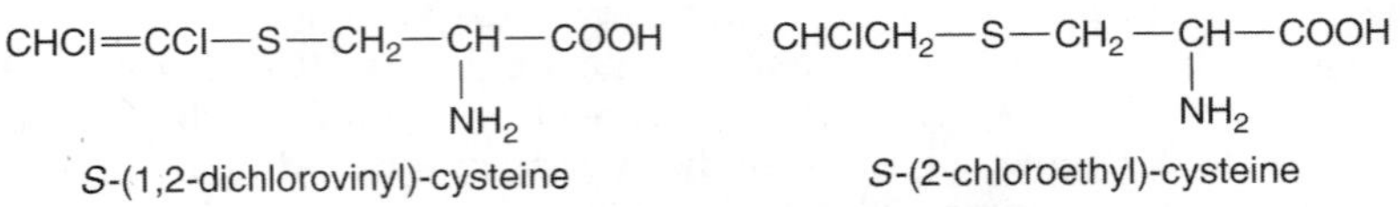

$$CHCl{=}CCl{-}S{-}CH_2{-}\underset{\underset{NH_2}{|}}{CH}{-}COOH \qquad CHClCH_2{-}S{-}CH_2{-}\underset{\underset{NH_2}{|}}{CH}{-}COOH$$

S-(1,2-dichlorovinyl)-cysteine   S-(2-chloroethyl)-cysteine

Figure 7.30 The structures of S-(1,2-dichlorovinyl)-L-cysteine and S-(2-chloroethyl)-DL-cysteine.

2-Bromohydroquinone   2-Bromohydroquinone
diglutathionyl conjugate

Figure 7.31 2-Bromohydroquinone and the diglutathionyl conjugate formed from it.

those in proteins. The one electron reduction yields a semiquinone that can generate superoxide, hydrogen peroxide, and hydroxyl radical. Furthermore, it seems that the ultimate toxic metabolite is a diglutathionyl conjugate formed from 2-bromohydroquinone (Fig. 7.31), as this will cause the same lesions in the kidney as bromobenzene, 2-bromophenol, and 2-bromohydroquinone when administered to animals.

The reasons for this are firstly the high activity of γ-glutamyl transpeptidase in the brush border membrane of the renal proximal tubular cells and secondly the activity of dipeptidases in the same region. These enzymes produce a cysteinyl-glycine conjugate and then a cysteine conjugate, respectively, from the original GSH conjugate produced in the liver. The cysteine conjugate is then taken up from the tubular lumen by the amino acid transporter system and concentrated in the proximal tubular cells.

It is these cells and the brush border membrane that sustains the damage, with the cell membrane and nucleus as early targets but mitochondrial damage as secondary target. Using the 2-bromohydroquinone diglutathionyl conjugate (Fig. 7.31) *in vitro* (renal proximal cells in culture), it was found to cause DNA single strand breaks mediated by ROS. The nephrotoxicity of bromobenzene seems therefore to be due to the damage to DNA in proximal tubular cells by the accumulated conjugate of the bromoquinone. The GSH conjugate (and breakdown products) is therefore a vehicle for localizing the hydroquinone in the kidney and so increasing the potential production of ROS.

Other quinones and and polyphenolic GSH conjugates are also toxic, such as benzohydroquinone. In the latter case, it is the triglutathionyl conjugate of benzohydroquinone that is the most toxic.

Thus, these similar examples illustrate that GSH conjugation may not always lead to a reduction in toxicity. Furthermore, they illustrate that several factors are responsible for the kidney being the target organ: active uptake processes, the presence of β-lyase and γ-glutamyltransferase, and possibly the importance of mitochondria in the brush border membranes, which are the ultimate target.

7.3.3 Antibiotics

Aminoglycosides

Aminoglycosides, such as **gentamycin**, are an important cause of acute proximal tubular necrosis and renal failure in humans. Chemically, they are carbohydrates with glycosidic linkages to side chains containing various numbers of amino groups (Fig. 7.32). Gentamycin, for example, has five amino groups. They are therefore cationic, and there is some correlation between the number of ionizable amino groups and the nephrotoxicity.

About 10% of patients suffer a moderate but significant decrease in glomerular filtration rate and increased serum creatinine. The drug may affect the structural integrity of the capillaries in the glomerulus, possibly by binding to ionic sites on the cell membrane. Alterations of glomerular structure have also been reported as another feature of the nephrotoxicity. Thus, a reduction in the glomerular filtration coefficient and the number and size of fenestrations in the glomerular endothelium has been observed. Also, the proximal tubular damage may cause feedback on the renin-angiotensin system, leading to angiotensin-induced vasoconstriction. This would also reduce the glomerular filtration.

Aminoglycosides have a low volume of distribution and are excreted by glomerular filtration, which is followed by active reuptake via a high-capacity, low-affinity transport system. The aminoglycosides remain in the kidney longer than in other exposed tissues such as the liver. Thus, with gentamycin, the time for the tissue concentration to be reduced by half is

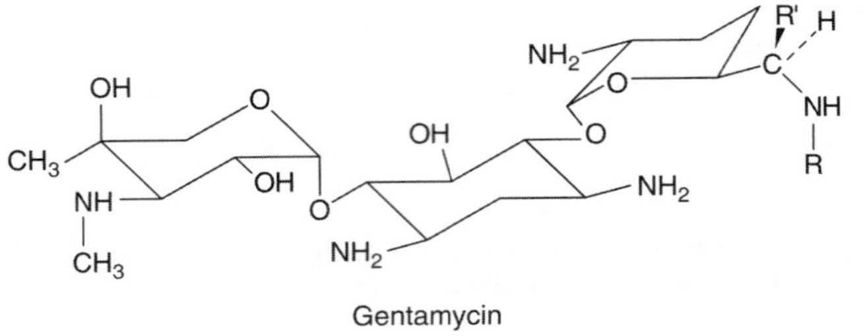

Figure 7.32 The structure of the aminoglycoside antibiotic gentamycin. The drug may be used as a mixture and R and R' = H or CH_3, depending on the particular component.

Gentamycin

five to six days. This suggests that gentamycin is binding to tissue components and will accumulate with repeated dosing. In the renal cortex, gentamycin achieves a concentration of two to five times that in the blood. It is believed that the cationic aminoglycosides bind to anionic phospholipids, especially phosphoinositides, on the brush border on the luminal side of the proximal tubule. The bound aminoglycoside is then engulfed by endocytosis into the proximal tubular cell as vesicles and are then incorporated into secondary lysosomes. The gentamycin thus achieves a high concentration in the proximal tubular cells, and it is these cells that are damaged and suffer necrosis. The accumulation of gentamycin in lysosomes means that the tubular cell concentration is 10 to 100 times that in plasma.

There are a number of possibilities for the underlying mechanism. Thus, the binding of the aminoglycoside can lead to overall impairment of the phosphatidyl inositol cascade and to inhibition of phospholipid breakdown because of effects on phospholipase action. The binding depends on the number and position of charged groups on the aminoglycoside. There is altered membrane phospholipid composition, and the activity of transport systems such as Na^+/K^+ ATPase, adenylate cyclase, and Mg^{2+}, K^+, and Ca^{2+} transport may be impaired. Oxidative phosphorylation in the mitochondria may be directly affected by the free gentamycin in the tubular cell, and if decreased, this will also adversely affect the transport systems that are so vital to the kidney. The accumulation in the lysosomes gives rise to phospholipidosis and myeloid inclusion bodies in the lysosomes by impairment of phospholipid degradation. This is a consistent feature of the toxicity. The lysosomes subsequently rupture and release hydrolytic enzymes, which contribute to the cell damage. Lipid peroxidation occurs and hydroxyl radicals have been implicated, although these may be secondary rather than primary events. The overall mechanism of toxicity is shown in Figure 7.33.

Some aminoglycosides also cause damage to the ear, in particular to the cochlear apparatus. This ototoxicity is irreversible and results from the progressive destruction of vestibular or cochlear sensory cells. This is associated with accumulation of the drug in the fluid (perilymph and endolymph) of the inner ear, where the half-life is much longer than in the plasma. The toxicity may be due to interference with the transport mechanism involved with maintaining ionic balance or interaction with membrane phospholipids (as in the kidney). Gentamycin, for example, damages the hair cells in the cochlear, a measure of the toxicity that can be detected by scanning electron microscopy. There is a clear dose-response relationship between the number of cells destroyed and the increasing dose of gentamycin. The ototoxicity can also be detected in the intact animal by the Preyer reflex test, which determines the threshold sound level for particular frequencies to which an animal responds by movement of the external ear. Patients suffer loss of balance, nausea, and vomiting and vertigo, associated with damage to the vestibular apparatus, whereas cochlear damage is manifested as tinnitus and progressive loss of hearing, ending in deafness.

7.3.4 Cephalosporins

These drugs, such as **cephaloridine** (Fig. 7.34), are β-lactam antibiotics, used therapeutically for the treatment of bacterial infections, but some cause kidney damage in human patients. Thus, acute doses of cephaloridine cause a dose-related proximal tubular necrosis, selectively damaging the S2 segment of the tubule. Cephaloridine accumulates in the kidney, especially the cortex, to a greater extent than in other organs. It is an organic anion, but also has a cationic (pyridyl) group (Fig. 7.34).

Cephaloridine is actively taken up from the blood into proximal tubular cells by the OAT 1 (Fig. 7.35). This requires a hydrophobic region as well as an anionic group, is specifically

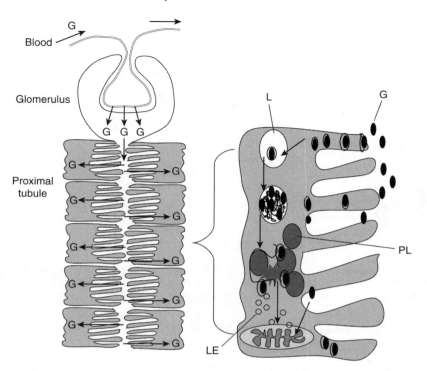

Figure 7.33 The renal accumulation and toxicity of gentamycin (G). Gentamycin is filtered in the glomerulus and enters the tubular lumen. Here, it is taken up by proximal tubular cells and in vesicles as part of the uptake process. These fuse with lysosomes (L) inside the cell. The accumulation of gentamycin inside the lysosome destabilizes it, causing it to rupture and release its hydrolytic enzymes (o). These cause damage within the cell. Also, gentamycin can directly damage mitochondria (M).

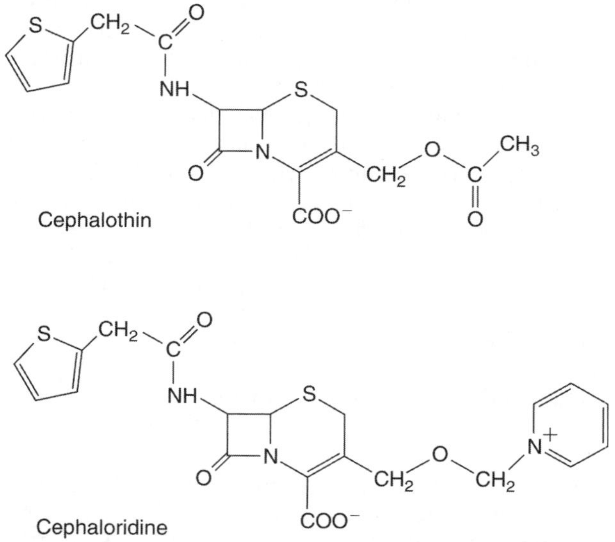

Cephalothin

Cephaloridine

Figure 7.34 The structures of the two cephalosporin antibiotics. Cephaloridine is nephrotoxic, cephalothin is not.

expressed in the kidney, and exports α-ketoglutarate in the opposite direction. Cephaloridine also seems to be secreted by the renal tubular cationic transport system [organic cation transporter (OCT)] into the tubular lumen. However, the anionic transport system for uptake into the tubular cell is more active, and the overall result is active accumulation. Binding to intracellular receptors may be involved. The movement of cephaloridine via diffusion out of

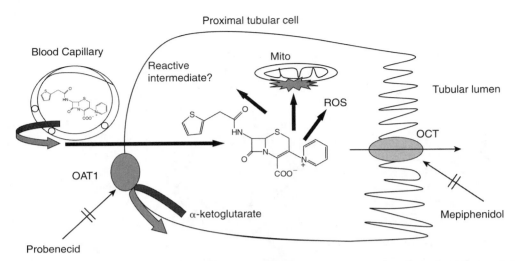

Figure 7.35 The uptake and elimination of cephaloridine by proximal tubular cells in the kidney and possible mechanisms of toxicity. The uptake can be inhibited (probenicid) and the elimination also inhibited (mepiphenidol). *Abbreviations*: OAT 1, organic anion transporter; OCT, organic cation transporter; ROS, reactive oxygen species.

Table 7.7 Accumulation of Two Cephalosporins in the Kidney

Drug	Concentration in renal cortex ($\mu g\ g^{-1}$)	Concentration in serum ($\mu g\ mL^{-1}$)	Cortex/serum	Nephrotoxic dose ($mg\ kg^{-1}$)
Cephaloridine	2576 ± 267	167 ± 12	15.1 ± 0.8	90
Cephalothin	431 ± 122	127 ± 16	3.2 ± 0.4	>1000

Source: Ref. 14.

the tubular cell into the lumen also seems to be limited, and hence the intracellular concentration remains high. Blocking active transport via OAT 1 with probenecid decreases both the nephrotoxicity and the concentration of cephaloridine in the proximal tubule. Conversely, inhibiting the export via OCT increases the nephrotoxicity.

Also, there is some correlation between accumulation and nephrotoxicity. For example, neonatal rabbits lacking a developed OAT 1 are less susceptible to the nephrotoxic effects of the drug. Similar drugs, such as cephalothin (Fig. 7.34), lacking the pyridine ring and, so, not cationic, are not nephrotoxic and do not accumulate (Table 7.7). This accumulation of cephaloridine is presumed to be due to the cationic group reducing efflux.

However, the nephrotoxicity involves more than this accumulation, as other cephalosporins such as cephalexin, which is not nephrotoxic, also accumulate and reach similar concentrations, although these are not sustained. Cephaloglycin is another nephrotoxic drug of the same class; this drug accumulates initially, but is also not sustained. So, clearly, other factors are also important.

The mechanism of toxicity is not completely clear, but a number of proposed hypotheses have been made (Fig. 7.36):

1. Metabolic activation via cytochromes P-450 to reactive metabolites. A reactive intermediate has been suggested as some inhibitors of cytochromes P-450 decrease the toxicity, and some inducers of the monooxygenases increase toxicity. However, other inducers and inhibitors do not change toxicity, and these treatments also affect the renal concentration of cephaloridine in a manner consistent with the effect on toxicity. As the β-lactam ring is unstable, a chemical rearrangement to produce a reactive intermediate is also possible.

2. Lipid peroxidation/oxidative stress (Fig. 7.36). Lipid peroxidation products (e.g., malondialdehyde) have been detected, and GSH is depleted seemingly by oxidation rather than conjugation, and prior depletion of GSH increases the toxicity. NADPH is

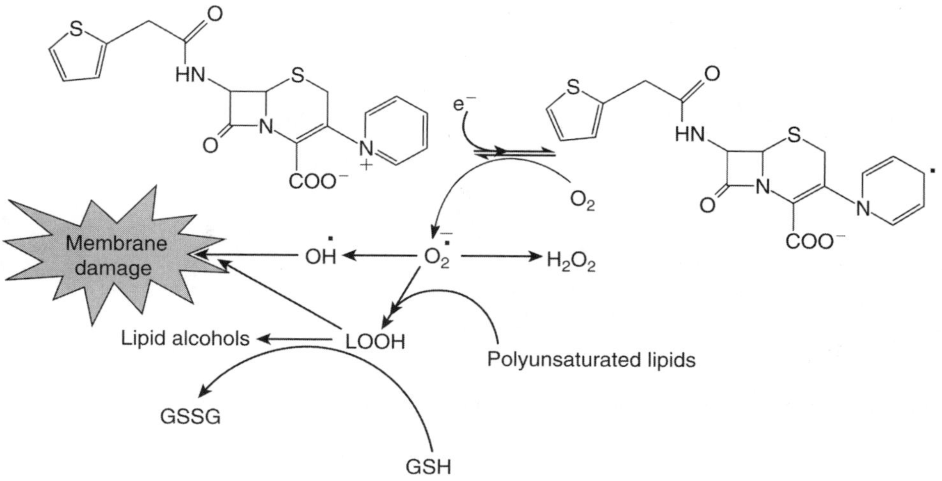

Figure 7.36 The interaction between a putative cephaloridine radical and oxygen to cause oxidative stress.

also depleted, and consequently, GSSG cannot be reduced back to GSH. Anaerobic metabolism of cephaloridine yields superoxide. Vitamin E- and selenium-deficient animals are more susceptible to the nephrotoxicity. Thus, it has been suggested that superoxide anion radical and hydroxyl radical may be formed and that lipid peroxidation could be responsible for the toxicity (Fig. 7.36). However, the role of the pyridine ring is uncertain, as cephalothin also produces malondialdehyde yet is not nephrotoxic.

3. Damage to the mitochondria and the intracellular respiratory processes. Damage to this organelle has been detected soon after exposure to the nephrotoxic cephalosporins along with reduced mitochondrial respiration (Fig. 7.35).

Cephaloridine will inhibit organic ion transport in the kidney and, this is preceded by the lipid peroxidation. Antioxidants block both events, suggesting that they are related. GSH depletion is a very early event, occurring before lipid peroxidation is detectable.

Thus, the susceptibility is the result of accumulation of the drug in the target organ to reach concentrations not achieved in other tissues. This is then followed by what is probably a combination of events such as formation of a reactive intermediate, possibly a free radical, stimulation of lipid peroxidation and depletion of GSH, and then peroxidative damage to cell membranes and mitochondria. Whether metabolic activation by cytochromes P-450, or chemical rearrangement, or reductive activation, or all the three are involved is not currently clear.

7.4 TISSUE LESIONS: LUNG DAMAGE

7.4.1 4-Ipomeanol

4-Ipomeanol (Fig. 7.37) is a pulmonary toxin produced by the mold *Fusarium solani*, which grows on sweet potatoes. The pure compound produces lung damage in a number of species when given intraperitoneally. This lung damage is manifested as edema, congestion, and hemorrhage. These are probably secondary or tertiary pathological changes resulting from the primary lesion, which is necrosis of the non-ciliated bronchiolar cells, also known as the Clara cells. Toxicologically, 4-ipomeanol is of particular interest as it is a specific lung toxin, which selectively damages one cell type, the Clara cell in the smaller bronchioles, although after high doses other cell types and those of the larger airways may also be damaged.

The elucidation of the mechanism has revealed that this specificity is due to a requirement for metabolic activation for which the Clara cell is particularly suited. Thus, early studies using radiolabeled 4-ipomeanol found that the compound was localized particularly in the lungs (when expressed as nmol g^{-1} wet weight of tissue), and was covalently bound to lung protein (Fig. 7.38). This binding was five times that seen in the liver. Furthermore, autoradiography revealed that the radiolabeled 4-ipomeanol was bound to the Clara cells,

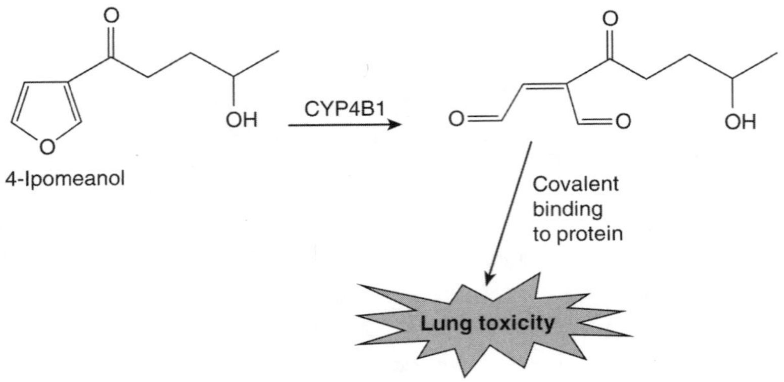

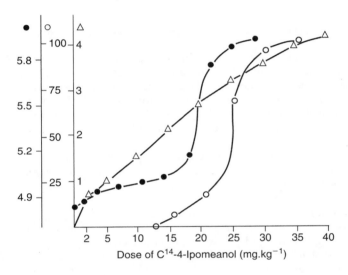

Figure 7.37 Structure and metabolic activation of 4-ipomeanol.

Figure 7.38 Dose-response curves for lethality (%, o), oedemagenesis (wet weight: dry weight ratios, •) and pulmonary covalent binding (nmol.mg^{-1} protein, △) of radio-labelled 4-ipomeanol given i.p. to the rat. *Source*: From Ref. 15.

which were necrotic, whereas ciliated and other cells lining the airways did not show necrosis or covalently bound radioactivity.

Various studies *in vitro and in vivo* indicated that 4-ipomeanol was metabolically activated by cytochrome P-450 to an alkylating species, which could covalently interact with protein, which was proportional to the lung damage (Figs. 7.37 and 7.38). Although the lung in general has a lower level of cytochrome than the liver, lung in rodents and other species has CYP4B1, which metabolizes 4-ipomeanol to a reactive intermediate. Thus, the rate of alkylation of and binding to protein is greater in the lung compared with the liver. It is of interest that in humans when 4-ipomeanol was evaluated as a potential treatment for lung cancer, liver toxicity occurred. This seems to be due to the fact that human CYP1A2 and CYP3A4 have similarities with rodent lung CYP4B1 and so can activate 4-ipomeanol. Perversely, human lung CYP4B1 does not!

Thus, the V_{max} for the covalent binding to protein is higher for the lung microsomal enzyme than for the liver, whereas the apparent K_m for the lung enzyme is about one-tenth that for the liver microsomal enzyme preparation. The lung enzyme, therefore, has a greater affinity for the ipomeanol and activates it more rapidly. GSH inhibited the binding to protein *in vitro*.

Studies *in vivo* indicated that the cause of death was probably pulmonary edema, as the time course for lethality and edemagenesis were similar (Fig. 7.39). Also, the pulmonary edema and lethality showed a very similar dose response, and the covalent binding to pulmonary protein was similarly dose dependent (Fig. 7.38).

Inhibitors of the microsomal enzymes decreased the covalent binding and the Clara cell necrosis and increased the LD$_{50}$, even though the blood and pulmonary levels of ipomeanol

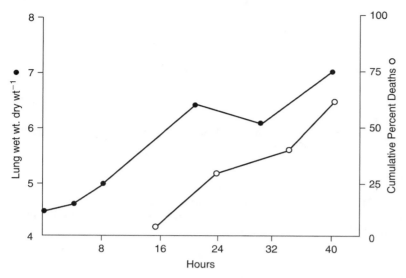

Figure 7.39 Dose-response curves for lethality (%), edemagenesis (wet weight-dry weight ratios), and pulmonary covalent binding (nmol mg^{-1} protein,) of radiolabeled 4-ipomeanol given i.p. to the rat. *Source*: From Ref. 15.

were increased. Recent studies with antibodies against CYP4B1 found that they blocked the covalent binding and toxicity.

The microsomal enzyme inducer phenobarbital decreased binding to both liver and lung protein and correspondingly increased the LD$_{50}$. In contrast, 3-methylcholanthrene induction increased binding in the liver and potentiated liver necrosis, but again decreased lung binding and lung damage. The LD$_{50}$ was again increased.

Diethyl maleate treatment, which depletes GSH, markedly increased covalent binding to protein and decreased the LD$_{50}$.

Analogues of 4-ipomeanol in which the furan ring was replaced by a methyl or phenyl group were very much less toxic and did not covalently bind to the lung or liver to any great extent. There is good correlation between the toxicity as measured by lethality or LD$_{50}$, edemagenesis, and covalent binding of radiolabeled ipomeanol to protein (Fig. 7.38).

The susceptibility of the lung is probably due to the location of the CYP4B1, which is particularly expressed in Clara cells. An additional reason could be the relative lack of protection such as lower levels of GSH than the liver, for instance.

Liver necrosis will occur in experimental animals treated with 3-methylcholanthrene. It seems likely that the metabolic activation takes place in situ rather than the reactive metabolite being transported from the liver.

In conclusion, the pulmonary toxicity of 4-ipomeanol seems to be due to metabolic activation probably by formation of an unstable epoxide on the furan ring catalyzed by CYP4B1 in the Clara cell. The epoxide can rearrange to an α,β unsaturated dialdehyde (Fig. 7.37), which can interact with macromolecules and covalently bind, leading to cellular necrosis and edema.

7.4.2 Paraquat

Paraquat is a bipyridylium herbicide, which is widely sold and used, but unfortunately has been involved in both accidental and intentional cases of poisoning. This usually occurs after oral ingestion rather than percutaneous absorption, although it will damage the skin and cause inflammation as it is a strong irritant. After absorption of a toxic dose, there will be abdominal pain, vomiting, and diarrhea. The main target organs are the lungs and kidneys, although there may also be cardiac and liver toxicity. Such multi-organ toxicity will occur with relatively large doses. Thus, the toxic effects are dose related, and so there may be minimal damage, which is reversible. Larger doses cause alveolar edema followed by pulmonary fibrosis if the patient

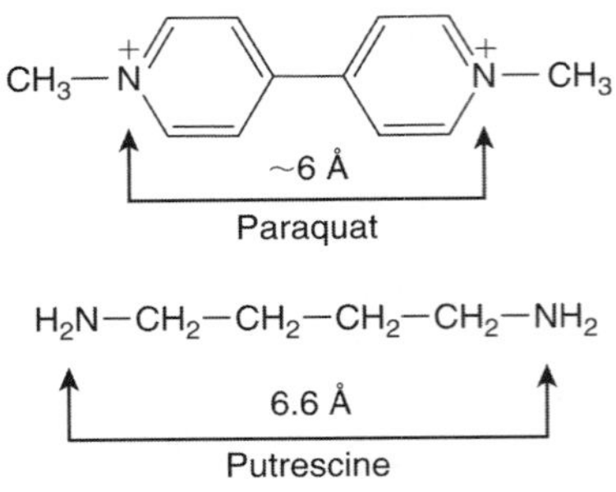

Figure 7.40 The structures of the herbicide paraquat and the polyamine putrescine.

survives beyond a few days. Pulmonary fibrosis and respiratory failure may develop up to six weeks after ingestion. Within 24 hours of a toxic dose, renal tubular function is affected, and so excretion is reduced, and this will exacerbate the toxicity.

Mechanism of Toxicity
The main target organ and that usually involved with lethality is the lung. The damage observed is destruction of the type I and type II alveolar epithelial cells. The lung is susceptible because paraquat is selectively taken up into the type I and type II cells by an active transport system. The lung-plasma ratio of paraquat in experimental animals dosed with the compound may reach values of 6 or 7:1 despite a fall in the plasma level after dosing. Paraquat thus reaches a higher concentration in the type I and type II alveolar cells than most other tissues, but it is not bound to tissue components or metabolized. The uptake system shows saturation kinetics, with an apparent K_m of 70 μM and V_{max} of 300 nmol g lung^{-1} h^{-1}. These kinetic constants are similar in the rat and human, and therefore the rat is a good model species. The system that takes up paraquat into the lung normally transports polyamines such as **putrescine** and **spermine** (Fig. 7.40). However, the size and shape of the paraquat molecule is similar to these amines. The preferred structure has two cationic nitrogen atoms separated by at least four methylene groups (6.6 Å), and it seems that the quaternary nitrogen atoms in paraquat have a separation that is sufficiently similar to allow it to be a substrate for this active transport system. The mechanism of toxicity involves an initial reaction between paraquat and an electron donor such as NADPH (Fig. 7.41). Paraquat readily accepts an electron and forms a stable radical cation. However, under aerobic conditions, this electron is transferred to oxygen, giving rise to superoxide. As there is a ready supply of oxygen in the lung tissue, this process can then be repeated, and a redox cycle is set up. This cycle causes two effects, both of which may be responsible for the toxicity:

1. it produces superoxide, which causes a variety of toxic effects; and
2. it depletes NADPH and so may compromise the alveolar cell and reduce its ability to carry out essential biochemical and physiological functions.

The superoxide produced can be detoxified by the action of the enzyme **superoxide dismutase**, which produces hydrogen peroxide. The hydrogen peroxide is then removed by catalase (Fig. 7.41)

However, the superoxide dismutase may be overwhelmed by the amount of superoxide being produced after a toxic dose of paraquat. The superoxide may then cause lipid peroxidation via the production of **hydroxyl radicals**. These may be produced from hydrogen peroxide in the presence of transition metal ions (see chap. 6).

The resulting lipid peroxides, if not detoxified, may give rise to lipid radicals and membrane damage. However, although there is experimental evidence that paraquat causes lipid peroxidation, there is little direct evidence from *in vivo* studies. This may reflect the fact that only a small proportion of lung cells are affected.

Lipid peroxidation will oxidize GSH, and the reduction of this oxidized glutathione (GSSG), catalyzed by GSH reductase, will also use NADPH and hence contribute to the depletion of this nucleotide (Fig. 7.41). The hexose monophosphate shunt will be stimulated to

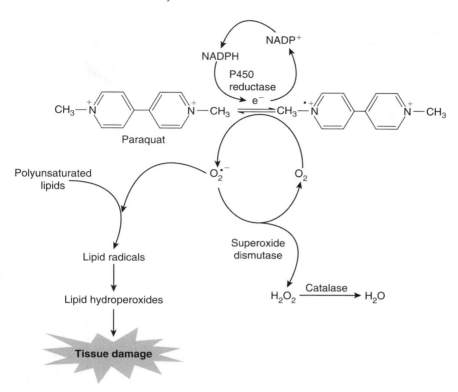

Figure 7.41   The proposed mechanism of toxicity of paraquat.

produce more NADPH, but this will be used for further reduction of paraquat. The lung is therefore a target for the toxicity of paraquat for two reasons:

1. Accumulation of the compound to toxic levels in certain lung cells.
2. Presence of high concentrations of oxygen.

The result is redox cycling, which produces active oxygen species, which can deplete NADPH and GSH and potentially cause peroxidation of membrane lipids. NADPH is generated by the hexose monophosphate shunt, and NADP can also be reduced by GSH (Fig. 6.18).

Treatment
There is no antidote to paraquat poisoning; therefore, prevention of absorption of paraquat from the gastrointestinal tract using **fuller's earth** as adsorbent and gastric lavage is usually employed. Experimental studies have shown that repeatedly giving an adsorbent and using gastric lavage can reduce the toxicity in rats, and this is used in patients. **Hemoperfusion** can also be used to reduce the blood concentration once absorption has started to take place. However, once the paraquat has accumulated in the lungs, little can be done to alter the course of the toxicity.

7.5   NEUROTOXICITY

7.5.1 Isoniazid
As well as being implicated as a hepatotoxin, the antitubercular drug isoniazid may also cause peripheral neuropathy with chronic use. In practice, this can be avoided by the concomitant administration of **vitamin B$_6$** (pyridoxine).

In experimental animals, however, chronic dosing with isoniazid causes degeneration of the peripheral nerves. The biochemical basis for this involves interference with vitamin B$_6$ metabolism.

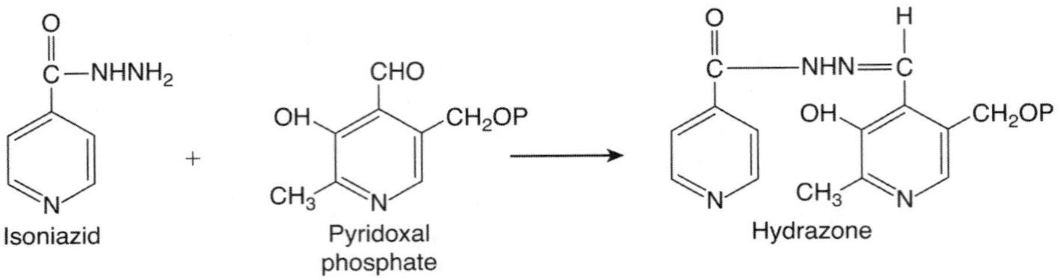

Figure 7.42 Reaction of isoniazid with pyridoxal phosphate to form a hydrazone.

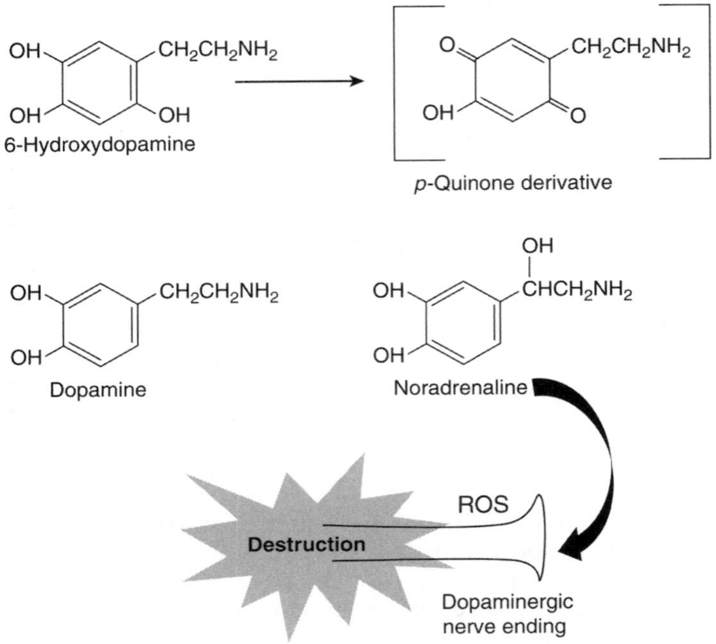

Figure 7.43 Structures of dopamine and noradrenaline, the analogue 6-hydroxydopamine and a possible oxidation product.

Isoniazid reacts with pyridoxal phosphate to form a hydrazone (Fig. 7.42), which is a very potent inhibitor of pyridoxal phosphate kinase. The hydrazone has a much greater affinity for the enzyme ($100-1000\times$) than the normal substratepyridoxal. The result of this is a depletion of tissue pyridoxal phosphate. This cofactor is of importance particularly in nervous tissue for reactions involving decarboxylation and transamination. The decarboxylation reactions are principally affected however, with the result that transamination reactions assume a greater importance.

In humans, peripheral neuropathy due to isoniazid is influenced by the acetylator phenotype (see chap. 5), being predominantly found in slow acetylators. This is probably due to the higher plasma level of isoniazid in this phenotype. In this case, therefore, acetylation is a detoxication reaction, removing the isoniazid and rendering it unreactive toward pyridoxal phosphate.

7.5.2 6-Hydroxydopamine

6-Hydroxydopamine is a selectively neurotoxic compound, which damages the sympathetic nerve endings. It can be seen from Figure 7.43 that 6-hydroxydopamine is structurally very similar to dopamine and noradrenaline, and because of this similarity it is actively taken up into the synaptic system along with other catecholamines. Once localized in the synapse, the 6-hydroxydopamine destroys the nerve terminal. A single small dose of 6-hydroxydopamine destroys all the nerve terminals and possibly the nerve cells as well.

The mechanism of the destruction of the nerve terminals is thought to involve oxidation of 6-hydroxydopamine to a *p*-quinone, the production of a **free radical** or of superoxide anion. It seems that a reactive intermediate is produced, which reacts covalently with the nerve terminal and permanently inactivates it.

Factors that influence the disposition of catecholamines will affect the toxicity. For instance, compounds that inhibit the uptake of noradrenaline reduce the destruction of adrenergic nerve terminals but not of dopaminergic ones. Interference with the oxidative metabolism of catecholamines also influences the toxicity of 6-hydroxydopamine.

This example of selective neurotoxicity particularly illustrates the potential importance of a similarity in chemical structure between an endogenous compound and a foreign compound in the distribution, localization, and type of toxic effect caused.

7.5.3 1-Methyl-4-Phenyl-1,2,3,6-Tetrahydropyridine (MPTP)

MPTP (Fig. 7.44) is a contaminant of a meperidine analogue, which was synthesized illicitly and used by drug addicts. Some of the drug addicts using this drug suffered a syndrome similar to Parkinson's disease. It was found that the contaminant MPTP was responsible for the symptoms. When monkeys were exposed to MPTP, they showed symptoms similar to the human victims. Furthermore, it was found that MPTP caused destruction of dopamine cell bodies in the substantia nigra of the brain. Although this example is similar to 6-hyroxydopamine, it also illustrates other important factors that determine the target organ and particular cell type that is affected by the toxicity. MPTP is not directly neurotoxic but must first be metabolically activated to MPP$^+$ in a two-step reaction (Fig. 7.44). As MPP$^+$ is charged, and therefore should not cross the **blood-brain barrier**, it is presumed to be formed in the brain itself. Indeed, when given to animals, MPP$^+$ itself is not neurotoxic. MPTP is metabolized by **monoamine oxidase B** to MPDP$^+$ in astrocytes and then undergoes autoxidation to MPP$^+$ (Fig. 7.44). Studies using inhibitors and more recently knockout mice

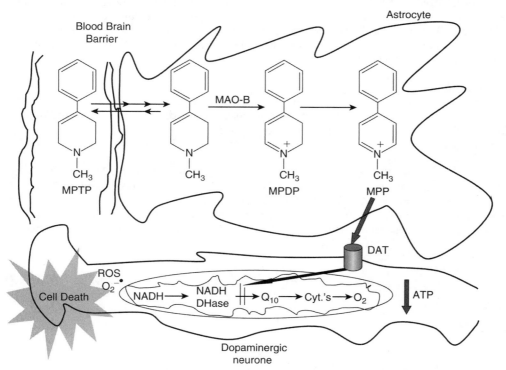

Figure 7.44 The metabolism and toxicity of MPTP. Diffusion into the brain is followed by metabolism in the astrocyte. The metabolite MPP$^+$ is actively transported into the dopaminergic neuron by DAT. It is accumulated there and is actively taken into mitochondria by another uptake system. Here, it inhibits mitochondrial electron transport between NADH dehydrogenase (NADH DHase) and coenzyme Q (Q10). Consequently, it blocks the electron transport system, depletes ATP, and destroys the neuron. *Abbreviations*: MPTP, 1-methyl-4-phenyl 1,2,3,6-tetrahydropyridine; DAT, dopamine transporter uptake system.

have shown that MAO B is necessary, which is why the rat is not sensitive to having little MAO B in astrocytes.

Although the MPP$^+$ is formed outside the neuron, it is then taken up by the **dopamine transporter uptake system** (DAT), where it accumulates and may associate with neuromelanin. DAT is located on the membrane of dopaminergic neurons, especially in the **substantia nigra**, and there is a correlation between the expression of DAT and the loss of the neurons.

However, as it is a charged molecule, it cannot cross the blood-brain barrier and be removed from the CNS. Furthermore, MPP$^+$ is concentrated in the matrix of mitochondria of the neuronal cells via an energy-dependent carrier. Consequently, high concentrations occur in the mitochondria. Here MPP$^+$ binds to and inhibits complex 1 of the electron transport chain (NADH dehydrogenase-coenzyme Q1 reductase), and this blocks oxidative phosphorylation. Therefore, mitochondrial ATP production stops, and cell ATP levels will decline dramatically (Fig. 7.44). GSH levels are also decreased, and there are changes in intracellular calcium concentration. These biochemical perturbations presage the death of the neuronal cell, and so the dopaminergic neurons are destroyed. It has been suggested that superoxide radicals and also other radical species are formed and such oxidative stress may contribute to the cytotoxicity of MPTP. Thus, superoxide dismutase and GSH peroxidase play a role in protection.

Also, research has recently indicated that **TNF-α** is involved in the cytotoxicity, and knockout mice studies have shown that absence of the two TNF-α receptors completely removes sensitivity to MPTP.

MPTP is also metabolized by other routes involving cytochromes P-450, FAD-dependent monooxygenases, and aldehyde oxidase. However, these seem to be detoxication pathways, as they divert MPTP away from uptake and metabolism in the brain. However, MPTP may inhibit its own metabolism by cytochromes P-450 and thereby reduce one means of detoxication. This example illustrates the importance of structure and physicochemical properties in toxicology. MPTP is sufficiently lipophilic to cross the blood-brain barrier and gain access to the astrocytes. The structure of the metabolite is important for uptake via the dopamine system, hence localizing the compound to a particular type of neuron. Again, uptake into mitochondria is presumably a function of structure, as a specific energy-dependent carrier is involved.

7.5.4 Domoic Acid

Domoic acid is a natural toxin, and poisoning results from eating shellfish, such as mussels, contaminated with the poison. It came to prominence after a disaster in Newfoundland in 1987, when 107 people were poisoned, and of these, 4 died.

It is an example of neurotoxicity due to a receptor interaction. The toxin is produced by **phytoplankton**, which then contaminates shellfish.

It causes gastrointestinal distress and disturbances and neurotoxicity. The neurotoxicity may be coma, convulsions, agitation, and loss of memory. If it is serious poisoning, the neurons in the hippocampus and amygdala may be destroyed. Domoic acid acts as an analogue of glutamate, having a similar structure to glutamate, and is believed to bind at the same sites on receptors (Fig. 7.45). It is believed to bind the "kainate receptor," one of the three types of receptor. Domoic acid is known as an **excitotoxin**.

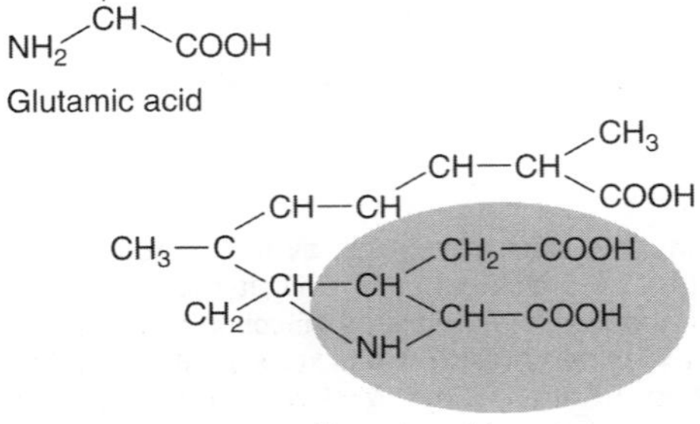

Figure 7.45 The structure of the excitotoxin domoic acid showing the similarity to glutamate, the normal agonist involved in brain function.

This binding leads to excitation of the glutamate receptors and excess release of glutamate. The postsynaptic membrane in the neuron is where the damage occurs, and this is where the glutamate receptors are. The result of the binding is excessive release of glutamate or the prevention of absorption of glutamate into the neuron. The excess glutamate then causes damage as a result of excessive simulation of receptors such as the NDMA receptor. This causes a rise in intracellular calcium, which, as previously discussed, can cause a number of adverse cellular events such as the production of ROS, activation of phospholipases, calpain, and protein kinase C.

Thus, this excess excitation leads to cell death by either apoptosis or necrosis in the neurons affected, which was found to be mostly in the hippocampus and amygdale, although some occurred in the cerebral cortex also.

7.5.5 Primaquine

Primaquine is an 8-aminoquinoline antimalarial drug (Fig. 7.46), effective against the exoerythrocytic forms of all four of the malarial species that infect humans, and is the only radically curative drug for the latent tissue forms of *Plasmodium vivax* and *Plasmodium ovale*. Unfortunately, despite being clinically important and effective, use of primaquine is limited, because it can cause hemolytic anemia, particularly in patients with **glucose-6-phosphate dehydrogenase** (G6PD) deficiency (see chap. 5). This deficiency is especially common in areas of the world where malaria is rife, and so this dose-limiting toxicity has a major impact on the

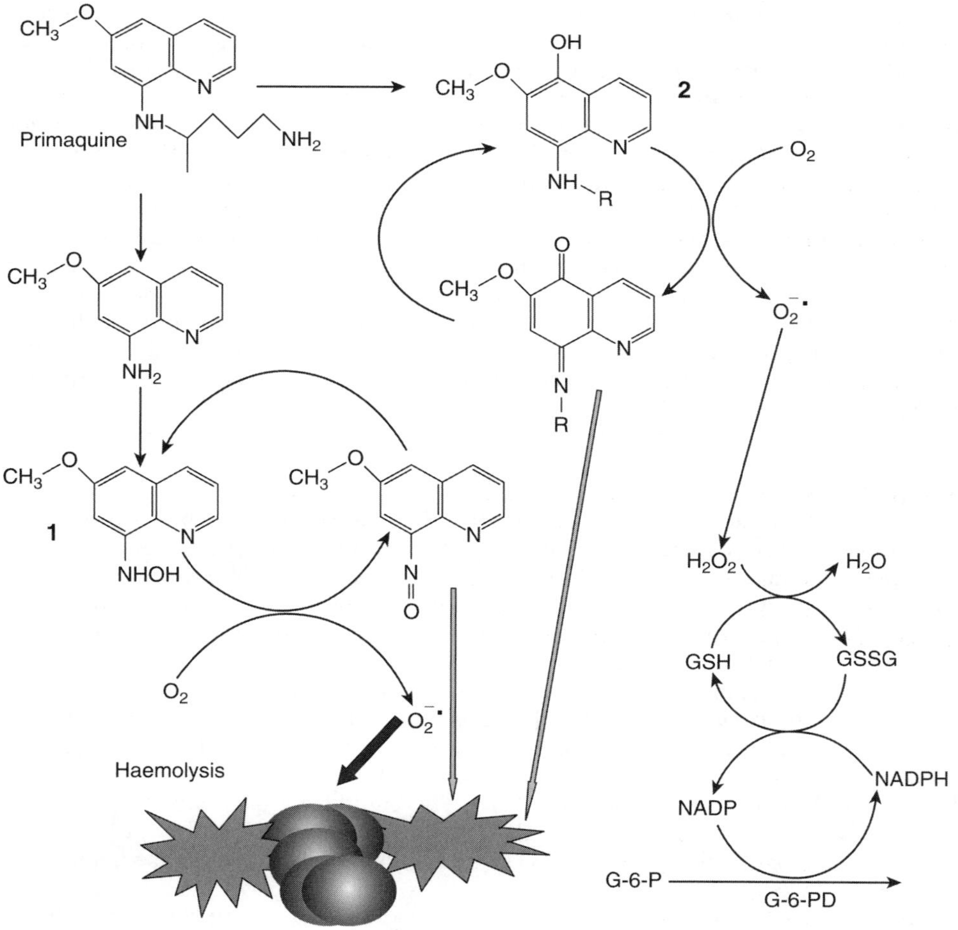

Figure 7.46 The metabolism and toxicity of primaquine. Two metabolites, (1) 6-methoxy-8-hydroxylaminoquinoline and (2) 5-hydroxyprimaquine, are known to be capable of causing oxidative stress and producing ROS. ROS and the metabolites can be removed by GSH, but when this is depleted, they may damage the red cell cytoskeleton and hemoglobin. *Abbreviations*: ROS, reactive oxygen species; GSH, glutathione.

usefulness of this drug. Indeed, there is a significant difference between the hemotoxic dose of primaquine in normal versus deficient individuals, being 20-fold lower in the deficient patients.

Studies have shown that primaquine causes toxicity to erythrocytes *in vivo* at concentrations used clinically, but this is believed to be due to a metabolite(s). However, the metabolite(s) responsible and the underlying mechanism(s) are still unclear.

6-Methoxy-8-hydroxylaminoquinoline, an N-hydroxylated metabolite of primaquine (Fig. 7.46), is directly toxic, causing hemolysis and methemoglobinemia in rats. However, there are several pathways of metabolism for primaquine and several potential toxic metabolites. Thus, hydroxylation of primaquine at the 5-position of the quinoline ring also forms redox-active derivatives able to cause oxidative stress within normal and G6PD-deficient human red cells as well as rat erythrocytes (Fig. 7.46).

5-Hydroxyprimaquine (Fig. 7.46) will produce ROS in red cells, causing oxidative injury to the cytoskeleton and the oxidation of hemoglobin. The oxidized hemoglobin will bind via disulfide bridges to cytoskeletal proteins (seen as **Heinz bodies** under the microscope). This and other damages to the red cell lead to their removal by the spleen, and therefore anemia develops. 5-Hydroxyprimaquine also causes a depletion of GSH and the formation of GSS-protein conjugates. This metabolite is considerably more potent than 6-methoxy-8-hydroxylaminoquinoline.

Both these compounds can also cause methemoglobin formation, stimulation of hexose monophosphate shunt activity in isolated suspensions of red cells, and reduced cell viability.

The hemolytic activity *in vivo* has not been *directly* linked to these metabolites. However, recent research has used a combination of *in vitro and in vivo* studies in which isolated red cells were exposed to a particular metabolite, then the cells were put back into the rats from whence they came, and their survival monitored. This research has confirmed the role of these two reactive metabolites as involved in the toxicity.

As illustrated (Fig. 7.46), both 5-hydroxyprimaquine and 6-methoxy-8-hydroxylamino-quinoline metabolites may undergo redox cycling to yield ROS, which can be detoxified by reduced GSH. In the absence of sufficiently reduced GSH, the ROS can react with and damage hemoglobin in the red cell. Also, reactive metabolites of the drug may be reduced by GSH and form more GSSG. Both the N-hydroxy metabolite and the 5-hydroxy metabolite will cause damage to red cells *in vitro* and damage to the cytoskeleton, including the formation of hemoglobin-protein adducts. The depletion of red cell GSH markedly increases the cytotoxicity of the N-hydroxy metabolite *in vitro* and also 5-hydroxyprimaquine, which is five times more potent in GSH-depleted erythrocytes. Removal of damaged red cells by the spleen will prematurely occur *in vivo*. Other drugs such as **dapsone** will also cause **hemolytic anemia** via the production of reactive metabolites.

7.5.6 Adriamycin/Doxorubicin

This anthracycline-type anticancer drug is very effective against various solid cancers but unfortunately is also cardiotoxic. It causes a dose-dependent **cardiomyopathy** (degeneration of the cardiac muscle), which leads to arrhythmias and congestive heart failure.

The drug interferes with the mitochondrial electron transport chain. It seems that this is due to a high affinity of the drug for lipids such as cardiolipin, a component of the mitochondrial inner membrane. It therefore accumulates there.

The structure of doxorubicin includes a quinone moiety; therefore, it can easily accept an electron and undergo redox cycling (Fig. 7.47). Because it accumulates in the mitochondria, it can accept electrons from the electron transport chain and divert them away from complex I. It becomes reduced to the semiquinone radical in the process. This will then reduce oxygen to superoxide and return to the quinone form (Fig. 7.47). This could lead to oxidation of GSH and mtDNA. The subsequent damage may lead to the opening of the mitochondrial permeability transition pore. Consequently, mitochondrial ATP production will be compromised, and ATP levels will decline.

The damage to mtDNA may persist in cardiac tissue and thereby compromise the ability of the cardiac mitchondria to function as efficient energy-producing units. This could cause the cardiac damage to persist. The mtDNA codes for mitochondrial enzymes, and therefore damage to it, will impact on the synthesis of new mitochondrial components. mtDNA is especially susceptible to damage. The mechanism is summarized in Figure 7.47.

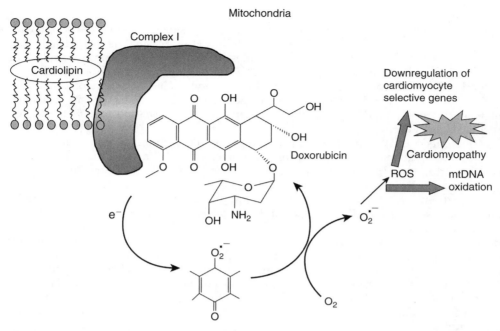

Figure 7.47 The mechanism of cardiotoxicity of doxorubicin. The drug can acquire electrons from mitochondrial complex I. The quinone thus produced can donate the electron to oxygen, and the superoxide produced damages heart tissues and mtDNA. *Abbreviation*: mtDNA, mitochondrial DNA.

The reasons for the heart being a target are threefold:

1. The drug interferes with the production of ATP, the supply of which is critical to cardiac function.
2. The heart has reduced levels of antioxidants such as GSH, GSH peroxidase, superoxide dismutase, and catalase and is therefore vulnerable to attack by ROS.
3. The drug indirectly alters the regulation of intracellular calcium levels via protein kinase C and can alter the transcription of gene- (downregulating) encoding enzymes (e.g., ATP/ADP translocase) and those involved in energy production and specific for cardiac tissue.

7.6 EXAGGERATED AND UNWANTED PHARMACOLOGICAL EFFECTS

The exaggerated or unwanted pharmacological responses are the most common toxic effects of drugs observed clinically, as opposed to direct toxic effects on tissues. However, this type of response may also be observed with other compounds such as the toxic cholinesterase inhibitors used as pesticides and nerve gases.

7.6.1 Organophosphorus Compounds

There are many different cholinesterase inhibitors that find use, particularly as insecticides but also as nerve gases for use in chemical warfare. Malathion is used for the treatment of head lice in humans, so in addition to occupational exposure, there is also exposure as a result of its use as a drug. Organophosphorus insecticides are the most widely used and the most frequently involved in fatal human poisonings. They may be absorbed through the skin, and there have been accidental poisoning cases arising from such exposure. Accidental contamination of food with insecticides such as parathion has led to a significant number of deaths. There are two types of toxic effects: **inhibition of cholinesterases** and **delayed neuropathy**.

Cholinesterase Inhibition

The inhibition of cholinesterases results in a number of physiological effects. The enzyme acetylcholinesterase, found in various tissues including the plasma, is responsible for hydrolyzing acetylcholine. This effectively terminates the action of the neurotransmitter at the synaptic nerve endings, which occur in the CNS, and other tissues such as glands and smooth muscle. The inhibition leads to accumulation of acetylcholine, and therefore, the signs of poisoning resemble excessive stimulation of cholinergic nerves. The toxicity becomes apparent when there is about 50% inhibition of acetylcholinesterase, and at 10% to 20% of normal activity, death occurs.

The toxic effects can be divided into three types as the accumulation of acetylcholine leads to symptoms that mimic the muscarinic, nicotinic, and CNS actions of acetylcholine. **Muscarinic receptors** for acetylcholine are found in smooth muscles, the heart, and exocrine glands. Therefore, the signs and symptoms are tightness of the chest, wheezing due to bronchoconstriction, bradycardia, and constriction of the pupils (miosis). Salivation, lacrimation, and sweating are all increased, and peristalsis is increased, leading to nausea, vomiting, and diarrhea.

Nicotinic signs and symptoms result from the accumulation of acetylcholine at motor nerve endings in skeletal muscle and autonomic ganglia. Thus, there is fatigue, involuntary twitching, and muscular weakness, which may affect the muscles of respiration. Hypertension and hyperglycemia may also reflect the action of acetylcholine at sympathetic ganglia.

Accumulation of acetylcholine in the CNS leads to a variety of signs and symptoms, including tension, anxiety, ataxia, convulsions, coma and, depression of the respiratory and circulatory centers.

The cause of death is usually respiratory failure due partly to neuromuscular paralysis, central depression, and bronchoconstriction.

The onset of symptoms depends on the particular organophosphorus compound, but is usually relatively rapid, occurring within a few minutes to a few hours, and the symptoms may last for several days. This depends on the metabolism and distribution of the particular compound and factors such as lipophilicity. Some of the organophosphorus insecticides such as **malathion**, for example (chap. 5, Fig. 12), are metabolized in mammals mainly by hydrolysis to polar metabolites, which are readily excreted, whereas in the insect, oxidative metabolism occurs, which produces the cholinesterase inhibitor. Metabolic differences between the target and nontarget species are exploited to maximize the selective toxicity. Consequently, malathion has a low toxicity to mammals such as the rat in which the LD_{50} is about 10 g kg^{-1}.

Mechanism of Inhibition of Cholinesterases

Inhibition of the cholinesterase enzymes depends on blockade of the active site of the enzyme, specifically the site that binds the ester portion of acetylcholine (Fig. 7.48). The organophosphorus compound is thus a pseudosubstrate. However, in the case of some compounds such as the **phosphorothionates** (parathion and malathion, for example), metabolism is necessary to produce the inhibitor.

In both cases, metabolism by the microsomal monooxygenase enzymes occurs in which the sulfur atom attached to the phosphorus is replaced by an oxygen (chap. 5, Fig. 12).

With malaoxon, the $P=O$ group binds to a serine hydroxyl group at the esteratic site of the cholinesterase enzyme in an analogous manner to the $C=O$ group in the normal substrate acetylcholine (Figs. 7.48 and 7.49). With acetylcholine, the enzyme–substrate complex breaks down to leave acetylated enzyme, which is then rapidly hydrolyzed to regenerate the serine hydroxyl group and hence the functional enzyme. With organophosphorus compounds such as malathion (Fig. 7.48), the bound organophosphorus compound also undergoes cleavage to release the corresponding thiol or alcohol, leaving phosphorylated enzyme. However, unlike the acetylated enzyme, intermediate produced with acetylcholine, the phosphorylated enzyme is only hydrolyzed very slowly, if at all. Consequently, the active site of the enzyme is effectively blocked. If the phosphorylated enzyme undergoes a change known as aging in which one of the groups attached to the phosphorus atom is lost, then the inhibition may become more permanent. However, the toxicity will ultimately depend on the affinity of the enzyme for the inhibitor and the rate of hydrolysis of the phosphorylated intermediate. These will in turn depend on the nature of the substituent groups on the phosphorus atom (Fig. 7.49).

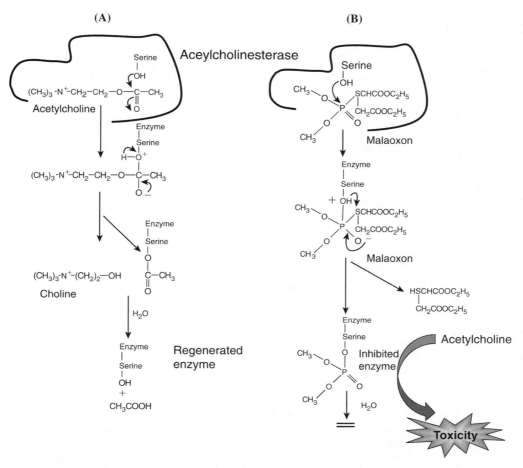

Figure 7.48 Mechanism of hydrolysis of acetylcholine by acetylcholinesterase (A) and the analogous reaction with malaoxon leading to blockade (=) of the enzyme.

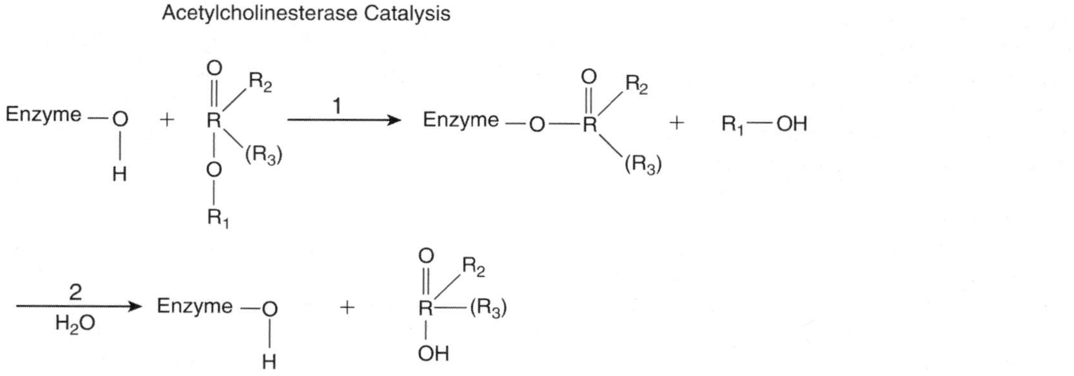

Acetylcholinesterase Catalysis

Figure 7.49 General scheme for acetylcholinesterase action. R may be equal to C or P. If R=P, then (R$_3$) is present. The group OR$_1$ may be replaced by SR$_1$, giving R$_1$SH on hydrolysis (reaction 1). IF R=P, reaction 2 is very slow, giving inactivated enzyme. The rate of hydrolysis or reactivation depends on the nature of R$_2$ and R$_3$.

Furthermore, the rates of metabolism to the active metabolite and via other routes will also be determinants of the toxicity.

Although a particular organophosphorus compound may be rapidly metabolized and therefore not accumulate in the animal, chronic dosing may cause a cumulative toxic effect because of the slow rate of reversal of the inhibition. Thus, the rate of regeneration may be slow with a half-life of the order of 10 to 30 days. Thus, in some cases, resynthesis of the enzyme may be the limiting factor. Thus, repeated, small nontoxic doses can cause eventual toxicity when sufficient of the cholinesterase is inhibited. Different cholinesterases have different half-lives. The acetylcholinesterases present in red blood cells, which are similar to that in the nervous tissue, remain inhibited after **di-isopropylfluorophosphate** for the lifetime of the red blood cell.

7.6.2 Delayed Neuropathy

Certain organophosphorus compounds, such as **tri-orthocresyl phosphate**, cause this toxic effect. The symptoms, which may result from a single dose, may not be apparent for 10 to 14 days afterward. The result is degeneration of peripheral nerves in the distal parts of the lower limbs, which may spread to the upper limbs. Pathologically, it is observed that the nerves undergo "'dying back" with axonal degeneration followed by myelin degeneration. The effect does not seem to be dependent on cholinesterase inhibition as tri-orthocresyl phosphate is not a potent cholinesterase inhibitor but causes delayed neuropathy. It seems that there is a covalent interaction with a membrane-bound protein, known as neuropathy target esterase, and the organophosphorus compound that may disturb metabolism in the neuron. This is followed by an aging process that involves loss of a group from the phosphorylated protein. This protein seems to have a function critical to the neuron. The reaction with the neuropathy target esterase occurs very rapidly, possibly within one hour of dosing, yet the toxicity is manifested only days later.

7.6.3 Treatment

Poisoning by organophosphorus compounds can be treated, and although the acute symptoms can be alleviated, the delayed neuropathy cannot. There are two treatments that may be used, both involving **antidotes**:

1. The compound **pralidoxime** (Fig. 7.50) is administered to regenerate the acetylcholinesterase. This it does by acting in place of the serine hydroxyl group in the enzyme and forming a complex with the organophosphorus moiety (Fig. 7.50). Pralidoxime must be administered quickly after the poisoning, because if the phosphorylated enzyme is allowed to age, then it will no longer be an effective antidote. The reactivation rate will depend on the particular organophosphorus compound.
2. The physiological effects of the accumulation of acetylcholine can be antagonized by the administration of **atropine**, and the symptoms alleviated.

If both atropine and pralidoxime are given together, there is a much greater effect than either given alone. Thus, when used after experimental parathion poisoning, each treatment

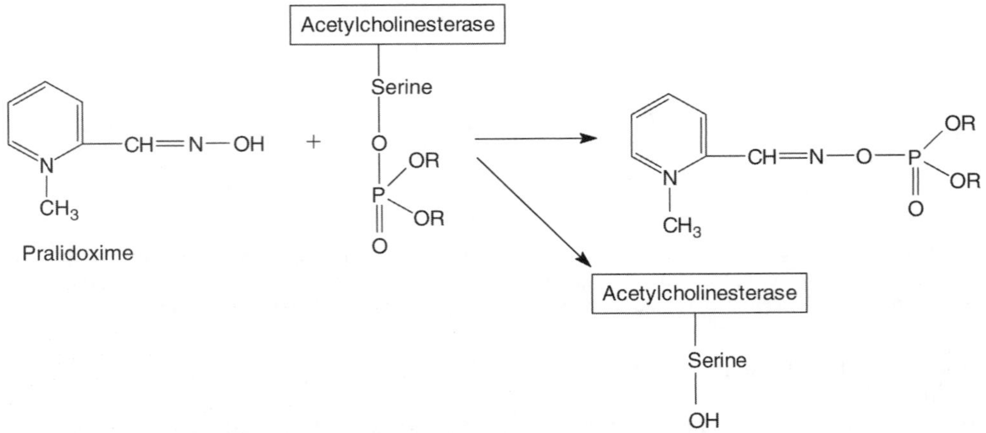

Figure 7.50 The mechanism of reaction of the antidote pralidoxime with phosphorylated acetylcholinesterase. The original acetylcholinesterase is thereby regenerated.

alone only increased the LD_{50} about two times, whereas in combination, the LD_{50} was increased 128 times. This is an example of synergy.

7.6.4 Cardiac Glycosides Digoxin/Digitoxin/Oubain

This group of natural substances includes drugs such as digoxin and digitoxin (from the foxglove) discovered by William Withering in 1785, which are now used to treat heart failure. There are also a number of plant toxins of this class such as convallotoxin from lily, hellebrin from henbane, and **oubain** from strophanthus. The compounds contain a steroid-like structure attached to one or more sugars (hence, the term "glycosides"). Digitoxin has three sugars (Fig. 7.51).

Digitoxin and related drugs are used as cardiac stimulants, causing a positive inotropic effect. Thus, they increase the strength and intensity of the contractions and so are used in the treatment of heart failure. Because they slow the electrical conduction between atria and ventricles, they can also be used in the treatment of atrial fibrillation, atrial tachycardia, and atrial fluter.

However, digitoxin and digoxin have a low therapeutic index, and low margin of safety and adverse effects are common even at therapeutic doses. Thus, nausea and vomiting are the most common adverse effects occurring at therapeutic doses, digitoxin causing vomiting in 3% of patients at the ED_{50} dose (Fig. 7.52). Sometimes, overdoses also occur, which result in serious toxicity. These generally affect the heart, and so, this is an example of exaggerated pharmacology at high dose.

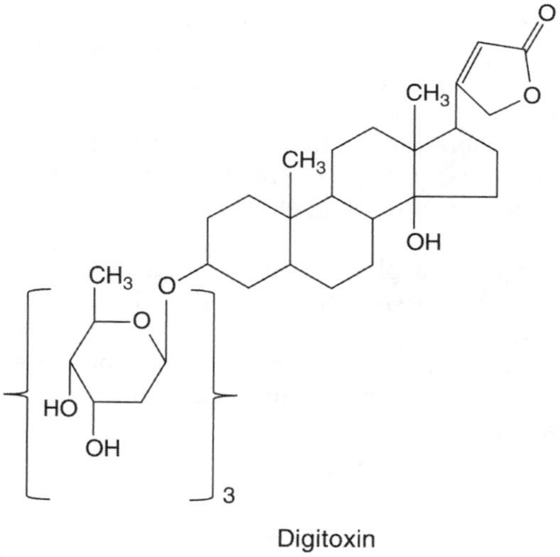

Digitoxin

Figure 7.51 The structure of the cardiac glycoside digitoxin.

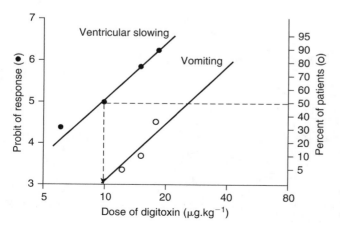

Figure 7.52 Comparison of the pharmacological and toxic effects of digitoxin in humans. The pharmacological effect was a 40% to 50% decrease in heart rate; the toxic effect, nausea, and vomiting, occurred after a single oral dose. The dotted line indicates that the dose required for 50% of patients to respond pharmacologically (*right hand scale*) is also the threshold for the toxic effect. *Source*: From Ref. 16.

    There is considerable patient variability in response to the drug, which has a slow distribution to its target sites, taking about 200 hours to reach steady-state levels and a long half-life (40 hours or possibly more). It has a high volume of distribution (420 L), being reasonably lipid soluble. The drug is also highly protein bound to plasma proteins. Therefore, it will take several hours even after an intravenous dose before a pharmacological effect is observed. Because of the low margin of safety, the dose is critical, and monitoring of the patients plasma level may be necessary.

    In addition to vomiting and diarrhea, severe toxicity can occur with erroneous or intentional overdoses. These include cardiac arrythmias and atrial tachycardia with atrial-ventricular (AV) block. These effects are due to increased intracellular calcium. Hyperkalemia also occurs. Disturbances of cognitive function, including visual disturbances, delirium, and convulsion are also adverse effects resulting from neuronal effects.

    In normal individuals, the effects of toxic doses of digitoxin are extreme slowing of the heart (bradycardia) and atrial fibrillation. In cardiac patients, however, the effects could be fatal with ventricular fibrillation, but certainly arrythmias and extra beats will occur.

    The mechanism of the pharmacological effect of digitoxin is inhibition of the Na^+/K^+ ATPase pump by binding to a specific amino acid in the protein (Fig. 7.53). This blocks the critical domain of the transporter, so it cannot transport the cations. The pump transports three Na^+ out and two K^+ into the cardiac cell, a Na^+ gradient is built up between the inside and outside (this is an example of an antiport; see chap. 3). Inhibition of the pump means that this gradient is reduced. Sodium is required in the extracellular milieu for other transporters such as the Na^+ Ca^+ exchanger in cardiomyocytes. This system transports Na^+ into the cardiomyocyte and Ca^+ out. So, inhibition of the Na^+/K^+ pump indirectly leads to increased Ca^+ and Na^+ in the heart cell. As Ca^+ in cardiomyocytes is the signal involved in contraction of cardiomyocytes, by activating the actin/myosin system, contractions are increased. This leads to the heart to beat more forcefully, known as a positive inotropic effect. However, this stimulation can become excessive and then result in arrythmias and fibrillation or heart block. For example, there may be ventricular extrasystoles, where the ventricle beats twice in rapid succession. The ventricle may start to beat independently of the atria, which is heart block. In serious cases of toxicity, ventricular fibrillation or rapid uncontrolled beating can occur, which is usually fatal.

    Reasons for variability in the dose required for pharmacological and adverse effects are partly pharmacokinetic. Thus, due to the lipid solubility, there is a big volume of distribution. Accumulation can occur because of the long half-life, and renal dysfunction is an important cause of variability in digitoxin toxicity. Thus, as renal function is an important factor in the half-life and blood level, half-lives of up to six days can occur in patients with compromised function (compared with 36–48 hours in normal patients).

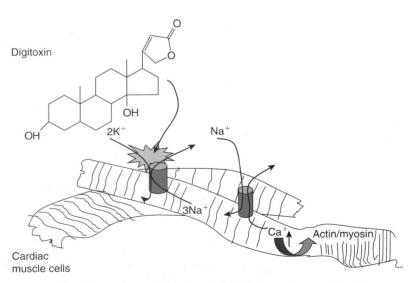

Figure 7.53   The mechanism of action of digitoxin in cardiac muscle cells.

Individual variability in gut microflora is another factor, as this reduces bioavailability.

Hepatic metabolism is a variable factor, and other drugs that induce or inhibit metabolism will alter the drug level in the body. As it has a low therapeutic index, this can have significant effects.

Thus, in patients taking digitoxin, who then start therapy with phenobarbital, (CYP3A4 inducer), the blood level of digitoxin decreases by about half. The dose would then need to be raised to compensate, but if the phenobarbital was stopped, the induction would decline and toxicity would ensue if there was no readjustment of the dose.

Coadministration of CYP3A4 inhibitors could cause a dangerous increase in the level of digitoxin also. Thus, quinidine increases the blood level and decreases clearance of digitoxin.

Because digitoxin is a substrate for p-glycoprotein, inhibitors of this transporter, such as atorvastatin, will increase drug levels and potentially cause toxicity. There have also been suggestions that the target receptor Na^+/K^+ ATPase shows variability between individuals.

7.6.5 Diphenylhydantoin

As with the digitalis glycosides, the toxic effects of the anticonvulsant drug diphenylhydantoin result from elevated plasma levels of the drug. This can simply be due to inappropriate dosage, but other factors may also be involved in the development of toxicity.

The toxic effects observed are nystagmus, ataxia, drowsiness, and sometimes more serious effects on the CNS. These toxic effects are clearly dose related and correlate well with the plasma levels of the unchanged drug. High plasma levels may be the result of defective metabolism as well as of excessive dosage (see chap. 5). Diphenylhydantoin is metabolized by hydroxylation of the aromatic ring (Fig. 7.54), and this is the major route of metabolism. However, a genetic trait has been described in humans in which there appears to be a deficiency in this metabolic route. The consequences of this are decreased metabolism and the appearance of toxicity after therapeutic doses due to the elevated plasma levels of the unchanged drug.

Another cause of toxicity with diphenylhydantoin can be the result of interactions with other drugs such as isoniazid. This drug is a noncompetitive inhibitor of the aromatic hydroxylation of diphenylhydantoin. Consequently, the elimination of diphenylhydantoin is impaired in the presence of isoniazid, and the plasma level is greater than anticipated for a normal therapeutic dose. Furthermore, it was found that there was a significant correlation

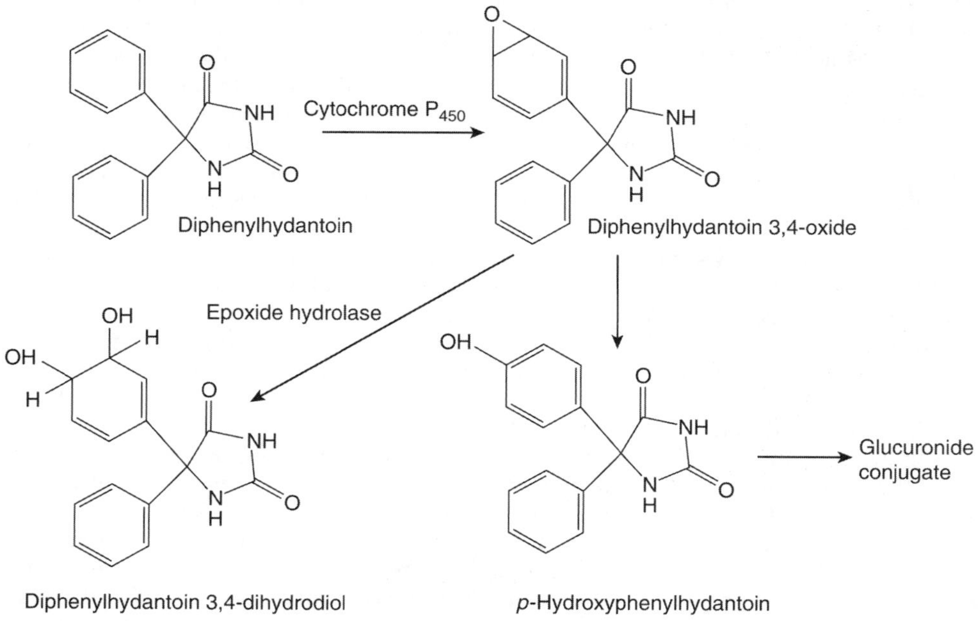

Figure 7.54   Metabolism of diphenylhydantoin.

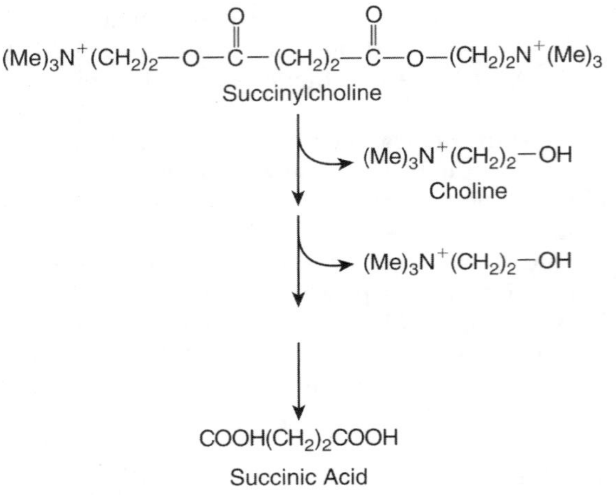

$$(Me)_3N^+(CH_2)_2-O-\overset{\overset{\displaystyle O}{\|}}{C}-(CH_2)_2-\overset{\overset{\displaystyle O}{\|}}{C}-O-(CH_2)_2N^+(Me)_3$$

Succinylcholine

$(Me)_3N^+(CH_2)_2-OH$

Choline

$(Me)_3N^+(CH_2)_2-OH$

$COOH(CH_2)_2COOH$

Succinic Acid

Figure 7.55 Metabolism of succinylcholine by pseudocholinesterases.

between the rate of acetylation of isoniazid and the appearance of adverse effects of diphenylhydantoin to the CNS.

This greater incidence in toxicity in slow acetylators corresponded with a greater increase in the plasma level of diphenylhydantoin when the drugs were taken in combination. This was due to higher plasma levels of isoniazid in slow acetylators, causing a greater degree of inhibition of diphenylhydantoin metabolism.

7.6.6 Succinylcholine

Succinylcholine is a neuromuscular blocking agent, which is used clinically to cause muscle relaxation. Its duration of action is short due to rapid metabolism—hydrolysis by cholinesterases (pseudocholinesterase or acylcholine acyl hydrolase)—in the plasma and liver to yield inactive products (Fig. 7.55). Thus, the pharmacological action is terminated by the metabolism. However, in some patients, the effect is excessive, with prolonged muscle relaxation and apnea lasting as long as two hours compared with the normal duration of a few minutes.

This occasional toxicity is due to a deficiency in the hydrolysis of succinylcholine; therefore, the parent drug circulates unchanged for longer periods of time, and consequently, the pharmacological effect is prolonged. This lack of hydrolysis is due to the presence of an atypical pseudocholinesterase. This aberrant enzyme hydrolyzes various substrates, including succinylcholine, at greatly reduced rates, and its affinity for both substrates and inhibitors is markedly different from that of the normal enzyme. The occurrence of this abnormal pseudocholinesterase is genetically determined and is controlled by two alleles at a single genetic locus. Homozygotes for the abnormal gene produce only the abnormal enzyme; heterozygotes produce a mixture of normal and abnormal enzymes. Homozygotes for the normal gene produce the normal pseudocholinesterase; the heterozygotes show an intermediate response. The abnormality shows a gene frequency of about 2% in certain groups (British, Greeks, Portuguese, North Africans), is rare in others (Australian aborigines, Filipinos), and is absent from some (Japanese, South American Indians).

There are other substrates for the enzyme such as diacetylmorphine and substance P, and other variants have been described, some of which may lack enzymic activity.

7.6.7 Botulism and Botulinum Toxin

Food may be contaminated with toxins produced by bacteria, such as botulinum toxin. This is produced by the bacterium *Clostridium botulinum* and is one of the two most potent toxins known to humans (the other being ricin). As little as one hundred-millionth of a gram (1×10^{-8} g) of the toxin would be lethal for a human. Fortunately, the toxin is destroyed by heat so that cooked food is unlikely to be contaminated (although the bacterial spores are quite resistant). The bacteria grow in the absence of air (they are *anaerobic*), and consequently, the foodstuffs most likely to be contaminated are those that are bottled or canned and eaten without cooking, for example, raw or lightly cooked fish.

For example, in October and November 1987, eight cases of botulism occurred, two in New York and six in Israel. All the victims had eaten Kapchunka, air-dried, salted whitefish, which had been prepared in New York, and then some had been transported by individuals to Israel. All the patients developed the symptoms of botulism within 36 hours, and one died. Some were treated with antitoxin and two received breathing assistance.

Botulinum toxin is a mixture of six large molecules, each of which consists of two components. The two components, a heavy (100 kDa) and light (50 kDa) polypeptide chain, are connected via a disulfide bond. The toxin may be associated with and protected from stomach acid by proteins such as hemogglutinins, which dissociate in the more alkaline intestine.

The toxin is absorbed from the intestine by transcytosis via epithelial absorptive cells, binding to the apical surface and being released on the basolateral side. Once in the blood (or lymph), the toxin can distribute to the presynaptic membrane of cholinergic nerve endings. The heavy chain binds to the walls of nerve cells, which then allows the whole toxin to be transported into the cell inside a vesicle (rather like a Trojan horse) via receptor-mediated endcytosis. Once inside the nerve cell, the light chain translocates into the cytosol, and acting as a peptidase destroys a synaptosomal protein, which in the case of toxin A and E is SNAP-25. By destroying the protein, the toxin prevents the release of the neurotransmitter acetylcholine from small packets at the ends of nerves by exocytosis. These nerves, attached to voluntary muscles, need acetylcholine to allow the flow of signals (impulses) between the nerve and the muscle (Fig. 7.56). By preventing the release of acetylcholine, botulinum toxin blocks muscle contraction, causing paralysis. Other botulinum neurotoxins target different proteins, e.g., toxin B targets VAMP/*synaptobrevin*. If the muscles affected are vital to life, such as those involved with breathing, the outcome is fatal. There are seven distinct neurotoxins.

Probably, the first cases of poisoning associated with the toxin were described in the 18th century in Germany where there was an outbreak caused by eating contaminated sausage in which many died. The syndrome caused by the toxin, known as botulism (derived from the Latin *botulus*, a sausage), has a high mortality rate. Furthermore, as the effect is essentially irreversible, victims who do survive can suffer paralysis for months. The symptoms can appear within a few hours of eating the contaminated food or may be delayed for several days.

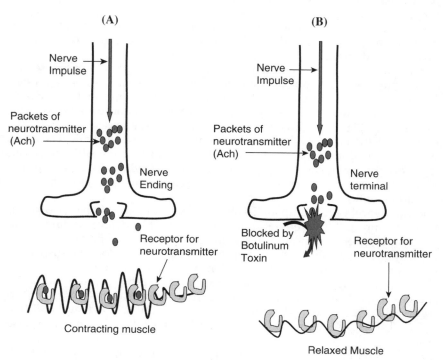

Figure 7.56 **(A)** Normal cholinergic nerve ending. The acetylcholine is released from the nerve ending and passes to the receptors at the neuromuscular junction. This causes the muscle to contract. **(B)** In botulium poisoning, the release of acetylcholine is blocked and so the muscle remains in the relaxed state.

The function of nerves, which cause contraction of muscles, is attacked by the toxin, and this leads to weakness of muscles causing blurred vision and difficulty in swallowing and breathing, and breathing may stop completely with fatal result if sufficient toxin has been absorbed into the body.

Cases of botulism are rare, and there is an antidote available (known as an *antitoxin*). Refrigeration of food and the use of preservatives such as sodium nitrite clearly have reduced the likelihood of contamination, but cases still occur despite this.

Surprisingly perhaps, given its extreme toxicity, botulinum toxin was introduced into medical practice in 1983 to treat patients with squint. Since then, its use has been expanded to include other disorders of muscle control suffered by patients with cerebral palsy or after a severe stroke where the brain cannot control the muscles, which may remain permanently contracted. Tiny amounts of the toxin are injected into the affected muscle, which then becomes paralyzed and, so, relaxed. There are several forms of the toxin that are now marketed as "Botox" (type A toxin) and Myobloc (type B toxin). More recently, Botox has been used in "cosmetic" medicine as a way of reducing lines and wrinkles in the face due to aging.

However, the toxin must be used with great care, and there have been cases of unwanted and long-term effects after its repeated cosmetic use and evidence that botox can move away from the injection site.

7.7 PHYSIOLOGICAL EFFECTS

7.7.1 Aspirin

Aspirin (acetylsalicylic acid) and other salicylates are still a common cause of human poisoning, both therapeutic and suicidal, and account for a significant number of deaths each year. Although the toxic effects have a biochemical basis, some of the effects caused are clearly physiological, and consequently, it has been used as an example in this category.

A common side effect of aspirin experienced by some patients at therapeutic doses is stomach irritation, bleeding, and in extreme cases, ulceration. This is related to the pharmacological effect of aspirin, which is inhibition of the enzyme cyclooxygenase (COX-1). As aspirin is nonionized in the acidic environment of the stomach, it can enter the lining cells. Here, it inhibits COX-1, and this leads to a reduction in the synthesis of prostaglandins. These mediators limit the secretion of acid and mucus by the stomach. The mucus helps protect the stomach lining from the damaging effects of the acid. Thus, aspirin reduces the protection and so increases the likelihood of damage.

Although aspirin is well absorbed from the gastrointestinal tract, after an overdose, the plasma levels may rise for as long as 24 hours, because if there is a large mass of tablets present in the stomach, this reduces their dissolution rate. Aspirin undergoes metabolism, mainly hydrolysis to **salicylic acid**, followed by hydroxylation to gentisic acid and conjugation with glucuronic acid and glycine (Fig. 7.57). The formation of the glycine and glucuronic acid conjugates is saturable, and therefore the half-life for elimination increases with increasing doses, and the steady-state blood level therefore increases disproportionately with increasing dose. When this saturation of conjugation occurs, there is more salicylic acid present in the blood, and therefore renal excretion of salicylic acid, which is sensitive to changes in the pH of urine, becomes more important.

The therapeutic blood level is greater than 150 mg L^{-1} but symptoms of toxicity occur at blood levels of around 300 mg L^{-1}. Therefore, knowledge of the blood level is important, particularly when aspirin is given in repeated doses. In children, therapeutic overdosage is responsible for the majority of fatalities from aspirin. When an overdose is suspected, measurement of the plasma level on two or more occasions will allow an estimate to be made of the severity of the overdose and whether the plasma level has reached its maximum. For interpretation of the blood level, the Done nomogram can be used. An overdose of 50 to 300 mg tablets in adults will give rise to moderate to severe toxicity and a blood level of 500 to 750 mg L^{-1} at 12 hours. The blood level must be interpreted with caution, however, because

1. the presence of metabolic acidosis will complicate the interpretation, because this will alter the distribution of salicylic acid (see below), and
2. with therapeutic overdose after repeated dosing, tissue levels may be higher than expected from the blood level.

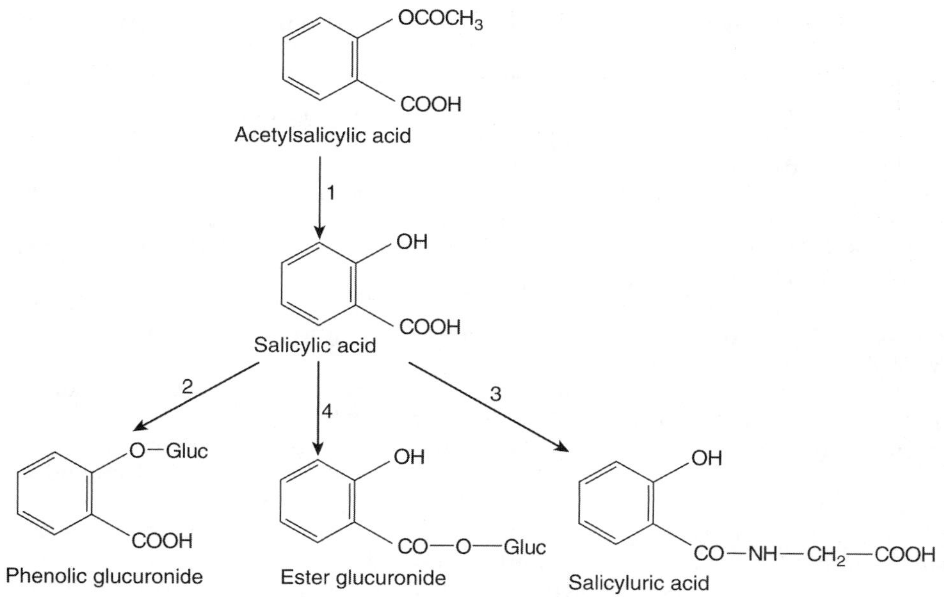

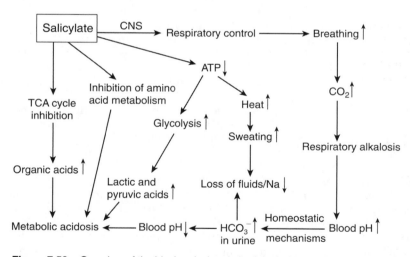

Figure 7.57 The metabolism of aspirin (acetylsalicylic acid). Step 1 (hydrolysis) yields the major metabolite, salicyclic acid, which is conjugated with glucuronic acid (pathways 2 and 4) or glycine (pathway 3). All these three pathways (2, 3, and 4) are saturable.

Figure 7.58 Overview of the biochemical and physiological interrelationships and changes in salicylate poisoning.

Salicylic acid is mainly responsible for the toxic effects of aspirin, and it has a number of metabolic and physiological effects some of which are interrelated (Fig. 7.58):

1. The initial effect of salicylate is stimulation of the depth of respiration due to increased production of carbon dioxide and greater use of oxygen. This is due to the uncoupling of oxidative phosphorylation in the mitochondria, which increases the rate of oxidation of substrates (see below).
2. As salicylate enters the CNS, it stimulates the respiratory center to give further increased depth and also rate of respiration. This causes the blood level of carbon dioxide to fall, as there is greater alveolar permeability to carbon dioxide than oxygen.

3. The loss of carbon dioxide causes the blood pH to rise, and this results in **respiratory alkalosis**.
4. The homeostatic mechanisms in the body respond, and the kidneys are stimulated to excrete large amounts of bicarbonate, so the blood pH falls.
5. The blood pH may return to normal in adults with mild overdoses. However, the blood pH can drop too far in children or more severely poisoned adults, resulting in metabolic acidosis. A lower buffering capacity or plasma protein-binding capacity may underlie the increased susceptibility to acidosis in children. Excretion of bicarbonate also means the bicarbonate in the blood is lower, and hence, there is an increased likelihood of metabolic acidosis.
6. As well as the increased excretion of bicarbonate, Na^+ and K^+ excretions are also increased. Because of the solute load, water excretion also increases. This results in hypokalemia and dehydration.
7. As well as these effects, salicylate also affects Krebs' cycle, carbohydrate metabolism, lipid metabolism, and protein and amino acid metabolism.
8. Krebs' cycle is inhibited at the points where α-ketoglutarate dehydrogenase and succinate dehydrogenase operate. This causes an increase in organic acids and an accumulation of glutamate.
9. ATP is depleted because of the **uncoupling of oxidative phosphorylation**, and hence, fat metabolism and glycolysis are stimulated.
10. Increased glycolysis leads to an increase in lactic and pyruvic acids. This contributes to the **metabolic acidosis**.
11. Salicylate also inhibits aminotransferases, leading to increased amino acid levels in blood and aminoaciduria. This also increases the solute load and contributes to the dehydration.
12. The uncoupling of oxidative phosphorylation stops ATP being produced by this process, and so the energy is dissipated as heat. The patient therefore suffers hyperpyrexia and sweating. Again, this produces dehydration.
13. Increased glycolysis to compensate for the loss of ATP leads to first hyper- and then hypoglycemia, especially in the brain. Mobilization of hepatic glycogen stores occurs through glucose-6-phosphatase activation via epinephrine release, possibly because of an effect on the CNS. This causes an initial hyperglycemia, but then when the glucose is completely used, hypoglycemia occurs.
14. The metabolic acidosis, if it occurs, allows salicylate to distribute to the CNS and other tissues more readily because of an increase in the proportion of the nonionized form (Fig. 7.59). Hence, the metabolic acidosis will exacerbate the toxicity to the CNS. This results in irritability, tremor, tinnitus, hallucinations, and delirium.

The basic mechanism underlying the toxicity of salicylate is the uncoupling of oxidative phosphorylation. For oxidative phosphorylation to take place, there is a requirement of a charge difference between the intermembrane space and the matrix of the mitochondria (Fig. 7.60). This is achieved when electrons move down the chain of multienzyme complexes and electron carriers (the electron transport chain), causing protons to move from the mitochondrial matrix to the intermembrane space. Consequently, a pH difference builds up, which is converted into an electrical potential across the membrane of approximately 200 mV over 8 nm.

When protons move back into the matrix along with phosphate through the pore of **ATP synthetase**, ATP is synthesized. The energy is derived from the potential difference.

As salicylate is a weak acid and has sufficient lipid solubility, it is able to diffuse across membranes. Thus, it can cross the mitochondrial membranes in its protonated form, releasing the proton into the matrix of the mitochondrion. By increasing the proton concentration, this dissipates the proton gradient and thus halts ATP production (Fig. 7.60).

The result of this is that the pH difference and therefore potential difference across the membrane collapses, and there is no energy to drive the ATP synthesis. However, electrons can still flow through the chain, which drives the citric acid cycle to produce them, and so cellular metabolism still continues, and indeed, without the constraint of ATP synthesis, the electron flow is more rapid and so other metabolic rates also increase in rate. Oxygen can still be reduced to water, and so this will be used up also and require replacement.

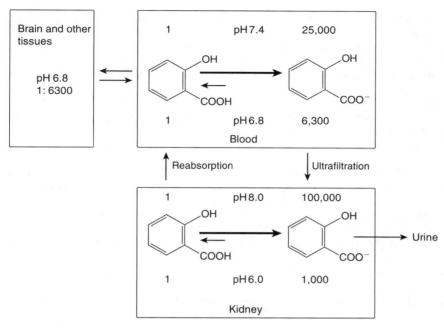

Figure 7.59 The effect of pH on the dissociation, distribution, and excretion of salicylic acid. The numbers represent the proportions of ionized and nonionized salicylic acid. The small horizontal arrows indicate the situation after overdose and the lower pH.

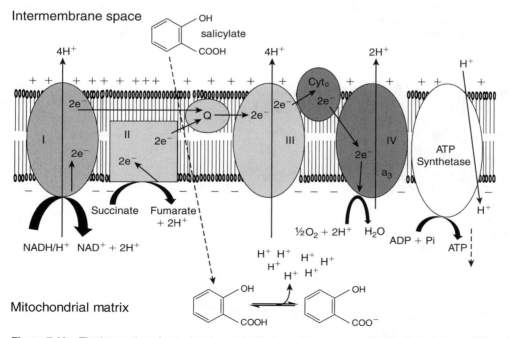

Figure 7.60 The interaction of salicylate ion and mitochondrial energy production. Salicylate can diffuse into the mitochondrial matrix, where it ionizes and loses a proton. This changes the predominantly negative charge and reduces the flow of protons through ATP synthetase. So ATP levels fall (↓).

The change in the acid-base balance in the body also alters the urine pH, making it more acidic (Fig. 7.59). This alters the excretion dynamics of salicylate, because more salicylate becomes nonionized and so is reabsorbed from the kidney tubules into the blood stream rather than being excreted into the urine. As can be seen from the Figure 7.59, lowering the pH of the urine to 6 results in a dramatic decrease in the ionization of the salicylate. Hence, elimination from the body is reduced.

The specific treatment for salicylate poisoning, apart from gastric lavage and aspiration to remove the drug from the stomach, is based on a knowledge of the biochemical mechanisms underlying the toxicity. There is no antidote, but treatment may be successful. Thus, treatment involves the following:

1. Correction of the metabolic acidosis with intravenous bicarbonate. This will also increase urine flow and cause it to become more alkaline (**alkaline diuresis**) and therefore facilitate excretion of salicylic acid and its conjugated metabolites. As the blood pH rises, the ionization of the salicylic acid increases, causing a change in the equilibrium and distribution of salicylate, which diffuses out of the CNS (Fig. 7.59).
2. Correction of solute and fluid loss and hypoglycemia with intravenous K^+ and dextrose and oral fluids.
3. Sponging of the patient with tepid water to reduce the hyperpyrexia.

Hemoperfusion and hemodialysis may be used in very severely poisoned patients.

7.8 BIOCHEMICAL EFFECTS: LETHAL SYNTHESIS AND INCORPORATION

7.8.1 Fluoroacetate

Monofluoroacetic acid (fluoroacetate) (Fig. 7.61) is a compound found naturally in certain South African plants, which causes severe toxicity in animals eating such plants. The compound has also been used as a **rodenticide**. The toxicity of fluoroacetate was one of the first to be studied at a basic biochemical level, and Peters coined the term "lethal synthesis" to describe this biochemical lesion.

Fluoroacetate does not cause direct tissue damage and is not intrinsically toxic but requires metabolism to fluoroacetyl CoA (Fig. 7.61). Other fluorinated compounds that are metabolized to fluoroacetyl CoA therefore produce the same toxic effects. For instance,

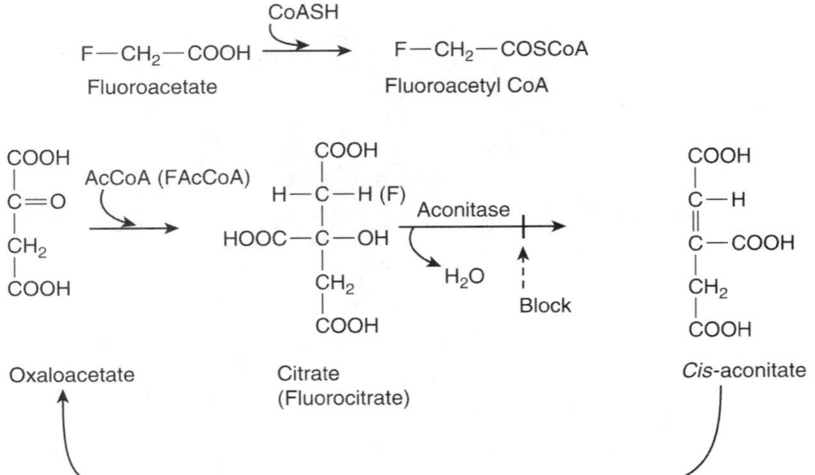

Figure 7.61   Metabolism of fluoroacetic acid to fluoroacetyl CoA (FAcCoA) and mechanism underlying blockade of the tricarboxylic acid cycle. Fluorocitrate cannot be dehydrated to *cis*-aconitate by aconitase and therefore blocks the cycle at this point.

compounds such as fluoroethanol and fluorofatty acids with even numbers of carbon atoms may undergo β-oxidation to yield fluoroacetyl CoA.

Fluoroacetyl CoA is incorporated into the tricarboxylic acid cycle (TCA cycle) in an analogous manner to acetyl CoA, combining with oxaloacetate to give fluorocitrate (Fig. 7.61). However, **fluorocitrate** inhibits the next enzyme of the TCA cycle, aconitase, and there is a build-up of both fluorocitrate and citrate. The TCA cycle is blocked, and the mitochondrial energy supply is disrupted. The inhibition arises from the fact that the fluorocitrate is a substrate for aconitase but binds strongly because the electronegative fluorine interacts with an iron atom ([4Fe-4S] cluster) in the enzyme. Furthermore, the aconitase cannot carry out the dehydration to *cis*-aconitate as the carbon-fluorine bond is stronger than the carbon-hydrogen bond. (Fig. 7.61).

Fluorocitrate is therefore a pseudosubstrate. As well as inhibiting cellular respiration, inhibition of the TCA cycle will also reduce the supply of 2-oxoglutarate. This may decrease the removal of ammonia via formation of glutamic acid and glutamine, and this might account for the convulsions seen in some species after exposure to fluoroacetate. The toxicity is manifested as a malfunction of the CNS and heart, giving rise to nausea, apprehension, convulsions, and defects of cardiac rhythm, leading to ventricular fibrillation. Fluoroacetate and fluorocitrate do not appear to inhibit other enzymes involved in intermediary metabolism, and the di- and trifluoroacetic acids are not similarly incorporated and therefore do not produce the same toxic effects.

7.8.2 Galactosamine

D-Galactosamine is an amino sugar (Fig. 7.62) normally found *in vivo* only in acetylated form in certain structural polysaccharides. Administration of single doses of this compound to certain species results in a dose-dependent hepatic damage resembling viral hepatitis, with focal necrosis and periportal inflammation. As well as a reversible hepatitis after acute doses, galactosamine may also cause chronic progressive hepatitis, cirrhosis, and liver tumors. As all the liver cells are affected, the damage tends to diffuse throughout the lobule.

A number of biochemical events have been found to occur. RNA and protein synthesis are inhibited, and membrane damage can be observed two hours after dosing. There is also an increase in intracellular calcium, which may be the crucial event that leads to cell death

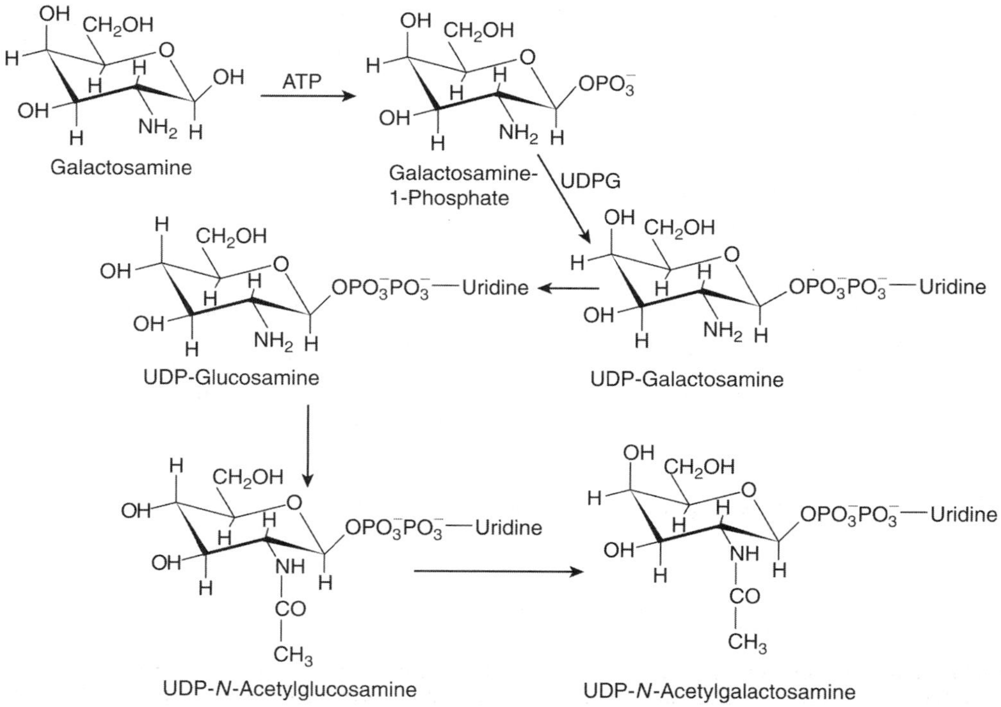

Figure 7.62   Metabolism of galactosamine.

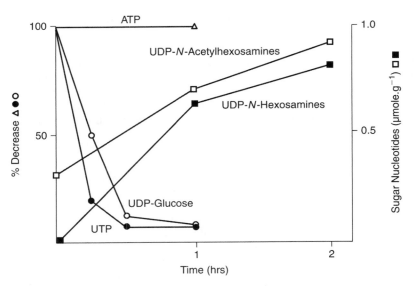

Figure 7.63 The effect of galactosamine on biochemical parameters. *Source*: From Ref. 17.

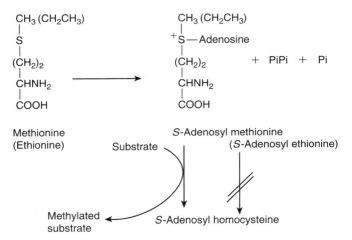

Figure 7.64 The role of methionine in methylation reactions and the mechanisms underlying ethionine hepatotoxicity. After the substrate is methylated, the *S*-adenosyl homocysteine remaining is broken down into homocysteine and adenine, both of which are reused. When *S*-adenosyl ethionine is formed, however, this recycling is reduced ($=$), and a shortage of adenine and hence ATP develops.

(see chap. 6). Also, levels of uridine triphosphate (UTP) and uridine diphosphate glucose (UDP-glucose) fall dramatically within the first hour after the administration of galactosamine. There is also a concomitant rise in UDP-hexosamines and UDP-*N*-acetylhexosamines (Fig. 7.63).

The rapid and extensive depletion of UTP is thought to be the basic cause of the toxicity. Galactosamine combines with UDP to form UDP-galactosamine (Fig. 7.62), which sequesters UDP, and therefore does not allow cycling of uridine nucleotides. Consequently, synthesis of RNA and hence protein are depressed, and sugar and mucopolysaccharide metabolism are disrupted. The latter effects may possibly explain the membrane damage that occurs. The toxic effects of galactosamine can be alleviated by the administration of UTP or its precursors.

Abnormal incorporation of hexosamines into **membrane glycoproteins** or glycolipids and a concomitant decrease in glucose and galactose incorporation may contribute to membrane damage, and the abnormal entry of calcium in the cell may play a role in the pathogenesis of the lesion. It is of interest, however, that although the mechanism may be compared with that of ethionine hepatotoxicity as described below, the eventual lesions are different. With ethionine, the lesion is fatty liver in contrast to the hepatic necrosis caused by galactosamine, although fatty liver may also occur with galactosamine.

7.8.3 Ethionine

Ethionine is a hepatotoxic analogue of the amino acid methionine (Fig. 7.64). Ethionine is an antimetabolite, which has similar chemical and physical properties to the naturally occurring amino acid. After acute doses, ethionine causes fatty liver, but prolonged administration results in liver cirrhosis and hepatic carcinoma. Some of the toxic effects may be reversed by the administration of methionine. The effects may be produced in a variety of species, although there are differences in response. The rat also shows a sex difference in susceptibility, the female animal showing the toxic response rather than the male.

After a single dose of ethionine, **triglycerides** accumulate in the liver, the increase being detectable after four hours. After 24 hours, the accumulation of triglycerides is maximal, being 15 to 20 times the normal level. Initially, the fat droplets accumulate on the endoplasmic reticulum in periportal hepatocytes and then in more central areas of the liver. Some species develop hepatic necrosis as well as fatty liver, and nuclear changes and disruption of the endoplasmic reticulum may also be observed.

Chronic administration causes proliferation of bile duct cells leading to hepatocyte atrophy, fibrous tissue surrounding proliferated bile ducts, and eventually cirrhosis and hepatocellular carcinoma.

The major biochemical changes observed are a striking depletion of ATP, impaired protein synthesis, defective incorporation of amino acids, and the appearance of RNA and proteins containing the ethyl rather than the methyl group. The plasma levels of triglycerides, cholesterol, lipoprotein, and phospholipid are all decreased.

The mechanism underlying this toxicity is thought to involve a deficiency of ATP. Methionine acts as a methyl donor *in vivo*, in the form of S-adenosyl methionine, and ethionine forms the corresponding S-adenosyl ethionine. However, the latter analogue is relatively inert as far as recycling the adenosyl moiety is concerned, and this is effectively trapped as S-adenosyl ethionine (Fig. 7.64). The resulting lack of ATP leads to inhibition of protein synthesis and a deficiency in the production of the apolipoprotein complex responsible for transporting triglycerides out of the liver. Consequently, there is an accumulation of triglycerides.

The reduction in protein synthesis obviously has other ramifications, such as a deficiency in hepatic enzymes and a consequent general disruption of intermediary metabolism. Methylation reactions are presumably also affected.

S-Adenosyl ethionine carries out ethylation reactions or ethyl transfer, and this is presumably involved in the carcinogenesis. Administration of ethionine to animals leads to the production of ethylated bases such as ethyl guanine. This may account for the observed inhibition of RNA polymerase and consequently of RNA synthesis. Incorporation of abnormal bases into nucleic acids and the production of impaired RNA may also lead to the inhibition of protein synthesis and misreading of the genetic code.

The depletion of ATP is the preliminary event leading to the pathological changes and the ultrastructural abnormalities of the nucleus and cytoplasmic organelles. Administration of ATP or precursors reverses all of these changes.

The exact mechanism underlying the carcinogenesis is less clear, but presumably involves inhibition of RNA synthesis or the production of abnormal ethylated nucleic acids and hence disruption of transcription, translation, or possibly replication. It is of interest that ethionine is not mutagenic in the Ames test, with or without rat liver homogenate. However, ethionine may be carcinogenic after metabolism to vinyl homocysteine (in which vinyl replaces ethyl), which is highly mutagenic.

7.9 BIOCHEMICAL EFFECTS: INTERACTION WITH SPECIFIC PROTEIN RECEPTORS

7.9.1 Carbon Monoxide

Despite the fact that carbon monoxide is no longer present in the gas used for domestic cooking and heating, it is still a major cause of poisoning. Indeed, it is one of the most important single agents involved in accidental and intentional poisoning, resulting in the deaths of several hundred people in Britain each year. There are in fact many sources of carbon monoxide: improperly burnt fuel in domestic fires, car exhausts, coal gas, furnace gas, cigarette smoke, and burning plastic. Thus, traffic policemen, firemen, and those trapped in fires and certain types of factory workers may all be potentially at risk. Because carbon monoxide is not

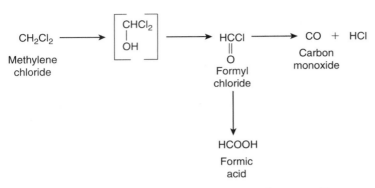

Figure 7.65   Metabolism of methylene chloride to carbon monoxide.

irritant and has no smell, a dangerous concentration may be reached before the victim is aware of anything untoward.

The pure gas is also used in laboratories, and it may be formed endogenously from the solvent **methylene chloride** by metabolism (Fig. 7.65). In the latter case, people using the solvent as a paint stripper in the home or for degreasing machinery in industry, and who inhale sufficient quantities, may be overcome some time after the exposure because of the slow release of methylene chloride from adipose tissue and metabolism to carbon monoxide. There is also a small amount of carbon monoxide produced normally from metabolism of the protoporphyrin ring of hemoglobin. As a result of the pioneering studies of Haldane in the 19th century and subsequent case studies, the symptoms are well documented. The concentration in the blood is normally measured as percent carboxyhemoglobin, as carbon monoxide binds to hemoglobin (see below).

A level of 60% carboxyhemoglobin will normally be fatal even for a few minutes, and if not fatal, may cause permanent damage; it is rare to find a blood level of greater than 80% carboxyhemoglobin. At a level of 20% carboxyhemoglobin, there may be no obvious symptoms but the ability to perform tasks can be impaired. When the level reaches 20% to 30%, the victim may have a headache, raised pulse rate, a dulling of the senses, and a sense of weariness. At levels of 30% to 40%, the symptoms will be the same but more pronounced, the blood pressure will be low and exertion may lead to faintness. At 40% to 60% carboxyhemoglobin, there will be weakness and incoordination, mental confusion, and a failure of memory. At concentrations of 60% carboxyhemoglobin and above, the victim will be unconscious and will suffer convulsions. There are many other clinical features: nausea, vomiting, pink skin, mental confusion, agitation, hearing loss, hyperpyrexia, hyperventilation, decrease in light sensitivity, arrhythmias, renal failure, and acidosis.

The spectrum of pathological effects includes peripheral neuropathy, brain damage, myocardial ischemia and infarction, muscle necrosis, and pulmonary edema. However, the main target organs are the brain and heart. This is because these organs have a relative inability to sustain an oxygen debt, and they use aerobic metabolic pathways extensively. The brain damage may be due to several mechanisms, including metabolic acidosis, hypotension, metabolic inhibition, and decreased blood flow and oxygen availability. The progressive hypotension, which is observed, may be an important contributor to the ischemia that occurs. Neural damage may follow both acute and chronic exposure. Death is due to brain tissue hypoxia, and respiratory failure may also occur.

The mechanism underlying carbon monoxide poisoning is well understood at the biochemical level. Carbon monoxide binds to the iron atom in hemoglobin at the same binding site as oxygen, but it binds more avidly, indeed about 240 times more strongly. The product is carboxyhemoglobin, which may contain one or more carbon monoxide molecules.

The hemoglobin molecule has four polypeptide chains, 2a and 2b, and each has a porphyrinic heme group with one iron atom at the center, which has five bonds used in the porphyrin structure, the sixth ligand being free for oxygen, (Fig. 7.66). Thus, hemoglobin has four binding sites for oxygen, and therefore, carbon monoxide can take up some of these. Loss of oxygen from the hemoglobin molecule causes a change in the tertiary structure of the whole molecule, a phenomenon known as cooperativity. Loss of molecules of oxygen from a fully

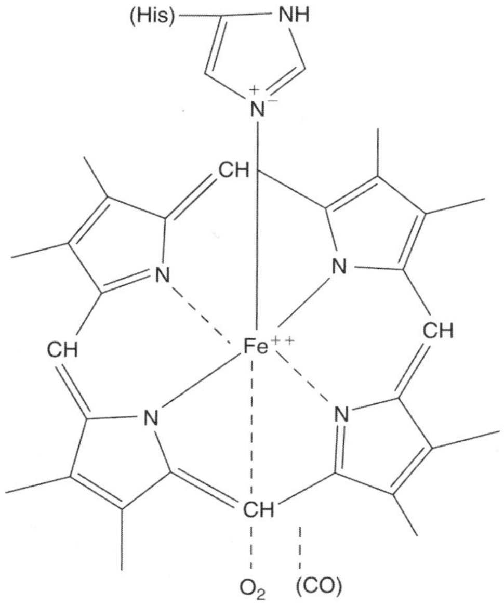

Figure 7.66 The heme moiety of the hemoglobin molecule showing the binding of the oxygen molecule to the iron atom. As shown in the diagram, carbon monoxide binds at the same site. *Abbreviation*: His, side chain of the amino acid histidine. *Source*: From Ref. 18.

oxygenated hemoglobin complex takes place in four steps, each with a different dissociation constant:

$$Hb_4O_8 \rightarrow Hb_4O_6 + O_2$$
$$K_1$$

where dissociation constant,

$$K_1 = \frac{[Hb_4O_6][O_2]}{[Hb_4O_8]}$$

$$K_2 = \frac{[Hb_4O_4][O_2]}{[Hb_4O_6]}$$

$$K_3 = \frac{[Hb_4O_2][O_2]}{[Hb_4O_4]}$$

$$K_4 = \frac{[Hb_4][O_2]}{[Hb_4O_2]}$$

(The smaller the dissociation constant, the more tightly the oxygen is bound.) When all four oxygen molecules are bound, each is equivalent. Loss of the first oxygen occurs when the partial pressure of the ambient oxygen falls. This loss of oxygen triggers a cooperativity change that facilitates the loss of the second oxygen molecule; thus, $K_2 > K_1$. Similarly, loss of the second oxygen facilitates loss of the third oxygen ($K_3 > K_2$). Loss of the fourth oxygen does not normally occur. Loss of the first oxygen requires a change in the partial pressure of O_2 of 60 mm Hg, the second requires a drop of 15 mm Hg, and the third a drop of 10 mm Hg. With a level of 50% carboxyhemoglobin, a fall in the partial pressure of O_2 of about 90 mm Hg is required for the loss of the first oxygen. This is due to the fact that the most common species will be $Hb_4 (O_2)_2 (CO)_2$, so the cooperativity normally expected is confined to one oxygen. Thus, binding of carbon monoxide causes a shift in the oxygen dissociation curve, resulting in more avid binding of oxygen, and so it is very much less readily given up in the tissues. Also, the binding of one or more carbon monoxide molecules to hemoglobin results, just as does the binding of oxygen, in an **allosteric change** in the remaining oxygen-binding sites.

The affinity of the remaining heme groups for oxygen is increased, and so the dissociation curve is also distorted. The consequence of this is that the tissue anoxia is very

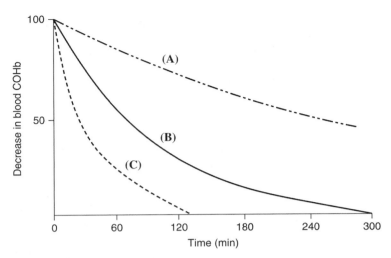

Figure 7.67 The dissociation of carboxyhemoglobin in the blood of a patient poisoned with carbon monoxide. The graph shows the effects of **(A)** breathing air **(B)**, oxygen, or **(C)** oxygen at increased pressure (2.5 atmospheres) on the rate of dissociation. *Source*: From Ref. 19.

much greater than would be expected from a simple loss of oxygen-carrying capacity by replacement of some of the oxygen molecules on the hemoglobin molecule.

The binding of carbon monoxide to hemoglobin is described by the Haldane equation:

$$\frac{[\text{COHB}]}{[\text{HbO}_2]} = \frac{M\ [P_{co}]}{[P_{o_2}]}$$

where $M = 220$ for human blood.

Therefore, when $P_{CO} = 1/220 \times P_{O2}$ the hemoglobin in the blood will be 50% saturated with carbon monoxide. Since air contains 21% oxygen, approximately 0.1% carbon monoxide will give this level of saturation. Hence, carbon monoxide is potentially very poisonous at low concentrations. The rate at which the arterial blood concentration of carbon monoxide reaches an equilibrium with the alveolar concentration will depend on other factors such as exercise and the efficiency of the lungs. Other factors will also affect the course of the poisoning.

Thus, physiological changes in blood flow through organs, such as shunting, may occur. Peripheral vasodilation due to hypoxia may exceed the cardiac output and hence fainting and unconsciousness can occur. Lactic acidemia will also result from impaired aerobic metabolism.

Whether the toxic effects are mainly due to **anemic hypoxia** or to the **histotoxic** effects of carbon monoxide on tissue metabolism is a source of controversy. Carbon monoxide will certainly bind to myoglobin and cytochromes such as cytochrome oxidase in the mitochondria and cytochrome P-450 in the endoplasmic reticulum, and the activity of both of these enzymes is decreased by carbon monoxide exposure. However, the general tissue hypoxia will also decrease the activity of these enzymes.

Treatment for the poisoning simply involves removal of the source of carbon monoxide and a supply of either fresh air or oxygen. Use of hyperbaric oxygen (2.5 atmospheres pressure) will facilitate the rate of dissociation of carboxyhemoglobin. Thus, the plasma half-life of carboxyhemoglobin can be reduced from 250 minutes in a patient breathing air to 23 minutes in a patient breathing hyperbaric oxygen (Fig. 7.67). Adding carbon dioxide may also be useful, as this reduces the half-life to 12 minutes at normal pressure. This is due to the stimulation by carbon dioxide of alveolar ventilation and acidemia, which increases the dissociation of carboxyhemoglobin.

7.9.2 Cyanide

Poisoning with cyanide may occur in a variety of ways: accidental or intentional poisoning with cyanide salts, which are used in industry or in laboratories; as a result of exposure to hydrogen cyanide in fires when polyurethane foam burns; from **sodium nitroprusside**, which is used therapeutically as a muscle relaxant and produces cyanide as an intermediate product; and from the natural product **amygdalin**, which is found in apricot stones, for example.

Amygdalin is a glycoside that releases glucose, benzaldehyde, and cyanide when degraded by the enzyme β-glucosidase in the gut flora of the gastrointestinal tract. There are many other natural sources of cyanide such as **Cassava tubers**, for example. There are various reports of human victims poisoned with cyanide, the most famous of which was the case of the influential Russian monk **Rasputin**, although this is disputed. From such case studies, the approximate lethal dose and toxic blood level is known. Thus, a dose of around 250 to 325 mg of potassium or sodium cyanide is lethal in humans, and a lethal case would have a blood level of around 1 mg mL^{-1}. A blood level of greater than 0.2 mg mL^{-1} is toxic. The symptoms of poisoning include headache, salivation nausea, anxiety, confusion, vertigo, convulsions, paralysis, unconsciousness coma, cardiac arrhythmias, hypotension, and respiratory failure. Both venous blood and arterial blood remain oxygenated, and hence the victim may appear pink.

The pathological effects seem to involve damage to the gray matter, and possibly the white matter, in the brain.

The mechanism underlying the toxicity of the cyanide ion is relatively straightforward and involves reversible binding to and inhibition of the cytochrome a-cytochrome a_3 complex (**cytochrome oxidase**) in the mitochondria.

This is the terminal complex in the electron transport chain, which transfers electrons to oxygen, reducing it to water (Fig. 7.68). Cyanide binds to the Fe^{3+} form of iron (Fig. 7.69), which is found in cytochromes such as cytochrome a_3, which undergo redox cycling. Thus, oxidized hemoglobin, cytochrome P-450, and cytochrome c are all targets. However, the effects on the mitochondria are the most significant because of the rapid effects on cell metabolism.

The binding of cyanide to cytochrome a_3 therefore blocks electron transport, which is coupled to ATP production, which requires the movement of protons into the mitochondrial intermembrane space (Fig. 7.68). Therefore, blockade of the movement of electrons past complex IV will stop the movement of protons, and so ATP synthesis will cease. It is the rapid depletion of ATP, which underlies the toxicity. Also, oxygen is not used, as the reduction to water occurs at complex IV and does not occur if the supply of electrons is interrupted. So, the peripheral oxygen tension rises, and the unloading gradient for oxyhemoglobin is decreased. Thus, higher circulating venous levels of oxyhemoglobin cause the victim to appear pink.

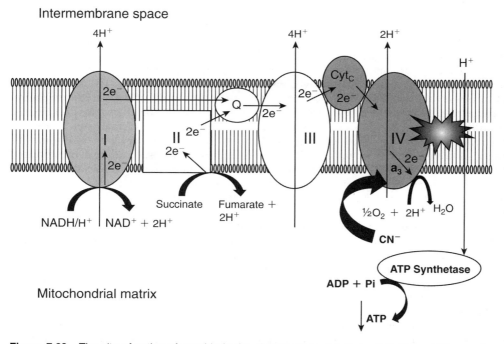

Figure 7.68 The site of action of cyanide in the electron transport chain. I, II, III, and IV: complexes in the electron transport chain. Cyanide blocks the action of a_3 and stops the reduction of water and the movement of electrons and protons. Therefore, ATP production stops (↓). *Abbreviations*: Q, coenzyme Q; cyt_c, cytochrome c; a_3: cytochrome a_3.

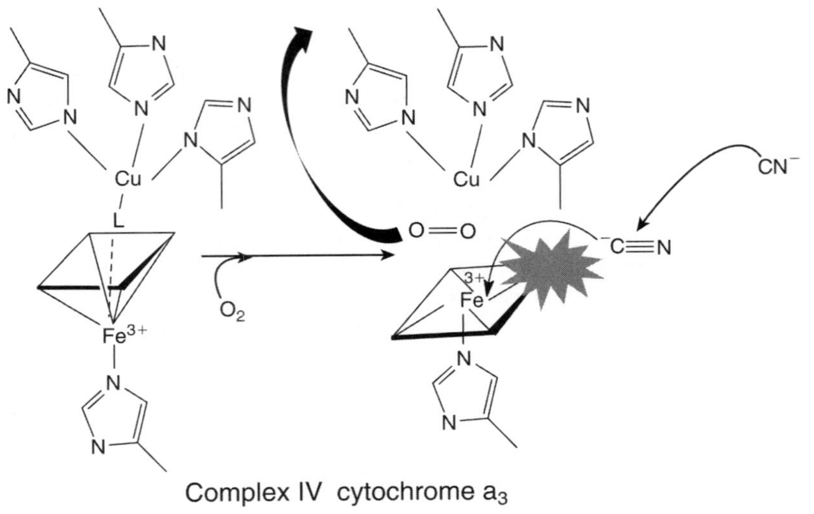

Complex IV cytochrome a₃

Figure 7.69 The details of interaction of cyanide with cytochrome a₃.

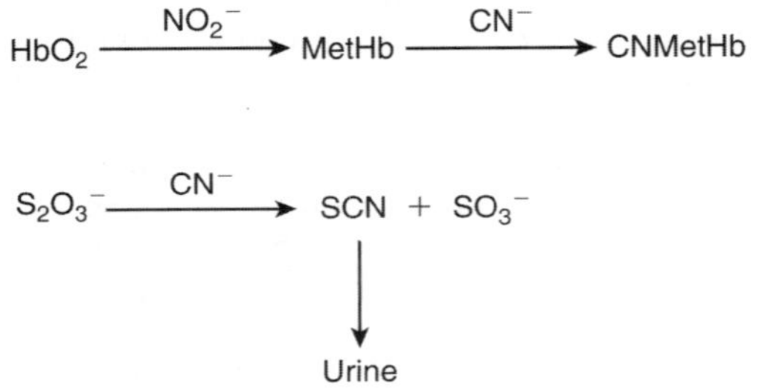

$$HbO_2 \xrightarrow{NO_2^-} MetHb \xrightarrow{CN^-} CNMetHb$$

$$S_2O_3^- \xrightarrow{CN^-} SCN + SO_3^-$$

Urine

Figure 7.70 The metabolism and detoxication of cyanide. *Abbreviations*: HbO₂, oxyhemoglobin; MetHb, methemoglobin; CNMetHb, cyanomethemoglobin.

The toxic effect is known as **histotoxic hypoxia**. Cyanide also directly stimulates chemoreceptors, causing hyperpnea. Lack of ATP will affect all cells, but heart muscle and brain are particularly susceptible. Therefore, cardiac arrythmias and other changes often occur, resulting in circulatory failure and delayed tissue ischemic anoxia. Death is usually due to respiratory arrest resulting from damage to the CNS, as the nerve cells of the respiratory control center are particularly sensitive to hypoxia. The susceptibility of the brain to pathological damage may reflect the lower concentration of cytochrome oxidase in white matter.

There are several antidotal treatments, but the blood level of cyanide should be determined, if possible, as treatment may be hazardous in some cases. Cyanide is metabolized in the body, indeed up to 50% of the cyanide in the circulation may be metabolized in one hour. This metabolic pathway involves the enzyme **rhodanese** and **thiosulfate** ion, which produces thiocyanate (Fig. 7.70). However, the crucial part of the treatment is to reduce the level of cyanide in the blood as soon as possible and allow the cyanide to dissociate from the cytochrome oxidase. This is achieved by several means. **Methemoglobin** will bind cyanide more avidly than cytochrome oxidase and therefore competes for the available cyanide. Converting some of the hemoglobin in the blood to methemoglobin will therefore decrease the circulating cyanide level and reduce the inhibition of the mitochondrial enzyme. Inhalation of amyl nitrite, which is volatile, may be used initially rapidly to oxidize hemoglobin to methemoglobin, and this is followed by sodium nitrite given intravenously to continue the production of methemoglobin. When methemoglobin binds cyanide, cyanomethemoglobin is formed, which cannot carry oxygen but will release its cyanide. Therefore, the administration of thiosulfate will facilitate the conversion of cyanide to thiocyanate and dissociation of cyanomethemoglobin. The thiocyanate is excreted in urine. An alternative treatment, which is now usually used when possible, is the administration of **cobalt edetate**, which is a chelating agent that binds cyanide and is excreted into the urine.

7.10 TERATOGENESIS

7.10.1 Actinomycin D
Actinomycin D is a complex chemical compound produced by the *Streptomyces* species of fungus and is used as an antibiotic. It is a well-established and potent teratogen and is also suspected of being carcinogenic.

The teratogenic potency of actinomycin D shows a marked dependence on the time of administration, being active on days 7 to 9 of gestation in the rat, when a high proportion of surviving fetuses show malformations (chap. 6, Fig. 22). Administration of the compound at earlier times, however (Fig. 7.71), results in a high fetal death and resorption rate. This falls to about 10% on the 13th day of gestation. The malformations produced in the rat are numerous, including cleft palate and lip, spina bifida, ecto- and dextrocardia, anencephaly, and disorganization of the optic nerve. Abnormalities of virtually every organ system may be seen at some time. Actinomycin D is particularly embryolethal, unlike other teratogens such as thalidomide (see below). This embryolethality shows a striking parallel with the incidence of malformations (Fig. 7.72). Although this has been observed to a certain extent with some other teratogens, in other cases, embryolethality and malformations seem to be independent variables.

It is well established that actinomycin D inhibits DNA-directed RNA synthesis by binding to guanosyl residues in the DNA molecule. This disrupts the transcription of genetic information and thereby interferes with the production of essential proteins. DNA synthesis may also be inhibited, being reduced by 30% to 40% in utero. It is clear that in the initial stages of embryogenesis, synthesis of RNA for protein production is vital, and it is not surprising that inhibition of this process may be lethal.

It has been shown that radiolabeled actinomycin D is bound to the RNA of embryos on days 9, 10, and 11 of gestation. Using incorporation of tritiated uridine as a marker for RNA synthesis, it was shown that only on gestational days 9 and 10 was there significant depression of incorporation in certain embryonic cell groups. On days 7,8, and 11 no significant depression of uridine incorporation was observed. The depression of RNA synthesis as measured by uridine incorporation therefore correlates approximately with teratogenicity. However, actinomycin D is also cytotoxic, and this can be demonstrated as cell damage in embryos on days 8, 9, and 10 but not at earlier times or on day 11. Therefore, the teratogenesis shows a correlation with cytotoxicity as well as with inhibition of RNA synthesis. Certainly,

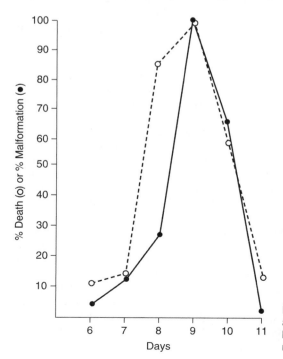

Figure 7.71 Embryolethality and teratogenicity of actinomycin D. This graph shows the relationship between the time of dosing and susceptibility to malformations (●) or death (○). *Source*: From Ref. 20.

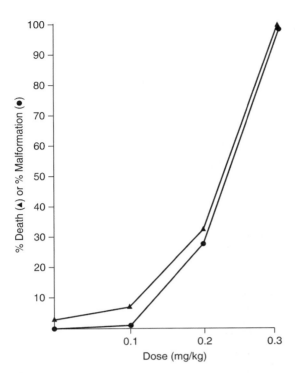

Figure 7.72 Embryolethality (▲) and teratogenicity (●) of actinomycin D. This graph shows the dose-response relationship for these two toxic effects. *Source*: From Ref. 20.

the concentration of actinomycin D in the embryo after administration on day 11 was too low for effective inhibition of RNA synthesis.

Although it is clear that actinomycin D is cytotoxic and inhibits RNA synthesis, the role of these effects in teratogenesis is not yet clear. RNA synthesis is a vital process in embryogenesis, preceding all differentiation, chemical and morphological, and therefore inhibition of it would be expected to disturb growth and differentiation. This may account for the production of malformations after administration on gestational day 7 in the rat, a period not normally sensitive in this species for the production of malformations. Excessive embryonic cell death may also be teratogenic.

7.10.2 Diphenylhydantoin

Diphenylhydantoin is an anticonvulsant drug in common use, which is suspected of being teratogenic in humans. There are at present insufficient data available to specify the exact type of malformation caused in humans, but in a few cases, craniofacial anomalies, growth retardation, and mental deficiency have been documented. Heart defects and cleft palate have also been described in some cases in humans. In experimental animals, however, diphenylhydantoin is clearly teratogenic, and this has been shown repeatedly in mice and rats. The defects most commonly seen are orofacial and skeletal, the orofacial defect usually described being cleft palate. Rhesus monkeys, however, are much more resistant to the teratogenicity, only showing minor urinary tract abnormalities, skeletal defects, and occasional abortions after high doses.

Although the teratogenicity of diphenylhydantoin in humans has not been demonstrated so clearly as that of thalidomide, the levels of drug to which experimental animals were exposed were not excessively high. The plasma levels of the unbound drug in the maternal plasma of experimental animals after teratogenic doses were only two to three times higher than those found in humans after therapeutic doses. The teratogenic effects in experimental animals were found to occur near the maternal toxic dose.

The types of defect produced in mice by diphenylhydantoin at various times of gestation are shown in Table 7.8. There is good correlation between the timing of these defects and the known pattern of organogenesis in the mouse.

Table 7.8 Timing of Teratogenic Effect of Diphenylhydantoin in the Mouse

Malformation	Treatment on gestational day					
	6–8	9–10	11–12	13–14	15–16	17–19
Orofacial	0	24	63	57	4	0
Eye defects	0	19	0	0	0	0
Limb defects	0	12	0	0	0	0
CNS defects	0	28	0	0	0	0
Skeletal defects	0	44	52	57	3	0
Kidney defects	0	25	17	7	4	0

Dose of diphenylhydantoin: 150 mg kg^{-1}. Figures are percentages of all surviving fetuses displaying the malformation.
Abbreviation: CNS, central nervous system.
Source: From Ref. 21.

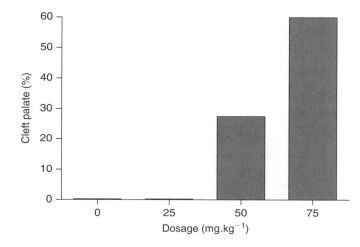

Figure 7.73 Dose-response relationship for diphenylhydantoin teratogenesis. The incidence of cleft palate in the surviving embryos is plotted against the dose of diphenylhydantoin given to pregnant mice on days 11, 12, and 13 of gestation. *Source*: From Ref. 21.

It can be seen that the greatest number of malformations of any sort occur on gestational days 11 to 12, with these being mainly orofacial and skeletal.

Diphenylhydantoin-induced malformations show a clear dose response, as can be seen from Figure 7.73 for pregnant mice treated on the 11th, 12th, and 13th days of gestation. No significant increase in fetal deaths was observed below 75 mg ·kg^{-1}, but above this dose level, more than 60% embryolethality was observed in utero, indicating a very steep dose-response curve. The mechanism underlying diphenylhydantoin teratogenicity is not fully understood, but enough is known to make it an interesting example. In both rats and humans, diphenylhydantoin is known to undergo aromatic hydroxylation, presumably catalyzed by the microsomal cytochromes P-450 system. However, pretreatment of pregnant mice with inducers or inhibitors of microsomal drug oxidation decreases or increases the teratogenicity, respectively. This paradoxical effect may simply be due to increases or decreases in the removal of the drug from the maternal circulation by metabolism and excretion following pretreatment.

It has been proposed that metabolic activation of diphenylhydantoin may be responsible for the teratogenicity. After the administration of radioactively labeled diphenylhydantoin to pregnant mice, radioactive drug or a metabolite was found to be covalently bound to protein in the embryo. It was shown that both the teratogenicity and embryolethality of diphenylhydantoin could be increased by using an inhibitor of epoxide hydrolase (see chap. 4), **trichloropropene oxide**. Similarly, the covalent binding of radiolabeled diphenylhydantoin to protein was also increased by this treatment.

Metabolism by epoxide hydrolase is effectively a detoxication pathway for reactive epoxides such as that proposed as an intermediate in diphenylhydantoin metabolism (Fig. 7.54), and inhibition of this enzyme therefore blocks the detoxication. It was postulated that the epoxide of diphenylhydantoin was the reactive and teratogenic intermediate produced by metabolism. The maternal plasma concentrations of drug in the mice used in this study were similar to those measured in humans after therapeutic doses of diphenylhydantoin had been given.

Although changes in the covalent binding of labeled diphenylhydantoin to fetal protein correlated with changes in teratogenicity, this does not prove a direct relationship. Indeed, other evidence suggests that diphenylhydantoin itself is the teratogenic agent. *In vitro*, using micromass cell culture, the drug was directly inhibitory, and toxicity was increased in the presence of various cytochromes P-450 inhibitors. Studies in mice *in vivo* also showed a correlation between increased maternal plasma levels of the unchanged drug and increased teratogenicity. Thus, there is conflicting evidence at present, and whether the putative reactive intermediate of diphenylhydantoin, which binds to protein, is the ultimate teratogen, or whether it is the parent drug, awaits clarification, but the embryo has limited metabolic capabilities as far as the necessary CYP is concerned. Metabolites could of course be generated in the maternal liver.

Alternatively, there is evidence that interaction with the **retinoic acid receptor** may be involved. Diphenylhydantoin is an agonist for this receptor, and exposure of mice in utero show highly upregulated expression of three RAR receptors: RAR α, β, and γ.

It is of interest that other antiepileptic drugs are also teratogenic such as valproic acid and **carbamezepine** and phenobarbital. It has therefore been suggested that the teratogenicity may be related to a common pharmacological action, resulting in cardiac arrhythmia and hypoxia in the embryo. Several of the drugs bind to a potassium ion channel, which results in inhibition of a potassium ion current at clinically relevant concentrations of the drug. A component of the potassium ion current plays a major role in cardiac rhythm regulation in the embryonic heart, and so inhibition causes arrhythmias and interruption of oxygen supply (hypoxia). It is significant that cardiovascular defects feature in the spectrum of teratogenic effects. Such events are known to be teratogenic when there are other causes. Also, after such an event, ROS are generated. Other drugs that inhibit the same potassium ion channel are also teratogenic, such as antiarrhythmic drugs and **erythromycin**, also causing cardiovascular defects.

7.10.3 Thalidomide

Despite the great interest in and notoriety of thalidomide as a teratogen, the underlying mechanisms of its teratogenicity are still not clear. It is, however, of particular interest in being a well-established human teratogen.

Thalidomide was an effective sedative sometimes used by pregnant women for the relief of morning sickness, and the drug seemed remarkably nontoxic. However, it eventually became apparent that its use by pregnant women was associated with characteristic deformities in the offspring. These deformities were **phocomelia** (shortening of the limbs), and malformations of the face and internal organs have also occurred. It was the sudden appearance of cases of phocomelia that alerted the medical world, as this had been a hitherto rare congenital abnormality.

It became clear that in virtually every case of phocomelia, the mother of the malformed child had definitely taken thalidomide between the third and eighth week of gestation. In some cases, only a few doses had been taken during the critical period. Analysis of the epidemiological and clinical data suggested that thalidomide was almost invariably effective if taken on a few days or perhaps just one day between the 20th and 35th gestational days. It has been possible to pinpoint the critical periods for each abnormality. Lack of the external ear and paralysis of the cranial nerve occur on exposure during the 21st and 22nd days of gestation. Phocomelia, mainly of the arms, occurs after exposure on days 24 to 27 of gestation with the legs affected one to two days later. The sensitive period ends on days 34 to 36 with production of hypoplastic thumbs and anorectal stenosis.

Initially, these malformations were not readily reproducible in rats or other experimental animals. It was later discovered that limb malformations could be produced in certain strains of white rabbits if they were exposed during the 8th to 16th days of gestation. Several strains of monkeys were found to be susceptible and gave similar malformations to those seen in humans. It was eventually found that malformations could be produced in rats but only if they were exposed to the drug on the 12th day of gestation. At teratogenic doses, thalidomide has no embryolethal effect, doses several times larger being necessary to cause fetal death.

The mechanism underlying thalidomide teratogenesis has been the subject of considerable research. Thalidomide exists as two isomers, one intriguing finding is that there is a difference in the toxicity of the two isomers of thalidomide, with only the S- enantiomer being teratogenic and not the R+ enantiomer (Fig. 7.74).

Initial research suggested that metabolism of thalidomide and possibly a reactive metabolite might be involved. The metabolism is complex, involving both enzymic and nonenzymic pathways. A number of metabolites arise by hydrolysis of the amide bond in the piperidine ring (Fig. 7.75), of which about 12 have been identified *in vivo*, and aromatic ring hydroxylation may also take place. The hydrolytic opening of the piperidine ring yields phthalyl derivatives of glutamine and glutamic acid. **Phthalylglutamic acid** may also be decarboxylated to the monocarboxylic acid derivative. Other monocarboxylic and dicarboxylic acid derivatives are also produced, some by opening of the phthalimide ring.

However, none of these metabolites was found to be teratogenic in the rabbit, although it has been subsequently found that they were unable to penetrate the embryo. Studies *in vivo* revealed that after the administration of the parent drug, the major products detectable in the fetus were the monocarboxylic acid derivatives. It therefore seems that the parent drug may cross the placenta and enter the embryo and undergo metabolism.

Subsequently, the phthalylglutamic acid (Fig. 7.75), a dicarboxylic acid, was shown to be teratogenic in mice when given on days 7 to 9 of pregnancy. However, other work has indicated that the phthalimide ring is of importance in the teratogenicity. Other studies have indicated that thalidomide acylates aliphatic amines such as putrescine and spermidine, histones, RNA, and DNA and that it may also affect ribosomal integrity.

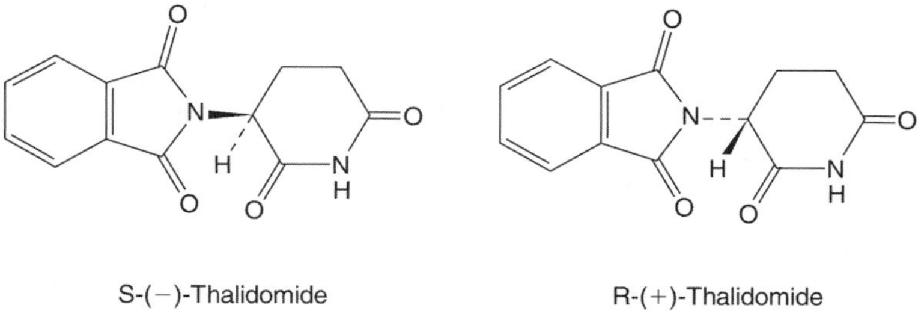

S-(−)-Thalidomide   R-(+)-Thalidomide

Figure 7.74   Structure of the isomers of thalidomide.

Thalidomide

Phthalylglutamine   Phthalylglutamic acid

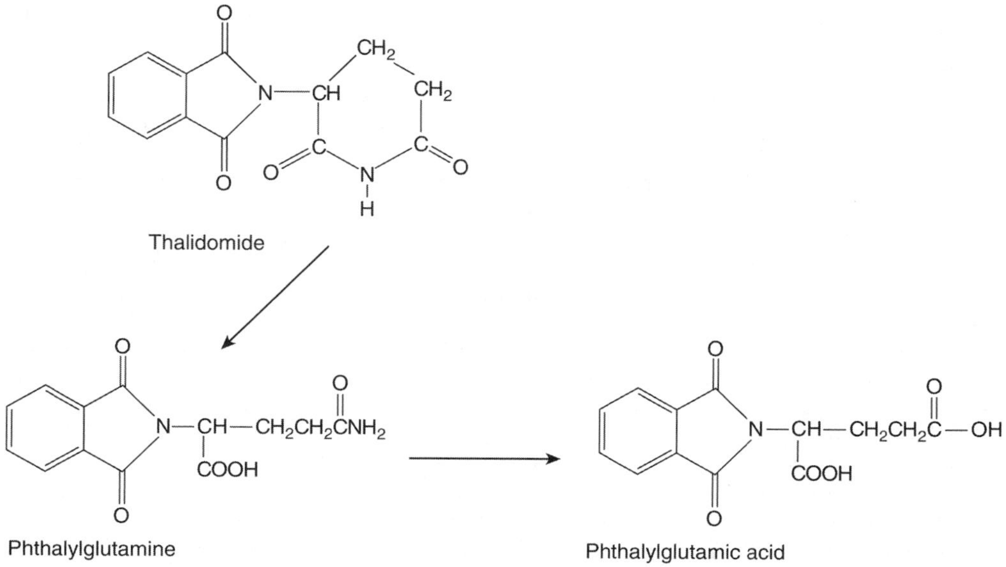

Figure 7.75   The structure of thalidomide and its two hydrolysis products.

Evidence was accumulated mainly from *in vitro* studies that a reactive metabolite(s), generated by cytochrome P-450, might be involved in the teratogenicity. This was derived using various *in vitro* systems such as tumor cells, human lymphocytes, limb bud, and embryo culture and a variety of techniques. The data suggested that an epoxide might be involved in the toxicity. However, other experimental data using an *in vitro* system involving tumor cells were inconsistent with these data and suggested that an epoxide intermediate was not involved.

More recent studies have indicated a probable mechanism that relies on the interaction of thalidomide with DNA. The phthalimide double-ring structure (Fig. 7.74) is relatively flat and can intercalate with DNA forming a stacked complex. Furthermore, observations were made that thalidomide

1. inhibits the development of new blood vessels from microvessels (angiogenesis) in the embryo;
2. inhibits the expression of specific cell adhesion molecules (**integrins**), which are intimately involved in cell growth and differentiation;
3. inhibits the action of certain growth factors; and
4. binds specifically to the promoter regions of genes involved in regulatory pathways for angiogenesis and integrin production.

Growth factors are clearly important in the development of the embryo as is growth of new blood vessels. Integrins are involved in differentiation having an important role in cell-cell communication.

The specific integrin ($\alpha v \beta 3$ subunits) is stimulated by two growth factors, insulin-like growth factor (IGF-I) and fibroblast growth factor type 2 (FGF-2). The integrin ($\alpha v \beta 3$) then stimulates angiogenesis, which is an integral part of the development of limb buds.

The transcription and therefore production of the integrin αv and $\beta 3$ subunits are each regulated by a promoter-specific transcription factor Sp1. A crucial step is binding of this transcription factor to guanine-rich promoter regions, which are found in many genes, including those for both αv and $\beta 3$. The genes for IGF-I and FGF also contain **polyG** sequences, and similarly, the gene expression is influenced by Sp1. Consequently, polyG sections are important for the process of differentiation and limb bud development (Fig. 7.76).

It has been possible to combine the observations and basic underlying molecular biology into a reasonable working hypothesis. It has been found that thalidomide not only intercalates into the DNA molecule but also binds to the polyG sites in the promoter regions. Furthermore, only the S-enantiomer, the teratogenic form (Fig. 7.76), fits properly into the major groove of the double helix and binds to the polyG sequences. Thus, it seems that a very specific stereochemical interaction may be necessary for this toxic effect to occur. It would seem that thalidomide competes for binding with Sp1 at these polyG sequences.

Because most genes (more than 90%) have **TATAA boxes**, the effect of competition for binding between Sp1 and thalidomide at polyG sequences would be minimal, and therefore, the development of many of the organs and tissues in the embryo would not be affected. Therefore, it would be only those genes *without* TATAA boxes that would be most affected, and these are the ones that include those coding for tissue-specific integrins and growth factors. Therefore, it is these that are affected, leading to a reduced level of integrin $\alpha v \beta 3$ and therefore reduced limb bud development (Fig. 7.76). This hypothesis is very plausible and shows how specific and intricate toxicodynamic mechanisms underlying toxicity may be.

Whether any of the metabolites or breakdown products of thalidomide are also active at these sites remains to be seen. Perhaps the particular lesson to be learnt from the tragedy of thalidomide is that a drug with low maternal and adult toxicity, which is similarly of low toxicity in experimental animals, may have high teratogenic activity. The human embryo seems to have been particularly sensitive to normal therapeutic doses in this case. However, it has resulted in new drugs now being more rigorously tested and also tested in pregnant animals.

As a postscript, it should be mentioned that thalidomide now has been investigated for a number of therapeutic indications such as certain forms of cancer, leprosy, and severe ulcers. Unfortunately, more cases of malformations have appeared despite warnings and controls on its use.

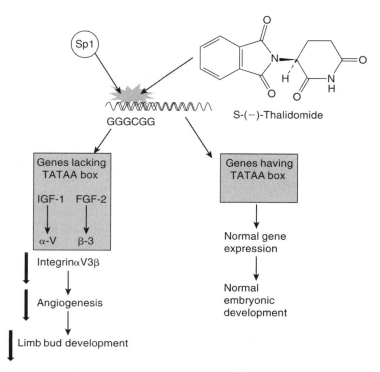

S-(−)-Thalidomide

Figure 7.76 The mechanism underlying the teratogenesis of thalidomide. IGF-1 and FGF-1 are growth factor genes. Sp1 is a specific transcription factor. Integrin αVβ3 is a growth factor composed of the subunits αV and β3, which increases angiogenesis. In the presence of thalidomide, the expression of the growth factor genes is reduced, hence less integrin is produced and less angiogenesis occurs.

7.11   IMMUNOTOXICITY

7.11.1 Halothane

Halothane is a very widely used anesthetic drug, which may cause hepatic damage in some patients. It seems that there are two types of hepatic damage, however. One is a very rare reaction, idiosyncratic, resulting in serious liver damage with an incidence of about 1 in 35,000. The other form of hepatotoxicity is a mild liver dysfunction, which is more common and occurs in as many as 20% of patients receiving the drug. The two different types probably involve different mechanisms.

The more common mild liver dysfunction is thought to be due to a direct toxic action of one of the halothane metabolites on the liver. It is manifested as raised serum transaminases (AST and ALT). Studies in experimental animals have indicated that under certain conditions, such as in phenobarbital-induced male rats exposed to halothane with a reduced concentration of oxygen, halothane is directly hepatotoxic. This is believed to be due to the reductive pathway of metabolism (Fig. 7.77), which produces a free radical metabolite. This metabolite will bind covalently to liver protein and can initiate lipid peroxidation, both of which have been detected in the experimental rat model. The other metabolites and fluoride ion have also been detected in experimental animals in support of this reductive pathway. Covalent binding to liver protein and the metabolic products of the reductive pathway have both been detected in a brain-dead human subject exposed to radiolabeled halothane during a transplant operation. The level of plasma fluoride correlated with the covalent binding. Thus, this reductive pathway, which would be favored in rats, anesthetized with halothane, and reduced oxygen concentration may be the cause of the direct toxic effect on the liver. The severe and rare hepatic damage has clearly different features from the mild hepatotoxicity. As well as centrilobular hepatic necrosis, patients suffer from fever, rash, and arthralgias and have serum

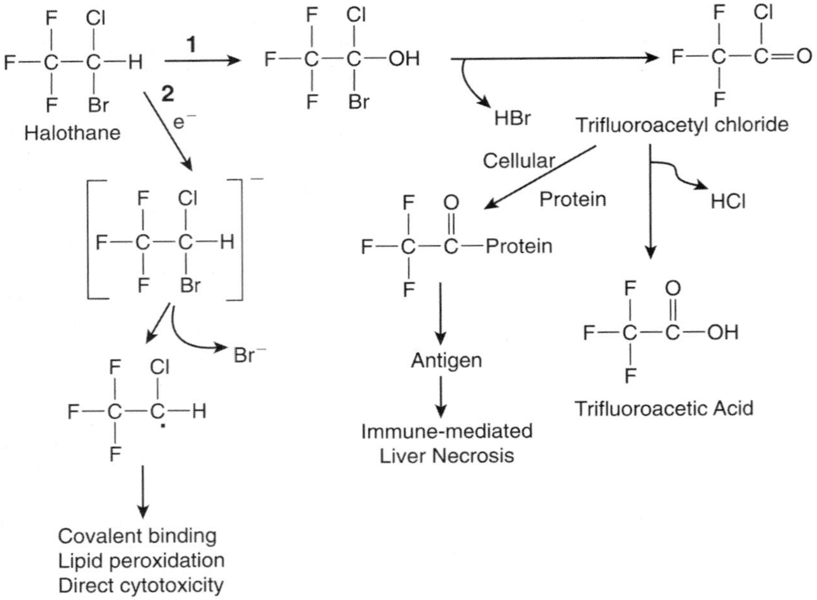

Figure 7.77 The metabolism of halothane and its proposed involvement in the toxicity. Pathway 1 (oxidative) and pathway 2 (reductive) are both catalyzed by cytochrome P-450.

tissue autoantibodies. About 25% of patients have antimicrosomal antibodies. There are also several predisposing factors, and those so far recognized are

1. multiple exposures, which seem to sensitize the patient to future exposures;
2. sex, females being more commonly affected than males in the ratio 1.8:1;
3. obesity, 68% of patients in one study were obese; and
4. allergy, a previous history of allergy was found in one third of patients. After multiple exposures the incidence increases from 1 in 35,000 patients to 1 in 3700.

Halothane is believed to cause this severe hepatic damage via an immunological mechanism whereby antibodies to altered liver cell membrane components are generated by repeated exposure to a halothane-derived antigen. The immunological reaction that ensues is directed at the liver cells of the patient. These are destroyed, resulting in hepatic necrosis. Initial studies using rabbits anesthetized with halothane revealed that lymphocytes from patients with halothane-induced hepatic damage were sensitized against the liver homogenate from these rabbits. Furthermore, lymphocytes from these patients were cytotoxic toward the rabbit hepatocytes. Serum from patients suffering from halothane-induced hepatic damage contained antibodies of the IgG type, directed against an antigenic determinant on the surface of rabbit hepatocytes when these were derived from rabbits that had been anesthetized with halothane. A subpopulation of lymphocytes (T lymphocytes, killer lymphocytes) from normal human blood attacked rabbit hepatocytes after incubation with sera from patients, showing the antibodies were specific for cells altered by halothane.

These studies suggested that patients had an immunological response to a halothane-derived antigen. Controlling the level of oxygen during anesthesia indicated that the oxidative metabolic pathway was involved rather than the reductive pathway. Only sera or lymphocytes from those patients with the severe type of damage, not the more common hepatic damage, reacted to rabbit hepatocytes in this way, and only hepatocytes from rabbits exposed to halothane were targets. Thus, these and later studies distinguished between the mild and severe hepatic damage.

The aerobic pathway of metabolism (pathway 1) (Fig. 7.77) produces **trifluoroacetyl chloride**, a highly reactive acyl chloride, which can react with nucleophiles such as amino groups similar to those on proteins. Alternatively, reaction with water yields trifluoroacetic acid. Trifluoroacetylchloride is the probable reactive metabolite that trifluoroacylates protein, most probably at lysine residues (Fig. 7.77). Removal of the trifluoroacyl moiety from the antigenic protein by chemical means abolished most of the interactions with antibodies from patients. Preparation of the suspected hapten, trifluoroacyl-lysine, inhibited the interaction between the antibody and altered cell protein. There are many target proteins that have been isolated, and the majority seems to be from the membrane of the endoplasmic reticulum, and some, including cytochrome P-450, are derived from the smooth endoplasmic reticulum. The majority of the antigenic activity is present in this fraction, although it is also detectable on the surface of the hepatocyte. The antigens correspond to at least five polypeptide fractions in rat, rabbit, and human liver, and they appear to be similar, although not identical in each species. These polypeptides have been characterized as having the following molecular weights: 100, 76, 59, 57, and 54 kDa, detected by immunoblotting with human serum from patients and rat, human, or rabbit liver microsomal fraction. The 100- and 76-kDa antigens are the most common. The 59-kDa protein has been identified as a microsomal carboxyl-esterase, and the 54-kDa protein may be cytochromes P-450. Other proteins are calreticulin and disulfide isomerase. Each of the antigens that are recognized by patients' sera are trifluoroacylated. However, antibodies recognize epitopes, which are a combination of the trifluoroacyl group and determinants of the polypeptide carrier. The trifluoroacyl group is essential and is recognized in the liver antigens by antitrifluoroacyl antibodies. However, there are other determinants as trifluoroacyl-lysine, the putative hapten, would not totally inhibit the binding of antibody from patients' serum to the liver antigen. This phenomenon is known as hapten inhibition. Also, the sera from different patients varied in the recognition of the different antigens. The antibodies also recognized native proteins.

The production of the antigens was reduced in rats in which a deutero analogue of halothane was given and the metabolism of which would be reduced by an isotope effect. The antigens could be formed *in vitro*, but only under aerobic conditions and not under anaerobic conditions, implicating the oxidative pathway rather than the reductive (Fig. 7.77).

Halothane-induced hepatitis seems to be mediated by both humoral and cell-mediated aspects of the immune system. Thus, both specific circulating antibodies and cytotoxic T lymphocytes are involved. One suggestion is that the mechanism involves **antibody-dependent cell cytotoxicity** (ADCC).

As described above, the haptenised protein or hapten protein complex has to be presented to the immune system for the system to be aware of it as an antigen. With intracellular antigens, this requires expression of fragment of the hapten-protein complex on the cell surface via major histocompatibility complex (MHC) molecules. These can bind almost any peptide (Fig. 7.78).

The MHC complex and antigen can then be detected such as by CD8 cytotoxic T cells and the cells destroyed. However, the fragments of the hapten complex can also be detected by other cells such as APC, and with the collaboration of T-helper cells, B cells would become activated and antibodies produced. These could then bind to the surface antigen as described for a type II reaction.

Furthermore, certain cells such as cytotoxic T lymphocytes and macrophages have Fc receptors, which bind to the Fc portion of the bound antibodies (e.g., IgG). As a result, the cytotoxic T cell or macrophage is activated, and lyses the cell to which the antibody is bound.

It is also likely that direct cytotoxicity of halothane metabolites is important in the process, causing cellular stress by lipid peroxidation and covalent binding to target proteins (Fig. 7.77). Then the cytokines produced by this stress cause upregulation of costimulatory molecules, which provides the second signal for the immune response to occur.

It may be that several mechanisms operate depending on exposure and host factors. A further twist is the suggestion that the reason so few individuals suffer the immune response is a result of tolerance. Clearly, from the "**danger hypothesis**," absence of the second signal, i.e., direct cytotoxicity from the free radical metabolite, because of better protective measures or lack of anaerobic conditions, then means that the immune response would not occur.

Another means of tolerance is based on molecular mimicry whereby a native protein has a similar epitope to the antigen, so the immune system is tolerant to this in most individuals. The suggestion is that the patients who are susceptible have some variation in this native protein so that tolerance has not been acquired.

Trifluoroacetylchloride is very reactive, and therefore, if any of it escapes from the vicinity of the cytochromes P-450 in the smooth endoplasmic reticulum, it might be expected to react with those proteins in highest concentration. The 59-kDa polypeptide, a microsomal carboxylesterase, constitutes 1.5% of the total microsomal protein, and therefore it fulfils this

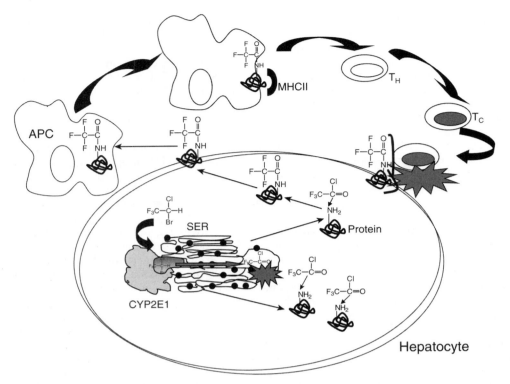

Figure 7.78 Postulated mechanism of halothane immune-mediated hepatotoxicity. This figure is only a partial explanation, involving Tc cells (cytotoxic lymphocytes). See text for complete description. CYP2E1 in liver cell activates the halothane to a reactive acyl chloride (*shown*), which reacts with proteins (e.g., enzymes in the SER). These are transported to cell surface and presented to immune system by APC. *Abbreviations*: APC, antigen-presenting cell; SER, smooth endoplasmic reticulum; MHCII, major histocompatability complex.

criterion. The rate of degradation of the carrier molecule may also be an important factor in determining whether an antigen becomes an immunogen. The concentration of antigen that is presented to the immune system may similarly be a very important factor. Some of the antigenic polypeptides detected seem to be long lived.

The resulting epitope density may depend on the number of lysine groups in the particular protein, and this will in turn affect the immunogenicity of the antigen. Trifluoroacyl adducts have been detected on the outer surface of hepatocytes, presumably as a result of the hapten-complex processing and delivery by MHC I, which is described above. The fact that the production of the trifluoroacetyl chloride is part of the major metabolic pathway and that the majority of patients produce trifluoroacylated proteins suggests that it is differences in the immune surveillance system or immune responsiveness, which determine which patients will succumb to the immunotoxic effect.

Clearly, an understanding of why when so many individuals will produce the hapten so few mount a severe immune response is crucial for the development of drug safety evaluation.

7.11.2 Practolol

Practolol (Fig. 7.79) is an antihypertensive drug, which had to be withdrawn from general use because of severe adverse effects that became apparent in 1974, about four years after the drug had been marketed. The toxicity of practolol was unexpected, and when it occurred, it was severe. Furthermore, it has never been reproduced in experimental animals, even with the benefit of hindsight.

The available evidence points to the involvement of an immunological mechanism, the basis of which is probably a metaboliteprotein conjugate acting as an antigen. It seems unlikely that the pharmacological action of this drug, β-blockade, is involved, as other β-blocking drugs have not shown similar adverse effects. The syndrome produced by this drug features lesions

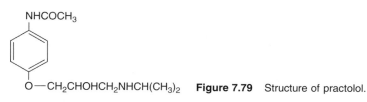

Figure 7.79 Structure of practolol.

to the eye, peritoneum, and skin. Epidemiological studies have firmly established practolol as the causative agent in the development of this toxic effect, described as the **oculomucocutaneous syndrome**. This involved extensive and severe skin rashes, similar to psoriasis, and keratinization of the cornea and peritoneal membrane.

Studies *in vivo* showed that practolol was metabolized to only a limited extent in human subjects and experimental animals. Thus, 74% to 90% of a dose of practolol is excreted in the urine in humans, and of this, 80% to 90% is unchanged, and less than 5% of the dose is deacetylated. No differences in metabolism were detected between patients with the practolol syndrome and those not suffering from it. However, in patients with the syndrome, there were no antibodies detected to practolol itself. Hamsters treated with practolol showed a marked and persistent accumulation of the drug in the eye. Furthermore, in a hamster liver microsomal system *in vitro*, it was shown that practolol could be covalently bound to microsomal protein and also added to serum albumin. It was further shown that human sera from patients with the practolol syndrome contained antibodies that were specific for this practolol metabolite generated *in vitro*. Factors affecting the covalent binding of the practolol metabolite to protein *in vitro* also influenced the reaction between the patient's sera and the protein conjugate. The absence of NADPH, the presence of GSH, and the use of 3-methylcholanthrene-induced hamster liver microsomes, all reduced the binding and the immunological reaction. These results indicated a requirement for NADPH-dependent microsomal enzyme-mediated metabolism and binding to protein to produce a hapten recognizable by the antibodies in the patients' sera. However, patients without the syndrome also had antibodies, although these were present at a lower level.

The identity of the metabolite has not yet been established, but it is neither the deacetylated nor 3-hydroxymetabolite found *in vivo*. However, deacetylation may be a prerequisite for the N-hydroxylation reaction, which has been suggested. Although the exact antigenic determinants of the metabolite have not been elucidated, it is known that the complete isopropanolamine side chain is necessary.

Therefore, it seems likely that practolol causes an adverse reaction that has an immunological basis because of the formation of an antigen between a practolol metabolite and a protein. The reason for the bizarre manifestations of the immunological reaction and the metabolite responsible has not yet been established.

7.11.3 Penicillin

Penicillin and its derivatives are very widely used antibiotics, which are well tolerated and have a low acute toxicity. Doses as high as 1 g kg^{-1} day^{-1} can be given by intravenous injection. However, penicillin and related derivative drugs cause more allergic reactions than any other class of drug. The incidence of allergic reactions to such drugs occurs in at least 1% of recipients.

The newer derivatives seem less likely to cause hypersensitivity reactions, perhaps because the protein adducts generated are shorter lived. All four types of hypersensitivity reaction have been observed with penicillin. Thus, high doses may cause hemolytic anemia and immune complex disease and cell-mediated immunity may give rise to skin rashes and eruptions, and the most common reactions are urticaria, skin eruptions, and arthralgia. Anti-penicillin IgE antibodies have been detected consistently with an anaphylactic reaction. The anaphylactic reactions (type 1; see above), which occur in 0.004% to 0.015% of patients, may be life threatening.

Pencillin is a reactive molecule both *in vitro and in vivo*, where it undergoes a slow transformation to a variety of products. The parent drug can react with protein, but some of the breakdown products (e.g., penicillenic acid) are more reactive than the parent drug and can

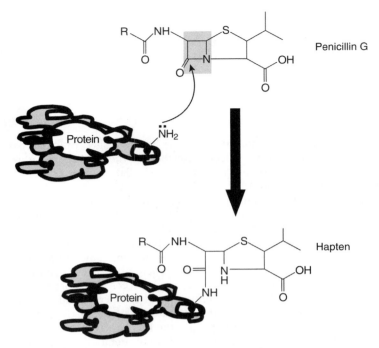

Figure 7.80 The reaction of protein with the penicillin molecule (hapten) to form a possibly antigenic conjugate. Breakdown products may also react with protein. The protein might be either soluble or cellular. For penicillin G, R=benzyl. The strained β-lactam ring is indicated.

bind covalently to protein, reacting with nucleophilic amino, hydroxyl, mercapto, and histidine groups. Therefore, a number of antigenic determinants may be formed from a single penicillin derivative, which may vary between individuals and cross-reactivity with other penicillins may also occur. Also, IgM and IgG can act as blocking antibodies toward IgE, in which case the severe anaphylactic reactions do not occur. This makes it difficult to predict the outcome of penicillin allergy.

The mechanism of penicillin immunotoxicity relies on the formation of a covalent conjugate with soluble or cellular protein. There is a clear relationship between conjugation to proteins and cells and the ability to cause hypersensitivity in humans. However, the immunogen(s) is still not known with certainty. There may be many different protein-drug conjugates with various breakdown products of penicillin, bound to varying extents. Thus, the conjugate may have many drug molecules bound per protein molecule. The conjugate most commonly formed, a benzyl penicilloyl derivative, involves formation of an α-amide derivative with the ε-amino group of the amino acid lysine via the strained β-lactam ring in the penicillin molecule (Fig. 7.80). This may also occur after reaction of penicillenic acid with protein.

The thiazolidine ring may also break open and give rise to conjugates through the sulfur atom. However, it seems that the penicilloyl derivative is the major antigenic determinant, and the disulfide-linked conjugate is the minor antigenic determinant. Thus, benzylpenicillenic acid is 40 times more reactive toward cellular proteins than benzylpenicillin itself and is highly immunogenic in the rat. This may be the ultimate immunogen. The importance of the benzyl penicilloyl amide conjugate was established using the technique of hapten inhibition where the penicillin derivative combines with the antibody and therefore inhibits the antibody-antigen reaction. This is normally visualized experimentally as a precipitation or agglutination of the antibody-antigen complex. Formation of penicilloic acid results in loss of immunogenicity. Not all of the drug-protein conjugates are immunogenic, however. For example, small amounts of penicillin are irreversibly bound to serum proteins such as albumin, but the penicilloyl groups do not seem to be accessible to antibodies. One of the immune responses leads to hemolytic anemia, and the mechanism underlying this seems reasonably well understood. This is shown in the Figure 7.81.

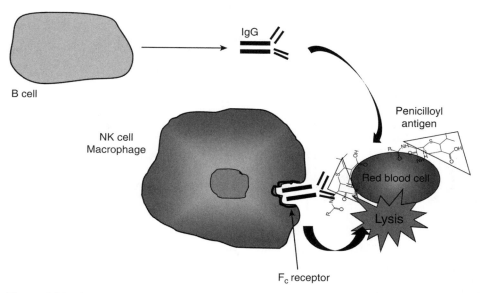

Figure 7.81   Antibody-dependent phagocyte-mediated lysis of red cells in penicillin immunotoxicity (type II hypersensitivity). *Abbreviation*: NK, natural killer cell.

Type II hypersensitivity occurs after high doses of penicillin, which lead to adducts on red cells (Fig. 7.81). These are recognized by the immune system via APC and T-helper cells to activate B cells to produce and retain a memory of the antigen. Subsequent exposure and appearance of the antigen on red cells once recognized is followed by the rapid production of antibodies of the IgG or IgM type by B cells from the original template. These are produced with a memory of the hapten and so recognize and bind to the hapten on the red cell. Antibodies have an F_c portion, which can be recognized by the F_c receptor on natural killer cells (NK) or macrophages. These phagocytic cells bind to the antibody-labeled red cells, release cytotoxic mediators, and so lyse the red cell (Fig. 7.81). This can lead to the destruction of significant numbers of red cells and therefore **hemolytic anemia**. This is a type II immune response but is antibody-dependent cell mediated. It is not unlike the immune response after halothane.

Again, the reasons why only some patients show this response are unclear. All patients on high doses of penicillin will have penicillin-derived adducts to their red cells, but only a small proportion will respond with anti-penicilloyl IgG antibodies. The majority are tolerant.

7.11.4 Hydralazine
The drug hydralazine is a vasodilator used for the treatment of hypertension. In a significant proportion of individuals, it causes a serious adverse effect, drug-induced **systemic lupus erythematosus** (SLE). This is a *systemic* kind of toxic effect, as there is no particular target organ or tissue. It is an example of immune-mediated toxicity that involves autoimmunity and shows a number of interesting features.

Thus, there are a number of predisposing factors that make it possible to identify patients at risk, although the mechanism(s) underlying some of these predisposing factors are unknown. When the occurrence in the exposed population is examined in the light of some of these factors, the incidence can be seen to be extremely high. The predisposing factors identified to date are

1.   dose,
2.   duration of therapy,
3.   acetylator phenotype,
4.   HLA type, and
5.   sex.

Thus, if the exposed population is divided into males and females and by the various doses administered, the incidence in susceptible populations can be appreciated (Table 7.9).

Table 7.9 Incidence of Hydralazine-Induced Systemic Lupus Erythematosus

Patients	Dose	Incidence over 3 yr
All patients	All doses	6.7%
	25	0%
	50	5.4%
	100	10.4%
Men	All doses	2.8%
	25 or 50	0%
	100	4.9%
Women	All doses	11.6%
	25 or 50	5.5%
	100	19.4%

Source: From Ref. 22.

Thus, the highest incidence recorded in one study was over 19% in the most susceptible population—females taking 100 mg of hydralazine twice daily.

We examine each of these factors in turn.

Dose
There is an increase in the incidence of hydralazine-induced lupus erythematosus (LE) in the exposed population with increasing dose as can be seen in Table 7.9. However, patients who develop LE do not have a significantly different cumulative intake of hydralazine from those patients who do not develop the syndrome. This latter observation is consistent with the absence of a clear dose-response relationship in many cases of toxicity with an immunological basis.

Duration of Therapy
The adverse effect develops slowly, typically over many months, and there is a mean development time of around 18 months.

Acetylator Phenotype
The LE syndrome only develops in those patients with the slow acetylator phenotype. Metabolic studies have shown that the metabolism of hydralazine involves an acetylation step (Fig. 7.82), which is influenced by the acetylator phenotype.

HLA Type
The tissue type seems to be a factor, as there is a preponderance of the **HLA-type DR4** in patients who develop the syndrome. Thus, in one study, an incidence of 73% for this HLA type was observed in patients suffering from the LE syndrome compared with an incidence of 33% in controls and 25% in patients not developing the LE syndrome.

Sex
Females seem to be more susceptible to hydralazine-induced LE than males. The ratio may be as high as 4:1.

Hydralazine-induced LE causes inflammation in various organs and tissues, giving rise to a number of different symptoms. Thus, patients suffer from arthralgia and myalgia, skin rashes, and sometimes vasculitis. There may also be hepatomegaly, splenomegaly, anemia, and leucopenia. Two particular characteristics detectable in the blood are antinuclear antibodies and LE cells. **Antinuclear antibodies** (ANA) are directed against single-stranded DNA and deoxyribonucleoprotein such as histones and occur in at least 27% of patients taking the drug, but not all of these develop the LE syndrome. Indeed, in one study of over three years, 50% of patients had an ANA titer of 1:20 or more. Antibodies directed against other

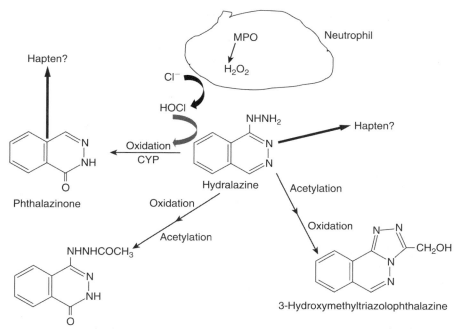

Figure 7.82 Some of the major routes of metabolism of hydralazine. As well as cyp-mediated oxidative metabolism, myeloperoxidase (MPO) in neutrophils will also oxidize the drug as shown.

cellular and tissue constituents such as DNA and immunoglobulins are also present, and antibodies against synthetic hydralazine-protein conjugates have also been detected.

There are thus various autoantibodies present, and if the auto-antigens are released by cellular breakdown, a **type III immune reaction** can occur where an immune complex is formed, which is deposited in small blood vessels and joints, giving rise to many of the symptoms. The immunoglobulins IgG and IgE act as both autoantibody and antigen, and hence immune complexes form. Such complexes stimulate the complement system leading to inflammation, infiltration by polymorphs and macrophages, and the release of lysosomal enzymes.

LE cells are neutrophil polymorphs, which have phagocytosed the basophilic nuclear material of leucocytes, which has been altered by interaction with antinuclear antibodies. The development of ANA requires a lower intake of hydralazine and occurs more quickly in slow acetylators than in rapid acetylators, and rapid acetylators have significantly lower titers of ANA than slow acetylators. There is also a significant correlation between the cumulative dose of hydralazine and the development of ANA, but as indicated above, patients who develop LE do not have a significantly different cumulative intake of hydralazine from those patients who do not develop the syndrome.

The mechanism of hydralazine-induced LE is not currently understood, but the evidence available indicates that it has an immunological basis. Hydralazine is a chemically reactive molecule, and it is also metabolized to reactive metabolites, possibly free radicals, by the cytochromes P-450 system (Fig. 7.83), which bind covalently to protein. The production of the metabolite phthalazinone correlates with the binding *in vitro*. However, no antibodies against such conjugates with human microsomal protein were detected in the sera of patients with the LE syndrome. Hydralazine may also be a substrate for the benzylamine oxidase system found in vascular tissue, and for a peroxidase-mediated metabolic activation system, which occurs in cells such as activated leucocytes. Thus, a **myeloperoxidase**/H_2O_2/Cl-system will metabolize hydralazine to phthalazinone and phthalazine, and this may also involve the production of reactive intermediates (Fig. 7.82). This system has also been suggested to be involved in the activation of the drug procainamide, which similarly causes an LE syndrome. Metabolic studies have shown that slow acetylators excrete more unchanged hydralazine and metabolize more via oxidative pathways (Fig. 7.82). Patients with the LE syndrome excrete more

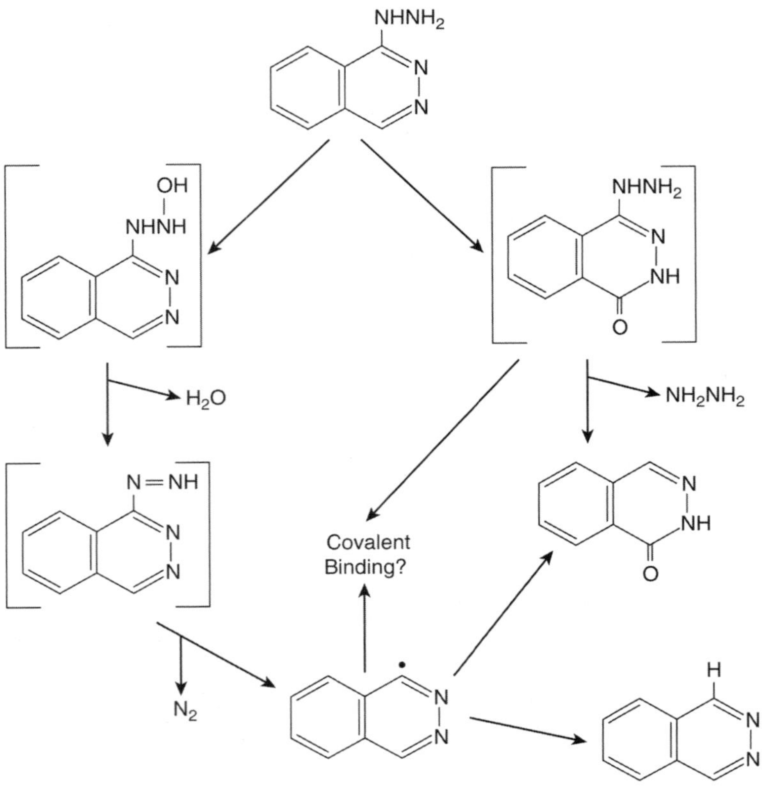

Figure 7.83 Possible routes for the metabolic activation of hydralazine. The oxidation of the hydrazine group may also involve the formation of a nitrogen-centered radical, which could also give rise to phthalazine with loss of nitrogen.

phthalazinone than control patients, although this is not statistically significant. There is no difference between males and females in the nature or quantities of urinary metabolites of hydralazine detected however, and so metabolic differences do not currently explain the sex difference in susceptibility. However, there is clearly scope for the formation of protein-drug conjugates, which may be antigenic, and hydralazine also reacts with DNA. Synthetic hydralazine-protein conjugates will stimulate the production of antibodies in rabbits, and antibodies in human sera from patients with the LE syndrome will recognize and agglutinate rabbit red blood cells to which hydralazine has been chemically attached. Hydralazine will abolish this reaction *in vitro*, indicating that it is the hapten or is similar to it.

Thus, interaction of hydralazine or a metabolite with macromolecules may underlie the immune response. An alternative or additional hypothesis involves inhibition of the complement system. The complement system helps remove immune complexes by solubilization, but if it is inhibited, deposition and accumulation of such complexes would be increased. Hydralazine and some of its metabolites interfere with part of the complement system, inhibiting the covalent binding of complement C4 by reaction with the thioester of activated C4. However, the concentrations required are highly relative to the normal therapeutic concentration. More recently, it has been shown that hydralazine inhibits DNA methylation in the T cell. The inhibition of DNA methyl transferase may initiate immune reactions via activation of genes as a result of this interference with DNA methylation. There is also another possibility suggested by analogy with the drug **procainamide**, which also causes SLE and has many similarities with hydralazine. This putative mechanism involves interference by a metabolite or the parent drug, with the development and maturation of T cells in the thymus. Here, the T-lymphocyte precursors develop into mature T lymphocytes and are selected for **tolerance** to self-peptides. The non-tolerant cells, which could react with "self"-antigens, are removed by apoptosis. This then guarantees a level of tolerance in the system. If a drug or metabolite interferes with this development of nonresponsiveness (as is the case with procainamide), non-tolerant lymphocytes may develop. Hence, antibodies to self-proteins and macromolecules such as nuclear material could occur as are indeed detected. These autoantibodies could in turn lead to an autoimmune reaction.

However, although the mechanism of hydralazine-induced LE is not yet understood, it is an important example of drug-induced toxicity for two reasons:

1. It illustrates the role and possibly the requirement for various predisposing factors in the development of an adverse drug reaction in a human population. An understanding of this should allow reduction of such adverse drug reactions by improved surveillance and prescription.
2. It reveals the difficulties of testing for this type of reaction in experimental animals when the various predisposing factors may not be present. However, the LE syndrome does occur in certain strains of mice and acetylation rates do vary between strains of laboratory animals. Using such specific models might therefore allow improved prediction.

7.12 MULTI-ORGAN TOXICITY

7.12.1 Ethylene Glycol

This substance is a liquid used in antifreeze, paints, polishes, and cosmetics. As it has a sweet taste and is readily available it has been used as a poor man's alcohol, but it may also be ingested accidentally and for suicidal purposes. Diethylene glycol was once used as a vehicle for the drug sulfanilamide, and when used for this, it caused some 76 deaths.

The minimum lethal dose of ethylene glycol is about 100 mL, and after ingestion, death may occur within 24 hours from damage to the CNS or more slowly (8–12 days) from renal failure.

There seem to be three recognizable clinical stages:

1. Within 30 minutes and lasting for perhaps 12 hours, there is intoxication, nausea, vomiting, coma, convulsions, nystagmus, papilledema, depressed reflexes, myo-clonic jerks, and tetanic contractions. Permanent optic atrophy may occur.
2. Between 12 and 24 hours, there is tachypnea, tachycardia, hypertension, pulmonary edema, and congestive cardiac failure.
3. Between 24 and 72 hours, the kidneys become damaged, giving rise to flank pain and acute renal tubular necrosis.

The clinical biochemical features reflect the biochemical and physiological effects. Thus, there is reduced plasma bicarbonate, low plasma calcium, and raised potassium. Crystals, blood, and protein may all be detected in the urine (crystalluria, hematuria, and proteinuria, respectively), and the urine may have a low specific gravity.

The mechanism of toxicity of ethylene glycol involves metabolism, but unlike previous examples, this does not involve metabolic activation to a reactive metabolite. Thus, ethylene glycol is metabolized by several oxidation steps eventually to yield oxalic acid (Fig. 7.84). The first step is catalyzed by the enzyme **alcohol dehydrogenase**, and herein lies the key to treatment of poisoning. The result of each of the metabolic steps is the production of NADH. The imbalance in the level of this in the body is adjusted by oxidation to NAD coupled to the production of lactate. There is thus an increase in the level of lactate, and lactic acidosis may result. Also, the intermediate metabolites of ethylene glycol have metabolic effects such as the inhibition of oxidative phosphorylation, glucose metabolism, Krebs' cycle, protein synthesis, RNA synthesis, and DNA replication.

The consequences of this are as follows:

1. acidosis due to lactate, oxalate, and the other acidic metabolites; this results in metabolic distress and physiological changes;
2. loss of calcium as calcium oxalate;
3. deposition of crystals of calcium oxalate in the renal tubules and brain;

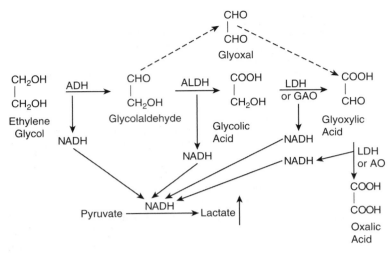

Figure 7.84 The metabolism of ethylene glycol. The NADH produced is used in the production of lactate, the level of which will rise dramatically in poisoning cases. *Abbreviations*: ADH, alcohol dehydrogenase; ALDH, aldehyde dehydrogenase; LDH, lactate dehydrogenase; GAO, glycolic acid oxidase; AO, aldehyde oxidase.

4. inhibition of various metabolic pathways leading to accumulation of organic acids;
5. impairment of cerebral function by oxalate and damage by crystals; also some of the aldehyde metabolites may impair cerebral function; and
6. damage to renal tubules by oxalate crystals, leading to necrosis.

Thus, the pathological damage includes cerebral edema, hemorrhage, and deposition of calcium oxalate crystals. The lungs show edema, and occasionally calcium oxalate crystals and degenerative myocardial changes may also occur. There is degeneration of proximal tubular epithelium, with calcium oxalate crystals and fat droplets detectable in tubular epithelial cells. The degeneration of distal tubules may also be seen.

Ethylene glycol is more toxic to humans than animals, and in general, the susceptible species are those that metabolize the compound to **oxalic acid**, although this is quantitatively a minor route. The treatment of poisoning with ethylene glycol reflects the mechanism and biochemical effects. Thus, after standard procedures such as gastric lavage to reduce absorption and supportive therapy for shock and respiratory distress, patients are treated with the following:

1. Ethanol; this competes with ethylene glycol for alcohol dehydrogenase, but as it is a better substrate, the first step in ethylene glycol metabolism is blocked—animal studies have shown that this doubles the LD_{50}.
2. Intravenous sodium bicarbonate; this corrects the acidosis—animal studies have shown that this increases the LD_{50} by around four times.
3. Calcium gluconate; this corrects the hypocalcemia.
4. Dialysis; this removes ethylene glycol.

Thus, the treatment of poisoning with ethylene glycol is a logical result of understanding the biochemistry of the toxicity.

7.12.2 Methanol

Methanol is widely used as a solvent and as a denaturing agent for ethanol and is also found in antifreeze. Mass poisonings have occurred because of ingestion in alcoholic drinks made with contaminated ethanol as well as from accidental exposure. Inhalation and skin absorption may cause toxicity. In humans, about 10 mL can cause blindness and 30 mL is potentially fatal, but there is variation in the lethal dose.

The clinical features are an initial mild inebriation and drowsiness, followed, after a delay of 8 to 36 hours, by nausea, vomiting, abdominal pain, dizziness, headaches, and

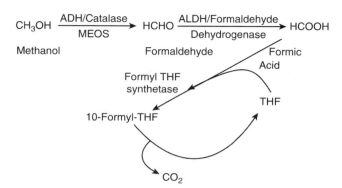

Figure 7.85 The metabolism of methanol. *Abbreviations*: ADH, alcohol dehydrogenase; MEOS, microsomal ethanol-oxidizing system; ALDH, aldehyde dehydrogenase; THF, tetrahydrofolate.

possibly coma. Visual disturbances may start within six hours and include blurred vision, diminished visual acuity, and dilated pupils, which are unreactive to light. These effects indicate that blindness may occur, and changes to the optic disc can be seen with an opthalmoscope. Severe depression of the CNS may occur at later stages because of accumulation of metabolites.

The mechanism underlying methanol poisoning again involves metabolism, with **alcohol dehydrogenase** catalyzing the first step in the pathway (Fig. 7.85), although the involvement of other enzymes such as catalase and a **microsomal ethanol oxidation system** have been proposed in some species. The second step, oxidation of formaldehyde to formic acid, is catalyzed by several enzymes: hepatic and erythrocyte aldehyde dehydrogenase; the folate pathway; and formaldehyde dehydrogenase (Fig. 7.85). The products of metabolism, formaldehyde and formic acid, are both toxic. However, formaldehyde has a very short half-life (about 1 minute), whereas formic acid accumulates. Formic acid is further metabolized to carbon dioxide or other products in a reaction that *in vivo* involves combination with **tetrahydrofolate** in a reaction catalyzed by formyl tetrahydrofolate synthetase (Fig. 7.85). It seems that the hepatic concentration of tetrahydrofolate regulates the rate of metabolism and hence removal of formic acid and that this may be a factor in the accumulation of formate. Those species that are less sensitive to methanol poisoning, such as the rat, remove formate more rapidly than humans and other sensitive primates.

The result of formate accumulation is **metabolic acidosis.** However, at later stages, the acidosis may also involve the accumulation of other anions such as lactate. This may be a result of inhibition of cytochrome oxidase and hence of mitochondrial respiration, tissue hypoxia due to reduced circulation of blood, or an increase in the NADH/NAD ratio. The acidosis that results from methanol poisoning will result in more formic acid being in the nonionized state and hence more readily able to enter the CNS. This will cause central depression and hypotension and increased lactate production. This situation is known as the "**circulus hypoxicus.**"

The other major toxic effect of methanol is the ocular toxicity. Although formaldehyde might be formed locally in the retina, this seems unlikely, whereas formate is known to cause experimental ocular toxicity. The mechanism suggested involves inhibition by formate of cytochrome oxidase in the optic nerve. As the optic nerve cells have few mitochondria, they are very susceptible to this "**histotoxic hypoxia.**"

The inhibition will result in a decrease in ATP and hence disruption of optic nerve function. Thus, stasis of axoplasmic flow, axonal swelling, optic disc edema, and loss of function occur. Studies have shown that formate alone will cause toxicity in the absence of acidosis, although this will exacerbate the toxicity. The liver, kidney, and heart may show pathological changes, and pulmonary edema and alveolar epithelial damage can occur. In severe cases of methanol poisoning, death may occur because of respiratory and cardiac arrest. The treatment of methanol poisoning involves firstly the administration of an antidote, ethanol, which blocks metabolism. Ethanol competes with methanol for alcohol dehydrogenase, as the enzyme has a greater affinity for ethanol. Methanol metabolism can be reduced by as much as 90% by an equimolar dose of ethanol, and the half-life becomes extended to 46 hours. **4-Methylpyrazole**, which also binds to alcohol dehydrogenase, has been used successfully in monkeys to treat methanol poisoning, as has folic acid.

Secondly, i.v. sodium bicarbonate is given for correction of the metabolic acidosis. Hemodialysis may be used in very serious cases.

7.13 MULTI-ORGAN TOXICITY: METALS

7.13.1 Cadmium

Cadmium is a metal that is widely used in industry in alloys, in plating, and in batteries and in the pigments used in inks, paints, plastic, rubber, and enamel. It is an extremely toxic substance, and the major hazard is from inhalation of cadmium metal or cadmium oxide. Although it is present in food, significant oral ingestion is rare, and absorption from the gut is poor (5–8%). However, various dietary and other factors may enhance absorption from the gastrointestinal tract. In contrast, up to 40% of an inhaled dose may be absorbed, and hence its presence in cigarettes is a significant source of exposure.

Cadmium is bound to proteins and red blood cells in blood and transported in this form, but 50% to 75% of the body burden is located in the liver and kidneys. The half-life of cadmium in the body is between 7 and 30 years, and it is excreted through the kidneys, particularly after they become damaged.

Cadium has many toxic effects, primarily causing kidney damage, as a result of chronic exposure, and testicular damage after acute exposure, although the latter does not seem to be a common feature in humans after occupational exposure to the metal. It is also hepatotoxic and affects vascular tissue and bone. After acute inhalation exposure, lung irritation and damage may occur along with other symptoms such as diarrhea and malaise. Chronic inhalation exposure can result in progressive fibrosis of the lower airways, leading to emphysema. This results from necrosis of alveolar macrophages and hence release of degradative enzymes, which damage the basement membranes of the alveolus. These lung lesions may occur before kidney damage is observed. Cadmium can also cause disorders of calcium metabolism, and the subsequent loss of calcium from the body leads to osteomalacia and brittle bones. In Japan, this became known as **Itai-Itai** ("Ouch-Ouch!") disease when it occurred in women eating rice contaminated with cadmium. The raised urinary levels of **proline** and **hydroxyproline** associated with chronic cadmium toxicity may be due to this damage to the bones.

Kidney damage is a delayed effect even after single doses, being due to the accumulation of cadmium in the kidney, as a complex with the protein metallothionein. **Metallothionein** is a low molecular weight protein (6500 Da) containing about 30% cysteine, which is involved with the transport of metals, such as zinc, within the body. Because of its chemical similarity to zinc, cadmium exposure induces the production of this protein and 80% to 90% of cadmium is bound to it *in vivo*, probably through SH groups on the protein. Cadmium induces metallothionein mRNA so that synthesis of the protein is increased. Thus, exposure to repeated small doses of cadmium will prevent the toxicity of large acute doses by increasing the amount of metallothionein available. The protein is thus serving a protective function. The cadmium-metallothionein complex is synthesized in the liver and transported to the kidney, filtered through the glomerulus, and reabsorbed by the proximal tubular cells, possibly by endocytosis. Within these cells, the complex is taken up into lysosomes and degraded by proteases to release cadmium, which may damage the cells or recombine with more metallothionein. The cellular damage caused by cadmium may be at least partly a result of its ability to bind to the sulfydryl groups of critical proteins and enzymes. Cadmium causes oxidative stress via the production of ROS, and in turn, this causes various effects including DNA damage.

This may be due to the interference with the mitochondrial electron transport chain. Thus, cadmium binds to complex III at the Q_0 site between semi-ubiquinone and heme b_{566}. This stops delivery of electrons to the heme and allows accumulation of semi-ubiquinone, which in turn transfers the electrons to oxygen and produces superoxide.

However, *in vitro* studies in hepatoma cells showed that lysosmal damage precedes the DNA and mitochondrial damage.

There seems to be a critical level of cadmium in the kidney when the kidney metallothionein is saturated and the free cadmium causes toxicity. The damage to the kidney occurs in the first and second segments of the proximal tubule. This can be detected biochemically as glucose, amino acids, and protein in urine. The proteins are predominantly of low molecular weight, such as β_2-microglobulin, which are not reabsorbed by the proximal tubules damaged by cadmium. Larger proteins in the urine indicate glomerular damage. The vasoconstriction caused by cadmium may affect renal function, and cadmium may cause fibrotic degeneration of renal blood vessels. The tubular cells degenerate, and interstitial fibrosis can occur. The proximal tubular damage can progress in chronic cadmium toxicity to distal tubular dysfunction, loss of calcium in urine giving rise to renal calculi and osteomalacia.

The binding of cadmium to metallothionein decreases toxicity to the testes but increases the nephrotoxicity, possibly because the complex is preferentially, and more easily, taken up by the kidney than the free metal. Dosing animals with the cadmium–metallothionein complex leads to acute kidney damage, whereas exposure to single doses of cadmium itself does not.

The **testicular damage** occurs within a few hours of a single exposure to cadmium and results in necrosis, degeneration, and complete loss of spermatozoa. The mechanism involves an effect on the vasculature of the testis. Cadmium reduces blood flow through the testis, and ischemic necrosis results from the lack of oxygen and nutrients reaching the tissue. In this case, cadmium is probably acting mainly indirectly by affecting a physiological parameter. However, pretreatment of animals with zinc reduces the testicular toxicity of cadmium by inducing the synthesis of metallothionein and hence reducing the free cadmium level.

The vasoconstriction, which is caused by cadmium, may underlie the hypertension observed in experimental animals. Cadmium is also carcinogenic in experimental animals, causing tumors at the site of exposure. Also, Leydig cell tumors occur in the testis of animals after acute doses of cadmium sufficient to cause testicular necrosis. This seems to be an indirect effect due to the reduced level of testosterone in the blood, which follows testicular damage. This causes Leydig cell hyperplasia and tumors to occur.

Thus, cadmium causes multi-organ toxicity, and at least some of the toxic effects are due to it being a divalent metal similar to zinc and able to bind to sulfydryl groups.

7.13.2 Mercury

Mercury can exist in three forms, elemental, inorganic, and organic, and all are toxic. However, the toxicity of the three forms of mercury are different, mainly as a result of differences in distribution. Some of these toxic properties have been known for centuries.

Elemental mercury (Hg^0) may be absorbed by biological systems as a vapor. Despite being a liquid metal, mercury readily vaporizes at room temperature and in this form constitutes a particular hazard to those who use scientific instruments containing it, for example.

Elemental mercury vapor is relatively lipid soluble and is readily absorbed from the lungs following inhalation and is oxidized in the red blood cells to Hg^{2+}. Elemental mercury may also be transported from red blood cells to other tissues such as the CNS. Elemental mercury readily passes across the **blood-brain barrier** into the CNS and also into the fetus. The metallic compound is only poorly absorbed from the gastrointestinal tract, however.

Inorganic mercury, existing as monovalent (mercurous) or divalent (mercuric) ions is relatively poorly absorbed from the gastrointestinal tract (7% in humans). After absorption, inorganic mercury accumulates in the kidney. Organic mercury is the most readily absorbed (90–95% from the gastrointestinal tract), and after absorption, distributes especially to the brain, particularly the posterior cortex. All the forms of mercury will cross the placenta and gain access to the fetus, although elemental mercury and organic mercury show greater uptake. The concentrations in certain fetal tissues, such as red blood cells, are greater than in maternal tissue.

Mercury is eliminated from the body in the urine and feces, with the latter being the major route. Thus, with methyl mercury, 90% is excreted into the feces. Methyl mercury is secreted into the bile as a cysteine conjugate and undergoes extensive **enterohepatic recirculation.**

The half-life of mercury is long, but there are two phases, the first being around 2 days, then the terminal phase, which is around 20 days. However, the half-life will depend on the form of mercury. Thus, methyl mercury has a half-life of about 70 days, whereas for inorganic mercury, this is about 40 days.

Organic mercury compounds, especially **phenyl** and **methoxyethyl mercury** may also be biotransformed into inorganic mercury by cleavage of the carbon-mercury bond. Although such compounds are more readily absorbed than inorganic mercury compounds, the toxicity is similar.

Elemental mercury is oxidized *in vivo* to inorganic mercury, a biotransformation that is probably catalyzed by catalase. It is selectively accumulated in the kidney and also by lysosomes. Inorganic mercury (Hg^{2+}) will induce the synthesis of metallothionein. Mercury binds to cellular components such as enzymes in various organelles, especially to proteins containing sulfydryl groups. Thus, in the liver, cysteine and GSH will react with mercury to produce soluble products, which can be secreted into the bile or blood.

Toxic Effects

Elemental Mercury Vapor. Although there may be toxic effects to the respiratory system from the inhalation of mercury vapor, the major toxic effect is to the CNS. This is especially true after chronic exposure. There are a variety of symptoms such as muscle tremors, personality changes, delirium, hallucination, and gingivitis.

Inorganic Mercury. Mercuric chloride and other mercuric salts will, when ingested orally, cause immediate acute damage to the gastrointestinal tract. This may be manifested as bloody diarrhea, ulceration, and necrosis of the tract. After 24 hours, renal failure occurs, which results from necrosis of the pars recta region of the proximal tubular epithelial cells. The epithelial cells show damage to the plasma membrane, endoplasmic reticulum, mitochondria, and effects on the nucleus. The result of this damage is excretion of glucose (glycosuria), amino acids (aminoaciduria), appearance of proteins in the urine (proteinuria), and changes in various metabolites excreted into urine. After an initial diuresis, there is a reduction in urine (oliguria), possibly developing into complete lack of urine (anuria). The effect on renal function can also be detected by determination of blood urea (BUN), which will be elevated in renal failure.

Chronic low-level exposure to inorganic mercury may lead to a glomerular disease, which has an immunologic basis. This type of nephropathy is accompanied by proteinuria and may involve glomerular damage due to immune complexes. Also, chronic exposure can give rise to salivation and gingivitis and erethism, which involves psychological effects such as nervousness and shyness.

Mercurous salts are less toxic than mercuric salts, probably as a result of lower solubility. Exposure of human subjects to mercurous chloride (calomel) may result in hypersensitivity reactions.

Organic Mercury. Mercury in this form, such as methyl mercury, is extremely toxic, mainly affecting the CNS. However, some organomercury compounds such as phenyl and methoxyethyl mercury cause similar toxic effects to inorganic mercury. There have been a number of instances in which human exposure to methyl mercury has occurred, and consequently, data are available on the toxic effects to humans as well as experimental animals. **Methyl mercury** was responsible for the poisoning that occurred in Japan, known as **Minamata disease**. This resulted from industrial effluent containing inorganic mercury contaminating the water of Minamata Bay in Japan. The microorganisms in the sediments at the bottom of the bay biotransformed the inorganic mercury ions into methyl and dimethyl mercury. As this form of mercury is lipid soluble, it was able to enter the **food chain** and so become concentrated in fish as a result of their eating small organisms that had absorbed the methyl mercury. The local population who consumed the fish therefore became contaminated with methyl mercury. Another episode occurred in Iraq when seed grain treated with a methyl mercury fungicide was used to make bread. Over 6000 people were recorded as exposed and more than 500 died. The major features of methyl mercury poisoning are paresthesia, ataxia, dysarthria, and deafness. There is a clear dose-response relationship for each of these toxic effects in exposed humans, which has been derived from the poisoning episode in Iraq. Thus, there is a linear relationship between body burden and the frequency of cases in the exposed population. However, the occurrence of each of these toxic effects shows a different profile and a different threshold (Fig. 7.86). The pathology involves degeneration and necrosis of nerve cells in the cerebral cortex, and particularly those areas dealing with vision. The blood-brain barrier is also disrupted.

Exposure of pregnant women to methyl mercury caused cerebral palsy and mental retardation in the offspring, despite lack of symptoms in the mothers.

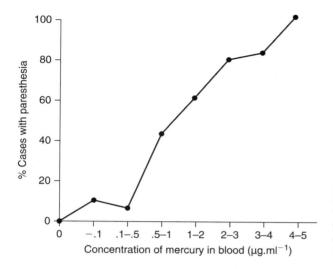

Figure 7.86 The dose-response relationship for methyl mercury in exposed humans. The concentration of mercury in the blood and the incidence of paresthesias are used as dose and response, respectively. *Source*: From Ref. 23.

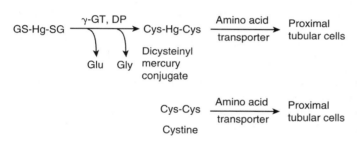

Figure 7.87 The formation of dicysteinyl mercury conjugates from the diglutathionyl conjugate and comparison with cystine, the natural substrate for the transporter. *Abbreviations*: γGT, γ glutamyl transferase; DP, dipeptidase.

Mechanism

Mercury is a reactive element, and its toxicity is probably due to interaction with proteins. Mercury has a particular affinity for sulfydryl groups in proteins and consequently is an inhibitor of various enzymes such as membrane ATPase, which are sulfydryl dependent. It can also react with amino, phosphoryl, and carboxyl groups. Brain pyruvate metabolism is known to be inhibited by mercury, as are lactate dehydrogenase and fatty acid synthetase. The accumulation of mercury in lysosomes increases the activity of lysosomal acid phosphatase, which may be a cause of toxicity, as lysosomal damage releases various hydrolytic enzymes into the cell, which can then cause cellular damage. The kidney is a target organ because mercury accumulates in the kidney. Indeed the renal uptake of Hg is very rapid despite the fact that Hg is bound to proteins and peptides in blood via SH groups. Uptake into the kidney therefore probably involves transporters. Uptake of mercury conjugates from the blood through the basolateral membrane of the proximal tubular cells is believed to involve an OAT. Uptake of diglutathionyl conjugates (Fig. 7.87) from the tubular lumen filtered out of the bloodstream involves apical enzymes and a transporter. Thus, first γ-glutamyltranspeptidase and then dipeptidase cleave off the glutamyl and glycinyl moieties and then transport of the cysteine conjugate is affected by one or more amino acid transporters. The cysteine conjugate is an analogue of the amino acid cystine, which is carried by the transporter (Fig. 7.87). Hence, the mercury conjugate mimics the naturally occurring amino acid.

Mitochondrial dysfunction is thought to be an early event in cell damage in the proximal tubules of the kidney caused by inorganic mercury. Oxidative stress is believed to play a part in this damage to the mitochondria, as there is evidence of lipid peroxidation. As mercury binds avidly to SH groups, the binding to critical protein SH groups is probably a major part of the toxicity. Indeed, mitochondrial enzymes are known to be inhibited by mercury. Calcium levels also rise in proximal tubular cells exposed to $HgCl_2$, and this will initiate mitochondrial and other cellular changes.

These effects on the mitochondria will lead to a reduction of respiratory control in the renal cells, and their functions, such as solute reabsorption, will be compromised.

The mechanism underlying the neurotoxicity of organic mercury is similar. Thus, methyl mercury can be conjugated with GSH and degraded to the cysteine conjugate in the gut after biliary excretion followed by reabsorption. Alternatively, the cysteine conjugate could be formed directly. The methyl mercury cysteine conjugate is lipophilic but is also an analogue of the amino acid methionine and is therefore a substrate for a specific transporter. Consequently, it is readily taken up into the astrocytes in the brain where interaction with critical thiol groups in proteins leads to toxicity and damage to certain neurons.

Mercury poisoning is usually treated with chelating agents such as **dimercaprol** or **penicillamine**, or hemodialysis may be used in severe cases. These help to decrease the body burden of mercury. However, chelating agents are not very effective after alkyl mercury exposure.

7.13.3 Lead

Lead has been known to be a poisonous compound for centuries and indeed was described as such in 300 BC. Consequently, many workers involved in lead mining and smelting and the preparation of lead-containing products such as paint have been occupationally exposed to the metal. It has even been suggested that lead poisoning may have been one of the causes of the fall of the Roman empire, probably resulting from the use of lead utensils for eating and, especially, drinking liquids, which would leach the lead from the vessel. Thus, high lead levels have been detected in the skeletons of Romans dating from the period.

Exposure to lead can occur in a variety of ways, via food and water and inhaled through the lungs, but for the general population, currently, the most important exposure is inhalation of airborne lead, which mainly derives from the combustion of leaded petrol, which contains the organic lead additive **tetraethyl lead**. Although the amount of lead in food may be greater than that in the air, absorption from the lungs is greater. Other sources in the environment are lead smelters, batteries, paints, lead water pipes, and insecticides such as lead arsenate. Paint used to be a significant source of lead, and lead poisoning in children, even recently, may have been due to flakes of old lead paint still being present in slum housing areas. During the time of Prohibition in the United States earlier this century, those drinking illicit Moonshine whisky suffered renal damage as a result of lead poisoning, which derived from the solder used to construct the stills. Lead poisoning has also resulted from the use of medicinal agents from certain countries. Industrial poisoning with lead became common in the industrial revolution with 1000 cases per year occurring in the United Kingdom alone at the turn of century, and exposure to lead still occurs in industry. Exposure to lead may be to the metal, lead salts, and organic lead.

Lead causes damage to a variety of organs and also causes significant biochemical effects. Thus, the kidneys, testes, bones, gastrointestinal tract, and the nervous system are all damaged by lead. The major biochemical effect is interference with heme synthesis giving rise to anemia.

Acute exposure to inorganic lead causes renal damage, in particular, damage to the proximal tubules. This is detectable biochemically as amino aciduria and glycosuria. Lead adversely affects **reproductive function** in both males and females, and recent studies in men occupationally exposed to lead have indicated that testicular function is adversely affected by lead. Animal studies have shown that lead is gametotoxic. Lead is also neurotoxic to the developing nervous system.

After absorption, lead enters the blood, and 97% is taken up by red blood cells. Here, lead has a half-life of two to three weeks during which there is some redistribution to the liver and kidney, then excretion into bile or deposition in **bone**. After an initial, reversible, uptake into bone, lead in bone becomes incorporated into the hydroxyapatite crystalline structure. Because of this, past exposure to lead is possible to quantitate using X-ray analysis. It is also possible to detect lead exposure and possible poisoning from urine and blood analysis, and the amount in blood represents current exposure. However, as lead is taken up into the red blood cell, both the free blood lead level and that in the erythrocytes needs to be known.

"Normal" blood levels in adults not occupationally exposed, in the United States, are in the region of 0.15 to 0.7 mg mL^{-1} with the average at 0.3 mg mL^{-1}. The threshold for toxicity is 0.8 mg mL^{-1}.

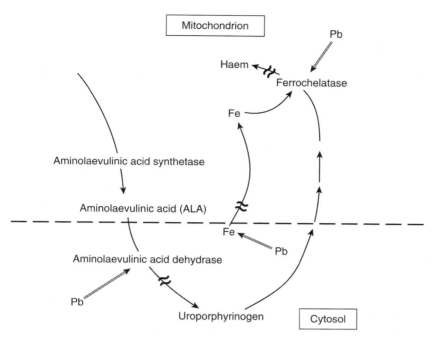

Figure 7.88   The synthesis of heme in the mammalian erythrocyte. The points of interference by lead are shown. $\approx$ Indicates inhibition of a metabolic pathway.

Interference with Heme Synthesis

Lead interferes with the synthesis of **heme**, and its inhibitory effects on the enzymes of this pathway can be readily detected in humans.

Myoglobin synthesis and cytochromes P-450 may also be affected. The mechanism of the interference of heme synthesis by lead can be seen in Figure 7.88. Lead inhibits heme synthesis at several points: aminolaevulinate synthetase, aminolaevulinate dehydrase, ferrochelatase, heme oxidase, and coproporphyrinogen oxidase. Thus, there is excess protoporphyrin available, but the inhibition of ferrochelatase means that protoporphyrin rather than heme is inserted into the globin moiety, and zinc replaces the iron. Lack of heme causes negative feedback control on aminolaevulinate synthetase, leading to increased synthesis of aminolaevulinate. The final result of the inhibition of these various enzymes is a decrease in the level of hemoglobin and hence anemia. The effects may be detected as coproporphyrin excretion in the urine, free erythrocyte protoporphyrin is increased and excess aminolaevulinate is excreted in urine. Significant inhibition of the enzyme aminolaevulinate dehydrase is detectable in volunteers with blood lead levels of 0.4 mg mL^{-1} resulting from normal exposure in an urban environment. With extensive exposure to lead such as may occur following industrial exposure or accidental poisoning, morphological changes in red blood cells may be seen. Indeed, the anemia caused by lead is due to both shortened lifetime of red blood cells as well as decreased synthesis of heme. The reduced lifetime of the red blood cell is due to increased fragility of the red cell membrane.

Renal Toxicity

Acute exposure to inorganic lead can cause reversible damage to the kidneys, manifested as tubular dysfunction. Chronic exposure to lead, however, causes permanent interstitial nephropathy, which involves tubular cell atrophy, pathological changes in the vasculature, and fibrosis. The most pronounced changes occur in the proximal tubules. Indeed, lead–protein complexes are seen as inclusion bodies in tubular cells, and the mitochondria in such cells have been shown to be altered with impaired oxidative phosphorylation. Clearly, this will influence the function of the proximal tubular cells in reabsorption and secretion of solutes and metabolites. Consequently, one indication of renal dysfunction is amino aciduria, glycosuria, and impairment of sodium reabsorption.

Neurotoxicity

The effects on the CNS are perhaps the most significant as far as humans are concerned, and children are especially vulnerable. The neurotoxicity is observed as **encephalopathy** and **peripheral neuropathy**. Encephalopathy is accompanied by various pathological changes, including cerebral edema, degeneration of neurons, and necrosis of the cerebral cortex. It is observed particularly in children and can be manifested as ataxia, convulsions, and coma. Cerebral palsy may also occur along with mental retardation and seizures. The mechanism of the neurotoxicity may involve direct effects of lead on neuronal transmission, which are thought to occur at blood lead levels of 0.3 to 0.5 mg mL^{-1} as assessed by behavioral studies, brain wave patterns detectable by electroencephalogram, and CNS-evoked potentials. Inhibition of cholinergic function via interference with extracellular calcium may underlie these changes in neurotransmission. The functioning of other neurotransmitters such as dopamine and aminobutyric acid are also affected by lead. There are also indirect effects via cerebral edema and cellular hypoxia. Peripheral neuropathy involves the degeneration of peripheral, especially motor, nerves. Dysfunction of motor nerves is detectable as a decrease in nerve conduction velocity at blood lead levels of 0.5 to 0.7 mg mL^{-1}.

Lead will also cause cancers in the kidney and lung in experimental animals. Furthermore, in humans exposed to lead similar tumors have been detected in epidemiological studies. Although the mechanism is not known, it is known that lead interferes with calcium-dependent protein kinases. The activation of PKC-mediated pathways leads to increased DNA synthesis. Thus, lead can cause increased cell replication and, so, hyperplasia. This may underlie the carcinogenicity. Lead may also act as a tumor promoter.

Manifestations of Lead Toxicity

Biochemical changes such as increased aminolaevulinate excretion and inhibition of aminolaevulinate dehydrase may be detected in urine and blood, respectively, at blood lead levels of 0.4 to 0.6 mg mL^{-1}. Anemia is a late feature, however. Neurotoxicity may be detectable at blood lead levels of 0.8 to 1.0 mg mL^{-1}. At blood lead levels greater than 1.2 mg mL^{-1}, encephalopathy occurs. Peripheral nerve palsies are rare, and the foot and wrist drop, which were once characteristic of occupational lead poisoning, only occur after excessive exposure and are now rarely seen. Similarly, seizures and impaired consciousness may result from involvement of the CNS. Bone changes are usually seen in children and are detected as bands at the growing ends of the bones and a change in bone shape.

Testicular dysfunction in humans occupationally exposed has been detected as a decrease in plasma testosterone and other changes indicative of altered testicular function.

Renal function changes are readily detectable as amino aciduria and glycosuria.

The treatment of lead poisoning involves the use of chelating agents to remove the lead from the soft tissues of the body. Thus, agents such as sodium calcium edetate may be used.

Organic lead is probably more toxic than inorganic lead, as it is lipid soluble. For example, triethyl lead, which results from breakdown of tetraethyl lead, is readily absorbed through the skin and into the brain and will cause encephalopathy. Symptoms are delusions, hallucinations, and ataxia, and the effects are rapid. Organic lead, however, has no effect on heme synthesis.

SUMMARY

Chemical Carcinogenesis

Acetylaminofluorene

Acetylaminofluorene is one of the most widely studied carcinogens and a very potent mutagen. It causes tumors in the liver, bladder, and kidney. Metabolism is important, hence, conjugates (sulfate and acetyl) of N-hydroxyacetylaminofluorene are more potent carcinogens than the parent compound. These may give rise to nitrenium and carbonium ions. Sulfate conjugation leads to cytotoxicity and tumorigenicity and DNA adducts. Some species (guinea pig) are resistant because of lack of formation of the N-hydroxy metabolite and low sulfotransferase activity.

Benzo[a]pyrene

Benzo[a]pyrene is a polycyclic aromatic hydrocarbon found in cigarette and other smoke and is a potent carcinogen. A planar molecule, it undergoes many metabolic transformations, including a variety of hydroxylations. The ultimate carcinogen seems to be $(-)$-7R,8S-dihydrodiol, 9S,10R-epoxide. The epoxide is adjacent to the bay region. This metabolite, produced by cytochrome P-450 1A1-mediated oxidation, is mutagenic and binds to DNA *in vitro* more extensively than the parent compound and yields the same conjugate as benzo[a] pyrene in cells in culture. Adducts are formed with the exocyclic amino group of guanine (N^2) and the O^6 and the ring nitrogen atoms. Other metabolites are also mutagenic and cytotoxic (e.g., quinones and other diol epoxides).

Dimethylnitrosamine

Dimethylnitrosamine is a hepatotoxic, mutagenic carcinogen. A single dose causes kidney tumors, repeated exposure causes liver tumors. The toxicity and carcinogenicity are due to metabolism to a methyldiazonium ion, which produces a methyl carbonium ion, which is a reactive methylating agent. DNA is methylated (N^7 of guanine), and the degree of methylation of DNA correlates with the tumorigenicity.

Vinyl Chloride

Vinyl chloride is a chemical used in the manufacture of plastics, which is carcinogenic and causes various toxic effects, including liver injury and damage to the bones and skin. Liver hemangiosarcomas are produced in animals and humans. Vinyl chloride undergoes metabolic activation by cytochrome P-450 to an epoxide, which may interact with DNA and form adducts (ethenodeoxyadenosine and ethenodeoxycytidine), which leads to mutations. These can be detected in white cells, and a mutant p21 ras protein can be detected in the serum of exposed workers. Also, reaction with GSH occurs.

Tamoxifen

This is used prophylactically to prevent breast cancer and as an extremely effective anticancer drug. But tamoxifen can cause liver tumors in rats. The rat conjugates the metabolite α-hydroxytamoxifen with sulfate, and this generates a reactive intermediate, which reacts with DNA and is genotoxic. Mice and humans mostly conjugate the α-hydroxytamoxifen with glucuronic acid. Therefore, it is not genotoxic or carcinogenic in mice or humans.

Clofibrate

This is a hyperlipdemic drug, which causes liver tumors in rats and peroxisome proliferation, hypertrophy, and hyperplasia. Chemicals causing this effect act on a receptor, called the PPAR. PPARα induces changes in the gene-encoding enzymes involved in lipid and lipoprotein metabolism. Transgenic mice lacking PPARα-given clofibrate do not show increased peroxisomes, liver size, or liver tumors.

The critical events thought to be involved in the hepatic carcinogenesis are (*i*) receptor binding and activation; (*ii*) induction of key target genes; (*iii*) increased cell proliferation and inhibition of apoptosis; (*iv*) oxidative stress causing DNA damage and/or increased signaling for cell proliferation; and (*v*) clonal expansion of initiated liver cells. Excess hydrogen peroxide produced in peroxisomes is not adequately detoxified and so can cause oxidative stress and DNA damage.

PPARα activation may lead to DNA replication, cell growth, and inhibition of apoptosis via TNF-α. Alternatively or additionally, the oxidative stress could activate the MAP kinase pathway, which will also influences cell growth, but this might be independent of PPARα activation.

Tissue Lesions: Liver Necrosis
Carbon Tetrachloride

Carbon tetrachloride is a hepatotoxic solvent, which causes centrilobular necrosis and fatty liver, liver cirrhosis, and tumors and kidney damage after chronic exposure. It is metabolized in liver by cytochrome P-450 by reduction to a free radical, which causes lipid peroxidation, which destroys membranes and produces toxic aldehydes such as 4-hydroxynonenal. The rough endoplasmic reticulum is damaged, hence protein synthesis is disrupted, and this may cause the fatty liver. The free radical also binds to protein and lipid. CYP2E1 is destroyed, whereas other isozymes, e.g., CYP1A1, are unaffected. So, a small dose protects against the hepatotoxicity of a subsequent large dose.

Valproic Acid

Valproic acid is a widely used antiepileptic drug, which occasionally causes liver dysfunction; mild transient elevation of the transaminases (ALT, AST), which resolves or severs with damage with fatty liver, jaundice, and necrosis, may lead to fatal liver failure.

As it is similar to a fatty acid, it forms an acyl CoA and a carnitine derivative. This depletes CoA from the intramitochondrial pool and carnitine. Also, valproic acid is metabolized to an unsaturated fatty acid analogue, which is incorporated into β-oxidation in mitochondria. The reactive analogue depletes GSH and damages mitochondria. Function is compromised and ATP depleted.

Nucleoside Analogues

These cause similar effects as valproic acid by damaging mitochondrial DNA and so reducing mitochondrial function.

Paracetamol

Paracetamol is a widely used analgesic, which causes liver necrosis and sometimes renal failure after overdoses in many species. The half-life is increased after overdoses because of impaired conjugation of the drug. Toxicity is due to metabolic activation and is increased in patients or animals exposed to microsomal enzyme inducers. The reactive metabolite (NAPQI) reacts with GSH, but depletes it after an excessive dose and then binds to liver protein. Cellular target proteins for the reactive metabolite of paracetamol have been detected, some of which are enzymes that are inhibited. Therefore, a number of events occur during which ATP is depleted, Ca levels are deranged, and massive chemical stress switches on the stress response.

Metabolic activation is catalyzed by cytochrome P-450, and the particular isoform (2E1, 1A2, or 3A4) depends on the dose.

Antidote is *N*-acetylcysteine, which promotes the synthesis of new GSH and may also be involved in the detoxication.

Bromobenzene

Bromobenzene is a hepatotoxic industrial solvent, which causes centrilobular liver necrosis. It is metabolized in the liver to a reactive epoxide (3,4), which is detoxified by conjugation with GSH, leading to excretion of a mercapturic acid. Depletion of GSH with an excess dose leads to binding of the reactive epoxide to liver protein and necrosis. GSH conjugate is excreted as a mercapturic acid in urine. Epoxide hydrolase detoxifies, and induction may also detoxify by increasing metabolism to a less reactive epoxide (2,3). Bromobenzene can also cause kidney damage, which may be due to diglutathionyl conjugate of 2-bromohydroquinone.

Isoniazid and Iproniazid

Both these substituted hydrazine drugs may cause liver damage after therapeutic doses. With isoniazid, a mild hepatic dysfunction may occur in 10% to 20% of patients and a more severe type in less than 1%. Both isoniazid and iproniazid yield hydrazine metabolites (acetylhydrazine and isopropylhydrazine, respectively), which are responsible for the hepatotoxicity after activation by cytochrome P-450. Isoniazid undergoes acetylation, which in humans is polymorphic. Slow acetylators are more at risk from the hepatotoxicity because acetylhydrazine is detoxified by acetylation.

Microcystins

These natural toxins are heptapeptides produced by cyanobacteria and have unusual structural features, incorporating three D-amino acids. Microcystin LR is hepatotoxic, as a result of inactivation of protein phosphatases, which leads to breakdown of the cytoskeleton and cell death.

Tissue Lesions: Kidney Damage
Chloroform

Chloroform is an anesthetic and solvent, which may be nephrotoxic and hepatotoxic. It requires metabolic activation by cytochrome P-450, and male mice are more susceptible to the nephrotoxicity than females, which are more likely to suffer hepatic damage. The renal damage, proximal tubular necrosis, is accompanied by fatty infiltration. The metabolic activation, which may take place in the kidney, produces phosgene, which is reactive and can bind to critical proteins.

Haloalkanes and Alkenes

Hexachlorobutadiene is a nephrotoxic industrial chemical, damaging the pars recta of the proximal tubule. Initial conjugation with GSH is necessary, followed by biliary secretion and catabolism resulting in a cysteine conjugate. The conjugate is reabsorbed and transported to the kidney where it can be concentrated and becomes a substrate for the enzyme β-lyase. This metabolizes it into a reactive thiol, which may react with proteins and other critical macromolecules with mitochondria as the ultimate target. The kidney is sensitive because the metabolite is concentrated by active uptake processes (e.g., OAT 1), which reabsorb the metabolite into the tubular cells.

Other halogenated compounds such as trichloroethylene may be metabolized to similar cysteine conjugates, which are also nephrotoxic but may not all require β-lyase activation.

Antibiotics

Gentamycin is a nephrotoxic antibiotic (aminoglycoside) having five cationic amino groups. The proximal tubular cells are damaged, the function of the mitochondria decreased, and the function of the kidney is compromised. Hydrolytic enzymes are released from lysosomes in which the gentamycin attached to phospholipids is stored, leading to further damage. Gentamycin has a long half-life, accumulates in the kidney (2–5 times blood concentration), and after initial glomerular filtration, is taken up and sequestered by lysosomes in the kidney. The concentration in proximal tubular cells reaches levels 10 to 100 times that in blood.

The drug binds to anionic phospholipids in the proximal tubule, may alter phospholipids metabolism, and cause phospholipidosis. Gentamycin also damages the hair cells in the ear.

Cephalosporins

These drugs (e.g., cephaloridine) may be nephrotoxic causing proximal tubular necrosis. Cephaloridine is actively taken up from blood into proximal tubular cells by OAT 1. The drug therefore accumulates in the kidney. Metabolic activation via cytochrome P-450 may be involved. GSH is oxidized, and as NADPH is also depleted, the GSSG cannot be reduced back to GSH. As vitamin E-depleted animals are more susceptible, it has been suggested that lipid peroxidation may be involved. Damage to mitochondria also occurs.

Tissue Lesions: Lung Damage
4-Ipomeanol

This pulmonary toxin is produced by a mold that grows on sweet potatoes. The toxin produces edema, congestion, and hemorrhage resulting from necrosis of the Clara cells (non-ciliated bronchiolar cells). Lung tissue (e.g., in rodents) has CYP4B1, which metabolizes 4-ipomeanol to a reactive intermediate. This binds to macromolecules in these cells, causing necrosis. Induction and inhibition of cytochrome P-450 may increase and decrease toxicity, and depleting GSH increases the toxicity. Paraquat, a bipyridylium herbicide, causes lung fibrosis and sometimes kidney failure. It is an irritant chemical, exposure to which has a high fatality rate. It is actively accumulated by the lung tissue via the polyamine uptake system. By accepting an electron from NADPH, it readily forms a stable free radical, which in the presence of oxygen produces superoxide anion radical. This may overwhelm the superoxide dismutase available and then hydroxyl radicals and hydrogen peroxide may be formed, leading to lipid peroxidation and tissue damage.

Neurotoxicity
Isoniazid
This drug may cause degeneration of peripheral nerves after repeated exposure as a result of the depletion of vitamin B_6. This is because isoniazid reacts with pyridoxal to form a hydrazone that inhibits pyridoxal phosphate kinase, so blocking formation of pyridoxal phosphate. As this effect is due to the parent drug, slow acetylators are more at risk, but the adverse effect can be prevented by supplying vitamin B_6 to the patient.

6-Hydroxydopamine
This selectively neurotoxic compound damages the sympathetic nerve endings. Because of its structural similarity to dopamine, it is actively taken up into the synaptic system. Once taken up, it oxidizes to a reactive quinone, which may bind to protein and produce reactive free radicals and superoxide. These events destroy the nerve terminals and nerve cells also in some cases.

1-Methyl-4-Phenyl-1,2,3,6-Tetrahydropyridine (MPTP)
This contaminant of a meperidine analogue, synthesized illegally, causes a syndrome similar to Parkinson's disease in both humans and monkeys. It does this by destroying the dopamine cell bodies in the substantia nigra area of the brain. After absorption into the brain, MPTP is metabolized first to $MPDP^+$ then to MPP^+ by the enzyme monoamine oxidase B. These compounds, being charged, do not go back across the blood-brain barrier and remain trapped in the brain. MPP^+ is taken up by the DAT and accumulates in neurons where it associates with neuromelanin. MPP^+ becomes concentrated in the mitochondrial matrix via an energy-dependent carrier and inhibits complex 1 of the electron transport chain. This leads to inhibition of oxidative phosphorylation, depletion of GSH, and increased cellular calcium. The result is the destruction of the dopaminergic neurons.

Domoic Acid
This is a natural toxin, and poisoning results from eating contaminated shellfish. It causes gastrointestinal distress and disturbances and neurotoxicity. Domoic acid acts as an analogue of glutamate and as an excitotoxin; excess excitation leading to neuronal cell death.

Primaquine
This is an 8-aminoquinoline antimalarial drug, which can cause hemolytic anemia, particularly in patients with G6PD deficiency causing toxicity to erythrocytes. Two metabolites, 6-methoxy-8-hydroxylaminoquinoline and 5-hydroxyprimaquine, are both cyotoxic to red cells. Both metabolites can undergo redox cycling to give ROS detoxified by reduced GSH. In the absence of sufficient reduced GSH, the ROS can react with and damage red cell hemoglobin.

5-Hydroxyprimaquine depletes GSH and is especially toxic in red cells depleted of GSH. It would be more toxic in G6PD deficient patients who have a deficiency in red cell GSH.

Adriamycin/Doxorubicin
This anthracycline type anticancer drug has a quinone moiety so can easily accept an electron and undergo redox cycling. As a result, it interferes with the mitochondrial electron transport chain, damages mitochondrial DNA, and leads to ATP depletion. The result is a dose-dependent cardiomyopathy.

Exaggerated and Unwanted Pharmacological Effects
Organophosphorus Compounds
This group of compounds is used as pesticides and nerve gases. The structure and therefore metabolism and potency varies. However, they all act in a similar manner. There are two toxic effects, cholinesterase inhibition and delayed neuropathy, but all OPs do not necessarily cause both. The cholinesterase inhibition results from the similarity between the organophosphorus compound and acetylcholine. The organophosphorus compound therefore acts as a pseudosubstrate but blocks the enzyme, in some cases, permanently. This is because the

organophosphate intermediate binds to the active site irreversibly or very strongly, and the phosphorylated enzyme, unlike the natural acetylated enzyme, is only hydrolyzed slowly. This enzyme blockade allows acetylcholine to build up, and so the toxicity and symptoms are a result of excessive stimulation of receptors: muscarinic receptors, leading to salivation, lacrimation, urination, defecation, bronchoconstriction, bradycardia, and miosis; nicotinic receptors, leading to fatigue, muscular twitches, and weakness, including respiratory muscles and hypertension; CNS receptors, leading to anxiety, tension, convulsions, coma, and depression of respiratory and circulatory centers. Treatment involves giving atropine and then pralidoxime. Delayed neuropathy results from the interaction between the organo-phosphorus compound and a neuropathy target esterase, which disturbs the metabolism in the neuron, leading to dying back of the long peripheral nerves.

Cardiac Glycosides

Digoxin/digitoxin and related drugs are used as cardiac stimulants, causing a positive inotropic effect. This group of substances, used as drugs, has a narrow margin of safety or low therapeutic index, and there is wide individual variation in response among patients.

Digitoxin causes inhibition of the Na^+/K^+ ATPase, this reduces the sodium gradient and leads to increased intracellular calcium. This causes the adverse effects, including vomiting, diarrhea, visual disturbances, hypotension, and ventricular tachycardia, leading to fibrillation.

Diphenylhydantoin

The toxic effects of this anticonvulsant drug, which result from elevated plasma levels, are dose related and are mainly effects on the nervous system (ataxia, drowsiness, nystagmus). The high plasma levels may be due to deficiencies in metabolism as well as excessive dosage. Deficiencies in metabolism may be genetic or due to coadministration of other drugs (e.g., isoniazid, especially in slow acetylators).

Succinylcholine

This muscle relaxant drug normally has a short duration of action because of metabolism (hydrolysis by pseudocholinesterase), which may occur in the plasma. This metabolism may be deficient in some patients leading to apnea (prolonged relaxation of respiratory muscles; 2 hours vs. a few minutes). This is due to an abnormal pseudocholinesterase, a genetically determined trait (2 alleles at a single locus occurring at a frequency of about 2% in some groups). Therefore, there are homozygotes for the normal gene and heterozygotes or homozygotes for the abnormal gene, leading to normal, intermediate, and deficient enzyme and metabolism.

Botulism and Botulinum Toxin

The toxin produced by the bacterium *C. botulinum* is a mixture of six large molecules and is one of the two most potent toxins known to humans. Each consists of two components, a heavy (100 kDa) and light (50 kDa) polypeptide chain. The toxin molecule is transported into nerve cells and destroys a synaptosomal protein, which prevents the release of the neurotransmitter acetylcholine. This blocks muscle contraction, causing paralysis. This can be fatal.

Physiological Effects

Aspirin

Aspirin is a major analgesic, which is responsible for many cases of therapeutic, accidental, and intentional poisoning. A common side effect is stomach irritation, bleeding, and sometimes, ulceration. This is due to inhibition of the enzyme cyclooxygenase. It is metabolized by hydrolysis to salicylic acid, which is conjugated. Conjugation pathways are saturable so that the half-life increases with dosage. Therefore, blood level of salicylate rises disproportionately with dosage. Salicylate is a lipid-soluble weak acid, which is able to diffuse across mitochondrial membrane and release the proton into the matrix. This uncouples oxidative phosphorylation, causing loss of ATP, heat generation, increased cellular metabolism and production of carbon dioxide, increased requirement for oxygen, and increased rate and depth of breathing via the effect on central respiratory control. This leads to metabolic alkalosis, and overcompensation leads to metabolic acidosis. Acidosis allows more nonionized salicylate to enter tissues, including CNS, leading to more metabolic effects and greater acidosis and decreased urinary excretion. Effects on CNS may lead to death. Treatment involves alkaline diuresis, reduction of temperature, and replacement of ions and glucose.

Biochemical Effects: Lethal Synthesis and Incorporation

Fluoroacetate

This naturally occurring toxicant is an analogue of acetate and is incorporated into acetyl CoA (fluoroacetate) and hence into Krebs' cycle (TCA cycle) as fluorocitrate. This blocks the enzyme aconitase, as the fluorine atom cannot be removed. The TCA cycle is blocked, and citrate accumulates. The mitochondrial energy supply is disrupted, hence cardiac damage occurs. Lack of oxaloacetate will allow ammonia to accumulate leading to convulsions.

Galactosamine

Galactosamine is an amino sugar, which causes hepatic damage similar to viral hepatitis (non-zonal) and focal necrosis. It may also cause cirrhosis and tumors. It causes inhibition of DNA and protein synthesis. Levels of UTP and UDP-glucose fall and other UDP-hexosamines and acetylhexosamines rise. Galactosamine forms UDP-galactosamine, which leads to depletion of UTP, so synthesis of RNA and protein are disrupted. Abnormal membrane glycoproteins may be produced, leading to membrane damage.

Ethionine

Ethionine is a hepatotoxic analogue of methionine causing fatty liver (accumulation of triglycerides). Chronic exposure causes cirrhosis, bile duct proliferation, and heptatocellular carcinoma. It forms *S*-adenosyl ethionine, which traps adenosyl leading to ATP depletion, which reduces triglyceride export from the liver. It also leads to ethylated bases in DNA.

Biochemical Effects: Interactions with Specific Protein Receptors

Carbon monoxide (CO)

Carbon monoxide is one of the most important agents involved in poisoning cases with many sources (fires, car exhausts, solvent metabolism) and difficult to detect. It binds avidly to iron atoms in hemoglobin-forming carboxyhemoglobin (COHb), which may be determined in the blood: 20% COHb may lead to impairment of normal function and 60% to death. The main target organs are the brain and heart; death is due to brain hypoxia. Binding of carbon monoxide to hemoglobin is stronger than O_2 (240×). Binding of carbon monoxide affects the cooperativity of oxyhemoglobin, so making loss of oxygen from carboxyhemoglobin more difficult. Treatment involves removal of the victim from the source of the carbon monoxide and provision of oxygen or oxygen at increased pressure.

Cyanide

Cyanide has many sources: natural (plant-Cassava), industrial (cyanide salts and nitriles), and accidental (fires). The target organ is the brain; death is from respiratory arrest. Cyanide blocks cytochrome a-a_3 (cytochrome oxidase) in mitochondria. The toxic level is 1 mg mL^{-1} in blood. Treatment involves giving dicobalt edetate (chelation). Alternatively, by giving $NaNO_2$, levels of methemoglobin are increased, and this binds cyanide. Detoxication is catalyzed by the enzyme rhodanese, and this pathway may be increased by giving NaS_2O_7.

Teratogenesis

Actinomycin D

This is a potent teratogen, at days 7 to 9 of gestation in rats, giving malformations (cleft palate, spina bifida, anencephaly) and embryolethality (both show a similar, steep dose response). It inhibits RNA synthesis and hence essential protein synthesis; it is also cytotoxic.

Diphenylhydantoin

This is an anticonvulsant drug causing orofacial (cleft palate) and skeletal malformations, especially on days 11 to 12 in mice. Teratogenesis shows a steep dose response. The mechanism may involve metabolism (epoxidation), and covalent binding to protein or interaction with the retinoic acid receptor may be involved.

Thalidomide

This is a sedative drug with low adult toxicity, which proved to be a very potent human teratogen, causing phocomelia (shortening of the limbs) and other defects when taken between the third and eighth week. In some cases, only a few doses were taken, but on the critical days (e.g., days 24–27 for phocomelia of arms). It is not readily reproducible in laboratory animals (e.g., rats). Mechanism is unknown, but a metabolite suspected, possibly produced by cytochrome P-450. A number of metabolites are produced and some chemical breakdown occurs. Phthalylglutamic acid metabolite is teratogenic in mice. Thalidomide may acylate nucleic acids and polyamines. The S-enantiomer is more embryotoxic than the R-enantiomer.

More recent studies have indicated a probable mechanism, which relies on the interaction of thalidomide with DNA. The phthalimide double-ring structure of *S*-thalidomide can intercalate with DNA and binds to the polyG sites in the promoter regions coding for growth factors and integrins.

Immunotoxicity

Halothane

This is an anesthetic drug, which causes hepatic damage. Mild hepatic dysfunction is common. Occasionally, serious hepatic necrosis occurs, which is an idiosyncratic reaction (factors: multiple exposures; sex—female>male; obesity; history of allergy). A reactive metabolite is produced, which binds to many target proteins in the liver cell, but mostly from the membrane of the endoplasmic reticulum, including cytochrome P-450. Expression of fragments of the hapten-protein complex on the cell surface is via MHC molecules. This stimulates an immunological reaction involving T lymphocytes and antibodies (antibody-dependent cell cytotoxicity). Hepatocytes are destroyed by lymphocyte-mediated cytolysis. It may be that several mechanisms operate depending on exposure and host factors.

Practolol

This antihypertensive drug was withdrawn from use after a number of cases of occulomucocutaneous syndrome (skin rashes, keratinization of the eye, and other tissues). Mechanism is unknown, but there is evidence for metabolism (microsomal enzyme mediated) to a reactive intermediate capable of reacting with protein and forming antigenic conjugates. Antipractolol antibodies are detected in patients. Syndrome is not reproducible in animals, but in hamsters, practolol accumulates in the eye.

Penicillin

This antibiotic drug and related derivatives are responsible for more allergic reactions than any other class of drug (1–10% of recipients). All four types of hypersensitivity reactions are caused depending on circumstances of exposure (dose etc.). Life-threatening anaphylactic reactions, urticaria, hemolytic anemia, and immune complex disease may all occur. The penicillin molecule is reactive and can bind covalently to various proteins (e.g., amino group of lysine) in various ways, but penicilloyl derivative is the major antigenic determinant (benzylpenicillenic acid 40 × more antigenic than benzylpenicillin). Cross-reactivity occurs between different penicillins. Type II hypersensitivity seems to be reasonably well understood and involves the lysis of red cells by NK cells.

Hydralazine

This is a vasodilator drug, which causes SLE in a significant proportion of patients. Several predisposing factors have been identified: dose (>25 mg); duration of therapy (mean 18 months); acetylator phenotype (slow); HLA type (DR4); and gender (females: males, 4:1). Antinuclear antibodies and antihydralazine antibodies are detected in serum. This causes a type III immune reaction. Mechanism is unclear but may involve interference with the development and maturation of T cells in the thymus, leading to an alteration of tolerance toward self-antigens. Interference with the complement system and interaction with nucleic acids also occur. Metabolism also may be mediated by myeloperoxidase in activated neutrophils.

Multi-Organ Toxicity

Ethylene glycol is a solvent used in antifreeze, paints etc. When ingested, it causes multi-organ toxicity and death. Target organs are the kidneys, optic system, brain, and heart. Toxicity is due to metabolism initially via alcohol dehydrogenase, leading to toxic aldehyde metabolites and eventually oxalic acid. Metabolites inhibit many metabolic pathways, but the overall effect is metabolic acidosis due to metabolites, excess NADH, and oxalate crystals in target organs. Treatment involves blocking metabolism with ethanol and correcting acidosis ($NaHCO_3$).

Methanol

Methanol is a solvent, which is added to ethanol and sometimes used in antifreeze. The main target organ is the optic system resulting from metabolic inhibition and systemic toxicity due to metabolic acidosis from formate and lactate. Toxicity is due to metabolism to formic acid via alcohol dehydrogenase and insufficient detoxication via tetrahydrofolate. The overall result is circulus hypoxicus. Treatment involves blockade of metabolism with ethanol and treatment of metabolic acidosis ($NaHCO_3$).

Multi-Organ Toxicity: Metals
Cadmium

Cadmium is a widely used metal responsible for multi-organ toxicity, some of which may occur over extended periods of time and may depend on the route of exposure. Thus, it causes acute testicular damage in rodents and kidney damage after chronic exposure or delayed after acute exposure. It may also be hepatotoxic, effects vascular tissue and bone, and causes Leydig cell tumors. It is detoxified by being bound to metallothionein, and binds to other proteins containing available SH groups. Enzyme inhibition may underlie some toxicity. Also, cadmium causes interference with the mitochondrial electron transport chain, leading to the production of ROS, oxidative stress DNA damage. Testicular toxicity is partly due to ischemic necrosis following vascular constriction. Kidney damage results from release of metal from metallothionein complex in kidney.

Mercury

Mercury exists in three forms: elemental, inorganic, and organic with different toxic effects. Elemental mercury is absorbed as a vapor and may enter the CNS and cause toxicity there. Inorganic mercury is poorly absorbed, but the cysteine conjugate of mercury is concentrated in the kidney by active transport. The kidney is the main target organ (also gastrointestinal tract if exposure by that route).

Organic mercury is very lipid soluble and is therefore well absorbed. It is toxic to the CNS and also teratogenic as in Minamata disease. The mechanism of toxicity involves binding to SH groups and inhibiting enzymes such as ATPase and uncoupling oxidative phosphorylation. It is treated with chelating agents (dimercaprol, penicillamine).

Lead

Lead is a toxic metal to which there is wide exposure. Exposure is via inhalation (main source, leaded petrol) and ingestion (water, old paint). Multi-organ toxicity occurs with the kidneys, central and peripheral nervous system, testes, red cells, bones, and gastrointestinal tract all damaged. After initial distribution into red blood cells, it is eventually deposited in bone. The main biochemical effect is interference with heme synthesis at several points. Kidney toxicity may be due to lead-protein complexes and inhibition of mitochondrial function. Damage to nerves leads to peripheral neuropathy.

Treatment involves use of chelating agents (EDTA).

Organic lead is also toxic to CNS (encephalopathy).

REVIEW QUESTIONS

1. Benzo[a]pyrene is a carcinogen. What is believed to be the metabolite responsible, and what is the evidence?

2. Dimethylnitrosamine is both a hepatic carcinogen and a hepatotoxicant. Does either effect require metabolism, and if so, what is (are) the metabolite(s) responsible for the carcinogenicity?

3. List the toxic effects of vinyl chloride. One of these shows an unusual dose-response relationship. Briefly describe and explain this. Name any toxic metabolites produced by metabolism of this compound.

4. What are the toxic effects of carbon tetrachloride? Briefly describe the mechanism underlying the acute toxic effects.

5. Which of the following are true?
   a. Paracetamol metabolism changes after large doses.
   b. Paracetamol toxicity does not require metabolism.
   c. When paracetamol is given to animals, it results in depletion of glutathione.
   d. There is no antidote for paracetamol poisoning.

6. Bromobenzene is toxic to the liver. It produces two reactive metabolites. Which one is thought to be responsible for the hepatotoxicity and why? Are there any routes of detoxication, and if so, what are they? What effect would treating with the enzyme inducer 3-methylcholanthrene have?

7. The drug isoniazid causes two different toxic effects. What are they? Are either of these effects due to a metabolite, and if so, which one? Which genetic factor is important in the toxicity and why?

8. What are the target organs for toxicity of chloroform and why?

9. Hexachlorobutadiene is toxic to which organ. Briefly explain the role of metabolism in the mechanism.

10. What is the underlying basis of the toxicity of carbon monoxide? What would you consider to be a dangerous concentration of carbon monoxide in the atmosphere in a factory? Show your calculations. (Haldanes constant=240; oxygen concentration in air=21%.)

11. Briefly describe the important aspects of mechanisms underlying the hepatotoxic effects of the anesthetic drug halothane.

12. Aminoglycoside drugs such as gentamycin may show toxic effects in which organs? Describe how these toxic effects may be detected.

13. What is the major target organ for paraquat toxicity, and what are the two main reasons for this?

14. Why is MPTP toxic to dopaminergic neurons in the brain?

15. Organophosphate insecticides such as parathion may lead to a variety of symptoms in mammals poisoned by them. List five such symptoms and explain these in terms of the mechanism of action.

16. What is the basic underlying biochemical effect caused by aspirin, and how does this lead to the various symptoms observed?

17. Administration of fluoroacetate to animals leads to inhibition of a particular enzyme. Name the enzyme that is inhibited, and briefly describe the mechanism and the biochemical and pathological consequences.

18. Galactosamine and ethionine both cause liver dysfunction. Comment on the similarities and differences between these two hepatotoxicants.

19. Why is nitrite sometimes administered as an antidote for cyanide poisoning?

20. If diphenylhydantoin is administered to pregnant mice on some days of gestation, there are no effects, whereas on other gestational days, there is more than 50% malformation of the fetuses. Explain this.

21. What are the predisposing factors identified in human patients that are important in the immune-mediated adverse effect lupus erythematosus caused by the drug hydralazine?

22. Ethylene glycol and methanol may both be used in antifreeze and are therefore sometimes accidentally or intentionally ingested together. What would be the

important biochemical features of poisoning with such a mixture, and what would the treatment be?

23. The heavy metal lead has a number of toxic effects but one is an effect on heme synthesis. How may this be detected, and what are the consequences?
24. Explain the mechanism(s) of carcinogenicity of fibrates. Why are these drugs not thought to be a risk for humans?
25. Valproate toxicity is manifested as fatty liver, but what is the underlying mechanism?
26. Describe the cellular target and the specific mechanism of toxicity of microcystin LR.
27. The structure of domoic acid is crucial to its toxicity. Explain this, and describe the toxicity.
28. The drug primaquine is toxic to a particular cell type. What is it? Why is there a genetic component to this toxicity?
29. Botulinum is highly toxic but can be used as a drug. Describe the therapeutic action and the mechanism.
30. Only the S-isomer of thalidomide is teratogenic. Explain this in terms of the mechanism.

REFERENCES

1. Gehring PJ, Watanabe PG, Park CN. Resolution of dose-response toxicity data for chemicals requiring metabolic activation: example–vinyl chloride. Toxicol Appl Pharmacol 1978; 44(3):581–591.
2. Greaves P, Goonetilleke R, Nunn G, et al. Two-year carcinogenicity study of tamoxifen in Alderley Park Wistar-derived rats. Cancer Res 1993; 53(17):3919–3924.
3. Peters JM, Cheung C, Gonzalez FJ. Peroxisome proliferators-activated receptor-α and liver cancer: where do we stand? J Mol Med 2005; 83:774–785.
4. Prescott LF, Wright N. The effects of hepatic and renal damage on paracetamol metabolism and excretion following overdosage. A pharmacokinetic study. Br J Pharmacol 1973; 49(4):602–613.
5. Potter WZ, Thorgeirsson SS, Jollow DJ, et al. Acetaminophen-induced hepatic necrosis. V. Correlation of hepatic necrosis, covalent binding and glutathione depletion in hamsters. Pharmacology 1974; 12 (3):129–143.
6. Mitchell JR, Jollow DJ, Potter WZ, et al. Acetaminophen-induced hepatic necrosis. I. Role of drug metabolism. J Pharmacol Exp Ther 1973; 187(1):185–194.
7. Jollow DJ, Mitchell JR, Potter WZ, et al. Acetaminophen-induced hepatic necrosis. II. Role of covalent binding *in vivo*. J Pharmacol Exp Ther 1973; 187:195–202.
8. Mitchell JR and Jollow DJ. Progress in hepatology. Metabolic activation of drugs to toxic substances. Gastroenterology 1975; 68(2):392–410.
9. Mitchell JR, Potter WZ, Hinson JA, et al. Toxic drug reactions. In: Gilette JR, Mitchell JR, eds. Handbook of Experimental Pharmacology, Vol. 28, Part 3, Concepts in Biochemical Pharmacology. Berlin: Springer-Verlag, 1975.
10. Gillette. 5th Int Congr Pharmacology, Vol. 2. Basel: Karger, 1973.
11. Zampaglione N, Jollow DJ, Mitchell JR, et al. Role of detoxifying enzymes in bromobenzene-induced liver necrosis. J Pharmacol Exp Ther 1973; 187(1):218–227.
12. Mitchell JR, Zimmerman HJ, Ishak KG, et al. Isoniazid liver injury: clinical spectrum, pathology, and probable pathogenesis. Ann Intern Med 1976; 84(2):181–192.
13. Timbrell JA, Mitchell JR, Snodgrass WR, et al. Isoniazid hepatoxicity: the relationship between covalent binding and metabolism *in vivo*. J Pharmacol Exp Ther 1980; 213(2):364–369.
14. Tune BM. The renal toxicity of beta-lactam antibiotics: mechanisms and clinical implications. In: DeBroe ME, Porter GE, Bennett WM, et al. eds. Clinical Nephrotoxins: Renal Injury from Drugs and Chemicals. Dordrecht: Kluwer Academic, 1998.
15. Boyd MR, Burka LT. *In vivo* studies on the relationship between target organ alkylation and the pulmonary toxicity of a chemically reactive metabolite of 4-ipomeanol. J Pharmacol Exp Ther 1978; 207(3):687–697.
16. Marsh DF. Outline of Fundamental Pharmacology. Springfield: Charles C.Thomas, 1951.
17. Decker K, Keppler D. Galactosamine hepatitis: key role of the nucleotide deficiency period in the pathogenesis of cell injury and cell death. Rev Physiol Biochem Pharmacol 1974; 71:77–106.
18. Timbrell JA. Introduction to Toxicology. London: Taylor & Francis, 1989.
19. Vale JA, Meredith TJ. Poisoning-Diagnosis and Treatment. London: Update Books, 1981.
20. Wilson JG. Embryological considerations in teratology. Ann N Y Acad Sci 1965; 123:219–227.

21. Harbison RD, Becker BA. Relation of dosage and time of administration of diphenylhydantoin to its teratogenic effect in mice. Teratology 1969; 2(4):305–311.

22. Cameron HA, Ramsay LE. The lupus syndrome induced by hydralazine: a common complication with low dose therapy. Brit Med J 1984; 289:410–412.

23. Bakir F, Damluji SF, Amin-Zaki L, et al. Methylmercury poisoning in Iraq. Science 1973; 181(196): 230–241.

BIBLIOGRAPHY

General

Aldridge WN. Mechanisms and Concepts in Toxicology. London: Taylor & Francis, 1996.

Boelsterli UA. Mechanistic Toxicology. Boca Raton, FL: CRC Press, 2007.

Caldwell J, Mills JJ. The Biochemical basis of toxicity. In: Ballantyne B, Marrs TC, Syversen T, eds. General and Applied Toxicology, vol. 1. London: Macmillan, 2000.

Cotgreave IA, Morgenstern R, Jernstrom B, et al. Current molecular concepts in toxicology. In: Ballantyne B, Marrs TC, Syversen T, eds. General and Applied Toxicology, Vol 1. 2nd ed. London: Macmillan, 1999:155–174.

Fenton JJ. Toxicology, A Case Oriented Approach. Boca Raton FL: CRC Press, 2002.

Flanagan RJ, Jones AL. Antidotes. London: Taylor & Francis, 2001.

Gregus Z, Klaassen CD. Mechanisms of toxicity. In: Klaassen CD, ed. Cassarett and Doull's Toxicology, The Basic Science of Toxicology. 6th ed. New York: McGraw Hill, 2001.

Hodgson E, Smart RC. In: Hodgson E, Smart RC, eds. Introduction to Biochemical Toxicology. 3rd ed. New York: Wiley, 2001. Various chapters.

Hodgson E, Bend JR, Philpot RM. Review in Biochemical Toxicology, Vols. 1. New York: Elsevier-North Holland, 1971. Various

Klaassen CD, ed. Cassarett and Doull's Toxicology: The Basic Science of Toxicology. 6th ed. New York: McGraw Hill, 2001. Various chapters.

Kocsis JJ, Jollow DJ, Witmer CM, et al, eds. Biological Reactive Intermediates III. Mechanisms of Action in Animal Models and Human Disease. New York: Plenum Press, 1986.

Monks TJ, Lau SS. Reactive intermediates and their toxicological significance. Toxicology 1988; 52:1.

Nagelkerke JF, van der Water B. Molecular and cellular mechanisms of toxicity. In: Mulder GJ, Dencker L, eds. Pharmaceutical Toxicology. London: Pharmaceutical Press, 2006.

Nelson SD, Pearson PG. Covalent and noncovalent interactions in acute lethal cell injury caused by chemicals. Annu Rev Pharmacol Toxicol 1990; 30:169.

Parkinson A. Biotransformation of Xenobiotics. In: Klaassen CD, ed. Cassarett and Doull's Toxicology, The Basic Science of Toxicology. 6th ed. New York: McGraw Hill. . 2001. This chapter has many examples of toxicity due to metabolism and metabolic activation.

Pratt WB, Taylor P, eds. Principles of Drug Action, The Basis of Pharmacology. New York: Churchill Livingstone, 1990.

Reed DJ. Mechanisms of chemically induced cell injury and cellular protection mechanisms. In: Hodgson E, Smart RC, eds. Introduction to Biochemical Toxicology. 3rd ed. Connecticut: Appleton-Lange, 2001.

Slater TF, ed. Biochemical Mechanisms of Liver Injury. London: Academic Press, 1978.

Snyder R, Parke DV, Kocsis J, et al. , eds. Biological Reactive Intermediates 2: Chemical Mechanisms and Biological Effects. New York: Plenum Press, 1981.

Witschi H, Haschek WM. Some problems correlating molecular mechanisms and cell damage. In: Bhatnagar RS, ed. Molecular Basis of Environmental Toxicity, Ann Arbor: Ann Arbor Science Publications 1980.

Chemical Carcinogenesis

Boocock DJ, Maggs JL, Brown K, et al. Major interspecies differences in the rates of O-sulphonation and O-glucuronylation of α-hydroxytamoxifen *in vitro*: a metabolic disparity protecting human liver from the formation of tamoxifen DNA adducts. Carcinogenesis 2000; 21:1851–1858.

Cattley RE, Deluca J, Elcombe C, et al. Do peroxisome proliferating compounds pose a hepatocarcinogenic hazard to humans?Regul Toxicol Pharmacol. 1998; 27:47.

Cooper CS, Grover PL, eds. Chemical Carcinogenesis and Mutagenesis II. Handbook of Experimental Pharmacology, Vol. 94. Berlin: Springer, 1990.

Esteller M. Aberrant DNA methylation as a cancer inducing mechanism. Ann Rev Pharmacol Toxicol 2005; 45:605–656.

Goodman JI, Watson RE. Altered DNA methylation: A secondary mechanism involved in carcinogenesis. Ann Rev Pharmacol Toxicol 2002; 42:501–525.

Haggerty HG, Holsapple MP. Role of metabolism in dimethylnitrosamine-induced immunosuppression: a review. Toxicology 1990; 63:1.

Harris CC. p53 Tumour suppressor gene: At the crossroads of molecular carcinogenesis, molecular epidemiology and cancer risk assessment. Environ Health Persp 1996; 104:435.

Hoivik DJ, Qualls CW, Mirabile RC, et al. Fibrates induce hepatic peroxisome and mitochondrial proliferation without overt evidence of cellular proliferation and oxidative stress in cynomologous monkeys. Carcinogenesis 2004; 25:1757–1769.

Jernstrom B, Graslund A. Covalent binding of benzo[a] pyrene 7,8-dihydrodiol 9,10-epxides to DNA: molecular structure, induced mutations and biological consequences. Biophys Chem 1994; 49:185–199.

Kim D, Guengerich FP. Cytochrome P450 activation of arylamines and heterocyclic amines. Annu Rev Pharmacol Toxicol 2005; 45:27–49.

Klaunig JE, Kamendulis LM. The role of oxidative stress in carcinogenesis. Ann Rev Pharmacol Toxicol 2004; 44:239–267.

Kozack R, Seo KY, Jelinsky SA, et al. Toward an understanding of the role of DNA adduct conformation in defining mutagenic mechanism based on studies of the major adduct (formed at N^2-dG) of the potent environmental carcinogen benzo[a]pyrene. Mutat Res 2000; 450(1–2):41–59.

Mulder GJ, Kroese ED, Meerman JHN. The generation of reactive intermediates from xenobiotics by sulphate conjugation and their role in drug toxicity. In: Gorrod JW, Oelschlager H, Caldwell J, eds. Metabolism of Xenobiotics. London: Taylor & Francis, 1988.

McGregor D. Carcinogenicity and genotoxic carcinogens. In: Ballantyne B, Marrs TC, Syversen T, eds. General and Applied Toxicology, vol. 2. 2nd ed. London: Macmillan, 2000.

Miller EC, Miller JA. The metabolism of chemical carcinogens to reactive electrophiles and their possible mechanism of action in carcinogenesis. In: Searle CE, ed. Chemical Carcinogens. Washington D.C.: American Chemical Society, 1976. A seminal article by the original authors of this important concept.

Moggs JG, Orphanides G. The role of chromatin in molecular mechanisms of toxicity. Toxicol Sci 2004; 80:218–224.

Moggs JG, Goodman JJ, Trosko JE, et al. Epigenetics and cancer: implications for drug discovery and safety assessment. Toxicol Appl Pharmacol 2004; 196:422–430.

Morimura K, Cheung C, Ward JM, et al. Differential susceptibility of mice humanised for peroxisome proliferator-activated receptor α to Wy-14643 induced liver tumorigenesis. Carcinogenesis 2005; 27:1074–1080.

Pegg AE. Alkylation and subsequent repair of DNA after exposure to dimethylnitrosamine and related carcinogens. In: Hodgson E, Bend JR, Philpot RM, eds. Reviews in Biochemical Toxicology, Vol. 5. New York: Elsevier-North Holland, 1983.

Pitot HC, Dragan YP. Chemical carcinogenesis. In: Klaassen CD, ed. Cassarett and Doull's Toxicology, The Basic Science of Toxicology. 6th ed. New York: McGraw Hill, 2001.

Ruddon RW. Chemical carcinogenesis. In: Pratt WB, Taylor P, eds. Principles of Drug Action, The Basis of Pharmacology. New York: Churchill Livingstone, 1990.

Sayer JM, Whalen DL, Jerina DM. Chemical strategies for the inactivation of bay-region diol-epoxides, ultimate carcinogens derived from polycyclic aromatic hydrocarbons. Drug Metab Rev 1989; 20:155.

Searle CE, ed. Chemical Carcinogens. 2nd ed. Washington D.C.: American Chemical Society, 1984. A reference text.

Smart RC, Akunda JK. Carcinogenesis. In: Hodgson E, Smart RC, eds. Introduction to Biochemical Toxicology. 3rd ed. New York: Wiley Interscience, 2001.

Smith SJ, Li Y, Whitley R, et al. Molecular epidemiology of p53 protein mutations in workers exposed to vinyl chloride. Am J Epidemiol 1998; 147:302–308.

Stanley LA. Molecular aspects of chemical carcinogenesis: The role of oncogenes and tumour suppressor genes. Toxicology 1995; 96:173.

Swenberg JA, Ham A, Koc H, et al. DNA adducts: effects of low exposure to ethylene oxide, vinyl chloride and butadiene. Mutat Res 2000; 464:77–86.

Thorgiersson SS, Glowinski IB, Mcmanus ME. Metabolism, mutagenicity and carcinogenicity of aromatic amines. In: Hodgson E, Bend JR, Philpot RM, eds. Reviews in Biochemical Toxicology, Vol. 5. New York: Elsevier-North Holland, 1983.

Waalkes MP, Ward JM, eds. Carcinogenesis. New York: Raven Press, 1994.

Williams GM. Mechanisms of chemical carcinogenesis and applications to human cancer risk assessment. Toxicology 2001; 166:3–10.

Wiseman H. Tamoxifen. Chichester: John Wiley and Sons, 1994.

Direct Toxic Action: Tissue Lesions

Anders MW, Elfarra AA, Lash LH. Cellular effects of reactive intermediates: Nephrotoxicity of S-conjugates of amino acids. Arch Toxicol 1987; 60:103.

Bannon MJ. The dopamine transporter: role in neurotoxicity and human disease. Toxicol Appl Pharmacol 2005; 204:355–360.

Bessems JGM, Vermeulen NPE. Paracetamol (acetaminophen) induced toxicity: molecular and biochemical mechanisms, analogues and protective approaches. Crit Rev Toxicol 2001; 31:55–138.

Bolchoz LJC, Morrow JD, Jollow DJ, et al. Primaquine induced hemolytic anaemia: effect of 6-methoxy-8-hydroxylaminoquinoline on rat erythrocyte sulphydryl status, membrane lipids, cytoskeletal proteins and morphology. J Pharmacol Exp Ther 2002; 303:141–148.

Bowman ZS, Morrow JD, Jollow, DJ, et al. Primaquine induced hemolytic anaemia: role of membrane lipid peroxidation and cytoskeletal protein alterations in the hemotoxicity of 5-hydroxyprimaquine. J Pharmacol Exp Ther 2005; 314:838–845.

Boyd MR. Biochemical mechanisms in chemical induced lung injury: Roles of metabolic activation. CRC Crit Rev Toxicol 1980; 7:103.

Cohen GM. Pulmonary metabolism of foreign compounds: Its role in metabolic activation. Environ. Health Perspect 85; 31: 1990.

Dai Y, Cederbaum AI. Inactivation and degradation of human cytochrome P4502E1 by CCl4 in a transfected HepG2 cell line. J Pharmacol Exp Ther 1995; 275; 1614–1622.

Dekant W, Vamvakas S. Biotransformation and membrane transport in Nephrotoxicity. Crit Rev Toxicol 1996; 26(3):309.

Fromenty B, Pessayre D. Impaired mitochondrial unction in microvesicular steatosis. J Hepatol 1997; 26:43–53.

Fukuda T. Neurotoxicity of MPTP. Neuropathol 2001; 21:323–332.

Gibson FD, Pumford NR, Samokyszyn VM, et al. Mechanism of acetaminophen-induced hepatotoxicity: Covalent binding versus oxidative stress. Chem Res Toxicol 1996; 9:580.

Gram TE. Chemically reactive intermediates and pulmonary xenobiotic toxicity. Pharmacol Rev 1997; 49:297–341.

Halmes NC, Hinson JA, Martin BM, et al. Glutamate dehydrogenase covalently binds to a reactive metabolite of acetaminophen. Chem Res Toxicol 1996; 9:541.

Huang Y-S, Chern HD, Su WJ, et al. Polymorphism of the N-acetyltransferase 2 gene as a susceptibility risk factor for antituberculosis drug-induced hepatitis. Hepatology 2002; 35:883–889.

Jaeschke K, Bajt ML. Intracellular signaling mechanisms of acetaminophen induced liver cell death. Toxicol Sci 2006; 89; 31–41.

Kiyomiya K, Matsushita N, Kurebe M, et al. Mitochondrial cytochrome c oxidase as a target site for cephalosporin antibiotics in renal epithelial cells (LLC-PK$_1$) and renal cortex. Life Sci 2002; 72: 49–57.

Landin JS, Cohen SD, Khairallah EA. Identification of a 54kDa mitochondrial acetaminophen binding protein as aldehyde dehydrogenase. Toxicol App Pharmacol 1996; 141:299.

Laurent G, Kishore BK, Tulkens PM. Aminoglycoside-induced renal phospholipidosis and nephrotoxicity. Biochem Pharmacol 1990; 40:2383.

Maret G, Testa B, Jenner P, et al. The MPTP story: MAO activates tetrahydropyridine derivatives to toxins causing parkinsonism. Drug Metab Rev 1990; 22:291.

Monks TJ, Lau SS. Reactive intermediates and their toxicological significance. Toxicology 1988; 52:1.

Monks TJ, Anders MW, Dekant W. Contemporary issues in toxicology: Glutathione conjugate mediated toxicities. Toxicol Pharmacol 1990; 106:1–19.

Monks TJ, Lau SS. The pharmacology and toxicology of polyphenolic glutathione conjugates. Ann Rev Pharmacol Toxicol 1998; 38:229–255.

Park BK, Kitteringham NR, Maggs JL, et al. The role of metabolic activation in drug induced hepatotoxicity. Ann Rev Pharmacol Toxicol 2005; 45:177–202.

Pohl LR. Biochemical toxicology of chloroform. In: Bend J, Philpot RM, Hodgson E, eds. Reviews of Biochemical Toxicology, Vol. 1. Amsterdam: Elsevier-North Holland, 1979.

Prescott LF. Paracetamol. London: Taylor & Francis, 1996.

Recknagel RO, Glende EA. Carbon tetrachloride hepatotoxicity: An example of lethal cleavage. CRC Crit Rev Toxicol 1973; 2:263.

Rush GF, Hook JB. The kidney as a target organ for toxicity. In: Cohen GM, ed. Target Organ Toxicity, Vol. 2. Boca Raton, FL: CRC Press, 1986.

Rush GF, Smith JH, Newton JF, et al. Chemically induced nephrotoxicity: role of metabolic activation. CRC Crit Rev Toxicol 1984; 13:99.

Smith LL. Paraquat. In: Roth RA, ed. Comprehensive Toxicology: Toxicology of the Respiratory System. Vol. 8. Oxford, U.K.: Elsevier, 1997.

Smith LL, Nemery B. The lung as a target organ for toxicity. In: Cohen GM, ed. Target Organ Toxicity, Vol. 2. Boca Raton, FL: CRC Press, 1986.

Takeda M, Babu E, Narikawa S, et al. Interaction of human organic anion transporters with various cephalosporin antibiotics. Eur J Pharmacol 2002; 438:137–142.

Tarloff JB, Goldstein RS, Hook JB. Xenobiotic biotransformation by the kidney: pharmacological and toxicological aspects. In: Gibson GG, ed. Progress in Drug Metabolism, Vol. 12. London: Taylor & Francis, 1990.

Timbrell JA. The role of metabolism in the hepatoxicity of isoniazid and iproniazid. Drug Metab Rev 1979; 10:125.

Timbrell JA. Acetylation and its toxicological significance. In: Gorrod JW, Oelschlager H, Caldwell J, eds. Metabolism of Xenobiotics. London: Taylor & Francis, 1988.

Van Bladeren PJ. Glutathione conjugation as a bioactivation reaction. Chem Biol Interact 2000; 129:61–76.

Walker RJ, Duggin GG. Drug nephrotoxicity. Annu Rev Pharmacol Toxicol 1988; 28:331–345.

Zhou S, Palmeria CM, Wallace KB. Doxorubicin induced persistent oxidative stress to cardiac myocytes. Toxicol Lett 2001; 121; 151–157.

Zimmerman HJ. Hepatotoxicity: The Adverse Effects of Drugs and Other Chemicals on the Liver. 2nd ed. Lippincott-Williams & Wilkins, 1999.

Pharmacological, Physiological, and Biochemical Effects

Cantilena LR. Clinical toxicology. In: Klaassen CD, ed. Cassarett and Doull's Toxicology, The Basic Science of Toxicology. 6th ed. New York: McGraw Hill, 2001.

Casida JE, Toia RF. Organophosphorus pesticides: their target diversity and bioactivation. In: Dekant W, Neumann HG, eds. Tissue Specific Toxicity: Biochemical Mechanisms. London: Academic Press, 1992.

Coburn RF, Forman HJ. Carbon monoxide toxicity. In: Fahri LE, Tenney SM, eds. Handbook of Physiology, The Respiratory System, Section 3, Vol. IV, Bethesda, MD: American Physiology Society, 1987.

Ellenhorn MJ, Barceloux DG. Medical Toxicology, New York: Elsevier, 1988.

Fosslien E. Review: mitochondrial medicine-molecular pathology of defective oxidative phosphorylation. Ann Clin Lab Sci 2001; 31:25–67.

Goldfrank LR, Flomenbaum NE, Lewin NA, et al. , eds. Goldfrank's Toxicologie Emergencies, 4th ed. Norwalk, Connecticut: Appleton & Lange, 1990.

Gossel TA, Bricker JD. Principles of Clinical Toxicology, 3rd edn. New York: Raven Press, 1994.

Hathway DE. Molecular Aspects of Toxicology. London: Royal Society of Chemistry, 1984.

Johnson MK. The target for initiation of delayed neurotoxicity by organophosphorus esters: Biochemical studies and toxicological applications. In: Bend J, Philpot RM, Hodgson E, eds. Reviews of Biochemical Toxicology, Vol. 4. Amsterdam: Elsevier-North Holland, 1982.

Nebert DW, Weber WW. Pharmacogenetics. In: Pratt WB, Taylor P, eds. Principles of Drug Action, The Basis of Pharmacology. New York: Churchill Livingstone, 1990.

Penney DG. Acute carbon monoxide poisoning: animal model: A review. Toxicology 1990; 62:123.

Penny DG. A review: Hemodynamic response to carbon monoxide. Environ Health Perspect 1988; 77:121.

Rocco TP, Fang JC. Pharmacological treatment of heart failure. In: Brunton L, Lazo J, Parker K, eds. Goodman & Gilman's The Pharmacological Basis of Therapeutics. New York: McGraw Hill, 2005.

Simpson LL. Identification of the major steps in botulinum toxin action. Ann Rev Pharmacol Toxicol 2004; 44:167–193.

Wallace KB, Starkov AA. Mitochondrial targets of drug toxicity. Ann Rev Pharmacol Toxicol 2000; 40:353–388.

Lethal Synthesis and Incorporation

Lauble H, Kennedy MC, Emptage MH, et al. The reaction of fluorocitrate with aconitase and the crystal structure of the enzyme inhibitor complex. Proc Natl Acad Sci U S A 1996; 93:13699–13703.

Peters RA. Biochemical Lesions and Lethal Synthesis. Oxford: Pergamon Press, 1963. The classic text first describing the concepts.

Teratogenesis

Juchau MR. Bioactivation in chemical teratogenesis. Annu Rev Pharmacol Toxicol 1990; 29:165.

Martz F, Failinger C, Blake DA. Phenytoin teratogenesis: correlation between embryopathic effect and covalent binding of a putative arene oxide metabolite in gestational tissue. J Pharmac Exp Ther 1977; 203:321.

Rogers JM, Kavlovck RJ. Developmental toxicology. In: Klaassen CD, ed. Cassarett and Doull's Toxicology, The Basic Science of Toxicology. 6th ed. New York: McGraw Hill, 2001.

Ruddon RW. Chemical teratogenesis. In: Pratt WB, Taylor P, eds. Principles of Drug Action, The Basis of Pharmacology. New York: Churchill Livingstone, 1990.

Stephens TD, Fillmore BJ. Hypothesis: thalidomide embryopathy- proposed mechanisms of action. Teratology 2000; 61:189–195.

Wilson JG, Fraser FC. Handbook of Teratology, Vol. 2. New York: Plenum Press, 1977. A reference text.

Witorsch RJ., ed. Reproductive Toxicology. New York: Raven Press, 1995.

Immunotoxicity

Amos HE. Immunological aspects of practolol toxicity. Int J Immunopharmacol 1979; 1:9.

Bourdi M, Amouzadeh HR, Rushmore TH, et al. Halothane induced liver injury in outbred guinea pigs: role of trifluoroacetylated protein adducts in animal susceptibility. Chem Res Toxicol 2001; 14: 362–370.

Chen ML, Gandolfi AJ. Characterisation of the humoral immune response and hepatotoxicity after multiple halothane exposures in guinea pigs. Drug Metab Rev 29:103–122. 1997.

Jiang X, Khursigara G, Rubin RL. Transformation of lupus inducing drugs to cytotoxic products by activated neutrophils. Science 1994; 266:810.

Park BK, Kiteringham N. Drug-protein conjugation and its immunological consequences. Drug Metab Rev 1990; 22:87.

Park BK, Kiteringham N, Powell H, et al. Advances in molecular toxicology: towards understanding idiosyncratic drug toxicity. Toxicology 2000; 153:39–60.

Perry HM. Late toxicity to hydralazine resembling systemic lupus erythematosus or rheumatoid arthritis. Am J Med 1973; 54:58. The first description of the phenomenon.

Pohl LR, Satoh H, Christ DD, et al. The immunologic and metabolic basis of drug hypersensitivities. Ann Rev Pharmacol Toxicol 28; 367: 1988.

Pohl LR, Kenna JG, Satoh H, et al. Neoantigens associated with halothane hepatitis. Drug Metab Rev 1989; 20:203.

Pratt WB. Drug allergy. In: Pratt WB, Taylor P, eds. Principles of Drug Action, The Basis of Pharmacology. New York: Churchill Livingstone, 1990.

Timbrell JA, Facchini V, Harland SJ, et al. Hydralazine-induced lupus: is there a toxic metabolic pathway? Eur J Clin Pharmacol 1984; 27:555.

Uetrecht J. Drug metabolism by leukocytes and its role in drug-induced lupus and other idiosyncratic drug reactions. CRC Crit Rev Toxicol 1990; 20:213.

Uetrecht J. Idiosyncratic drug reactions: Current understanding. Ann Rev Pharmacol Toxicol 2007; 47: 513–539.

Multi-Organ Toxicity

Fielder RJ, Dale EA. Toxicity Review: Cadmium and its Compounds. London: HMSO, 1983.

Goyer RA, Clarkson TW. Toxic effects of metals. In: Klaassen CD, ed. Cassarett and Doull's Toxicology, The Basic Science of Toxicology. 6th ed. New York: McGraw Hill, 2001.

Jacobsen D, Mcmartin KE. Methanol and ethylene glycol poisonings. Mechanism of toxicity, clinical course, diagnosis and treatment. Med Toxicol 1986; 1:309.

Park JD, Liu Y, Klaassen CF. Protective effects of metallothionein against the toxicity of cadmium and other metals. Toxicology 2001; 163:93–100.

Thorne PS. Occupational toxicology. In: Klaassen CD, ed. Cassarett and Doull's Toxicology, The Basic Science of Toxicology. 6th ed. New York: McGraw Hill, 2001.

Glossary

ABC: ATP-binding cassette. Also known as MDR (multidrug resistant) protein. A family of proteins located in the plasma membrane, which serve to pump drugs and other foreign chemicals out of cells.

α-carbon: first carbon after functional group.

acetylator status/phenotype: genetically determined difference in the acetylation of certain foreign compounds giving rise to rapid and slow acetylators.

acidemia: decrease in blood pH.

acidosis: the condition where the pH of the tissues falls below acceptable limits.

aciduria: decrease in urinary pH.

acro-osteolysis: dissolution of the bone of the distal phalanges of the fingers and toes.

ACTH: adrenocorticotrophic hormone.

actin: cytoskeletal protein.

acyl: group such as acetyl, propionyl, etc.

acylation: addition of an acyl group.

ADCC: antibody-dependent cell cytotoxicity.

adenocarcinoma: malignant epithelial tumor.

adenoma: benign epithelial tumor of glandular origin.

ADI: acceptable daily intake.

ADP ribosylation: transfer of ADP ribose from NAD+ to a protein.

adrenergic: nerves responding to adrenaline.

β-adrenoceptor: an autonomic receptor, which is of two types, β1 and β2.

aglycone: portion of molecule attached to glycoside as in a glucuronic acid conjugate.

AHH: aryl hydrocarbon hydroxylase.

Ah receptor: a protein, which binds polycyclic hydrocarbons such as dioxin (TCDD). Binding to this receptor is part of the process of induction of xenobiotic metabolizing enzymes.

AhR: Ah receptor.

alkalosis: the condition where the pH of the tissues rises above acceptable limits.

alkyl: group such as methyl or ethyl.

alkylation: addition of an alkyl group.

allosteric (change): alteration of protein conformation resulting in alteration in function.

allozymes: alternative electrophoretic forms of a protein coded by alternative alleles of a single gene.

allyl: the unsaturated group, $CH_2=CH-$.

allylic: containing the allyl group.

ALT: alanine transaminase; alanine aminotransferase; previously known as SGPT (serum glutamate pyruvate transaminase).

aminoaciduria: excretion of amino acids into the urine.

amphipathic (amphiphilic): molecules possessing both hydrophobic, nonpolar, and hydrophilic polar moieties.

ANA: antinuclear antibodies.

anaphylactic (anaphylaxis): a Type I immunological reaction.

aneuploidy: increase or decrease in the normal number of chromosomes of an organism (karyotype).

antihypertensive: drug used for lowering blood pressure.

antiport: membrane carrier system in which two substances are transported in opposite directions.

antitubercular: drug used to treat tuberculosis.

anuria: cessation of urine production.

APC: antigen presenting cell.

aplastic: absence of tissue such as bone marrow in aplastic anemia.

apoprotein: protein component of an enzyme, e.g. enzyme minus lipid, metal ions and cofactors etc.

apoptosis: programmed cell death.

apurinic: loss of a purine moiety.

arteriole: small branch of an artery.

Arthus (reaction): Type III immediate hypersensitivity reaction.

aryl: aromatic moiety.

arylamine: aromatic amine.

arylated: addition of aromatic moiety.

arylhydroxamic acid: *N*-hydroxy aromatic acetylamine.

AST: aspartate transaminase; previously known as SGOT (serum glutamate oxalate transaminase).

astrocytes: cells found in the central nervous system.

ATP: Adenosine triphosphate.

AUC: area under the plasma concentration versus time curve.

autoantibodies: antibodies directed against "self" tissues or constituents.

autoimmune: immune response in which antibodies are directed against the organism itself.

autoradiography: use of radiolabeled compounds to show distribution in a tissue, organ, or even whole animal.

axoplasm: cytoplasm of an axon.

azo: N=N group.

basophil: a granulocyte (type of white blood cell) distinguishable by Leishman's stain as containing purple blue granules.

basophilic: cells that stain readily with basic dyes.

Bcl-2: oncogene.

β-cells: insulin-producing cells of the pancreas.

bioactivation: metabolism of a foreign substance to a chemically reactive metabolite.

bioavailability: proportion of a drug or foreign compound absorbed by the organism.

biomarker: a biochemical or biological marker of exposure, response, or susceptibility to chemicals.

biosynthesis: synthesis within and by a living organism.

biotransformation: chemical change brought about by a biological system.

blebbing: appearance of blebs (protrusions) on outside of cells.

BNPP: bis-*p*-nitrophenyl phosphate.

bolus: a single dose.

bradycardia: slowing of heart beat.

bronchiole: small branch of the bronchial tree of the lungs.

bronchitis: inflammation of the bronchial system.

bronchoconstriction: constriction of the airways of the lungs.

bronchoscopy: visual examination of the bronchial system with a bronchoscope.

bronchospasm: spasms of constriction in the bronchi.

canalicular: relating to the canaliculi.

canaliculi: smallest vessels of the biliary network formed from adjoining hepatocytes.

carbanion: chemical moiety in which a carbon atom is negatively charged.

carbene: free radical with two unpaired electrons on a carbon atom.

carbinolamine: chemical moiety in which carbon to which amino group is attached is hydroxylated.

carbocation: a positively charged carbon atom.

carbonium (ion): chemical moiety in which a carbon atom is positively charged.

carcinogenic: a substance able to cause cancer.

cardiac arrhythmias: abnormal beating rhythms of the heart.

cardiolipin: double phospholipid that contains four fatty acid chains. Found mainly in mitochondrial inner membrane.

cardiomyopathy: pathological changes to heart tissue.

cardiotoxic: toxic to heart tissue.

caspases: cystein protease enzymes involved in cell signaling.

catabolized: metabolically broken down.

cDNAs: complementary DNA strands.

centrilobular: the region of the liver lobule surrounding the central vein.

chelation: specific entrapment of one compound, such as a metal, by another in a complex.

chemoreceptors: biological receptors modified for excitation by a chemical substance.

chiral: the presence of asymmetry in a molecule giving rise to isomers.

chloracne: a particular type of skin lesion caused particularly by halogenated hydrocarbons.

chloracnegenicity: ability to cause chloracne.

cholestasis: cessation of bile flow.

cholinergic: receptors that are stimulated by acetylcholine.

cis: when two groups in a chemical structure are on the same side of the molecule.

clastogenesis: occurrence of chromosomal breaks, which results in a gain, loss, or rearrangement of pieces of chromosomes.

clathrate: shape or appearance of a lattice.

clinical trials: initial studies carried out with a drug in human subjects.

coadministered: when two substances are given together.

co-oxidation: oxidation of two substrates in the same reaction.

coplanar: in the same plane.

cotransport: the transport of two substances together such as by a membrane active transport system.

CoA: Coenzyme A.

codominance: full expression in a heterozygote of both alleles of a pair without either being influenced by the other.

COHb: carboxyhemoglobin.

colorectal: colon and rectum portion of the gastrointestinal tract.

cooperativity: change in the conformation of a protein such as a hemoglobin or an enzyme leading to a change in binding of the substrate.

corneal: relating to the cornea of the eye.

corneum: horny layer of skin.

cristae: folds within the mitochondrial structure.

crystalluria: appearance of crystals in the urine.

cyanosis: condition when there is an excessive amount of reduced hemoglobin in the blood giving rise to a bluish coloration to skin and mucus membranes.

CYP: cytochrome P450.

cytolytic: causing lysis of cells.

cytoskeletal: relating to the cytoskeleton.

cytoskeleton: network of protein fibrils within cells.

cytotropic: a class of antibodies that attach to tissue cells.

dalton: unit of molecular weight.

DBCP: dibromochloropropane.

DCVC: dichlorovinyl cysteine.

DDE: p,p'-dichloro-diphenyl-dichloroethylene.

DDT: p,p'-dichloro-diphenyl-trichloroethane.

deacetylation: removal of acetyl group.

dealkylation: removal of alkyl group.

deaminate: removal of amine group.

dechlorination: removal of chlorine group.

de-ethylation: removal of ethyl group.

dehalogenation: removal of halogen atom(s).

dehydrohalogenation: removal of halogen atom(s) and hydrogen, e.g., removal of HCl or HBr from a molecule.

demethylation: removal of methyl group.

denervated: a tissue deprived of a nerve supply.

depolymerize: the unraveling of a polymer.

derepression: remove repression of nucleic acid to allow transcription of gene.

desulfuration: removal of sulfur from a molecule.

deuterated: insertion of deuterium in place of hydrogen in a molecule.

diastereoisomers: stereoisomeric structures that are not enantiomers and thus not mirror images are diastereoisomers. They have different physical and chemical properties. A compound with two asymmetric centers may thus give rise to diastereoisomers.

dienes: unsaturated chemical structure in which there are two double bonds.

dimer: a macromolecule such as a protein may exist as a pair of subunits or dimers.

distal: remote from the point of reference.

dopaminergic: receptors responsive to dopamine.

DR4: HLA (human lymphocyte antigen) type, DR4.

DT diaphorase: Dicoumarol-sensitive NADH/NADPH dye reductase.

ductule: small duct.

dyspnea: labored breathing.

ED$_{50}$: effective dose for 50% of the exposed population.

ED$_{99}$: effective dose for 99% of the exposed population.

electrophile: a chemical that is attracted to react with electron-rich center in another molecule.

electrophilic: having the property of an electrophile.

embryogenesis: development of the embryo.

embryolethal: causing death of embryo.

embryo toxic: toxic to embryo.

enantiomer: a compound with an asymmetric carbon atom yields a pair of isomers or enantiomers, which are nonidentical mirror images.

encephalopathy: degenerative disease of the brain.

endocytosis: uptake or removal of substance from a cell by a process of invagination of membrane and formation of a vesicle.

endogenous: from inside an organism.

endometrium: mucus membrane of the uterus.

endonucleases: enzymes involved in DNA fragmentation.

endothelial/endothelium: layer of epithelial cells lining cavities of blood vessels and heart.

enzymic: involving an enzyme.

Epigenetic: when used to describe a carcinogen, one that does not interact directly with genetic material. However, it may cause changes to DNA methylation, for example.

epinephrine: adrenaline.

epithelial/epithelium: tissue covering internal and external surfaces of mammalian body.

epitope: antigenic determinant.

exogenous: from outside the organism.

extracellular: outside the cell.

extrahepatic: outside the liver.

FAD: Flavin adenine dinucleotide.

favism: syndrome of poisoning by Fava beans.

Fc: one of the two segments of immunoglobulin molecule.

fenestrations: perforations.

FEV: forced expiratory volume.

fibroblast: connective tissue cell.

flavoprotein: protein in which the prosthetic group contains a flavin, e.g., FMN.

FMN: Flavin mononucleotide.

FMO: Flavin monooxygenase.

folate: the cofactor folic acid and precursor of tetrahydrofolic acid.

gametotoxic: toxic to the male or female gametes.

genome: complete set of hereditary factors as in chromosomes.

glial cells: supporting cells, such as astrocytes, found in the central nervous system.

a2μ-globulin: endogenous protein that may form complexes with exogenous chemicals, which may accumulate in the kidney resulting in damage.

glomerulonephritis: inflammation of the capillary loops in the glomerulus.

glomerulus: a functional unit of the mammalian kidney consisting of a small bunch of capillaries projecting into a capsule (Bowman's capsule), which serves to collect the filtrate from the blood of those capillaries and direct it into the kidney tubule.

gluconeogenesis: synthesis of glucose.

glucosides: conjugates with glucose.

glucuronidation: addition of glucuronic acid to form a conjugate.

glucuronide: glucuronic acid conjugate.

glycolipid: molecule containing lipid and carbohydrate.

glycoprotein: molecule containing protein and carbohydrate.

glycoside: carbohydrate conjugate.

glycosidic: linkage between xenobiotic and carbohydrate moiety in conjugate.

glycosuria: glucose in urine.

granulocyte: leukocyte containing granules.

granulocytopenia: deficiency of granulocytes.

GS-: glutathione moiety.

GSH: reduced glutathione (γ-glutamyl-cysteinyl-glycine).

GSSG: oxidized glutathione.

heme: iron protoporphyrin moiety as found in hemoglobin.

hemangiosarcoma: a particular type of tumor of the blood vessels.

hematuria: appearance of blood in urine.

hemodialysis: passage of blood through a dialysis machine in order to remove a toxic compound after an overdose.

hemolysis: breakdown/destruction of red blood cells and liberation of hemoglobin.

hemoperfusion: passage of blood through a resin or charcoal column to remove toxic compounds.

hemoprotein: protein containing heme.

haloalkanes: alkanes containing one or more halogen atoms.

Hazard: the intrinsic capability of a substance to cause an adverse effect.

Hb: hemoglobin as in Hb_4O_2, which is four subunits with one molecule of oxygen bound.

HCBD: hexachlorobutadiene.

heparin: endogenous compound that stops clotting.

hepatocarcinogen: substance causing tumors of the liver.

hepatocellular: relating to the hepatocytes.

hepatomegaly: increase in liver size.

hepatotoxin: substance causing liver injury.

heterocyclic: a ring structure containing a mixture of atoms.

heterozygote: possessing different alleles for a particular character.

histones: basic proteins associated with nucleic acids.

homocytotropic: having affinity for cells from the same species (see cytotropic).

homogenate: mixture resulting from homogenization of a tissue.

homolytic: equal cleavage of a chemical bond.

homozygote: possessing the same alleles for a particular character.

humoral: immune response that involves the production of specific antibodies.

Hormesis: the concept that at low doses of a toxicant there may be positive, beneficial effects, whereas at higher doses there will be toxic effects.

hydrolase: hydrolytic enzyme.

hydroperoxide: chemical compound containing the group $-OOH$.

hydrophilic: having an affinity for water.

hydrophobic: a substance that does not tend to associate with water.

hydropic degeneration: pathological process involving the accumulation of water in cells as a result of alteration of ion transport.

hydroxyalkenals: hydroxylated aliphatic aldehyde products of lipid peroxidation, e.g., 4-hydroxynonenal.

hydroxylases: oxidative enzymes that add a hydroxyl group.

hyper-: prefix meaning increased or raised.

hyperbaric: higher than normal atmospheric pressure.

hyperplasia: increase in the number of cells in a tissue or organ.

hyperpnea: increased breathing.

hyperpyrexia: raised body temperature.

hypersensitivity: increased sensitivity to immunogenic compounds.

hyperthermia: abnormally raised temperature.

hypothyroidism: decreased activity of the thyroid gland.

hypocalcaemia: low blood calcium level.

hypolipidemia: low blood lipid.

hypophysectomy: removal of pituitary gland.

hypothalamus: region of the brain lying immediately above the pituitary responsible for coordinating and controlling the autonomic nervous system.

hypoxia: low oxygen level.

hypoxic: tissue suffering a low oxygen level.

imine: the group $H-N=R'$.

immunogen: substance able to elicit a specific immune response.

immunogenic: description of an immunogen.

immunoglobulin (Ig): one of five classes of antibody protein involved in immune responses: IgA, IgD, IgE, IgG, and IgM.

immunohistochemical/-histochemistry: use of specific antibody labeling to detect particular substances in tissues.

immunosuppression: reduction in the function of the immune system.

immunotoxic: toxic to or toxicity involving the immune system.

infarction: loss of blood supply to a tissue due to obstruction and subsequent damage.

initiation: first stage in chemical carcinogenesis in which ultimate carcinogen reacts with genetic material and causes a heritable change.

Interferon: macromolecule produced in the body response to a stimulus such as an infection.

Ifn: interferon.

inulin: soluble polysaccharide.

intron: portion of a primary RNA transcript that is deleted during splicing.

invagination: infolding of biological membrane.

isoelectric point: pH at which there is no net charge on a molecule such as an amino acid and therefore there is no movement on electrophoresis.

Isoenzyme, isozyme: one of several forms of an enzyme where the different forms usually catalyze similar but different reactions.

karyolysis: loss of the nucleus during necrosis.

karyorrhexis: fragmentation of nucleus to form basophilic granules.

karyotype: the array of chromosomes carried by the cell.

kDa: kilodaltons; units of molecular weight.

kinase: enzyme catalyzing transfer of high energy group from a donor such as ATP.

kinins: polypeptides such as bradykinin found in the plasma and involved in the inflammatory response.

L-DOPA: dihydroxyphenylalanine, a drug used to treat Parkinson's disease.

lacrimation: promoting tear formation.

lactoperoxidase: peroxidase enzyme found in mammary glands.

lavage: washing with fluid such as isotonic saline as in gastric lavage.

LD$_{50}$ (LC$_{50}$): lethal dose (or concentration) for 50% of the exposed population.

LD$_1$: lethal dose for 1% of the exposed population.

leukocytopenia: deficiency of leukocytes.

leukopenia: see leukocytopenia.

LH: luteinizing hormone.

ligand: substance that binds specifically.

liganded: bound to ligand.

ligandin: binding protein identical with glutathione-*S*-transferase B.

ligase: enzyme catalyzing the joining together of two molecules and involving breakdown of ATP or other energy-rich molecule.

ligation: tying off a vessel or duct.

lipases: enzymes that degrade lipids.

lipoidal: lipid like.

lipophilic: lipid liking. A substance that associates with lipid.

lipophilicity: a term used to describe the ability of a substance to dissolve in or associate with lipid and therefore living tissue.

lipoprotein: macromolecule that is a combination of lipid and protein. Involved in transport of lipids out of liver cells and in blood.

lithocholate: secondary bile acid.

lobule: a unit of structure in an organ such as the liver.

LOOH: lipid hydroperoxide.

lyase: enzyme that adds groups to a double bond or removes groups to leave a double bond.

lymphocyte: type of white blood cell produced in the thymus and bone marrow. Two types, T and B.

lymphoid tissue: tissue involved in the immune system, which produces lymphocytes.

lymphokines: soluble factors that are associated with T lymphocytes and cause physiological changes such as increased vascular permeability.

lysis: breakage of a chemical bond or breakdown of a tissue or macromolecule.

macrophage: a phagocytic type of white blood cell.

MAO: Monoamine oxidase.

MDR: a family of proteins located in the plasma membrane, which serve to pump drugs and other foreign chemicals out of cells.

megamitochondria: exceptionally large mitochondria.

MEOS: microsomal ethanol oxidizing system.

mercapto-: −SH group.

metallothionein: metal-binding protein.

methylation: addition of a methyl group.

MHC: major histocompatibility complex.

Michaelis—Menten kinetics: kinetics describing processes such as the majority of Enzyme-mediated reactions in which the initial reaction rate at low substrate concentrations is first order but at higher substrate concentrations becomes saturated and zero order. Can also apply to excretion for some compounds.

microbodies: peroxisomes.

microflora: microorganisms such as bacteria.

midzonal: the region between the periportal and centrilobular regions of the liver.

miosis: constriction of the pupils.

mispairing: alteration of the pairing of the bases in DNA.

mitogens: stimulants of the immune system.

μm: micrometers.

MNNG: *N*-methyl-*N'*-nitro-*N*-nitrosoguanidine.

monomer: single subunit of a compound such as a protein.

monomorphic: with one form.

MPTP: 1-Methyl-4-phenyl-1,2,3,6-tetrahydropyridine.

MTP: mitochondrial permeability transition.

mucosa: mucus membrane.

muscarinic receptors: receptors for acetylcholine found in smooth muscle, heart, and exocrine glands.

mutagen: substance that causes a heritable change in DNA.

mutagenesis: process in which a heritable change in DNA is produced.

mtDNA: mitochondrial DNA.

myalgia: muscle pain.

myelin: concentric layers of plasma membrane wound round a nerve cell process.

myeloid: pertaining to bone marrow.

myelotoxicity: toxicity to the bone marrow.

myocardial: relating to heart muscle.

myoclonic: relating to myoclonus, a shock-like muscle contraction.

myoglobin: protein similar to hemoglobin.

myosin: protein found in most vertebrate cells and always when actin is present.

N-dealkylation: removal of an alkyl group from an amine.

NAD: Nicotinamide adenine dinucleotide.

NADP: Nicotinamide adenine dinucleotide phosphate.

NADPH: Reduced form of NADP.

NAPQI: N-acetyl-p-benzoquinone imine.

nephron: functional unit of the kidney.

nephropathy: pathological damage to the nephrons of the kidney.

nephrotoxicity: toxicity to the kidney.

neuronal: relating to neurones (nerve cells).

neuropathy: damage to the nervous system.

neurotoxic: toxic to the nervous system.

neurotransmission: passage of nerve impulses.

neurotransmitter: endogenous substance involved in the transmission of nerve impulses such as adrenaline.

neutrophil: phagocytic granulocyte (white blood cell).

NIH: National Institutes of Health, where the NIH shift was discovered.

nitroso: the group $-N\rightarrow O$.

nitroxide: the group $\equiv N\rightarrow O$.

nitrenium: the group $-N^{+}-H$.

nm: nanometers.

NOAEL: No observed adverse effect level. The dose or exposure level of a chemical at which no adverse effect is detected.

nomogram: graph such as blood concentration of a compound versus time used for the estimation of the toxicity expected from the compound as in overdose cases.

nonvascularized: lacking in blood vessels.

nucleoside: base-sugar (e.g., adenosine).

nucleotide: base-sugar phosphate (e.g., adenosine triphosphate)

OAT: organic anion transporter.

OCT: organic cation transporter.

oliguria: concentrated urine.

oncogene: gene that when activated in cells can transform them to neoplastic cells.

ontogenesis: development of an organism.

oocyte: an egg cell; may be primary (diploid) or secondary (haploid).

opacification: becoming opaque.

ophthalmoscope: instrument for examining the retina of the eye.

organogenesis: formation of organs and limbs in the embryo during the gestation period.

osteomalacia: increase in uncalcified matrix in the bone resulting in increased fragility.

β-oxidation: degradative metabolism of fatty acids leading to the production of acetyl-CoA.

oxirane: epoxide ring.

oxygenase: enzyme catalyzing oxidation reaction in which molecular oxygen is utilized.

p53: tumor suppressor gene. The "guardian of the genome." May activate apoptosis.

pachytene: one of the stages in meiosis and also of development of the mammalian spermatocyte.

pampiniform plexus: collection of vessels delivering blood to and draining blood from the mammalian testis.

papilledema: edema of the optic papilla.

parenchyma: tissue of a homogeneous cell type, such as the liver, which is composed mainly of hepatocytes.

parenteral: routes of exposure to a compound other than via the gastrointestinal tract.

paraesthesia: tingling sensations in the fingers and toes.

PBBs: polybrominated biophenyls.

PCBs: polychlorinated biphenyls.

pentose: carbohydrate with five carbon atoms.

peptidase: enzyme, which cleaves peptide bonds.

peptide: molecule made up of amino acids joined by peptide bonds ($-CO-NH-$). percutaneous: through the skin.

perfusion: passage of blood or other fluid through a tissue.

perinatal: the period between the end of pregnancy (7th month in humans) and the first week of life.

periportal: the region of the liver lobule surrounding the portal tract.

peroxidized: oxidized by the addition of $-O_2$.

peroxisome: intracellular organelle, which carries out fatty acid oxidation and other oxidative transformations.

Peroxisomal proliferators: chemicals that change the number and characteristics of peroxisomes (intracellular organelles which carry out oxidation of fatty acids). These chemicals also may cause liver cancer and induce a number of enzymes.

peroxy: the chemical group $-O-O-$

pGp: p-glycoprotein. One of the ABC (MDR) proteins.

phagocytic: a cell that is able to engulf foreign particles.

pharmacodynamic (toxicity): relating to the action of a compound with a receptor or enzyme for example.

pharmacokinetic (toxicity): relating to the concentration of a compound at a target site.

phospholipidosis: accumulation of phospholipids.

photosensitization: sensitization of skin to light.

physico-chemical characteristics: characteristics of a molecule such as lipid, solubility, size, polarity.

pinocytosis: uptake of a solution by a cell.

PKC: protein kinase c. Enzyme involved in cell signaling.

pleiotropic: ability of genes or proteins to evoke a series of downstream responses in a cell at the same time.

polyG sequences: stretches of DNA with a preponderance of guanine bases.

polymerase: enzyme that catalyses polymerization.

polypeptide: a chain of amino acids.

polysaccharide: a chain of carbohydrate residues.

porphyrin: ring composed of four pyrrole rings.

posttranscriptional: event occurring after transcription of DNA code to mRNA.

portal tract (triad): the group of vessels seen in sections of liver tissue consisting of a bile ductule, hepatic arteriole, and portal venule.

potentiation: when an effect due to two substances with different modes of action is greater than expected from the effects of the individual components.

PPAR: peroxisome proliferator-activated receptor. Receptor involved in the induction of peroxisomal enzymes other responses to certain chemicals.

PPi: diphosphate.

ppm: parts per million.

prostaglandins: endogenous chemical mediators derived from unsaturated fatty acids such as arachidonic acid.

proteases: enzymes that degrade proteins.

proteinuria: presence of protein in the urine.

proto oncogene: gene that may be converted into an oncogene.

protonated: molecule to which a proton is added.

pseudosubstrate: substance that mimics the normal substrate for an enzyme.

pyknosis: change occurring in a cell during necrosis in which there is shrinkage of the nucleus and increased staining intensity.

pyrolysis: breakdown of a substance or mixture by heat/burning.

racemate: mixture of isomers; racemic mixture.

ras: oncogene as in H-*ras*; K-*ras*.

Raynaud's syndrome: severe pallor of the fingers and toes.

redox: reduction and oxidation cycle.

reticuloendothelial: system of phagocytic cells found in the liver, bone marrow, spleen, and lymph nodes.

rhinitis: inflammation of the nasal passages.

ribosomes: particles found on the smooth endoplasmic reticulum involved with protein synthesis.

risk: a measure of the probability that an adverse effect will occur.

rodenticide: pesticide used specifically for killing rodents.

ROS: reactive oxygen species.

rRNA: ribosomal RNA.

RT: room temperature.

semiquinone: quinone reduced by the addition of one electron.

sensitization: sensitization of the immune system as a result of exposure to an antigen.

SER: smooth endoplasmic reticulum.

sinusoid: blood-filled space that may be a specialized capillary such as found in the liver in which the basement membrane is modified to allow for efficient passage.

sp^2: orbital.

sp^3: orbital.

spectrin: cytoskeletal protein associated with the cytoplasmic side of the membrane.

spermatocytes: cells in the testes that develop into sperm.

splenomegaly: enlarged spleen.

stasis: stoppage or slowing down of flow.

steatosis: fatty infiltration in organ or tissue.

steric factors: factors relating to the shape of a molecule.

sulfation: addition of a sulfate group.

superoxide: O_2^-; oxygen radical.

symport: membrane carrier system in which two substances are transported in the same direction.

synergism: when an effect due to two substances with similar modes of action is greater than expected from the effects of the individual components.

synergist: a compound that causes synergism with another compound.

$T_{1/2}$: half-life.

tachypnea: increased breathing rate.

TCDD: 2,3,7,8-tetrachlorodibenzo-p-dioxin; dioxin.

TD_{50}: toxic dose for 50% of the exposed population.

teratogen: compound that causes abnormal development of the embryo or fetus.

teratogenesis: development of abnormal embryo or fetus.

thiol: −SH group.

thiolate: −S−

thiyl: −S; sulfur free radical.

thrombocytopenia: low level of platelets in the blood.

thymocytes: cells found in the thymus.

thyroidectomy: removal of the thyroid gland.

tinnitus: "ringing" in the ears.

TLV: threshold limit value.

α-**TH:** α-tocopherol; vitamin E.

α-**TQ:** α-tocopherol quinone.

toxication: increase in toxicity.

Toxicodynamics: study of the effects of toxic substances on biological systems (e.g., interaction with receptors).

toxicokinetics: study of the kinetics of toxic substances.

trans: when two groups in a chemical structure are on the opposite side of the molecule.

transversion: transformation of a purine base into a pyrimidine or vice versa.

trifluoroacyl: the group CF_3-CO-.

triglycerides: lipids in which glycerol is esterified with three fatty acids.

trihydric: molecule having three alcoholic hydroxyl groups.

trimodal: frequency distribution, which divides into three groups.

trisomy: possession of an extra chromosome.

tumorigenic: able to cause tumors.

UDP: uridine diphosphate.

UDPGA: uridine diphosphate glucuronic acid; glucuronic acid donor.

ultrafiltrate: fluid formed in renal tubule from blood passing through the glomerulus/ Bowman's capsule in the kidney.

uncouplers: compounds that uncouple oxidative phosphorylation from ATP production in the mitochondria.

uniport: membrane carrier system in which one substance is transported.

urticaria: appearance of weals on the skin; may occur as part of an immunological reaction.

uterine: part of uterus.

vacuolated: cell showing vacuoles.

vacuoles: spaces inside cells.

vasculature: blood vessels in a tissue.

vasculitis: inflammation of the vascular system.

vasoconstriction: constriction of blood vessels.

vasodilator: compound that dilates the blood vessels.

VCM: vinyl chloride monomer.

veno-occlusive: occlusion of blood vessels.

venule: very small tributary of a vein.

VLDL: very low density lipoprotein.

wheals: raised patches on the skin; may be part of an immune reaction.

xenobiotics: substances foreign to living systems.

Answers to Review Questions

CHAPTER 1

1. Paracelsus. He recognized the importance of the relationship between the size of the dose and the effect. This underlies the dose-response relationship.
2. Cycasin is a glucose conjugate of methylazoxymethanol. When hydrolyzed by the enzymes found in the gut, it releases methylazoxymethanol, which is carcinogenic. When the cycasin is given by other routes, i.p., for example, it is not hydrolyzed and so is not carcinogenic.
3. Five orders of magnitude (10^5) or 100,000 times.
4. Orfila realized that as much as pathology, chemical analysis was important.

CHAPTER 2

1. The therapeutic index is the ratio between the toxic and effective doses of a drug. It is calculated as TD_{50}/ED_{50} or alternatively LD_{50}/ED_{50}. The larger the ratio, the greater the relative safety of the drug. The margin of safety is a similar parameter except that it is calculated as TD_1/ED_{99}. It is a more critical parameter than the therapeutic index.
2. Tolerance is the modification of the biological effect of a chemical as a result of repeated dosing. For example, repeated dosing with phenobarbital leads to a decrease in the anesthetic effect of the drug as a result of enzyme induction. Giving animals a small dose of carbon tetrachloride renders a second larger dose less toxic. This may be a result of induction of repair processes and destruction of cytochrome P-450 caused by the small first dose.
3. Synergism.
4. The dose response is predicated on the following assumptions:
   a. The toxic response is a function of the concentration of the compound at the site of action
   b. The concentration at the site of action is related to the dose
   c. The response is causally related to the dose.
5. Quantal types of toxic effects are known as "all or none" effects such as death, the presence of a tumor, or loss of consciousness.
6. The LC_{50} is the concentration of a chemical, which is lethal to 50% of the organisms exposed. It is usually applied to aquatic organisms exposed through the water but can also be used for air concentrations where the individual dose is unknown.
7. The NOEL is the "no observed effect level" and it is determined from the dose-response curve when the toxic effect being measured is plotted against the dose. It is the highest dose where no toxicity is observed. The ADI is the "acceptable daily intake" and is usually determined as NOEL/100. The arbitrary factor of 100 is applied to account for individual differences between humans (factor of 10) and for species differences (factor of 10) as the NOEL is derived from animal toxicity studies. The ADI is the amount of a food additive that is felt to be safe to be ingested on a daily basis.

8. The toxic effects of tri-*o*-cresyl phosphate are cumulative such that a dose of 1 mg kg^{-1} day^{-1} for 30 days will give the same toxic effects as a single dose of 30 mg kg^{-1}. See text in Chapter 2 for further details and a full explanation.

9. Biomarkers are parameters that can be measured to determine exposure, response, or effect and susceptibility.

10. Hormesis is the phenomenon whereby low doses of chemicals have positive, beneficial effects whereas higher doses are harmful.

CHAPTER 3

1.   a. The pH partition theory states that "only non-ionized lipid soluble compounds will be absorbed by passive diffusion down a concentration gradient."

b. Active transport is a mechanism for the transport of chemicals through biological membranes and has the following features: it has a specific carrier system, it utilizes metabolic energy, it may work against a concentration gradient, it may be inhibited by chemicals interfering with metabolic energy supplies or competitively, it may be saturated at high concentrations, and it is a zero-order process.

c. Plasma protein binding is the interaction between a chemical and a blood protein, usually albumin. This process may limit the distribution, metabolism, and excretion of a chemical. The process may be saturated at high doses and there may be competition between chemicals leading to displacement of one chemical by another.

d. Enterohepatic recirculation is the process whereby a chemical is eliminated into the bile and after reaching the small intestine is reabsorbed. This is usually a result of metabolism by the gut flora, which will cleave glucuronide conjugates, for example, allowing the aglycone released to be reabsorbed. The process prolongs exposure of the animal to the chemical and may occur repeatedly.

2. The abbreviation ADME stands for absorption, distribution, metabolism, excretion.

3. Passive diffusion, active transport, facilitated diffusion, phago-/pinocytosis, and filtration.

4. Lipophilicity, pK_a/charge/polarity, size/shape, and similarity to endogenous molecules.

5. Inhaled particles of 20 μm diameter or more tend to be retained in upper parts of the respiratory tract (trachea, pulmonary bronchi and terminal bronchioles) and then removed by ciliary action. In contrast, smaller particles of around 6 μm diameter reach all parts of the respiratory system including the alveolar sacs. Smaller (2 μm) and very small particles (0.2 μm) may not reach the alveolar sacs, only the terminal bronchioles and alveolar ducts. Particles of diameter 1 μm or less may be absorbed if they reach the alveolar sacs.

6. The statement is true. Passive diffusion is a first-order rate process as it is dependent on the concentration of the chemical. In contrast, active transport is a zero-order process as it is not dependent on the concentration.

7. The difference between the two drugs pentobarbital and thiopental is that thiopental has a sulfur atom at position 2 whereas pentobarbital has an oxygen atom. Consequently thiopental is more lipophilic and therefore will be more readily absorbed and will distribute more rapidly into fat tissue.

8. Log P is the logarithm (log$_{10}$) of the partition coefficient.

9. Surface area: skin, lungs, and gastrointestinal tract all have a large surface area.

Blood flow: the lungs have a better blood flow than the gastrointestinal tract and the skin has a poor blood flow.

Thinness of barrier between site of absorption and blood: this barrier is greatest for the skin and least for the lungs.

pH of absorption site: this is variable for the gastrointestinal tract and may be as low as 2, whereas for lungs this will be closer to 7. It is not really a factor in the case of skin. However, the presence of fur on skin may be important in retaining chemicals in contact with the skin or affording some protection. Similarly the presence of food in the gastrointestinal tract may either increase or decrease/delay absorption.

10. The first-pass effect is the phenomenon in which a drug or other chemical is metabolized or otherwise removed during the first pass through an organ, usually the liver.

11. A weak acid is most likely to be absorbed from the stomach where the ionization will be suppressed.

12. Albumin.

13. Barbiturates such as phenobarbital are weak acids. The toxicity of the barbiturate is mainly the result of the effects on the central nervous system. Only the nonionized form of the drug will distribute into the central nervous system. The proportion ionized will depend on the pK_a and the pH of the blood. By increasing the pH of the blood using sodium bicarbonate administration to the poisoned patient, ionization of the barbiturate will be increased and distribution to tissues such as the brain will be decreased. Urinary excretion of the barbiturate will also be increased because the urinary pH will be increased.

14. The plasma level of a chemical is an important piece of information because
    a. It reflects the absorption, distribution, metabolism, and excretion of the chemical.
    b. It may reflect the concentration of the chemical at the target site.
    c. It can be used to derive toxicokinetic parameters.
    d. It may indicate tissue exposure.
    e. It may indicate accumulation is occurring as doses are increased or repeated.
    f. It is central to any toxicokinetic study.

15. The volume of distribution of a chemical can be determined from the plasma level and dose as follows:
    Volume of distribution (liters)
    Alternatively it can be determined using the area under the plasma concentration–time curve (AUC).
    It should be noted that to determine the V_D, the dose must have been administered by intravenous injection or the bioavailability must be known to be 100%.

16. A chemical can be shown to undergo first-pass metabolism by comparing its blood concentration after it has been administered through two different routes of administration.

17. The biliary excretion of chemicals and their metabolites is influenced by (a) the molecular weight of the chemical, (b) the polarity of the molecule, and (c) the species of animal.

18. ABC transporter proteins are crucial elements in the detoxication process as they pump potentially toxic chemicals out of cells and tissues. Thus the transporters in the intestine pump chemicals back into the lumen of the gut, those in the brain form part of the blood-brain barrier, keeping chemicals out of the organ, as also occurs in the placenta where transporters move chemicals out of the fetal bloodstream and back into the maternal bloodstream.

CHAPTER 4

1. a. Cytochrome P-450 is a heme-containing enzyme system that is involved in the metabolism of chemicals, especially those foreign to the body. It is most abundant in the liver and is located in the smooth endoplasmic reticulum. It requires molecular oxygen and NADPH, and the complex, which contains cytochrome P-450 also contains NADPH cytochrome P-450 reductase.
    b. Glutathione is a tripeptide consisting of glutamate, cysteine, and glycine. It is found in many tissues but is most abundant in the liver. It is extensively involved in the detoxication of reactive metabolites either by forming conjugates or by reducing the reactive intermediate. The sulfur atom acts as a nucleophile,

reacting with electrophilic groups or atoms in foreign chemicals or their metabolites or free radicals.

c. N-acetyltransferase is the enzyme that catalyzes the addition of the acetyl group to appropriate molecules. This phase 2 metabolic reaction adds the acetyl group, derived from acetyl CoA, to amino, sulfonamido, and hydrazine groups. In humans, there are two N-acetyltransferase enzymes (NAT1 and NAT2). The activity of NAT2 varies between human individuals as a result of mutations in the gene coding for it. This leads to fast and slow acetylator phenotypes.

d. Glucuronic acid is a carbohydrate, derived from glucose by oxidation, which is utilized in the phase 2 conjugation of xenobiotics. It is first activated as UDP-glucuronic acid and then, in the presence of glucuronosyl transferase, it is added to oxygen in hydroxyl or carboxylic acid groups, sulfur, or nitrogen. In each case the relevant atom in the xenobiotic reacts with the C-1 carbon atom. The resulting conjugate is normally in the β configuration. The resulting conjugate is water soluble.

2. The consequences of metabolism are:
   a. The biological half-life is decreased,
   b. The duration of exposure is reduced,
   c. Accumulation of the compound in the body is avoided,
   d. The duration of the biological activity may be affected.

3. Cytochrome P-450 is mostly located in the smooth endoplasmic reticulum.

4. Some phase 1 metabolic transformations are aromatic hydroxylation, aliphatic hydroxylation, N-hydroxylation, N-oxidation, deamination, N-, S- and O-dealkylation, alcohol oxidation, azo and nitro reduction, ester and amide hydrolysis. Some phase 2 metabolic transformations are glucuronic acid conjugation of phenols and aromatic acids, N-glucuronidation, sulfate conjugation of hydroxyl groups, sulfate conjugation of amino groups, glutathione conjugation with epoxides, aromatic nitro and halogen compounds, acetylation of amines, sulfonamido and hydrazine groups, amino acid conjugation of aromatic acids, catechol methylation.

5. Cytochrome P-450 requires oxygen, NADPH, and NADPH cytochrome P-450 reductase.

6. Benzene can form five aromatic hydroxylated products. It may also be metabolized to the open-chain *trans, trans*-muconic acid.

7. N-demethylation is the result of carbon oxidation.

8. Sulfate conjugation requires 3'-phosphoadenosyl-5'-phosphosulphate (PAPS). This is synthesized from ATP and sulfate so both of these are necessary.

9. Glutathione conjugation can be the result of either an enzyme-mediated reaction or a chemical reaction.

10. A number of amino acids can be used for conjugation of aromatic acids but the most common is glycine. Others utilized are taurine, glutamine, ornithine, for example. However, the one used may depend on the species of animal.

11. Inorganic mercury may be methylated by biological systems.

12. CYP3A4 is the most important isoform of cytochrome P-450 in humans.

13. a. An electrophile
    b. A nucleophile
    c. A free radical
    d. A redox reagent

CHAPTER 5

1. a. Glucose-6-phosphate dehydrogenase (G6P dehydrogenase) deficiency is a sex-linked genetic disorder. Males who have this disorder are deficient in the enzyme G6P dehydrogenase, which is the first enzyme in the pentose phosphate shunt. Lack of this enzyme results in a deficiency in NADPH and therefore an inability to reduce oxidized glutathione (GSSG) to reduced glutathione (GSH) in the red

cell. Therefore, these individuals suffer when exposed to drugs and other chemicals, which can oxidize hemoglobin as there is insufficient GSH to protect the red cell.

b. Cytochrome P-450 2D6 is an isozyme that is involved in metabolizing nitrogen-containing xenobiotics. There is genetically determined variability, which affects this isozyme such that some individuals may have a mutation resulting in reduced amounts of enzyme. The individuals who have this are known as poor metabolizers and this is a recessive trait. The toxicological importance of this trait is that a number of toxic reactions are associated with the poor metabolizer status. Thus debrisoquine metabolism is decreased in poor metabolizers and these individuals suffer exaggerated hypotension after a standard therapeutic dose. Perhexiline may cause liver dysfunction and peripheral neuropathy in poor metabolizers, penicillamine may cause skin rashes, and phenformin may cause lactic acidosis in poor metabolizers.

c. The diet of an animal may influence the toxicity of a chemical in a number of ways

  i. By altering the metabolism as a result of reduced levels of enzymes or cofactors. This may be due to decreased protein intake or decreased intake of cofactors such as sulfate.

  ii. By altering the distribution of a chemical. For example, a low-protein diet may lead to low levels of plasma protein that will lead to decreased binding capacity and therefore a higher plasma level of unbound chemical.

  iii. By decreasing the level of protective sub-substances such as glutathione or vitamin E. This could be due to a decreased level of methionine or protein in the case of glutathione or of the vitamin itself in the case of vitamin E.

d. The age of an animal may be a factor affecting the toxicity of a chemical. Thus chemicals may be more or less toxic in young or old animals. Chemicals which require metabolic activation may be less toxic in neonatal animals as a result of reduced levels of enzyme activity. However, this may be offset by reduced levels of protective agents such as glutathione. Conversely, when the parent substance has biological activity, for example, hexobarbital, young animals are much more susceptible. Distribution of chemicals may be different in animals of different ages as a result of reduced levels of plasma proteins in both neonates and old animals or differences in the particular proteins in the blood in neonates. The blood-brain barrier is less effective in neonates and this is why morphine is much more toxic to the neonate than the adult animal. Renal excretion of chemicals may be reduced in old age, hence elimination from the body is reduced, the half-life is longer and repeated exposure may lead to accumulation and toxicity.

2. Chirality may feature in one of the following ways:
   (a) substrate stereoselectivity, (b) product stereoselectivity, (c) inversion of configuration, (d) loss of chirality.

3. Species differences in absorption may occur for any of the following reasons: differences in gastrointestinal pH or anatomy, differences in rate of blood flow to the absorption site, differences in breathing rate, differences in skin type, and the presence or absence of fur.

4. Biliary excretion varies between species as the molecular weight threshold for biliary excretion varies between species. Thus the threshold in rats is about 300 whereas it is 500 in humans.

5. The following are some examples: Phase 1 aromatic hydroxylation of aniline varies with species, the metabolism of malathion differs between mammals and insects, and the metabolism of amphetamine varies between different mammalian species.

6. The conjugation of phenol with glucuronic acid is zero in the cat but the facility is present in many other species. Conversely the conjugation of phenol with sulfate is zero in the pig but the facility is present in many other species. There are many

quantitative (and qualitative) differences in the conjugation of arylacetic acids with amino acids.

7. A sex difference in the toxicity of chloroform to the kidney occurs in mice. Male animals are more sensitive than females. The reason is that chloroform is metabolized more rapidly in male mice than females and this is affected by male hormones (androgens).

8. Genetic factors can affect the toxicity of chemicals either (a) by influencing the response to the chemical or (b) by influencing the disposition of the compound. An example of (a) is the deficiency of glucose-6-phosphate dehydrogenase, a sex-linked trait. This trait leads to increased sensitivity to a variety of drugs and chemicals resulting in hemolytic anemia. An example of (b) is the acetylator phenotype. This trait leads to differences in metabolism of amines, sulfonamides, and hydrazines. With the substituted hydrazine drug isoniazid, the slow acetylator phenotype are more susceptible to the peripheral neuropathy and the hepatic damage caused by this drug.

9. a. Cytochrome P-450 2D6, cytochrome P-450 2C, pseudocholinesterase.
   b. acetyltransferase, glucuronosyl transferase.

10. a. A protein-deficient diet would increase the toxicity of a chemical when the parent compound was responsible for the toxicity and the reduced level of enzymes likely to occur after a protein-deficient diet would decrease detoxication.
    b. A protein-deficient diet would decrease the toxicity of a chemical, which was metabolically activated as the likely lack of enzyme protein would decrease this activation. However protein-deficient diets may also decrease glutathione level and so lower protection.

11. Chemicals that require metabolic activation may be less toxic in neonatal animals as a result of reduced levels of enzyme activity. However levels of protection may also be reduced.

12. The barbiturate type, the clofibrate type, the polycyclic hydrocarbon type, the steroid hormone type, and alcohol/acetone/isoniazid type. For example, the polycyclic hydrocarbon type act through a receptor. The barbiturate type increase liver blood flow.

13. Enzyme induction can be detected as
    a. A decrease in half-life or plasma level of a chemical
    b. A change in proportion of metabolites
    c. A change in a pharmacological or toxic effect
    d. An increase in plasma γ-glutamyl transferase
    e. An increase in the excretion of D-glucaric acid
    f. An increase in urinary 6-β-hydroxycortisol excretion in humans

14. a. NADPH cytochrome P-450 reductase, epoxide hydrolase, cytochrome b_5.
    b. Glucuronosyl transferases, glutathione transferases.

15. A chemical may inhibit the enzymes of drug metabolism in the following ways:
    a. Interaction of the compound with the active site of the enzyme (e.g., piperonyl butoxide and cytochrome P-450)
    b. Competitive inhibition (e.g., methanol and ethanol with alcohol dehydrogenase)
    c. Destruction of the enzyme (e.g., vinyl chloride and cytochrome P-450)
    d. Reduced synthesis (e.g., cobalt and cytochrome P-450)
    e. Allosteric effects (e.g., carbon monoxide in its interaction with hemoglobin)
    f. Lack of cofactors (e.g., buthionine sulfoxide inhibits glutathione synthesis)

16. a. Enzyme induction can be important toxicologically because it may change the amount of a chemical that is metabolized to a toxic metabolite or conversely increase the detoxication of a chemical which is inherently toxic. Therefore enzyme induction may increase or decrease the toxicity of a chemical depending on the mechanism and role of metabolism.
    b. Biliary excretion may be important in the toxicity of a compound if that compound or a metabolite is excreted into the bile. Biliary excretion can have toxicological implications if there is (*i*) extensive enterohepatic recirculation or

(*ii*) if biliary excretion is the major route of excretion. Enterohepatic recirculation will repeatedly expose the liver and gastrointestinal tract to the compound or a metabolite possibly increasing the chances of toxicity. As biliary excretion is usually an active process, this may be saturated at high doses leading to accumulation in and toxicity to the liver.

c. The flow of blood through an organ may be a factor in the toxicity of a chemical to that organ. Thus the greater the blood flow, the greater the exposure of the organ to the chemical present in the bloodstream. Blood flow is also important in removing a chemical from a site of exposure such as the lungs. The greater this is, the more rapidly will the chemical be absorbed. Blood also delivers nutrients and oxygen to tissues and organs. If this is reduced by the action of the chemical causing vasoconstriction or swelling of a tissue through which the blood passes, the resulting ischemia may lead to tissue damage.

d. The lipophilicity of a chemical is an intrinsic physicochemical characteristic, which influences its absorption, distribution, metabolism, and excretion. Lipophilic compounds are more readily absorbed than hydrophilic compounds. Lipophilic compounds will distribute into tissues more readily especially those with high lipid content. With regard to the metabolism of chemicals by cytochrome P-450, for some reactions there is a correlation between lipophilicity and ease of metabolism, i.e., more lipophilic compounds seem to be better substrates for the enzyme for some reactions. Lipohilicity will tend to decrease excretion of chemicals at least in the urine (although if volatile they may be exhaled relatively easily).

17. A molecule will be amphipathic if it has a lipophilic moiety and a hydrophilic (e.g., charged) moiety. Thus chlorphentermine and chloroquine are amphipathic. The toxicological significance is that such molecules can cause the accumulation of phospholipids, a phenomenon known as phospholipidosis. This can be because of interference with the metabolism or disposition of phospholipids. This is sometimes observed during safety evaluation studies of drugs, often with cationic drugs, hence the term cationic amphiphilic drug (CAD).

18. The drug tamoxifen was found to cause cancers in rats during experimental studies but fortunately there are metabolic differences between rats and humans, and also mice. Tamoxifen undergoes several routes of oxidative metabolism. (*i*) Aromatic hydroxylation to 4-hydroxy tamoxifen eliminated after conjugation. (*ii*) Oxidation of alkyl groups attached to the nitrogen atom leads to dealkylation. These are detoxication pathways. (*iii*) Tamoxifen also undergoes oxidation on the α-carbon atom of the aliphatic chain and the α-hydroxylated product is conjugated. This pathway shows species differences and is crucially involved in the carcinogenicity. In rats, the preferred route is conjugation with sulfate. The conjugate formed readily loses the sulfate group forming a positively charged moiety, a carbocation, which is electrophilic. The mouse conjugates with glucuronic acid, a clear detoxication pathway, as a stable metabolite is formed. The mouse is much less susceptible. Humans have more glucuronosyl transferase activity than mice and very much more than rats. Furthermore, humans have much lower levels of sulfotransferase activity. Overall humans effectively detoxify tamoxifen.

CHAPTER 6

1. a. Direct toxic effects: tissue lesions;
    b. pharmacological, physiological, and biochemical effects;
    c. teratogenesis;
    d. immunotoxicity;
    e. mutagenesis; and
    f. carcinogenesis.

2. a. Its blood supply,
    b. the presence of a particular enzyme or biochemical pathway,
    c. the function or position of the organ,

    d. the vulnerability to disruption or degree of specialization,

    e. the ability to repair damage,

    f. the presence of particular uptake systems,

    g. the ability to metabolize the compound and the balance of toxication/detoxication systems, and

    h. binding to particular macromolecules.

3. Toxic responses may be detected by one of the following: death (cell or organism), pathological change, biochemical change, physiological change, or changes in normal status.

4. The liver shows the following types of toxic response: steatosis (fatty liver), cytotoxic damage, cholestatic damage, cirrhosis, vascular lesions, liver tumors, and proliferation of peroxisomes.

5. The area around the portal tract of the liver is (b) zone 1 or (c) the periportal region.

6. The correct answer is (b) and (d).

7. Fatty liver may be caused by mechanisms (a), (b), and (c).

8. Chemicals which can damage (a) the liver include carbon tetrachloride, paracetamol, bromobenzene, isoniazid, vinyl chloride, ethionine, galactosamine, halothane, dimethylnitrosamine; (b) the kidney include hexachlorobutadiene, cadmium and mercuric salts, chloroform, ethylene glycol, aminoglycosides, phenacetin; (c) the lung include paraquat, ipomeanol, asbestos, monocrotaline, sulfur dioxide, ozone, naphthalene; (d) the nervous system include MPTP, hexane, organophosphorus compounds, 6-hydroxydopamine, isoniazid; (e) the testes include cadmium, cyclophosphamide, phthalates, ethanemethane sulfonate, 1,3-dinitrobenzene; (f) the heart include allylamine, adriamycin, cobalt, hydralazine, carbon disulfide; (g) the blood include nitrobenzene, aniline, phenylhydrazine, dapsone.

9. The important primary events that underlie cellular injury are lipid peroxidation, covalent binding, changes in thiol status, enzyme inhibition, and ischemia.

10. This is an increase in the permeability of the inner mitochondrial membrane and then the opening of a mega channel or pore spanning both inner and outer mitochondrial membranes. This change will allow most solutes (of molecular weight less than 1500) into or out of the mitochondria. Thus protons enter, changing the import charge differential on the membranes, and ATP leaks out. Water also enters causing swelling. The ATP synthase enzyme changes and becomes an ATPase thus breaking down any remaining ATP. Any Ca^{++} in the mitochondria can leak out into the cytoplasm. ATP depletion becomes critical as there is not enough for glycolysis or for any energy requiring processes. This is the so-called mitochondrial permeability transition, and calcium ion activation of hydrolytic enzymes may underlie the progression to both apoptosis and necrosis.

11. (a) Kidney damage may be detected by measurement of serum creatinine, serum urea, urinary enzymes (γGT), urinary protein, urinary amino acids, and urinary glucose; (b) Liver damage may be detected by measurement of serum enzymes (ALT, AST), serum protein level, plasma bilirubin, and liver function (dye clearance).

12. During toxicity, damage to proteins, lipids, and DNA may occur, all of which can be repaired. Thus oxidation of SH groups in proteins to produce P-SS-P or P-SO, for example, can be repaired by reduction using the enzyme thioredoxin, which in turn is reduced by NADPH. GSH and thioltransferase can also be used. P-SS-G (GSH protein adducts) can be repaired by reduction using the enzyme glutaredoxin. GSH and NADPH are required. As already indicated, when a protein structure is damaged, heat shock proteins may repair it. As well as reduction by vitamin E, as described, lipids damaged by oxidation to peroxides can be broken down by phospholipase A_2 and the fatty acid peroxide is replaced. Although nuclear DNA is well protected by histones, there are a number of repair systems for DNA. These will be discussed in the section of genotoxicity. Mitochondrial DNA, however, does not have histones for protection or efficient repair mechanisms. Basically DNA repair is either error free, returning the DNA to its original undamaged state, or error prone giving improved but still altered DNA. The processes generally involve

    i. Recognition of damage

    ii. Removal of damage

    iii. Repair-DNA synthesis

    iv. Ligation-sticking together

This must occur before cell division to be effective, otherwise there may be mispairing of bases, rearrangements and translocations of sections of DNA.

13. The toxic products of lipid peroxidation are malondialdehyde, hydroxyalkenals, and lipid peroxides.
14. The main characteristic features of teratogenesis are
    a. Selectivity to embryo/fetus
    b. Involvement of genetic influences
    c. Susceptibility usually depends on the stage of development
    d. Specificity of effect (different types of teratogens give similar effects if given at the same time)
    e. The basic manifestations of teratogenic effects are death of the embryo/fetus, malformations, growth retardation, and functional disorders
    f. The access to the embryo/fetus is via the maternal bloodstream
    g. The dose-response curve may be steep and there is a threshold.

15. Some examples: *N*-methyl/*N*-nitro-*N*-nitrosoguanidine is a cytotoxic teratogen. Nitrophen (herbicide) is a teratogen that is believed to act through a receptor. Colchicine is a teratogen that interferes with spindle formation during cell division. 6-Aminonicotinamide is a teratogen that interferes with energy production from the tricarboxylic acid cycle. For other examples, see Chapter 6.
16. Chemicals can interact with the immune system
    a. By inhibiting or depressing immune function
    b. By elicitation of an immune response.

17. (a), (b), and (d) are involved in a type III response.
18. Characteristics of chemically induced immune reactions are (a) no true dose response, (b) repeated exposure to the chemical is necessary, (c) the reaction is not normally dependent on the chemical structure and so different chemicals can give similar reactions, and (d) the site of action may not always be the site of exposure.
19. It seems that usually the immune system, specifically the naïve T cells, need to get two signals for there to be an immune response, even when the antigenicity of the agent is high. Exposure seems to lead to a level of tolerance. One signal required is the APC-T cell receptor interaction, which is antigen specific. The second are signals indicating that the foreign substance is indeed potentially dangerous. This is the so-called "Danger hypothesis." Thus if the drug or a metabolite causes damage to tissue or cellular stress, for example, inflammatory cytokines will be released. These could form part of the second "danger" signal. The second signal also acts on the T cell via APC cells and co-stimulatory receptors (B7, for example, which interacts with CD28 on the T cell).
20. Interaction of chemicals with genetic material may cause aneuploidization, clastogenesis, and mutagenesis.
21. The four types of mutagenic change are base-pair transformations, base-pair additions or deletions, large deletions and rearrangements, and unequal partition or nondisjunction.
22. The Ames test using Salmonella bacteria is the most commonly used test.
23. The carcinogenic process involves at least three stages
    a. Initiation
    b. Promotion
    c. Progression

24. The three types of carcinogens are
    a. Genotoxic carcinogens (e.g., dimethylsulfate, 1,2-dimethylhydrazine)
    b. Epigenetic carcinogens (e.g., phorbol esters, asbestos, oestrogens)
    c. Unclassified carcinogens (e.g., peroxisome proliferators such as clofibrate) (note that these are now known to act through receptors and may be similar to oestrogens, i.e., epigenetic carcinogens).

25. The somatic cell mutation theory underlies chemical carcinogenesis. This theory presumes that there is an interaction between the carcinogen and DNA to cause a mutation. This mutation is then fixed when the DNA divides. Further replication then provides a clone of cells with the mutation, which may become a tumor. This theory fits much of the data available on carcinogenesis. Thus, such a mutation would account for the heritable nature of tumors, their monoclonal character, the mutagenic properties of carcinogens, and the evolution of malignant cells *in vivo*. However, although many carcinogens are also mutagens, by no means all of them.

CHAPTER 7

1. The metabolite responsible for the carcinogenicity of benzo[a]pyrene (i.e., the ultimate carcinogenic metabolite) is the 7,8-dihydrodiol-9,10-oxide. Some of the evidence is as follows: the 7,8-dihydrodiol metabolite of benzo[a]pyrene binds more extensively to DNA after microsomal enzyme activation than other metabolites or benzo[a]pyrene itself; the nucleoside adducts formed are similar to those formed from benzo[a]pyrene itself; the synthetic 7,8-dihydrodiol 9,10-oxides are highly mutagenic; the 7,8-dihydrodiol is carcinogenic whereas the 4,5- and 9,10-epoxides are not.

2. Metabolism is required for both the carcinogenicity and direct cytotoxicity of dimethylnitrosamine. The metabolism of dimethylnitrosamine produces methyl diazonium ion which can fragment into a methyl carbonium ion which is believed to be the ultimate carcinogen and responsible for the liver necrosis.

3. Vinyl chloride causes liver cancer (hemangiosarcoma), narcosis, and Raynaud's phenomenon that comprises scleroderma, acro-osteolysis, and liver damage. The appearance of liver tumors is dose dependent but reaches a maximum at about 22% incidence. This seems to be due to the fact that the metabolism that is necessary for the carcinogenicity, is saturated, and/or inhibited by vinyl chloride. The toxic metabolites produced are the epoxide and chloroacetaldehyde.

4. Carbon tetrachloride causes centrilobular liver necrosis and steatosis after acute exposure, and liver cirrhosis, liver tumors, and kidney damage after chronic administration. The mechanism underlying the acute toxicity to the liver involves metabolic activation by cytochrome P-450 to yield a free radical (trichloromethyl free radical). This reacts with unsaturated fatty acids in the membranes of organelles and leads to toxic products of lipid peroxidation including malondialdehyde and hydroxynonenal. This results in hepatocyte necrosis and inhibition of various metabolic processes including protein synthesis. The latter leads to steatosis as a result of inhibition of the synthesis of lipoproteins required for triglyceride export.

5. Statements (a) and (c) are true.

6. The metabolite of bromobenzene that is believed to be responsible for the hepatic necrosis is bromobenzene 3,4-oxide. This reacts with liver cell protein, which causes cell death. The reactive metabolite can be detoxified by conjugation with glutathione or be detoxified by metabolism to a dihydrodiol by epoxide hydrolase. Pretreatment of animals with the enzyme inducer 3-methylcholanthrene decreases the toxicity. This is because it increases metabolism to the 2,3-oxide. This reactive metabolite is not as toxic as the 3,4-bromobenzene oxide readily undergoing rearrangement to 2-bromophenol. 3-Methylcholanthrene also induces epoxide hydrolase and so increases detoxication.

7. Isoniazid causes peripheral neuropathy and liver damage (centrilobular necrosis). The liver damage is caused by a metabolite, acetylhydrazine, whereas the peripheral neuropathy is caused by the parent drug interacting with pyridoxal phosphate and thereby interfering with the metabolism of vitamin B_6. The genetic factor acetylator phenotype is important in both types of toxicity. Thus with peripheral neuropathy, the slow acetylators are more susceptible to the toxicity because they have higher

levels of unchanged drug. With the liver damage, the slow acetylators are more susceptible because they detoxify the acetylhydrazine less efficiently.

8. Chloroform causes kidney damage (proximal tubular necrosis) and liver damage (hepatic necrosis). The mechanism is believed to involve metabolic activation of the chloroform in the kidney to produce phosgene, which is probably responsible for the toxicity. Both liver and kidney damage may be modulated by treating animals with enzyme inducers and therefore it seems likely that the liver damage is also mediated by a reactive metabolite.

9. Hexachlorobutadiene is toxic to the kidney. It is metabolized by glutathione conjugation. The glutathione conjugate is excreted into the bile and while in the gut and in contact with the gut flora, the conjugate is broken down with loss of the glutamyl and glycinyl residues. The remaining cysteine conjugate is reabsorbed from the gastrointestinal tract. In the kidney, the cysteine conjugate may be degraded by the enzyme CS-lyase to give a thiol metabolite. This is toxic to the kidney, damaging the mitochondria. The concentrating ability of the kidney may lead to higher concentrations of this metabolite in the proximal tubular cells than in the blood.

10. Carbon monoxide is toxic as a result of its binding to hemoglobin. This binding both displaces oxygen and makes loss of oxygen bound to hemoglobin more difficult. The result is that the tissues become starved of oxygen and eventually susceptible tissue such as the heart and brain are unable to function and the animal dies. A dangerous concentration of carbon monoxide would be considered to give rise to 50% carboxyhemoglobin. Therefore the air concentration of carbon monoxide, which will give rise to this, can be calculated as follows:

$$\frac{COHb}{O_2Hb} = 240 \times \frac{[CO]}{[O_2]}$$

Therefore knowing that the concentration of carboxyhemoglobin chosen as dangerous is 50%, allows the following calculation

$$\frac{50}{50} = 240 \times \frac{[CO]}{21}; \text{Therefore} \frac{1 \times 21}{240} = [CO] = 0.0875\%.$$

11. Halothane is metabolized by cytochrome P-450 to a reactive intermediate, trifluroacetyl chloride. This acylates the amino groups of liver proteins, especially those containing lysine. The acylated, altered proteins are probably transported from the liver and presented to the immune system. The immune system recognizes these proteins as antigens and antibodies are generated in response. After a subsequent exposure to halothane, the altered proteins, which are expressed on the surface of the liver are targets for the antibodies, which bind to them. However, T-lymphocytes also become involved and these bind to the Fc portion of the antibodies attached to the altered liver protein. During this process the T-lymphocytes release cytolytic factors, which lyse the liver cells in the vicinity. Therefore, those areas of the liver where metabolism of halothane has produced the reactive intermediate, which binds to protein (the centrilobular area) is damaged by the T-lymphocytes resulting in necrosis.

12. Gentamycin damages the proximal tubular cells in the kidneys and also the hair cells in the cochlear apparatus in the ear. Kidney damage may be detected by histopathology and by biochemical abnormalities such as raised serum urea and creatinine levels and the presence of enzymes such as γ-glutamyl transferase, amino acids, protein, and glucose in the urine. The toxicity of gentamycin to the ear may be detected by histopathology or more easily by the Preyer reflex test, which measures hearing thresholds for particular sound frequencies.

13. Paraquat mainly damages the lungs. This is because (a) there is a specific uptake system for polyamines, which accumulates paraquat in lung tissue and (b) because there is a high level of oxygen in the lungs. This oxygen accepts electrons from the stable paraquat radical and the superoxide formed causes lipid peroxidation, which damages the tissue.

14. MPTP is a molecule, which is sufficiently lipophilic to cross the blood-brain barrier and enter the astrocyte cells. Once in these cells, it can be metabolized by monoamine oxidase B to MPDP and then MPP both of which are charged molecules. These metabolites are therefore not able to diffuse out of the astrocyte into the bloodstream and away from the brain. However, the structure of MPP allows it to be taken up by a carrier system and concentrated in dopaminergic neurones. In the neurone, it inhibits the mitochondrial electron transport chain leading to damage to the neurone.

15. (a) Bronchoconstriction, miosis, salivation, lacrimation, sweating, vomiting, diarrhea; (b) muscular weakness and fatigue, involuntary twitching; (c) tension, anxiety, convulsions. Symptoms (a) are due to the excess acetylcholine at muscarinic receptors. Symptoms (b) are due to excess acetylcholine at nicotinic receptors. Symptoms (c) are due to the accumulation of acetylcholine in the central nervous system. Acetylcholine accumulates because the organophosphates inhibit the enzyme acetylcholinesterase, which normally removes the neurotransmitter substance.

16. Salicylic acid, the major metabolite of aspirin, uncouples the electron transport chain in the mitochondria. This results in (a) increased use of oxygen and production of carbon dioxide, (b) lack of ATP, and (c) excess energy no longer utilized in ATP production. The result is increased respiration and raised temperature. The alterations in respiration lead to alkalosis followed by acidosis. The lack of ATP and loss of respiratory control will cause increased metabolic activity and hypoglycemia after an initial mobilization of glucose from glycogen.

17. Fluoroacetate causes inhibition of aconitase, an enzyme in the tricarboxylic acid cycle. This is due to the formation of fluorocitrate, which binds to aconitase and inhibits the enzyme. This is because the fluorine atom cannot be removed from the fluorocitrate unlike the hydrogen atom in the normal substrate, citrate. The result is complete blockade of the cycle and this means tissues become starved of ATP and other vital metabolic intermediates. This causes adverse effects in the heart as the organ is particularly sensitive to deficiency of ATP.

18. Similarities: Both galactosamine and ethionine are toxic as a result of interference with intermediary metabolism. Galactosamine depletes UTP, ethionine depletes ATP. Both form derivatives that effectively trap the cofactor precursor (uridine and adenine, respectively). Both may cause cirrhosis and tumors after repeated dosing.

    Differences: Galactosamine causes diffuse hepatic necrosis, which is not zonal, whereas ethionine causes steatosis, which is initially periportal.

19. Nitrite, which may be administered as sodium nitrite solution or inhaled as amyl nitrite, will oxidize hemoglobin to methemoglobin. This oxidized form of hemoglobin will bind cyanide more strongly than cytochrome a-a3 and therefore tend to remove the cyanide bound inside cells and sequester it in the blood from where it is more readily eliminated.

20. The developing embryo varies in its susceptibility to chemicals. Therefore depending on when a particular chemical is administered, the effects may range from nothing through malformations to death. In the case of diphenylhydantoin, gestational days 11–12 are the most sensitive for orofacial malformations, whereas days 13–14 are more sensitive for skeletal defects. These differences reflect the organ system under development at a particular time in gestation.

21. The predisposing factors for susceptibility to hydralazine-induced lupus identified in humans are (a) dose, (b) duration of therapy, (c) the acetylator phenotype, (d) HLA (tissue) type, and (e) gender.

22. Both ethylene glycol and methanol produce, as a result of metabolism, acidic metabolites and also increase production of NADH through the utilization of alcohol dehydrogenase in this metabolism. The increased amount of NADH leads to increased production of lactic acid. Therefore, the overriding biochemical toxicity of both solvents is acidosis. Treatment for poisoning for both compounds will involve (a) treatment of the acidosis with bicarbonate administration and (b) use of ethanol as a competitive inhibitor of alcohol dehydrogenase to block metabolism of both compounds, so allowing elimination of the unchanged solvents to occur.

23. Interference by lead of heme synthesis occurs at a number of points, including the enzymes aminolevulinic acid dehydrase (ALA dehydrase) and ferrochelatase and uptake of iron into the mitochondrion. Therefore, several biochemical changes can be used as biomarkers: inhibited activity of ALA dehydrase, increased aminolevulinic acid levels in urine, excess protoporphyrin in red cells, decreased red cell count. The consequences of the inhibition of heme synthesis are less hemoglobin, fragile red blood cells, and so reduced numbers of red blood cells leading to anemia. The individual animal or human will therefore have less oxygen-carrying capacity.

24. Fibrates (such as clofibrate) are hypolipidemic drugs. They cause a number of biological and biochemical effects, including enzyme induction, peroxisomal proliferation, liver hypertrophy and hyperplasia, and liver cancer. The mechanism is believed to be mediated through the PPARα receptor. Animals lacking this receptor do not show these effects.

    There are several critical events thought to be involved in the hepatic carcinogenesis caused by PPARα agonists: (i) receptor binding and activation, (ii) induction of key target genes, (iii) increased cell proliferation and inhibition of apoptosis, (iv) oxidative stress causing DNA damage and or increased signaling for cell proliferation, and (v) finally, clonal expansion of initiated liver cells.

    It is suggested that PPARα activation may lead to DNA replication, cell growth and inhibition of apoptosis via tumor necrosis factor α (TNF-α). Alternatively or additionally, the oxidative stress could activate the MAP kinase pathway, which will also influences cell growth, but this might be independent of PPARα activation. The role of these events is still unclear.

    Humans are not thought to be at risk because they do not have a functional receptor. There is no evidence of these effects in humans.

25. Valproic acid is similar to a fatty acid and therefore can become incorporated into fatty acid metabolism. This involves formation of an acyl CoA derivative and also a carnitine derivative. However, this depletes both acetyl-CoA from the intramitochondrial pool and carnitine and so compromises the mitochondria and reduces the ability of the cell to metabolize short, medium, and long-chain fatty acids via β oxidation.

    Furthermore valproic acid is metabolized by dehydrogenation, a reaction catalyzed by CYP 2C9, to a fatty acid analogue. This fatty acid analogue is then incorporated into the β oxidation pathway in the mitochondria. However, the metabolite analogue is reactive as it has conjugated double bonds. This reactive intermediate disrupts β oxidation by damaging the enzymes involved in the process. It also depletes mitochondrial glutathione, allowing still more damage to occur. Therefore overall, damage to the mitochondria and disruption of mitochondrial fatty acid metabolism so as to reduce or terminate β oxidation of fatty acids is the mechanism underlying the toxic effect of valproic acid.

26. Microcystins are produced by blue-green algae. Some of them are toxic. Microcystin LR is toxic to the liver. The structure of microcystin includes an electrophilic carbon atom, which is part of the Mdha amino acid. It is taken up into the liver by an organic anion transporter system (OAT) and therefore is concentrated in the liver. The structure of the microcystins means they are able to associate with the enzymes protein phosphatases, such as PP-1, PP-2A, and PP-2B via hydrophobic and ionic interactions. However, this association allows the electrophilic carbon to form a covalent bond with the SH group of cysteine in the structure of the enzyme. This inactivates the enzyme. Microcystins are extremely potent inhibitors of protein phosphatases. Protein phosphatases are enzymes involved in the regulation of proteins by the addition and removal of phosphate groups, which is an essential to the function and integrity of many proteins. When inactivated, certain proteins (cytoskeletal proteins) can be hyperphosphorylated leading to inactivation or structural breakup and cell death. Change in phosphorylation status also affects MAP kinases and, therefore, the MAPK cascade, and this may be responsible for the apoptosis, which is observed as another toxic effect. Lower concentrations of microcystin affect the protein dynein, which, with ATP, drives vesicles along microtubules in the cell.

27. Domoic acid is a natural toxin produced by phytoplankton, which then contaminates shellfish. It causes neurotoxicity via a receptor interaction. It also causes

gastrointestinal distress and disturbances. The neurotoxicity may be coma, convulsions, agitation and loss of memory. If it is serious poisoning, the neurones in the hippocampus and amygdala maybe destroyed.

Domoic acid, known as an excitotoxin, acts as an analogue of the amino acid glutamate, incorporating a similar structure. It is believed to bind at the same site on the "kainate receptor."

This binding leads to excitation of the glutamate receptors and excess release of glutamate. The post-synaptic membrane in the neurone is where the damage occurs and this is where the glutamate receptors are. The result of the binding is excessive release of glutamate or the prevention of absorption of glutamate into the neurone. The excess glutamate then causes damage as a result of excessive simulation of receptors such as the NDMA receptor. This causes a rise in intracellular calcium, which can cause a number of adverse cellular events such as the production of reactive oxygen species, activation of phospholipases, calpain, and protein kinase C.

This excess excitation leads to cell death either by apoptosis or necrosis in the neurones affected, mostly in the hippocampus and amygdala, although some occurs in the cerebral cortex also.

28. Primaquine is toxic to the red blood cell. The toxicity involves oxidative damage by a metabolite(s) of the drug to hemoglobin. Normally in the red cell oxidizing species such as drug metabolites and reactive oxygen are removed by reduced glutathione. This is maintained by glutathione reductase and NADPH. NADPH is generated by the pentose phosphate shunt. The first enzyme in this pathway is glucose-6-phosphate dehydrogenase. Some individuals have a genetic deficiency in this enzyme, a sex linked (male) trait. Lack of the enzyme makes individuals sensitive to drugs such as primaquine, due to lack of NADPH and reduced glutathione. They suffer red cell damage and hemolytic anemia.

29. *Botulinum toxin* is a mixture of six large molecules, each of which consist of two components, a heavy (100kDa) and light (50kDa) polypeptide chain. The heavy chain binds to the walls of nerve cells, which then allows the whole toxin molecule to be transported into the cell inside a vesicle via receptor-mediated endocytosis. Once inside, the light chain translocates into the cytosol and acting as a peptidase, destroys a synaptosomal protein.

By destroying the protein, the toxin prevents the release of the neurotransmitter acetylcholine from small packets at the ends of nerves by exocytosis. These nerves, attached to voluntary muscles, need acetylcholine to allow the flow of signals (impulses) between the nerve and the muscle. By preventing the release of acetylcholine, botulinum toxin blocks muscle contraction, causing paralysis and relaxation. The therapeutic action relies on relaxation of muscles, generally in the face. It is therefore used to treat blepharospasm (uncontrolled contractions) and stroke-induced permanent facial muscle contractions.

30. Only the S-isomer of thalidomide is teratogenic. Explain this in terms of the mechanism.

A probable mechanism underlying the teratogenicity of thalidomide relies on the interaction of thalidomide with DNA. The phthalimide double-ring structure is relatively flat and can intercalate with DNA forming a stacked complex. Furthermore, observations were made that thalidomide

   a. inhibits the development of new blood vessels from microvessels (angiogenesis) in the embryo,
   b. inhibits the expression of specific cell adhesion molecules (integrins), which are intimately involved in cell growth and differentiation,
   c. inhibits the action of certain growth factors, and
   d. binds specifically to the promoter regions of genes involved in regulatory pathways for angiogenesis and integrin production. These contain polyG sequences.

Only the S-enantiomer, the teratogenic form, fits properly into the major groove of the double helix and binds to the polyG sequences. It seems that a very specific stereochemical interaction may be necessary for this toxic effect to occur.

Growth factors and integrins are important in the development of the embryo. The specific integrin ($\alpha v\beta 3$ subunits) is stimulated by two growth factors, IGF-I and FGF-2. The integrin ($\alpha v\beta 3$) then stimulates angiogenesis, which is an integral part of the development of limb buds.

The transcription and therefore production of the integrin αv and $\beta 3$ subunits are each regulated by a promoter specific transcription factor Sp1. This binds to guanine-rich promoter regions, which are found in many genes including those for both αv and $\beta 3$, IGF-I, and FGF.

Thalidomide competes for binding with Sp1 at these polyG sequences. More than 90% of genes have TATAA boxes, so competition for binding between Sp1 and thalidomide is minimal. Therefore development of many of the organs and tissues in the embryo would be unaffected. Only those genes *without* TATAA boxes (i.e., integrin $\alpha v\beta 3$, IGF-1, FGF-1) are most affected therefore leading to reduced limb bud development.

Index

Pulmonary cancer, 205
Purine oxidation, 94
Puromycin, 245
Putrescine, 338
Pyridine, 88
Pyrolysis of food, during cooking, 154–155
Pyrrolizidine alkaloid monocrotaline
    metabolic activation, 200
    structure, 200
Pyrrolizidine alkaloids
    lung damage, 205

Quantal effects, 20
Quaternary ammonium compounds, 69
Quinoline, 88
Quinones, 97

Rate constants, 60
Rate of diffusion, 38
Raynaud's phenomenon, vinyl chloride, 300
Reactive intermediates, 123
Reactive oxygen species (ROS), 122–123
Receptor interactions, 15–20
Receptor-mediated teratogenicity, 243–244
Redox reagents, 122–123
Reductases, 96
Reduction, 96–99
Reductive dehalogenation, 98
Renal toxicity
    lead, 391
Reproductive systems
    vinyl chloride effects on, 300
Respiration, cellular
    inhibition of, 235
Respiratory alkalosis, 356
Respiratory failure, 235
Response biomarker, 8
Retinoic acid receptor
    diphenylhydantoin interaction with, 370
Reversibility, 9
Rifampicin
    cholestatic damage, 199
Ring hydroxylation, acetylaminofluorene, 295
Risk assessment, 27
RNA, 259
R's, concept of the, 13

Salicylate-induced alkalosis, 236
Salicylate poisoning, 235
Salicylic acid, physiological effects, 355–356
Salmonella typhimurium, 13
Saturation kinetics, 65
Scatchard plot, 55
Secobarbital, 85
Sedormid, 85
Selective estrogen receptor modulator (SERM), 303
Selectivity, teratogenesis, 238
Semiquinone radicals, 121
S(–) enantiomer of thalidomide, 131
Sensitivity
    heightened, 237
    to teratogens, 240

Sensitization, 9
SERM. *see* Selective estrogen receptor modulator (SERM)
Sex differences, in metabolism, 146–149
Short-lived reactive metabolites/intermediates, 123
Signal transduction pathways
    chemical inhibition of, 216
Sinusoidal surface, of the liver cell, 37
Sinusoids, liver, 196
SKF 525A, 180–181
Skin
    absorption through the, 44–45, 135
    foreign compound toxicity, 208
Smooth endoplasmic reticulum (SER), 76
Sodium nitroprusside, 364
Somatic cell mutations, 244
Somatic cell-mutation theory, 274
S-oxidation, 90–91
Species differences
    in disposition, 134–138
    in metabolism, 138–144
Specific, acquired immunity, 248
Specificity
    immunotoxicity, 257
    of teratogens, 239–240
Spectrin, 37
Spermatocyte death, 229
Spermine, 338
Sphingomyelin, 37
Starvation, effect on metabolism of foreign compounds, 161
Steatosis (fatty liver), 198
    mechanism, 224–225
Stereoselectivity, 131
Strain differences in disposition, 144–146
Streptozotocin, 207, 235
Streptomycin, 245
Stress, effect on disposition, 160
Stress proteins, 231–232
Substrate deficiency, in teratogenesis, 245
Succinylcholine, 75, 159
    toxic effects, 352
Sulfamethoxazole, 256
Sulfamethazine, 152
Sulfanilamide, 75, 111–112, 146
Sulfate conjugation reactions, 105–106
Sulfides, 98
Sulfisoxazole, 56
Sulfonamides
    drugs, 56
    kidney toxicity, 202
Sulfotransferase enzyme, 106
Sulfoxides, 98
Superoxide dismutase, 338
    protective role of, 232
Susceptibility
    of the organism, 8
    to teratogenesis, 239
Symport, 42
Synergism, 14
Synergy, 14–15
Systemic effects, of toxicity, 9, 35